Handbook on Psychopathy and Law

Oxford Series in Neuroscience, Law, and Philosophy

Series Editors
Lynn Nadel, Frederick Schauer, and Walter P. Sinnott-Armstrong

Conscious Will and Responsibility
Edited by Walter P. Sinnott-Armstrong and Lynn Nadel

Memory and Law
Edited by Lynn Nadel and Walter P. Sinnott-Armstrong

Neuroscience and Legal Responsibility
Edited by Nicole A. Vincent

Handbook on Psychopathy and Law
Edited by Kent A. Kiehl and Walter P. Sinnott-Armstrong

Handbook on Psychopathy and Law

Edited by
Kent A. Kiehl
Walter P. Sinnott-Armstrong

OXFORD
UNIVERSITY PRESS

Oxford University Press is a department of the University of Oxford.
It furthers the University's objective of excellence in research, scholarship,
and education by publishing worldwide.

Oxford New York
Auckland Cape Town Dar es Salaam Hong Kong Karachi
Kuala Lumpur Madrid Melbourne Mexico City Nairobi
New Delhi Shanghai Taipei Toronto

With offices in
Argentina Austria Brazil Chile Czech Republic France Greece
Guatemala Hungary Italy Japan Poland Portugal Singapore
South Korea Switzerland Thailand Turkey Ukraine Vietnam

Oxford is a registered trademark of Oxford University Press in the UK and certain other
countries.

Published in the United States of America by
Oxford University Press
198 Madison Avenue, New York, NY 10016

© Oxford University Press 2013

All rights reserved. No part of this publication may be reproduced, stored in a
retrieval system, or transmitted, in any form or by any means, without the prior
permission in writing of Oxford University Press, or as expressly permitted by law,
by license, or under terms agreed with the appropriate reproduction rights organization.
Inquiries concerning reproduction outside the scope of the above should be sent to the
Rights Department, Oxford University Press, at the address above.

You must not circulate this work in any other form
and you must impose this same condition on any acquirer.

Library of Congress Cataloging-in-Publication Data
Handbook on psychopathy and law / [edited by] Kent A. Kiehl, Walter Sinnott-Armstrong.
   pages cm
Includes bibliographical references and index.
ISBN: 978-0-19-984138-7
1. Antisocial personality disorders. 2. Psychopaths—Legal status, laws, etc. I. Kiehl, Kent A.
II. Sinnott-Armstrong, Walter, 1955–
RC555.H363 2013
616.85′82—dc23
2012046215

9 8 7 6 5 4 3 2 1
Printed in the United States of America
on acid-free paper

CONTENTS

	Foreword Robert D. Hare	vii
	Contributors	xi
1.	**Introduction** Kent A. Kiehl and Walter P. Sinnott-Armstrong	1

PART 1 DIAGNOSIS OF PSYCHOPATHY

2.	**Assessment of Psychopathy: The Hare Psychopathy Checklist Measures** Adelle Forth, Sune Bo, and Mickey Kongerslev	5
3.	**Alternatives to the Psychopathy Checklist—Revised** Katherine A. Fowler and Scott O. Lilienfeld	34

PART 2 DEVELOPMENTAL PERSPECTIVES ON PSYCHOPATHY

4.	**Developmental Conceptualizations of Psychopathic Features** Dustin A. Pardini and Amy L. Byrd	61
5.	**Adolescent Psychopathy and the Law** Michael J. Vitacco and Randall T. Salekin	78

PART 3 DECISION MAKING AND PSYCHOPATHY

6.	**The Decision-Making Impairment in Psychopathy: Psychological and Neurobiological Mechanisms** Michael Koenigs and Joseph P. Newman	93
7.	**Do Psychopaths Make Moral Judgments?** Jana Schaich Borg and Walter P. Sinnott-Armstrong	107

PART 4 NEUROSCIENCE AND PSYCHOPATHY

8.	**Functional Neuroimaging and Psychopathy** Nathaniel E. Anderson and Kent A. Kiehl	131

9. Structural Brain Abnormalities and Psychopathy 150
   Marina Boccardi

   PART 5 GENETICS OF PSYCHOPATHY

10. Quantitative Genetic Studies of Psychopathic Traits in Minors: Review and Implications for the Law 161
    Essi Viding, Nathalie M. G. Fontaine, and Henrik Larsson

11. The Search for Genes and Environments that Underlie Psychopathy and Antisocial Behavior: Quantitative and Molecular Genetic Approaches 180
    Irwin D. Waldman and Soo Hyun Rhee

    PART 6 TREATMENT OF PSYCHOPATHY

12. Treatment of Adolescents with Psychopathic Features 201
    Michael F. Caldwell

    PART 7 RECIDIVISM AND PSYCHOPATHY

13. Psychopathy and Violent Recidivism 231
    Marnie E. Rice and Grant T. Harris

14. Taking Psychopathy Measures "Out of the Lab" and into the Legal System: Some Practical Concerns 250
    John F. Edens, Melissa S. Magyar, and Jennifer Cox

    PART 8 RESPONSIBILITY OF PSYCHOPATHS

15. Criminal Responsibility and Psychopathy: Do Psychopaths Have a Right to Excuse? 275
    Paul Litton

16. Why Psychopaths Are Responsible 297
    Samuel H. Pillsbury

    PART 9 DETENTION OF PSYCHOPATHS

17. Preventive Detention of Psychopaths and Dangerous Offenders 321
    Stephen J. Morse

18. Some Notes on Preventive Detention and Psychopathy 346
    Michael Louis Corrado

19. Psychopathy and Sentencing 358
    Erik Luna

    Index 389

FOREWORD

Some 20 years ago I wrote that following a lecture in Florida on psychopathy and emotions a forensic psychiatrist approached me and said, "Your research implies that psychopaths may be mentally disordered, perhaps not as responsible for their behavior as we once thought. Until now, a diagnosis of psychopathy has been 'the kiss of death' for many murderers. Will it now become the 'kiss of life' for them?" I can't recall how I responded, but it is clear that his comment and question reflected dilemmas that persist to this day: What implications do the findings from risk assessment, behavioral genetics, and neuroscience have for the criminal justice system, particularly with respect to psychopathy and legal culpability? Is psychopathy a mitigating factor, an aggravating factor, neither, or dependent on the context?

There are no simple or generally satisfying answers to these questions, as the contents of this volume clearly indicate. First, there is disagreement among researchers and commentators about what the science actually tells us about the nature of psychopathy (e.g., genetically or neurologically "damaged," or just different?). Second, there is debate about the extent to which the legal system will be, or should be, influenced by what the science says about psychopathy and culpability.

My own view is that psychopathic individuals have an intellectual understanding of the rules of society and the conventional meanings of right and wrong and know enough about what they are doing to be held accountable for their actions. Like Iago in Shakespeare's *Othello*, they choose which rules to follow or to ignore, based on their own self-interest, a calculating appraisal of the circumstances, and a lack of concern for the feelings or welfare of others. They lack empathy, guilt, or remorse for their actions and are emotionally "disconnected" from others. But, they do not ignore or break every moral or legal code, nor do they make everyone they encounter a victim. There is little doubt that many psychopathic features are associated, in theoretically relevant ways, with a variety of brain structures and functions that differ from those of the majority of other individuals. But this does not necessarily mean that they suffer from a neurological *deficit* or *dysfunction*. Indeed, psychopaths might claim that because they are not encumbered by emotional baggage they are more rational than most people. As a psychopathic offender in one of our research projects put it, "The psychiatrist said that my problem is I think more with my head than with my heart." He did not see this as a problem, and went on to say that he was "a cat in a world of mice."

This unintended but succinct allusion to the evolutionary view of psychopathy as an adaptive life strategy implied that he merely was doing what nature intended him to do. Whatever the merits of this particular view, we should consider the possibility that the actions of psychopaths reflect cognitive, affective, and behavioral processes and strategies that are different from those of other people, but for reasons other than neuropathology or deficit, in the traditional medical and psychiatric sense of the terms.

I say this because it is tempting–for experts and laypersons alike–to explain the callous, manipulative, and remorseless behavior of psychopaths in terms of "something" that doesn't work properly. Such explanations are understandable when the observed differences between psychopathic and other individuals involve brain regions and circuitry that are related to emotional, social, and executive functions that characterize psychopathy. And it is not surprising that many observers view clinical descriptions and empirical findings through a prism of dysfunction when dealing with adjudicated criminals, particularly those who are violent. It is more difficult to do so with respect to psychopathic entrepreneurs, stockbrokers, financial consultants, politicians, clinicians, lawyers, academics, and so forth.

At a meeting of the MacArthur Foundation Law and Neuroscience Project at Stanford University in January 24–25, 2008, the topic of psychopathy and neuroscience was listed under a category labeled "Diminished Brains." I argued that the use of this label prejudged the issue; the label subsequently was changed to "Differing Brains." It may turn out that psychopathy *is* causally associated with functional and structural deficits or abnormalities, but, for now, it is difficult to differentiate correlation from causation. Are the brain structures and circuitry of psychopaths the cause of psychopathic behavior, correlates of such behavior, or the result of a life-long pattern of unusual cognitive and behavioral strategies? How do genetics and environment play into these issues?

Whatever the answers, some might argue that psychopaths lack the emotional wherewithal needed to translate intellectual, moral knowledge into behavior acceptable to society, and that this is a *deficit* that places them at a disadvantage when making crucial life decisions. That is, like Iago and the offender who thinks more with his head than with his heart, it is possible that their ability to make "calculating" decisions that primarily serve their own best interests (at least in the short term) reflects a deficiency in the emotional processes that contribute to "conscience" and that help others to make prosocial life decisions.

The public is becoming increasingly more fascinated with psychopaths, both as villains and antiheros. Unfortunately, much of the information it receives comes from what has become a "psychopathy industry," with dramatic and often uninformed portrayals of "psychopaths" in television programs and movies, magazine articles, newspaper reports, and popular books. Brain scans are great attention-grabbers, and even though their scientific meaning may be uncertain they tend to have considerable impact on the public—and no doubt legal—perceptions of psychopathy. "They must be mad or brain-damaged to do that," goes a popular refrain.

Perhaps, but among the issues that concern me are the following, listed in no particular order of importance.

- There needs to be agreement on the conceptualization and measurement of the psychopathy construct used in the legal system. Psychopathy overlaps with, but is not identical to, antisocial personality disorder as defined in DSM-IV. They may become more similar to one another in DSM-5, but it is unlikely that they will be interchangeable, at least at the measurement level. Self-report measures are important for research but in legal settings are not viable substitutes or proxies for carefully conducted clinical assessments.
- Scores on some measures of psychopathy, particularly the Psychopathy Checklist–Revised (PCL-R) and its derivatives, are generally reliable, but problems may arise when they are obtained by opposing sides in an adversarial legal system. The problems might be minimized by ensuring that those who conduct psychopathy assessments for the courts are qualified and trained to do so in accordance with the highest professional standards, and without regard for who pays the bill. I've been told that this is a naïve expectation.
- Current measures of psychopathy appear to identify a dimensional construct, although this does not rule out the possibility that individuals with extremely high scores on an instrument (e.g., on all four factors of the PCL-R) are qualitatively different from those with lower scores (i.e., members of a taxon). In any case, how high up the dimension must an individual be before being considered psychopathic

enough for purposes of determining culpability? Researchers use standardized thresholds (e.g., ≥ 30 on the PCL-R) for psychopathy, but what thresholds, cutoff points, or patterns of scores will be appropriate for use in the legal system?

- This brings up several related issues. How different from "normal" do brain structure and function, and cognitive, affective, and behavioral processes, need to be in order to be considered "abnormal" or "deviant" for legal purposes? What is "normal?" With respect to the dimensional/taxon issue, do differences from normality gradually appear as the measured level of psychopathy increases (suggesting dimension), or do they emerge only at a very high measured level of psychopathy (suggesting taxon)? What degree of difference from normality does a "psychopathic brain" represent? Can we have a "psychopathic brain" without a high psychopathy score? What about an individual who has a very high psychopathy score but a "normal" brain?
- At present we know little about the variability in brain structure and function in the general population, and even less about how such variability relates to differences in genetics, environment, personality, and behavior. What proportion of the general population has the structural and functional features found in psychopathy but without any indication of psychopathic behavior? What is the ecological validity of the laboratory tasks used in cognitive/affective laboratory paradigms?
- Besides the measurement error associated with the assessment of psychopathy, there are methodological, measurement, and statistical problems in acquiring and interpreting neuroimaging data. There also is uncertainty about what such data tell us about underlying cognitive and affective processes. As a well-known psychiatrist said after a presentation I had given on psychopathy and brain imaging, "Some pretty pictures, but what do they mean?" My response was that they may provide a neurological basis for understanding psychopathic behavior, to which he replied, "But, not necessarily a causal basis!" Similar considerations have been raised about the implications of neuroscience for criminal culpability in general.

The chapters in this volume provide a valuable framework for discussing the difficult scientific, philosophical, and legal issues that arise when science informs debates about criminal responsibility. The basic issues are not new, but the literature relevant to the issues has increased dramatically in recent years. For example, over the past 50 years the number of publications on psychopathy has increased from less than 15 per year to more than 250 per year, with a cumulative total approaching 3,000. The number of active researchers has increased from a dozen or so to many hundreds and is growing rapidly, with a healthy mix of those with basic and applied interests. Researchers now have their own professional organization, the Society for the Scientific Study of Psychopathy (SSSP; www.psychopathysociety.org). The increasing breadth, depth, and sophistication of the multidisciplinary thinking and research on psychopathy are truly impressive, but we have a lot to learn. For now, we should be judicious in drawing out the potential implications of this research for legal matters.

Robert D. Hare
University of British Columbia

CONTRIBUTORS

Nathaniel E. Anderson
University of New Mexico
Albuquerque, NM, USA

Sune Bo
Psychiatric Research Unit
Roskilde, Denmark

Marina Boccardi
Laboratory of Epidemiology, Neuroimaging, and Telemedicine (LENITEM)
Brescia, Italy

Jana Schaich Borg
Stanford University
Stanford, CA, USA

Amy L. Byrd
University of Pittsburgh
Pittsburgh, PA, USA

Michael F. Caldwell
University of Wisconsin—Madison
Madison, WI, USA

Michael Louis Corrado
University of North Carolina, Chapel Hill, Law School
Chapel Hill, NC, USA

Jennifer M. Cox
Texas A&M University
College Station, TX, USA

John F. Edens
Texas A&M University
College Station, TX, USA

Nathalie M. G. Fontaine
Department of Criminal Justice
Indiana University
Bloomington, IN, USA

Adelle Forth
Carleton University
Ottawa, ON, Canada

Katherine A. Fowler
National Institute of Mental Health
Bethesda, MD, USA

Grant T. Harris
Research & Academics, Waypoint Centre for Mental Health Care
Penetanguishene, ON, Canada

Kent A. Kiehl
The Nonprofit Mind Research Network & University of New Mexico
Albuquerque, NM, USA

Michael Koenigs
University of Wisconsin—Madison
Madison, WI, USA

Mickey Kongerslev
Psychiatric Research Unit
Roskilde, Denmark

Henrik Larsson
Department of Medical Epidemiology and Biostatistics
Karolinska Institute
Hagalund, Sweden

Scott O. Lilienfeld
Emory University
Atlanta, GA, USA

Paul Litton
University of Missouri School of Law
Columbia, MO, USA

Erik Luna
Washington and Lee University School of Law
Lexington, VA, USA

Melissa S. Magyar
Texas A&M University
College Station, TX, USA

Stephen J. Morse
University of Pennsylvania Law School
Philadelphia, PA, USA

Joseph P. Newman
University of Wisconsin—Madison
Madison, WI, USA

Dustin A. Pardini
University of Pittsburgh Medical Center
Pittsburgh, PA, USA

Samuel H. Pillsbury
Loyola Law School
Los Angeles, CA, USA

Soo Hyun Rhee
University of Colorado, Boulder
Boulder, CO, USA

Marnie E. Rice
Research & Academics, Waypoint Centre
for Mental Health Care
Penetanguishene, ON, Canada

Randall T. Salekin
Department of Psychology
University of Alabama
Tuscaloosa, AL, USA

Walter P. Sinnott-Armstrong
Mind Research Network
Duke University
Durham, NC, USA

Essi Viding
Division of Psychology and Language
Sciences University College London
London, UK

Michael J. Vitacco
Georgia Health Sciences University
Augusta, GA, USA

Irwin D. Waldman
Emory University
Atlanta, GA, USA

CHAPTER 1

Introduction

Kent A. Kiehl and Walter P. Sinnott-Armstrong

Experts estimate that psychopaths comprise less than 1% of the population but commit as much as 30% to 40% of the violent crime in the United States. Given that the cost of crime in the United States is estimated to be $2.3 trillion per year[1] and violent crime constitutes the vast proportion of those costs, psychopathy is likely the most expensive mental health disorder known to man. Of course, these financial estimates do not include the emotional damage to victims of crime or increased fear for everyone.

In addition to these practical problems, psychopaths pose theoretical challenges for legal theory. If psychopaths cannot appreciate that or why their acts are wrong, it might seem unfair to punish them in the same way as normal criminals. Yet every state and country holds them criminally responsible. Apart from responsibility, psychopathy is among the best available predictors of recidivism. This predictive power raises questions of whether psychopathy should be treated as an aggravating or mitigating factor in sentencing, whether psychopaths should be released by parole boards when they behaved well in prison, and whether they should be detained in prison after serving their sentences to prevent them from committing more crimes.

Outside of law, psychiatrists need to ask whether psychopathy should count as a mental illness, disorder, or disease[2] and how it can be treated, if at all. Philosophers need to ask whether psychopathy creates special problems for moral theories, including intuitionism[3] and motivational internalism about moral judgments.[4] Business leaders need to figure out how to detect psychopathy in their potential or current partners, employees, and customers. Of course, each of us also needs to figure out how to avoid psychopaths in our personal lives—or deal with them if we cannot avoid them.

To address these issues, we need to learn more about psychopathy. Unfortunately, the amount of funding for psychopathy research is minuscule compared to other social problems such as drug abuse. Fortunately, the pace of work on psychopathy has exploded recently. Aided by the standard diagnosis using the Hare Psychopathy Checklist-Revised (PCL-R), researchers have been able to compare disparate studies to attain new levels of precision and generality. New tools—especially structural and functional brain imaging as well as genetics—have provided surprising insights and hold out some hope for eventual treatments. Access to new populations in prisons has made it possible to study high-scoring psychopaths in depth. The legal system is beginning to show interest in this new science of psychopathy.

To capture this excitement, we organized a three-day workshop on psychopathy and the law in conjunction with a meeting of the Society for the Scientific Study of Psychopathy in New Orleans in 2009. The speakers at this workshop were then asked to compose chapters for the present volume in light of the feedback that they received on their talks. They were also asked to explain the main debates and views of other scholars instead of simply arguing for their own theses.

The resulting chapters present overviews of a wide variety of perspectives on psychopathy as well as debates about how the legal system should handle psychopaths. This volume should interest not only scientists who study psychopathy but also practitioners who need to deal with psychopaths in the legal system as well as students who are just entering the field.

NOTES

1. See Kiehl and Hoffman, 2011.
2. See Nadelhoffer and Sinnott-Armstrong, forthcoming.
3. See Sinnott-Armstrong, forthcoming-a.
4. See Sinnott-Armstrong, forthcoming-b.

REFERENCES

Kiehl, K. A., & Hoffman, M. B. (2011). The criminal psychopath: History, neuroscience and economics. *Jurimetrics: The Journal of Law, Science, and Technology*, Summer, *51*, 355–397.

Nadelhoffer, T., & Sinnott-Armstrong, W. (Forthcoming). Is psychopathy a mental disease? In: Vincent, N (Ed.), *Neuroscience and responsibility* (pp. 227–253). New York: Oxford University Press.

Sinnott-Armstrong, W. Forthcoming-a. Do psychopaths refute internalism? In: Schramme, T. (Ed.), *Being amoral: Psychopathy and moral incapacity*. Cambridge, MA: MIT Press.

Sinnott-Armstrong, W. Forthcoming-b. Moral disagreements with psychopaths. In: Bergmann, M. (Ed.), *Challenges to Religious and Moral Belief: Disagreement and Evolution*. New York: Oxford University Press, 2013.

PART ONE
DIAGNOSIS OF PSYCHOPATHY

CHAPTER 2

Assessment of Psychopathy: The Hare Psychopathy Checklist Measures

Adelle Forth, Sune Bo, and Mickey Kongerslev

The construct of psychopathy has a long history within the fields of clinical psychology and psychiatry. As Millon, Birket-Smith, and Simonsen (1998) have noted, psychopathy was the first specific personality disorder to be recognized in psychiatry. Although descriptions of psychopathy have varied over time, and controversies still exist, most descriptions have focused on a mixture of callous and unemotional personality traits coupled with antisocial and interpersonally problematic behaviors. However, until the 1980s, a fundamental problem plagued research on psychopathy: the lack of a reliable and valid measure of the construct. The Canadian psychologist Robert Hare addressed this deficiency in developing the Psychopathy Checklist (PCL; Hare, 1980). The PCL was an instrument for measuring psychopathy that incorporated descriptions of personality traits and antisocial behaviors based on clinical observations. Over the years, the PCL has been revised (PCL-R; Hare, 1991, 2003), and other instruments for assessment of psychopathy in different samples and settings have been derived from the PCL: the Psychopathy Checklist: Screening Version (PCL:SV; Hart, Cox, & Hare, 1995) and the Psychopathy Checklist: Youth Version (PCL:YV; Forth, Kosson, & Hare, 2003). The PCL-R and its derivatives are currently the most widely used measures for the assessment of psychopathy in research as well as in clinical, forensic, and pretrial and other legal settings.

This chapter discusses the development, assessment procedures, psychometric properties, strengths, and limitations of the PCL-R, PCL:SV, and PCL:YV. The discussion also touches on salient issues pertinent to the use of these instruments in clinical and legal settings, such as cutoff scores, categorical versus dimensional, misuses, labeling, and relationship with Antisocial Personality Disorder (*Diagnostic and Statistical Manual of Mental Disorders* [*DSM–IV* and *DSM–IV–TR*]; American Psychiatric Association, 1994, 2000).

HARE PSYCHOPATHY CHECKLIST-REVISED

Description of the PCL-R

No standard methods for assessing psychopathy existed before the 1980s. Researchers used their own clinical judgment along with various self-report scales to classify participants. This situation changed dramatically with the advent of the PCL developed by Robert Hare (Hare, 1980). Hare's work on the PCL at the University of British Columbia has had a far-reaching impact on psychopathy and its assessment. Hare and his colleagues originally developed the PCL to assess psychopathic traits in criminal offenders for research purposes. However, increasing interest in the clinical construct of psychopathy from both research and applied psychology and psychiatry perspectives prompted the publication of the Hare PCL-R in 1991. The PCL-R has played an essential role in the assessment of psychopathy during the past 20 years. The PCL-R provides researchers and practitioners with a uniform metric for the assessment of psychopathy, thereby enhancing both research and clinical work.

Hare and colleagues began the development of the PCL by reviewing the clinical literature, paying specific attention to the construct of psychopathy as described by the American psychiatrist Hervey Cleckley. Cleckley (1976) described 16 features ranging from positive features (e.g., good intelligence, social charm, and absence of delusions and anxiety), emotional-interpersonal features (e.g., lack of remorse, untruthfulness, unresponsiveness in interpersonal relations), and behavioral problems (e.g., inadequately motivated antisocial behavior, unreliability, failure to follow any life plan). This resulted in a pool of more than 100 items encapsulating both behavioral aspects and trait-like factors essential to the concept of psychopathy. Later, the item pool was reduced due to redundancies and problems with obtaining reliable ratings. A preliminary scoring procedure was developed and the results analyzed to ascertain which items could discriminate between individuals with high and low levels of psychopathy. Of the original 100 items, 22 remained, and the resulting scale was named the Psychopathy Checklist (PCL; Hare, 1980). The PCL was made available to researchers, and based on feedback, modifications were made to the PCL in 1985: two items were deleted, scoring criteria were made more explicit, the procedures for dealing with missing information clarified, and sources of information described in more detail. In 1991 the PCL-R (Hare, 1991) was published and in 2003, the second edition of the PCL-R was published. Although no changes were made to the item descriptions, the second edition included several other modifications: (1) a more detailed discussion of appropriate uses and misuses of the scale; (2) a summary of the factor structure, reliability, validity, and generalizability of the scale; and (3) comparison tables for a large sample of North American and European offenders and forensic patients.

The PCL-R (Hare, 1991, 2003) is a 20-item rating scale designed to measure the construct of psychopathy. Each item is rated on a 3-point scale (0 = the item definitely does not apply; 1 = the item applies to some degree; 2 = the item definitely applies). A total summed score can range from 0 to 40, reflecting the extent to which the person matches the prototypical psychopath. Table 2.1 lists the items and factors in the PCL-R and its derivatives, the PCL:SV and PCL-YV. No exclusion criteria are associated with the PCL-R, a feature that permits the assessment of individuals with different psychiatric diagnoses and comorbid diagnoses, such as psychiatric offenders.

The PCL-R is scored based on a semistructured interview and review of available file and collateral information (Hare, 2003). The interview takes approximately 90 to 120 minutes, may be spread over several sessions, and covers a variety of domains (educational, employment, substance use, family background, relationships, and criminal history). For clinical assessments it is required that the evaluator have access to collateral information in order to check the accuracy of the information given in the interview. This requirement makes the PCL-R assessment format different from self-report questionnaires designed to assess psychopathy (see Lilienfeld & Fowler, 2006). Psychopathic individuals can lie when answering self-report questionnaires but they can also lie in an interview. Thus, to assess the credibility of what they say and to know when they are distorting, minimizing, or omitting information, the evaluator must have access to information from other sources.

What Is the Factor Structure of the PCL-R?

Three models have been proposed to represent the factor structure underlying PCL-R total scores in adults. Although each model differs in the number of underlying factors proposed, they all assume that the factors of the PCL-R correlate and represent a high-order construct of psychopathy. The majority of the studies that have set out to investigate the structure of the PCL-R have used a variety of factor analytic techniques to explore and investigate the underlying facets of psychopathy: exploratory factor analysis, confirmatory factor analysis, and structural equation modeling (Bishopp & Hare, 2008).

Table 2.1 Items and Factors in the Hare Psychopathy Checklist (PCL) Scales

PCL-R	PCL:SV	PCL:YV
Factor 1	Part 1/Factor 1	
Facet 1: Interpersonal	Facet 1: Interpersonal	Factor 1: Interpersonal
1. Glibness—superficial charm	1. Superficial	1. Impression management
2. Grandiose sense of self-worth	2. Grandiose	2. Grandiose sense of self-worth
4. Pathological lying	3. Deceitful	4. Pathological lying
5. Conning, manipulative		5. Manipulation for personal gain
Facet 2: Affective	Facet 2: Affective	Factor 2: Affective
6. Lack of remorse or guilt	4. Lacks remorse	6. Lack of remorse
7. Shallow affect	5. Lacks empathy	7. Shallow affect
8. Callous, lack of empathy	6. Does not accept responsibility	8. Callous/lack of empathy
16. Failure to accept responsibility		16. Failure to accept responsibility
Factor 2	Part 2/Factor 2	
Facet 3: Lifestyle	Facet 3: Lifestyle	Factor 3: Behavioral
3. Need for stimulation	7. Impulsive	3. Stimulation seeking
9. Parasitic lifestyle	9. Lacks goals	9. Parasitic orientation
10. Lack of realistic, long-term plans	10. Irresponsibility	13. Lack of goals
14. Impulsivity		14. Impulsivity
15. Irresponsibility		15. Irresponsibility
Facet 4: Antisocial	Facet 4: Antisocial	Facet 4: Antisocial
11. Poor behavioral controls	8. Poor behavioral controls	10. Poor anger controls
12. Early behavioral problems	11. Adolescent antisocial behaviour	12. Early behavior problems
18. Juvenile delinquency	12. Adult antisocial behavior	18. Serious criminal behavior
19. Revocation of conditional release		19. Serious violations of release
20. Criminal versatility		20. Criminal versatility
Additional items		Additional items
11. Promiscuous sexual behavior		11. Impersonal sexual behavior
17. Many short-term marital relationships		17. Unstable interpersonal relationships

Notes: Items are from Hare (2003), Hart et al. (1995), and Forth et al. (2003), respectively. Table adapted from Hare and Neumann (2009). Reprinted by permission of R. D. Hare.

The most widely used model is the two-factor model originally reported by Harpur, Hakstian, & Hare (1988; Hare et al., 1990; Harpur, Hare, & Hakstian, 1989). In this model, 17 of the 20 items in the PCL-R load on two correlated dimensions. The first, commonly referred to as Factor 1, represents the interpersonal and affective features of psychopathy. The second, commonly referred to as Factor 2, reflects chronic impulsive, irresponsible, and antisocial tendencies. These factors show differential correlations with external criteria, including psychiatric diagnoses, emotional processing performance, and risk for recidivism and violence (Harpur et al., 1989). In 2001, Cooke and Michie proposed a three-factor model based on 13 of 20 PCL-R items. Arguing that the antisocial items were a consequence of other more personality-based items, they removed the antisocial items before conducting the factor analyses. In this model, the interpersonal and affective factor splits into an interpersonal and affective factor, with a third factor focusing on impulsive and irresponsible behaviors.

Recently, Hare & Neumann (2006) and Neumann, Hare, & Newman (2007) have argued for a four-factor model of psychopathy.[1] The four latent factors that represent the PCL-R construct are indicated in Table 2.1. The four-factor model has the same factors as the three-factor model but with the addition of an antisocial factor. Support for the four-factor model is present across various studies and in diverse samples: forensic and civil psychiatric patients (Vitacco, Rogers, Neumann, Harrison, & Vincent, 2005), youth offenders (Jones, Cauffman, Miller, & Mulvey, 2006; Neumann, Kosson, Forth, & Hare, 2006; Salekin, Brannen, Zalot, Leistico, & Neumann, 2006; Vitacco, Neumann, Caldwell, Leistico, & Van Rybroek, 2006), and additionally in samples from the general population (Hare & Neumann, 2006; Neumann & Hare, 2008).

What Cutoff Score to Use?

In North America a cutoff score of 30 is generally used to "diagnose" psychopathy and classify participants in empirical studies. However, in some predictive validity studies of reoffending, a lower cutoff, ranging from 15 to 27, has been used (Barbaree, 2005; Grann, Långström, Tengström, & Kullgren, 1999; Hare, Clark, Grann, & Thornton, 2000; Looman, Abracen, Maillet, & diFazio, 2005; Rice & Harris, 1997; Rice, Harris, & Cormier, 1992; Seto & Barbaree, 1999; Tengström, Grann, Långström, & Kullgren, 2000). In Europe the tradition among practitioners and researchers has been to use a lower cutoff score, often 25.

The PCL-R provides a dimensional score that represents the degree to which the offender resembles the prototypical psychopath. Recent empirical research supports the assumption that the construct underlying the PCL-R is dimensional in nature (Edens, Marcus, Lilienfeld, & Poythress, 2006; Guay, Ruscio, Knight, & Hare, 2007). This dimensionality extends to both the PCL:SV (Walters et al., 2007) and the PCL:YV (Murrie et al., 2007). However, some researchers may wish to use cutoff scores to group participants, such as classifying those scoring less than 20 as "low," 21 to 29 as "moderate," and 30 or above as "high." The PCL-R manual provides T-scores and percentiles for clinicians to compare the individual they are assessing to large samples of offenders or forensic psychiatric patients.

Judicial decision making often involves categorical decisions such as the following questions: Is the person guilty or not? Should a sentence of imprisonment be imposed or not? Should the offender be released or not? If clinicians and judicial decision makers mechanically/simplistically use 30 as the line "drawn in the sand," with a score of 29 meaning the person is not a psychopath and a score of 30 meaning the person is a psychopath, this could lead to problematic consequences. For example, a person scoring 30 may be described as having poor treatment responsivity and being at higher risk for institutional behavioral problems and for reoffending. In contrast, a person scoring 29 could potentially be described as not a psychopath and being amenable to treatment and at low risk for institutional misconduct and future violence. However, taking measurement error into account renders such assumptions unwarranted. Still, it is not uncommon for defense lawyers to cross-examine an expert witness vigorously for the prosecution on their PCL-R item scores with the intent to have the expert lower his or her scores. Once the scores are below the 30 threshold, some defense lawyers will then claim that since the defendant is now not a psychopath (i.e., score fails below the 30 cutoff), judicial decision makers can disregard the clinician's findings relating to treatability, management, and risk.

How Reliable Is the PCL-R?

Many studies confirm the internal consistency of the PCL-R total and factor scores (Hare, 2003). Alpha coefficients for total scores range from .81 in male forensic patients to .87 in studies using file review coding of male offenders (Hare, 2003). The standard error of measurement (SEM) takes into account both the standard deviation and the reliability of the scale. The SEM of the PCL-R for the total score is 3 for a single rating and 2 for the average of two ratings (Hare, 2003).

Do two evaluators agree on scores? Studies have consistently reported excellent interrater reliability for PCL-R total scores (Hare, 2003; single Intra-class Correlation Coefficient (ICCs) ranging from .86 to .94). Moreover, studies that have compared clinical ratings to independent research ratings have also found very strong agreement (e.g., ICC = .92; Woodworth & Porter, 2002; single ICC = .91, Brown & Forth, 1997) In short, these studies indicate that it is possible to obtain similar scores on the PCL-R in both applied and research settings. Nonetheless, the reliability of the raters likely depends on how similar the sources of information are, when the evaluations were conducted (within weeks, months, or years of each other), the training of the raters, and the context of the evaluation. Gacono and Hutton (1994) compared scores of clinicians conducting PCL-R training to the clinicians being trained. Trainers and trainees interviewed the forensic patients together and had access to the identical medical and institution files. Across 146 sets of PCL-R total scores, 92% of the cases were within 2 points of each other. In contrast, Edens, Boccaccini, and Johnson (2010) found poor interrater reliability for 20 sex offenders. The absolute single rater intraclass correlations for Total, Factor 1, and Factor 2 scores were .42, .16, and .56. It is not clear what accounted for the rater differences on Factor 1 scores, whether it was the inherent difficulty of assessing more subjective traits or whether evaluators found it particularly challenging to assess child molesters. See Table 2.2 for a summary of the reliability of PCL-R scores in adversarial contexts. Recently, and more specific to sexual violent predator evaluations, researchers have compared PCL-R scores from prosecution (petitioner) and defense (respondent) and found substantial discrepancies. Murrie, Boccaccini, Johnson, and Janke (2008) and Murrie et al. (2009) reported substantially lower levels of agreement from opposing sides of a case (ICCs ranging from .39 to .42). When comparing scores from the same side (two prosecution experts) inconsistent findings have emerged. Levenson (2004) found higher levels of agreement in state-contracted evaluators in Florida evaluating the same offender, whereas in Texas, Boccaccini, Turner, and Murrie (2008) found relatively poor level of agreement between state-contracted evaluators scoring the same offender. Boccaccini et al. (2008) concluded that "about 45% of the variance would be attributable to offenders' true standing on the PCL-R; about 30%, to evaluator differences; and about 20%, to

Table 2.2 PCL-R Evaluator Scores in Adversarial Contexts

Study	Sample/Context	Comparison	ICC_{A1}	Average difference
Boccaccini et al. (2008)	Sex offenders Texas civil commitment	22 Prosecution vs. prosecution expert	.47	—
Murrie et al. (2009)	Sex offenders Texas civil commitment	35 Prosecution vs. defense expert	.42	5.79
		13 Prosecution vs. prosecution experts	.24	6.37
		7 Defense vs. defense experts	.88	2.57
Murrie et al. (2008)	Sex offenders Texas civil commitment	23 Prosecution vs. defense expert	.39	7.81
Levenson (2004)	Sex offenders Florida civil commitment	69 Prosecution vs. prosecution expert	.84	—
Lloyd et al. (2010)	Violent offenders Canada dangerous offender/LTO hearings	15 Prosecution vs. defense expert	.67	2.55
		7 Court-appointed vs. defense expert	.82	2.28

Notes: ICC_{A1} absolute agreement, single rater; samples in Murrie et al. (2008) and Murrie et al. (2009) are overlapping samples. LTO, long-term offender.

adversarial allegiance" (p. 276). In Canada, only one study has examined the reliability of experts in an adversarial context (Lloyd, Clark, & Forth, 2010). Scores from prosecution and defense experts were available in only 15 cases and were not as blatantly different as seen in past studies. Moreover, an even stronger agreement was obtained between court-appointed experts and defense experts (although the sample size was very small). Until additional research is conducted in other jurisdictions using larger samples, it is difficult to know what is contributing to the very low level of interrater agreement in Texas for samples of sexual offenders.

How Prevalent Are Psychopathic Traits?

The PCL-R has proven applicable in multiple contexts. It has exhibited reliable findings in different cultures (Cooke, Kosson, & Michie, 2001; Sullivan & Kosson, 2006); among forensic populations including offender and forensic psychiatric; among various samples of psychiatric populations, including forensic (Gacono & Hutton, 1994); and among patients with substance abuse issues (Rutherford, Alterman, Cacciola, & McKay, 1997; Rutherford, Cacciola, Alterman, & McKay, 1996).

Although the PCL-R was developed mainly with data form Caucasian male offenders and forensic patients, the psychometric properties have been confirmed in many different offender settings and with other patients (Hare, 2006), including women, substance abusers, and sex-offenders, as well as in diverge cultural settings (Cooke, Kosson, & Michie, 2001). Table 2.3 lists the mean PCL-R scores across different settings and samples. People who have come into contact with the justice system (forensic psychiatric and offenders) have higher PCL-R scores than people in the community. The mean PCL-R scores in both the Vanman, Mejia, Dawson, Schell, and Raine (2003) and the DeMatteo, Heilbrun, and Marczyk (2006) studies of community samples are relatively high because of their recruitment methods. Vanman et al. (2003) recruited people from a temporary employment setting, and DeMatteo et al. (2006) placed advertisements in local newspapers asking for "charming, intelligent, adventurous, aggressive, and impulsive...get bored easily and like to live on the edge" (p. 136) persons to participate. Higher PCL-R scores have been found with male offenders as compared to female offenders. Mean PCL-R scores differ across sexual offenders, with child molesters scoring lower than rapists, and sexual offenders who offend sexually against both children and adults scoring the highest.

Is the PCL-R Valid?

To date there is an extensive literature confirming the validity of the PCL-R. In 1991 Hare concluded: "There is increasing evidence that PCL scores are related, in appropriate ways, to a variety of clinical, self-report, and demographic variables. At the same time, they are unrelated to, or only weakly related to, variables that theoretically should not be associated with psychopathy"

Table 2.3 PCL-R Scores across Different Samples

Study	Sample	N	Mean (SD)
Weiler & Widom (1996)	Community—Childhood abuse	629	9.2 (6.9)
	Community—No childhood abuse	440	6.8 (5.9)
DeMatteo et al. (2006)	Community men	54	14.0 (6.6)
Vanman et al. (2003)	Community men and women	80	18.5 (8.4)
Hare (2003)	Female offenders	1,218	19.9 (7.5)
Hare (2003)	Male offenders	5,408	22.1 (7.9)
Hare (2003)	Male forensic psychiatric patients	1,246	21.5 (6.9)
Hare (2003)	Child molesters	94	20.9 (6.3)
Hare (2003)	Rapists	162	25.5 (6.2)
Hare (2003)	Mixed sexual offenders	24	29.0 (6.6)

(Hare, 1991: 48). Twelve years later, in the second edition of the manual, Hare (2003) presented substantial references to studies underscoring the legitimacy of the PCL-R as a valid assessment instrument for psychopathy, including findings relating the PCL-R to other measurements of psychopathy, comparing it with other instruments and scales (e.g., self-report measurements), and linking it to violence and recidivism. The construct of psychopathy has also been studied in various basic research contexts that have included investigations of how psychopaths process information, as well as the physiological and anatomical characteristics of psychopathy (e.g., via functional magnetic resonance imaging and measurement of electrocortical activity and biochemical correlates) (Blair, 2007; Blair, Mitchell, & Blair, 2005; Glen, Raine, & Schug, 2009; Kiehl et al., 2004, 2006; Sadeh & Verona, 2008).

Although a detailed discussion of the predictive validity of the PCL-R for risk of violence and treatment response is beyond the scope of this chapter, an outline of some of the main findings shall be provided (see Rice and Harris, Chapter 13 for more detailed discussion). The PCL-R has been found to be useful as a predictor of behavior in a number of different studies: violent behavior in forensic hospitals (Hare et al., 2000; Kroner & Mills, 2001; Walters, 2003), engagement in violence after institutional release (Glover, Nicholson, Hemmati, Bernfeld, & Quinsey, 2002; Kroner & Mills, 2001; Serin, 1996; Serin & Amos, 1995), recidivism in female offenders (Salekin, Rogers, Ustad, & Sewell, 1998), criminal recidivism in male offenders (Douglas & Webster, 1999; Rice & Harris, 1995; Serin, 1996; Serin & Amos, 1995), sexual recidivism in psychopaths with deviant sexual interest (Hildebrand, de Ruiter, & de Vogel, 2004; Rice & Harris, 1997; Serin, Mailloux, & Malcolm, 2001), and sexual reoffending in released sexual offenders (Furr, 1993). In a meta-analysis of 68 studies using the PCL scales, Leistico, Salekin, DeCoster, and Rogers (2008) reported effects sizes of .50 and .47 for general and violent recidivism. In a recent survey of forensic clinicians, 54% reported regularly using the PCL-R or the PCL:YV when performing a violence risk assessment (Viljoen, McLachlan, & Vincent, 2010).

In general, offenders with psychopathic traits are more prone to engage in criminal behavior across the lifespan compared to offender groups without psychopathy (Porter, Birt, & Boer, 2001), engage in more criminal activity (Blackburn & Coid, 1998; Brown & Forth, 1997; Porter, Fairweather, Drugge, Hervé, Birt, & Boer, 2000), and exhibit more instrumental aggression (Cornell et al., 1996; Williamson, Hare, & Wong, 1987; Woodworth & Porter, 2002).

For years the efficacy of psychopathy treatment has been questioned, based on the assumption that individuals with psychopathic traits are highly treatment-resistant. Some researchers have also argued that treatment programs that are emotion focused; are psychodynamic or insight-oriented; or are aimed at developing empathy, enhancing conscience, and increasing interpersonal skills are not the appropriate type of treatment programs for psychopathic individuals (Harris & Rice, 2006; Thornton & Blud, 2007; Wong & Burt, 2007). Researchers have also reported that psychopaths exhibit poor motivation, show little treatment progress, and often terminate treatment prematurely (Hobson, Shine, & Roberts, 2000; Ogloff, Wong, & Greenwood, 1990). The interpersonal and affective features of psychopathy (Factor 1) especially seem to interfere with treatment adherence (Hobson et al., 2000). Moreover, inappropriate treatment approaches are contraindicative to the resocialization of offenders (Rice et al., 1992). Therapeutic community treatment programs, though very extensive and applied worldwide in diverse criminal settings, have failed to show positive findings in the treatment of psychopathy (Harris & Rice, 2006).

On the other hand, "therapeutic nihilism" is not warranted. Treatment programs targeted at reducing risk for violence and recidivism, including relapse-prevention techniques and those that target risk-needs-responsivity principles, have consistently been shown to be effective with offenders (Andrews, Bonta, & Wormith, 2006; Dowden & Andrews, 2000; Hanson, Bourgon, Helmus, & Hodgson, 2009). Cognitive-behavioral techniques have proven to be among the best treatment approaches, resulting in reduced violence and recidivism among offenders with many psychopathic traits (Caldwell, Skeem, Salekin, & Van Rybroek, 2006; Doren & Yates, 2008; Langton, Barbaree, Harkins, & Peacock, 2006; Olver & Wong, 2009; see Caldwell, Chapter 12 for a more detailed discussion).

Despite some success, caution is warranted when treating psychopaths because research has demonstrated how their manipulative skills can be improved as a result of inappropriate treatment (Hare et al., 2000; Hobson et al., 2000; Looman, Abracen, Serin, & Marquis, 2005; Rice et al., 1992). Thus, it is critical that future research investigate associations between psychopathy and treatment in order to disentangle how high levels of psychopathy are related to treatment outcome. Also, future research needs to identify the core psychological mechanisms, structures, and processes underlying psychopathy, assuming that treatment interventions become most effective when they are tailored to and directly target such core psychopathological mechanisms, structures, and processes (Allen, Fonagy, & Bateman, 2008; Livesley, 2003, 2010). Moreover, advances in the areas of neuroscience and genetic research may inform development of effective prevention and early intervention strategies.

Psychopathy and Antisocial Personality Disorder: Toward *DSM–V*

Although psychopathy is among the strongest predictors of risk for offending and risk for violence (e.g., Leistico et al., 2008), it is not an explicit and separate diagnosis in the DSM system. Additional confusion surrounds the diagnosis of Antisocial Personality Disorder (APD) and its relationship to both psychopathy and "sociopathy." The three terms are sometimes used interchangeably (e.g., Blackburn, 1988), whereas a general consensus among most researchers is that the constructs of APD, psychopathy, and "sociopathy" are related but distinct (Hare & Neumann, 2008).

Exacerbating this confusion is the statement found in *DSM–IV–TR* noting that APD "has also been referred to as psychopathy, sociopathy, or dissocial personality disorder" (APA, 2000: 702). This creates a potential pitfall by raising the possibility that some clinicians might equate APD with psychopathy, including utilizing research findings about treatment amenability and risk assessment related to psychopathy, when describing the implications of a person diagnosed with APD (Hare & Neumann, 2009). Anecdotal evidence comes from court testimony where one mental health expert stated that "persons with antisocial personality are psychopaths. They are some of the most dangerous people alive" (*R. vs. Talbot*, 1995).

The term "sociopath" was coined in 1930 by Partridge to describe those people who had problems with or refused to adapt to society. Lykken (2006) proposed that sociopaths manifest similar traits as psychopaths but develop these traits as a result of poor parenting and other environmental factors whereas psychopaths are genetically predisposed to a temperament that makes them difficult to socialize. The term "sociopath" is rarely used in the empirical literature and there are no assessment instruments developed to identify the construct.

In the *DSM–IV–TR* (APA, 2000), the criteria for APD are based largely on behavioral symptoms. By excluding the richer clinical descriptions reflecting the concept of psychopathy, a diagnosis of APD has more limited clinical utility, when it comes to predicting recidivism, institutional management, and treatment amenability. Ogloff (2006) describes APD as a diagnostic category merely for behavioral delinquency pertaining to criminality, and research confirms that the prevalence for APD in offenders is higher compared to that of psychopathy (50%–80% vs. 15%; Hare, 2003), making it an overinclusive concept with limited predictive value in correctional samples. When comparing the criteria defining APD in the DSM system with the psychopathy items reflecting Factor 1 and Factor 2, including the corresponding four facets, only three of the eight items from Factor 1 (two items from Facet 1—pathological lying and conning/manipulative—and one item from Facet 2—lack of remorse or guilt) are found in the criteria for APD. On the contrary, six out of ten of the items from Factor 2 (three items from Facet 3—need for stimulation and proneness to boredom, impulsivity, and irresponsibility—and three from Facet 4—poor behavioral controls, early behavioral problems, and criminal versatility) overlap with APD criteria. Therefore only 37.5% of the interpersonal/affective symptoms from the PCL-R and 60% of the social deviance symptoms are found in the criteria defining APD (Ogloff, 2006).

Various researchers have expressed concern about the conceptual development of *DSM–III* and *DSM–IV* criteria in relation to the diagnosis of APD, including the overinclusiveness of APD in offender populations, the lack of diagnostic validity, and limited predictive power when estimating reoffending behavior and recidivism (Hare, 2003; Ogloff, 2006; Rogers, Duncan, Lynett, & Sewell, 1994). Lykken (2006) concludes that "Identifying someone as having APD is about as nonspecific and scientifically unhelpful as diagnosing a sick patient as having a fever or an infectious or a neurological disorder" (p. 4).

In contrast to APD, research indicates that those who meet the criteria for psychopathy as measured by the PCL-R display distinct clinical features, including divergent patterns of cognitive processing (Blair et al., 2004) and an impaired ability to perceive the emotional content of language (Hare, 1996; Kosson, Lorenz, & Newman, 2006). Further, functional magnetic resonance imaging (fMRI) studies have revealed differences in brain structure between psychopaths and nonpsychopaths; for example, high scoring psychopaths have decreased gray matter in the prefrontal gray cortex (Raine, Lencz, Bihrle, LaCasse, & Coletti, 2000; Yang, Raine, Narr, Coletti, & Toga, 2009), different hippocampus structure (Boccardi et al., 2010; Laakso et al., 2001; Raine et al., 2004), and decreased gray matter near the limbic system (de Oliveira-Souza et al., 2008).

The *DSM–V* is scheduled for publication in 2013. The *DSM–V* task force should either (1) work toward a clear and explicit distinction between psychopathy and Antisocial Personality Disorder, and refrain from equating the two concepts in the introduction and in the associated features sections or (2) incorporate personality features and more stable traits, as diagnostic criteria defining APD, much in line with the descriptions in PCL-R, *DSM–II*, and International Classification of Diseases (ICD)-10, thereby enhancing the clinical and predictive relevance of APD and improving diagnostic validity.

How Generalizable Is the PCL-R?

Assessing psychopathy in female offenders and female forensic patients with the PCL-R has proven to be valid (Dolan & Völlm, 2009; Hare, 2003), and the descriptive properties very similar to those of the male population (e.g., Loucks & Zamble, 2000; McDermott et al., 2000; Richards, Casey, & Lucente, 2003; Richards, Casey, Lucente, & Kafami, 2003; Vitale & Newman, 2001; Vitale, Smith, Brinkley, & Newman, 2002; Warren et al., 2003; Windle & Dumenci, 1999). Support for the two-factor, three-factor, and four-factor models of psychopathy have been found for both males and females (Hare, 2003; Warren et al., 2003). Similarities in criminal correlates are reported (Richards et al., 2003; Vitale et al., 2002), as well as response to treatment (Richards et al., 2003), and the relation to psychopathology, including personality disorders (Salekin, Rogers, & Sewell, 1996; Warren et al., 2003).

Yet, several differences exist between male and female PCL-R results (Bolt, Hare, Vitale, & Newman, 2004). The PCL-R is not as strong a predictor of institutional aggression or post-release offending in women as in men (Dolan & Völlm, 2009). In samples of serious female offenders, the PCL-R does predict reoffending (Loucks & Zamble, 2000; Richards et al., 2003) but may not be as strong a predictor in samples of less serious female offenders (Dolan & Doyle, 2000; Salekin et al., 1996). The majority of studies have found lower PCL-R scores for females than for males in substance abusers (Alterman, Cacciola, & Rutherford, 1993), forensic psychiatric patients (Grann, 2000), and offenders (Vitale et al., 2002). In the PCL-R manual, Hare (2003) reported that 15% of male offenders had a PCL-R score above 30, and for women only 7% had a score of 30 or above (see section on PCL:SV for a comparison of men and women differences).

Studies investigating the psychometric properties of the PCL-R outside North America have reported lower scoring in the United Kingdom and other European countries. Across cultures, Factor 1 displays good correspondence, specifically Facet 2 (affective), indicating that this may be the pan-cultural core deficit of psychopathy (Cooke, Michie, Hart, & Clark, 2005). Yet, more cross-cultural studies are required to compare and adapt the metric properties of the PCL-R.

Based on a meta-analytic review, Skeem, Edens, Camp, and Colwell (2004) concluded that PCL-R scores were not influenced by ethnicity, and though the data provided to develop the PCL-R were from primarily Caucasian offenders, Item Response Theory (IRT) analysis has displayed psychometric correspondence for Caucasian and African American male offenders (Cooke et al., 2001). Data for the second edition of the PCL-R was obtained from diverse cultural settings and did not appear to affect the metric properties of the PCL-R (Hare, 2003).

Some studies, though, indicate cultural and ethnic differences in relation to the mean scores for the PCL-R, highlighting how cultural factors contribute to variations observed in expression of psychopathy (see Sullivan & Kosson, 2006 for review). More research is needed to clarify the similarities and differences surrounding the metric properties of the PCL-R, when applied in various cultural settings.

Forensic Uses of the PCL-R

Several studies have surveyed the use of expert testimony regarding the assessment of psychopathy, sociopathy, or APD in criminal and civil court proceedings. Table 2.4 lists the range of criminal and civil cases in which evidence concerning psychopathy or related constructs was introduced in Canada and the United States. Lyon and Ogloff (2000) and Zinger and Forth (1998) reviewed cases in which psychopathy, sociopathy, or APD were evaluated, whereas in more recent studies by DeMatteo and Edens (2006) and Walsh and Walsh (2006), only cases in which the PCL-R was used were included. Most recently, Viljoen, MacDougall, Gagnon, and Douglas (2010) examined adolescent criminal cases in Canada and the United States in which psychopathy was assessed. These studies show that psychopathy has played a role in a diverse range of criminal cases, with the majority of testimony regarding psychopathy being associated with an increased severity of disposition. For example, in death penalty hearings in the United States and dangerous offender proceedings in Canada, a diagnosis of psychopathy, sociopathy, and APD is an aggravating factor for the death penalty (in the United States), and considered to be associated with a higher risk of violent recidivism and a lack of treatment responsivity in dangerous offender hearings. With respect to the insanity defense, the diagnosis of psychopathy does fulfill the disease of the mind requirement but has not ever fulfilled the second requirement of not appreciating the nature or quality of the act or knowing that it is wrong. The most common use of psychopathy in adolescent criminal cases was to help determine whether the youth should or should not be transferred to adult court.

Two studies have examined the use of PCL-R–based psychopathy assessments in United States criminal proceedings between 1991 and 2004. DeMatteo and Edens (2006) and Walsh and Walsh (2006) studies were based on analyses of legal databases of written judicial judgments and thus included only information the judges felt was pertinent in making their decision. The majority of cases using the PCL-R occurred after 2000, with the prosecution being most likely to introduce PCL-R evidence. The PCL-R was used most often in sexual violent predator hearings where a higher PCL-R score was related to higher risk for sexual recidivism, a central issue in determining whether to declare someone a sexual violent predator. Written court decisions are a useful method of gauging what the judicial decision maker is relying on to support his or her final determinations. However, future research needs to code both the mental health professionals' written report and their court testimony to determine their clinical conclusions and whether any cautions were mentioned when testifying about psychopathy.

Concerns About Misuse

Assessment of psychopathy using the PCL-R requires that the prescribed method for administration be followed (see Hare, 2003). When the PCL-R is used to direct adjudication in criminal justice context, and treatment initiations in clinical settings, there are potentially serious consequences for misuse (Hare, 1998). The information provided by the PCL-R must be

Table 2.4 Contexts for the Use of Psychopathy Assessments in Canadian and United States Courts

Study	Data source/Country	Search terms/Years	Contexts used
Zinger & Forth (1998)	QuickLaw–Canada	Psychopathy, sociopathy, APD	• Sentencing hearings • Transfers of youth to adult court • Dangerous offender hearings
	Number of cases not specified	Prior to 1998	• Appeals on parole eligibility • Mental state at time of offence
Lyon & Ogloff (2000)	Legal databases—United States	Psychopathy, sociopathy, APD	• Witness credibility • Competency to stand trial • Mental state at time of offence
	Number of cases not specified	Prior to 1999	• Repeat/habitual offender laws • Sexual predator laws • Death penalty sentencing • Juvenile transfer to adult court • Child custody issues • Civil commitment
Walsh & Walsh (2006)[a]	LexisNexis—United States	PCL-R	• Sexual violent predator hearings • Parole determination
	76 cases	1991–2004	• Death penalty sentencing • Guilt determination • Sentencing • Juvenile transfer to adult court • Civil commitment • Termination of parental rights • Competency to stand trial
DeMatteo & Edens (2006)[a]	LexisNexis-United States	PCL-R	• Sexual violent predator hearings • Parole/probation determination
	87 cases	1991–2004	• Death penalty sentencing • Sentencing • Mental state at time of offense • Death penalty sentencing • Juvenile transfer to adult court
Viljoen et al. (2010)[a]	West Law—Canada	Psychopathy/PCL scales	• Juvenile transfer to/from adult court • Disposition
	Lexis-Nexis—United States	Prior to 2009	• Adult sentences (Canada) • Voluntariness of confession • Mental state at time of offense
	111 adolescent offenders		• Sex offender commitment • Death penalty sentencing • Dangerous offender legislation

[a]Rank ordered in order of frequency.

interpreted in line with the latest research findings and in accordance with the manual (see Edens, 2001; Hare, 1988 for examples of misuse). All available information must be incorporated in the assessment procedure in order to provide the most comprehensive and in depth analysis of the individual as possible. Clinicians who employ PCL-R assessment must be able to explain and justify the way items are interpreted and scored (Book et al., 2006).

The PCL-R has been subjected to extensive psychometric scrutiny due to the serious consequences related to an assessment of psychopathy, ranging from an application of the label "psychopath" to a designation of "dangerous offender" (or even a death sentence). Various studies underscore the importance of extensive education and specific training before conducting an assessment of psychopathy (Edens, Desforges, Fernandez, & Palac 2004; Hare, 1998, 2003). Hare (2003) explicitly lists a set of criteria practitioners must possess when employing the PCL-R in their work. Clinicians and researchers need to keep themselves updated on current empirical literature regarding the PCL-R. Also, they should regularly discuss and compare ratings to ensure validity and reliability.

PSYCHOPATHY CHECKLIST: SCREENING VERSION

The Psychopathy Checklist: Screening Version (PCL:SV; Hart et al., 1995) was derived from the PCL-R to provide researchers and clinicians with a shorter assessment instrument for the screening of psychopathic traits in nonforensic populations.

Development and Description of the PCL:SV

A detailed, comprehensive, and time-consuming (albeit necessary) evaluation of the individual is required when the PCL-R is used to assess psychopathy in offender populations. As a consequence, its applicability in nonforensic settings has sometimes been problematic. For example, criminal records are often unavailable outside forensic settings, making some of the PCL-R items unscoreable. Hart and colleagues (1995) developed the PCL:SV as a more cost-effective and appropriate instrument for the assessment of psychopathic traits in nonforensic populations.

The PCL:SV contains 12 items, which, as with the PCL-R, is scored on a 3-point scale. Ratings are summed to yield Total Scores ranging from 0 to 24, which can be interpreted both dimensionally and categorically. When interpreted categorically, a cutoff score for psychopathy of 18 or more is recommended. When used for clinical purposes, PCL:SV scores should not be used to diagnose psychopathy, but instead be used to determine if an individual warrants a comprehensive PCL-R assessment.

When developing the PCL:SV some of the original PCL-R items were shortened and simplified, whereas other items were combined. For example, item 3 in the PCL:SV (Deceitful) is derived from collapsing, shortening, and simplifying item 4 (Pathological Lying) and 5 (Conning and Manipulative) from the PCL-R. Also, for PCL-R items reflecting antisocial behaviors, scoring procedures and items were modified, so that access to formal criminal records was not required.

The factor structure of PCL:SV is isomorphic to the two-, three-, and four-factor solutions found with the PCL-R (Cooke, Michie, Hart, & Hare, 1999; Guy & Douglas, 2006; Skeem, Mulvey, & Grisso, 2003). However, recent work using confirmatory factor analysis suggests that the four-factor solution provides the best overall fit for the PCL:SV (Hill, Neumann, & Rogers, 2004; Neumann & Hare, 2008; Vitacco, Neumann, & Jackson, 2005). Accordingly the PCL:SV consists of two parts (Part 1 and Part 2) reflecting the two factor structure of the PCL-R, with four scales representing the four facets of the PCL-R.

Research indicates that the PCL:SV correlates highly with the PCL-R ($r \geq .80$), suggesting that the PCL:SV measures the same overall construct of psychopathy as measured by the PCL-R (Hart et al., 1995). Correlations between the Part 1/Factor 1 and Part 2/Factor 2 of the PCL:SV and PCL-R scores also revealed good correspondence (r .68 and .81, respectively; Hart et al., 1995).

The PCL:SV is often used in research studies, and overall the findings support the concurrent, discriminant, and predictive validity of the instrument (Gray et al., 2004; Douglas, Strand, Belfrage, Fransson, & Levander, 2005; Huchzermeier, Brub, Geiger, Kernbichler, & Aldenhoff, 2008; Nicholls, Ogloff, & Douglas, 2004), and show similar relations to external constructs and variables as does the PCL-R (Coid, Yang, Ullrich, Roberts, & Hare, 2009; Douglas et al., 2005; Forth, Brown, Hart, & Hare, 1996; Hill, Rogers, & Bickford, 1996; Monahan et al., 2000). For example, in a community sample in United Kingdom, Coid and colleagues (2009) found PCL:SV scores correlated positively with suicide attempts, violent behavior, criminal behavior, homelessness, substance abuse, APD, histrionic, and borderline personality disorders, as well as panic and obsessive–compulsive disorders. In civil psychiatric patients in the United States, PCL:SV scores are predictive of violent behavior in the community (AUC = .73, Skeem & Mulvey, 2001). Finally, in a Swedish forensic psychiatric and correctional sample, PCL:SV scores correlated positively with institutional aggression, the HCR-20 score (Webster, Douglas, Eaves, & Hart, 1997), substance use problems, a diagnosis of personality disorder, and negatively with a diagnosis of psychosis. Taken together, research on the instrument's factorial structure as well as correlations with the PCL-R and external variables indicate that the PCL:SV measures the same construct as the PCL-R.

Reliability of PCL:SV

The reliabilities of the PCL:SV in community, forensic psychiatric patients, and correctional offenders are good. The PCL:SV has good internal consistency and interrater reliability (Coid et al., 2009; Douglas et al., 2005; Forth et al., 1996).

How Prevalent Are PCL:SV Psychopathic Traits?

The prevalence of psychopathic traits depends on the sample and gender. Table 2.5 lists findings from seven studies that tested for differences in mean PCL:SV scores between men and women across different samples. In the general population psychopathy is rare. Coid et al. (2009) found that 0.6% of a representative general population sample in Great Britain had scores of 13 or greater on the PCL:SV (only one person scored above the cutoff score of 18), with 71% of the sample having no psychopathic traits (i.e., scoring 0 on the PCL:SV). In a community sample in United States (Neumann & Hare, 2008), about 75% of the sample had scores of 2 or less and only 1.2% had scores in the "potential psychopathic" range

Table 2.5 PCL:SV Scores Across Settings and Sexes

Study	Population	Male N Mean (SD)	Female N Mean (SD)	Significance
Forth et al. (1996)	Undergraduates: Canada	75 6.4 (5.1)	75 2.7 (2.5)	M > F***
Coid et al. (2009)	Community: United Kingdom	301 1.5 (0.16)	319 .54 (0.08)	M > F***
Neumann & Hare (2008)	Community: United States	196 3.5 (3.8)	318 1.7 (2.8)	Not reported
Douglas et al. (2005)[a]	Forensic psychiatric: Sweden	191 12.6 (5.5)	79 12.1 (5.4)	ns
	Offenders: Sweden	168 14.9 (6.1)	31 11.7 (6.4)	M > F
Nicholls et al. (2004)	Forensic psychiatric: Canada	146 8.6 (4.3)	90 6.5 (3.5)	M > F***
Vitacco et al. (2005)	Civil psychiatric: United States	483 9.3 (5.6)	357 7.3 (5.2)	M > F***
de Oliveira-Souza et al. (2008)	Civil psychiatric: Brazil	19 19.5 (4.2)	31 17.2 (3.5)	M > F*

[a]Significance level not reported.
* $p < .05$; *** $p < .001$.

(i.e., scores of 13 or greater on the PCL:SV). In forensic psychiatric and correctional samples substantially higher PCL:SV scores have been observed. Regardless of sample, females consistently score lower than males on the PCL:SV and other psychopathy measures (see Dolan & Völlm, 2009, for a review).

One of the challenges of assessing psychopathic traits in nonforensic populations is the lack of collateral information. It is possible that in the three large-scale studies of general and civil psychiatric populations (Coid et al., 2009; Neumann & Hare, 2008; Vitacco, et al., 2005), the prevalence of psychopathic traits was underestimated since no collateral information was obtained. One method of obtaining collateral information is to contact friends, relatives, or employers of the participants. This method was used in both the de Oliveira-Souza et al. (2008) and Forth et al. (1996) studies and may be partly why the average scores are somewhat higher in these studies than in other community samples.

PSYCHOPATHY CHECKLIST: YOUTH VERSION

As should be clear from the foregoing, the strong association between psychopathy and many forms of serious criminal behavior is well established (Cornell et al., 1996; Hare et al., 2000; Harris & Rice, 2006; Hemphill, Hare, & Wong, 1998; Leistico et al., 2008; Neumann & Hare, 2008; Woodworth & Porter, 2002). The construct of psychopathy identifies a small but distinct subgroup of antisocial adults with specific neurological, cognitive, and affective characteristics (Blair et al., 2005; Dolan & Fullam, 2009; Glenn & Raine, 2008) who tend to chronically commit more violent and severe crimes, often beginning at an early age (Hare, McPherson, & Forth, 1988; Marshall & Cooke, 1999). As a consequence, individuals with psychopathic traits require a costly and disproportionate amount of attention from the criminal justice system (Loeber & Farrington, 2000).

To understand the causes and complex developmental trajectories of psychopathic adult offenders, research on psychopathy must extend to children and adolescents (Forth & Book, 2007; Frick, 2002, 2009). Such research is important not only for understanding the causes of psychopathy and aggression, but also for identifying protective developmental factors. Indeed, research on psychopathy in children and adolescents is important for developing prevention and early intervention programs by identifying high-risk individuals, prodromal features, as well as bio-psycho-social mechanisms such programs should address. Treatment interventions targeting psychopathic traits in childhood and adolescence may be more effective because personality traits are generally more changeable at these developmental stages (Forth & Book, 2007, Frick, 2009; Roberts & DelVecchio, 2000; Salekin, Worley, & Grimes, 2010). Psychopathy may also be a valuable construct for understanding the rather small subgroup of adolescents (approximately 6%) who become persistent violent offenders accounting for the majority of violent acts (Moffitt, Caspi, Harrington, & Milne, 2002). It may also help with the subtyping of conduct disordered children and adolescents (Frick, 2009; Moffitt et al., 2008). Currently, only limited evidence exists for the longitudinal stability of psychopathic traits. Nevertheless, current data suggest an early emergence of psychopathic traits, and that these traits have modest stability (Frick, 2009; Frick & Marsee, 2006; Lynam, Derefinko, Caspi, Loeber, & Stouthamer-Loeber, 2007; Salekin, Rosenbaum, & Lee, 2008). Clearly, data from large-scale and extensive longitudinal follow-up studies are the best way to determine the stability of psychopathic traits over time.

A necessary requirement for research on psychopathy in children and adolescents is the development of a measure specifically designed for these populations. Several such assessment instruments have been constructed over the last two decades, such as the self-report Youth Psychopathic Traits Inventory (YPI; Andershed, Hodgins, & Tengström, 2007; Andershed, Kerr, Stattin, & Levander, 2002), the Inventory of Callous and Unemotional Traits (Kimonis et al., 2008), and scales derived from selected items in the Millon Adolescent Clinical Inventory (MACI; Millon, 1993; Murrie & Cornell, 2000; Salekin, Ziegler, Larrea, Anthony, & Bennett, 2003). Three other instruments designed to measure psychopathy in children or adolescents are all derived from the Hare PCL-R: the Antisocial Process Screening Device (APSD; Frick & Hare, 2001), the Child Psychopathy Scale (Lynam, 1997), and the Psychopathy Checklist: Youth Version (PCL:YV; Forth et al., 2003).

Development and Description of the PCL:YV

The PCL:YV was developed as an adaptation of the Hare PCL-R for the assessment of psychopathic traits in adolescent males and females aged 12 to 18 years, inclusive, both in forensic and nonforensic settings. Because the goal was to develop an instrument to measure the same construct as measured by the PCL-R in adults, Forth et al. (2003) used the same assessment format and item content found in the PCL-R items. Thus, the PCL:YV consists of 20 items measuring the interpersonal, affective, antisocial, and behavioral features of psychopathy (see Table 2.1), and uses an expert rater format to rate items on a 3-point scale, using information from interview and collateral information.

However, because adolescents differ from adults in many important ways, several crucial modifications also were made to take these differences into account. First, rating instructions were modified to make the instrument sensitive to the ways in which normative behavior changes with age. Accordingly, raters are instructed to evaluate the adolescent's behavior in the context of normative behavior of same-age peers, and with consideration of normative adolescent development. Second, a scoring system was developed that is tailored to the salient life experiences of adolescents. Thus, the scoring system reflects and emphasizes the greater importance that peers, family, and school have in the daily lives of adolescents compared to adults. Third, these modifications were also reflected at the item level, where several items had to be modified to permit their assessment and scoring in youth. This led to modifications of all the individual items titles, descriptions, and/or source of information (for a detailed outline of the scoring modifications at the item level, see Forth et al., 2003: 117–121). Fourth, a new interview guide was developed in light of the above-mentioned goal and modifications. The interview guide contains age-appropriate questions and is useful for obtaining the relevant interview-information needed to score the PCL:YV. Finally, because of its potential (adverse) implications for adolescents, in particular young offenders, it was decided not to provide a cutoff score for the diagnosis of psychopathy in youth. Instead, the PCL:YV provides a dimensional assessment of adolescent precursors and features of psychopathy.

The PCL:YV manual contains information on the instrument's psychometric properties based on data from 2,438 youth in three countries (Canada, the United Kingdom, and the United States; Forth et al., 2003). Like the PCL-R, the PCL:YV yields both a total score (with a maximum of 40), as well as separate scores for the four factors (see Table 2.1).

Use and Administration of the PCL:YV

The PCL:YV can be used in both clinical/applied and research settings, provided users have the appropriate qualifications and training. User qualifications and training requirements are generally the same as those for the PCL-R. In addition, professionals using the PCL:YV also need to have knowledge about adolescent development and normative behavior in order to rate the items in an age-appropriate and valid manner.

Users, especially those working in clinical and forensic settings, must consider the potentially negative consequences of an assessment of psychopathy for adolescents. Generally, the context of the evaluation should guide the choice of which instruments, PCL:YV or otherwise, to use (Beutler & Groth-Marnat, 2003). Forth et al. (2003) also stress that the PCL:YV should not be used as the sole foundation for describing or making recommendations for youth within the mental health and criminal justice system.

Administration of the PCL:YV requires assessment of file and collateral information, plus a semistructured interview. It is recommended that the rater begin by reviewing all available collateral information before interviewing the adolescent, so the interviewer knows where to probe for additional information, and also to be in a better position to judge the extent to which the youth lies, denies, distorts, or minimizes during the interview. The interview takes approximately 90 minutes to complete (perhaps spread out over more than one occasion), and the review of charts and collateral information takes between 1 and 2 hours.

Raters rate the 20 items on a 3-point ordinal scale (2, 1, or 0) based on the degree to which the assessed adolescent's personality and behavior resembles the item descriptions. Thus, in the scoring, a certain amount of inference and judgment is required.

Psychometric Properties of the PCL:YV

The PCL:YV manual summarizes the instrument's psychometric properties, based on data from 19 different samples of adolescents (institutionalized, probation, and community) from Canada, the United Kingdom, and the United States.

PCL:YV total scores were found to vary considerably as a function of setting. Institutionalized samples had the highest scores, and community and clinical samples had relatively low scores. The diversity in range of scores across the 19 samples is a reflection of differences in populations from which they were drawn. For example, the sample with highest mean score (31.8), consisted of violent offenders, with long criminal histories, whereas the sample with the lowest mean score (2.8), was a community sample, which consisted mostly of high school nonoffenders. Recently Sevecke, Pukrop, Kosson, and Krischer (2009) also reported low PCL:YV scores in a sample of German high school students (5.0).

In 3 of the 19 samples, 2 from institutionalized offenders and 1 probation sample, the PCL:YV scores were based on solely file information. This allowed for comparison of scores between this procedure and the standard assessment procedure (interview plus file information). As was the case with adult offenders (Hare, 2003), the results showed that file-only assessment resulted in somewhat lower scores compared to samples where standard assessment procedure was used. Unfortunately it is not clear from the data whether these lower scores reflect difficulty in scoring some of the items from file-only information or whether it was due to sample characteristics. Future research needs to address this question. Until then, these findings underscore the importance of attempting to obtain an interview when assessing adolescents with the PCL:YV.

In using an expert-rater format, a certain amount of inference and judgment is involved in rating of many of the items. Thus, studies of interrater reliability are important to see if the item descriptions are specific enough to allow for general consensus among different raters across items as well as factor and total scores.

Generally the PCL:YV shows good reliability (Forth et., 2003). Internal consistency, the extent to which the different items measure the same construct, is acceptable, ranging from .85 to .94 across settings.

Interrater reliability is high for total scores (ICC$_1$ of .90 to .96; Forth et al., 2003). Other researchers have obtained good to excellent interrater reliability for total scores (ICCs range from .82 to .98; see Andershed et al., 2007; Cauffman, Kimonis, Dmitrieva, & Monahan, 2009; Das, de Ruiter, Doreleijers, & Hillege, 2009). The interrater reliability for factor scores is more variable, ranging from a low of .43 to .86 in research studies (Forth et al., 2003; Skeem & Cauffman, 2003). It is likely that the interrater reliability of PCL:YV total and factor scores will be lower in clinical and forensic settings where youth may be more motivated to engage in impression management and where the amount of training is more variable.

Factor Structure of the PCL:YV

The different two-, three-, and four-factor models of the PCL-R (Hare, 1991, 2003; but see also Cooke, Michie, & Hart, 2006; Hare & Neumann, 2006) have also been tested on the PCL:YV. Use of confirmatory factor analytic techniques, however, has yielded only mixed support for the original two-factor model of the PCL-R (Cooke & Michie, 2001; Hare, 2003; Hill et al., 2004), pointing toward three- and four-factor models as a better fit.

Research on the PCL:YV shows that both the three- and four-factor models provide good fit (Forth et al., 2003; Jones, Cauffman, Miller, & Mulvey, 2006; Neumann et al., 2006;

Salekin, 2006; Salekin, Brannen, Zalot, Leistico, & Neumann, 2006). Determining which of these models is the most optimal remains an open question. Forth and colleagues (2003) and Neumann et al. (2006) have argued that for various statistical, methodological, and theoretical reasons, the four-factor model might be the preferred model. Others have argued that the three-factor model is a better fit (Cauffman et al., 2009; Jones et al., 2006). A recent study of the three- and four-factor models of the PCL:YV in a German sample of both male and female detainees and community youth found the four-factor model to be problematic in all samples, but the three-factor model to be adequate in incarcerated and community male (but not in females; Sevecke et al., 2009). Vitacco, Neumann, Caldwell, Leistico, and Van Rybroek (2006) have shown, using structural equation modeling, that the four-factor model accounts for more variance (20%) associated with instrumental violence than does the three-factor model (8%).

Stability of Psychopathic Traits

Edens, Skeem, Cruise, and Cauffman (2001) underscore the need for longitudinal research to determine if psychopathic juveniles become psychopathic adults, or as Seagrave and Grisso (2002) stated: "there must be a demonstration that psychopathy as measured in adolescence is predictive of serious delinquency in later adolescence and psychopathy in adulthood" (p. 233). More recently, Lynam and Gudonis (2005) concluded that juvenile and adult psychopathy appears to share many fundamental features, and that the phenotype expression of psychopathy remains more or less stable over time. Seven studies have examined the stability of psychopathic traits (Blonigen, Hicks, Krueger, Patrick, & Iacono, 2006; Burke, Loeber, & Lahey, 2007; Forsman, Lichtenstein, Andershed, & Larsson, 2008; Frick, Kimonis, Dandreaux, & Farrell, 2003; Lynam et al., 2007; Obradovic, Pardini, Long, & Loeber, 2007), and all found that psychopathic traits were stable from early adolescence into early adulthood. However, none of these studies used the PCL:YV to measure psychopathy, in part because (1) several studies were initiated before the publication of the PCL:YV, (2) self-report measures were used for ease of administration, and (3) several researchers used instruments they developed themselves.

Although psychopathic traits appear fairly stable across time, several studies have highlighted how diverse moderators affect the stability of psychopathy during development—demonstrating synergistic interactions, wherein risk factors combine in a more-than-additive fashion to produce offending (Cadoret, R. J., Yates, W. R., Troughton, E., Woodworth, G. et al., 1995; Cohen, Berliner, & Mannarino, 2003). Another line of research has investigated interaction effects between risk factors, where one variable reduces the impact of the other (the risk factor). An array of moderating, mediating, and buffering factors seems to reduce commitment to an antisocial lifestyle, including IQ, nurturing relationship with adults, good academic performance, and prosocial peer groups (Fergusson & Lynskey, 1996; Quinton & Rutter, 1988; Werner & Smith, 1992).

Generalizability of the PCL:YV

Within institutionalized and probations settings there is a small negative correlation between PCL:YV total scores and age at time of assessment. Younger youth scored slightly higher on the PCL:YV than their older counterparts. Important age-related changes in some of the features measured by the PCL:YV accordingly require further investigation (Salekin et al., 2008).

Ethnicity is another important variable to consider (Sullivan & Kosson, 2006). Correlations between PCL:YV Total score and ethnicity (Caucasian vs. non-Caucasian) were mostly small (Forth et al., 2003), and studies comparing Caucasians and African American adolescents have found no significant differences in PCL:YV Total Scores between these two groups (Murrie, Cornell, Kaplan, McConville, & Levy-Elkon, 2004). More recently, Schmidt, McKinnon, Chattha, and Brownlee (2006) also reported no differences between Caucasians and Aboriginal young offenders on the PCL:YV. However, research is needed to determine if the PCL:YV is metrically equivalent across different ethnic and cultural groups.

Few studies have assessed psychopathic traits in female adolescents. Forth et al. (2003) found that male adolescents scored slightly higher than female adolescents. More recent studies, however, suggest that the relationships between the PCL:YV and gender needs further investigation. For example, Sevecke et al. (2009) found that neither the two-, nor three-, nor four-factor model of psychopathy provided a consistently acceptable fit for German females. Moreover, research on the association between psychopathic traits and future recidivism in females is equivocal. Schmidt et al. (2006) found that the PCL:YV predicted both general and violent recidivism in males, but the effects were weaker in females. Both Vincent, Odgers, McCormick, and Corrado (2008) and Edens, Campbell, and Weir (2007) reported that PCL:YV scores did not predict violent reoffending in female adolescents. Such findings suggest that care should be taken when interpreting PCL:YV scores in females (see Vincent & Kinscherff, 2008 for a review).

PCL:YV, Aggression, and Reoffending

The PCL:YV is linked to general and violent reoffending in various types of adolescent offenders (Catchpole & Gretton, 2003; Edens et al., 2007; Forth & Book, 2010; Schmidt et al., 2006; Vincent et al., 2008), including sex offenders (Gretton, McBride, Hare, O'Shaughnessy, & Kumka, 2001). The PCL:YV is also associated with violence, especially instrumental violence (Flight & Forth, 2007; Murrie et al., 2004; Salekin, Neumann, Leistico, & Zalot, 2004; Vitacco et al., 2006). However, the predictive utility of the PCL:YV for long-term offending has been examined in only two studies. One study found that the PCL:YV predicted 10-year reoffending (Gretton, Hare, & Catchpole, 2004) and another found that it did not (Edens & Cahill, 2007). Moreover, in the largest study to date of 1,170 male adolescent offenders, PCL:YV scores were related to short-term reoffending (12 months or less) but only weakly related to self-reported and official recidivism at 36 months (Cauffman et al., 2009). Forth et al. (2003) warn that it is not appropriate to rely on PCL:YV scores to impose harsher sentences or to use scores in determining whether an adolescent offender should be tried as an adult in court.

The Potential Impact of the Psychopathic Label

Extending the construct of psychopathy to children and adolescents has created much debate and controversy (Salekin, 2006). Clearly one of the major concerns is the potential misuse of the term, especially in forensic settings, where use of the label "psychopathy" could stigmatize the youth, potentially resulting in harsher punishment and preventing them from gaining access to treatment and rehabilitative programs (Steinberg, 2002). Forth et al. (2003) provide no cutoff score for labeling youth as psychopathic and also explicitly warn against doing so.

Boccaccini, Murrie, Clark, and Cornell (2008) have argued that although concerns about labeling are legitimate, such concerns about the possible negative consequences of assessing psychopathy in youth should be considered a hypothesis to be empirically tested, rather than given truths. Table 2.6 summarizes the published studies that have examined the impact of the psychopathy traits or label in youth. Researchers have examined the influence of diagnostic criterion (i.e., describing the personality traits associated with psychopathy) and diagnostic labels of psychopathy or conduct disorder. Researchers have found that describing a youthful offender using either the psychopathy label or underlying traits sometimes results in more punitive decision making among undergraduates, juvenile probation officers, potential jurors, juvenile justice judges, and clinicians (Chauhan, Burnette, & Repucci, 2007; Edens, Guy, & Fernandez, 2003; Jones & Cauffman, 2008; Murrie, Cornell, & McCoy, 2005; Murrie, Boccaccini, & Cornell, 2007; Rocket, Murrie, & Boccaccini, 2007; Vidal & Skeem, 2007). More specifically, these studies have demonstrated that describing youth as psychopathic resulted in harsher sentence recommendations and higher ratings of risk for future violence. The effects for treatment access or amenability have been mixed. Perhaps contrary to what most would expect, three studies have shown that the diagnosis of psychopathy assigned to defendants was associated with perceptions of an increased need for treatment among juvenile probation officers (Murrie et al.,

Table 2.6 Summary of Youth Psychopathy Labeling Studies

Study	Sample	Manipulation	Labeling effects		
			Increased risk	Harsher punishment	Negative treatment beliefs
Edens et al. (2003)	Undergraduate students	Psychopathic traits vs. positive traits		Yes	Yes
Murrie et al. (2005)	Juvenile probation officers	Psychopathic traits vs. positive traits	Yes	No	No
		Psychopathy vs. conduct disorder vs. no disorder	No	No	No[a]
Murrie et al. (2007)	Judges	Psychopathic traits vs. positive traits	Yes	No	No[a]
		Psychopathy vs. conduct disorder vs. no disorder	No	No	No[a]
Vidal & Skeem (2007)	Juvenile probation officers	Is a psychopath and psychopathic traits vs. no disorder with positive traits	Yes	Yes	Yes
Chauhan et al. (2007)	Undergraduates, law students, judges, clinicians and developmental experts[b]	Psychopathic traits and no label vs. psychopathic traits and is a psychopath vs. no psychopathic traits and is a psychopath vs. no psychopathic traits and no label	Yes	Yes	Yes
Rockett et al. (2007)	Juvenile justice clinicians	Psychopathic traits vs. positive traits	Yes		No
		Psychopathy vs. conduct disorder vs. no disorder	Yes		No
Boccaccini et al. (2008)	Jury-eligible community	Psychopathic traits vs. positive traits	Yes	Yes	No
		Is a psychopath vs. psychopathy vs. conduct disorder vs. no disorder	Yes	Yes	No[a]
Jones & Cauffman (2008)	Judges	Psychopathic traits and no label vs. psychopathic traits and is a psychopath vs. no mental health information	Yes	Yes	Yes

Notes: Adapted from Boccaccini et al. (2008).
"Yes" in the effect column indicates that the psychopathic traits or label had a significant effect in at least one analysis. "No" indicates that there was no significant effect of the labels. Empty column indicates that the dependent variable was not measured in the study.
[a] In this case, the presence of the psychopathy label and traits were associated with an increased likelihood of recommending treatment.
[b] There were no trait or labeling effects for law students or judges.

2005), juvenile and family court judges (Murrie et al., 2007), and jury-eligible community participants (Boccaccini et al., 2008). Several major conclusions can be drawn from these studies: (1) describing psychopathic traits produces stronger effects compared to using the label themselves; (2) there is evidence for a general labeling effect but not a specific labeling effect (no difference between a diagnosis of conduct disorder or psychopathy); (3) the negative effects of psychopathic traits have been found for decisions regarding degree of risk for future offending and whether harsher sanctions are warranted; and (4) explicitly labeling young defendants "(as a) psychopath" led jurors to assume they posed greater risk for future crime (Boccaccini et al., 2008). Such results have important implications for clinical practice, and should alert practitioners to their use of language and diagnostic labels.

In summary, the PCL:YV appears to be a psychometrically sound instrument for the assessment of psychopathic traits in adolescence. Yet, both researchers and clinicians should be mindful of the potentially detrimental effects for youth when assessing psychopathy. If an adolescent is diagnosed as a psychopath this label may remain with the person and have ongoing negative repercussions. Instead of being viewed as simply a measure with a short-term goal of diagnosis, developing measures such as the PCL:YV to identify psychopathic traits in youth should instead be considered a necessary component of systematic research that will lead to the development of early intervention and prevention programs.

Conclusion

The PCL-R, PCL:SV, and PCL:YV represent substantial advances in the assessment of psychopathy. However, to understand this complex disorder successfully will take the sustained commitment from scientist and clinicians from diverse backgrounds, using a multitude of different techniques, and focusing on international collaborations. A major challenge for new generations of psychopathy researchers will be to contribute to the refinement of assessment measures for psychopathy and to decipher the etiological mechanisms underlying the development of this disorder. Given the enormous social, personal, and financial costs of psychopathy, not to invest heavily in understanding this disorder is simply not an option.

NOTE

1. In Table 2.1, the term "facet" is used to denote the four factors in the PCL-R and PCL:SV. This is done to show that the factors in the four-factor model are subsidiary factors of the original two-factor model.

REFERENCES

Allen, J. G., Fonagy, P. & Bateman, A. (2008). *Mentalizing in clinical practice*. Washington, DC: American Psychiatric Publishing.

Alterman, A. I., Cacciola, J. S., & Rutherford, M. J. (1993). Reliability of the Revised Psychopathy Checklist in substance abuse patients. *Psychological Assessment: A Journal of Consulting and Clinical Psychology, 5*(4), 442–448.

American Psychiatric Association (1994). *Diagnostic and statistical manual of mental disorders* (4th ed.). Washington, DC: Author.

American Psychiatric Association (2000). *Diagnostic and statistical manual of mental disorders* (4th ed.)—*Text Revision*. Washington, DC: Author.

Andershed, H., Hodgins, S., & Tengström, A. (2007). Convergent validity of the Youth Psychopathic Traits Inventory (YPI): Association with the Psychopathy Checklist: Youth Version (PCL:YV). *Assessment, 14*, 144–154.

Andershed, H., Kerr, M., Stattin, H., & Levander, S. (2002). Psychopathic traits in non-referred youths: A new assessment tool. In E. Blaauw & L. Sheridan (Eds.), *Psychopaths: Current international perspectives* (pp. 131–158). The Hague: Elsevier.

Andrews, D. A., Bonta, J., & Wormith, J. S. (2006). The recent past and near future of risk and/or need assessment. *Crime and Delinquency, 52*, 7–27.

Barbaree, H. E. (2005). Psychopathy, treatment behavior, and recidivism: An extended follow-up of Seto and Barbaree (1999). *Journal of Interpersonal Violence, 20*, 1115–1131.

Beutler, L. E., & Groth-Marnat, G. (2003). *Integrative assessment of adult personality* (2nd ed.). New York: Guilford.

Bishopp, D., & Hare, R. D. (2008). A multidimensional scaling analysis of the Hare PCL-R: Unfolding the structure of psychopathy. *Psychology, Crime, and Law, 14,* 117–132.

Blackburn, R. (1988). On moral judgements and personality disorders: The myth of psychopathic personality revisited. *British Journal of Psychiatry, 153,* 505–512.

Blackburn, R. & Coid, J. W. (1998). Psychopathy and the dimensions of personality disorders in violent offenders. *Personality and Individual Differences, 25,* 129–145.

Blair, R. J. R. (2007). The amygdala and ventromedial prefrontal cortex in morality and psychopathy. *Opinion, Trends in Cognitive Sciences, 11(9),* 387–392.

Blair, J., Mitchell, D., & Blair, K. (2005). *The psychopath: Emotion and the brain.* Oxford: Blackwell.

Blair, R. J. R., Mitchell, D. G. V., Peschardt, K. S., Colledge, E., Leonard, R. A., Shine J. H., Murray, L. K., & Perret, D. I. (2004). Reduced sensitivity to others' fearful expressions in psychopathic individuals. *Personality and Individual Differences, 37,* 1111–1122.

Blonigen, D. M., Hicks, B. M., Krueger, R. F., Patrick, C. J., & Iacono, W. G. (2006). Continuity and change in psychopathic traits as measured via normal-range personality: A longitudinal-biometric study. *Journal of Abnormal Psychology, 115,* 85–95.

Boccaccini, M. T., Murrie, D. C., Clark, J. W., & Cornell, D. G. (2008). Describing, diagnosing, and naming psychopathy: How do youth psychopathy labels influence jurors? *Behavioral Sciences and the Law, 26,* 487–510.

Boccaccini, M. T., Turner, D. B., & Murrie, D. C. (2008). Do some evaluators report consistently higher or lower PCL-R scores than others? Findings from a statewide sample of sexually violent predator evaluations. *Psychology, Public Policy, and Law, 14,* 262–283.

Boccardi, M., Ganzola, R., Rossi, R., Sabattoli, F., Laakso, M. P., Repo-Tiihonen, E., et al. (2010). Abnormal hippocampal shape in offenders with psychopathy. *Human Brain Mapping, 31,* 438–447.

Bolt, D. M., Hare, R. D., Vitale, J. E., & Newman, J. P. (2004). A multigroup item response theory analysis of the Psychopathy Checklist-Revised. *Psychological Assessment, 16,* 155–168.

Book, A., Clark, H. Forth, A., & Hare, R. D. (2006). The Psychopathy Checklist- Revised and the Psychopathy Checklist: Youth Version. In R. Archer (Ed.), *Forensic uses of clinical assessment instruments* (pp. 147–179). Mahwah, NJ: Lawrence Erlbaum.

Brown, S. L., & Forth, A. E. (1997). Psychopathy and sexual assault: Static risk factors, emotional precursors, and rapist subtypes. *Journal of Consulting and Clinical Psychology, 65,* 848–857.

Burke, J. D., Loeber, R., & Lahey, B. B. (2007). Adolescent conduct disorder and interpersonal callousness as predictors of psychopathy in young adults. *Journal of Clinical Child and Adolescent Psychology, 36,* 334–346.

Cadoret, R. J., Yates, W. R., Troughton, E., Woodworth, G., & Stewart, M. A. (1995). Genetic environmental interaction in the genesis of aggressivity and conduct disorders. *Archives of General Psychiatry, 52,* 916–924.

Caldwell, M. F., Skeem, J., Salekin, R., & Van Rybroek, G. (2006). Treatment response of adolescent offenders with psychopathy features: A 2-year follow-up. *Criminal Justice and Behavior, 33,* 571–596.

Catchpole, R. E. H., & Gretton, H. M. (2003). The predictive validity of risk assessment with violent young offenders: A 1-year examination of criminal outcome. *Criminal Justice and Behavior, 30,* 668–708.

Cauffman, E., Kimonis, E. R., Dmitrieva, J., & Monahan, K. C. (2009). A multimethod assessment of juvenile psychopathy: Comparing the predictive utility of the PCL:YV, YPI, and NEO PRI. *Psychological Assessment, 21,* 528–542.

Chauhan, P., Reppucci, N. D., & Burnette, M. L. (2007). Application and impact of the psychopathy label to juveniles. *International Journal of Forensic Mental Health, 6,* 3–14.

Cleckley, H. (1976). *The mask of sanity,* (5th ed.). St. Louis, MO: Mosby.

Cohen, J. A., Berliner, L., & Mannarino, A. P. (2003). Psychosocial and pharmacological interventions for child crime victims. *Journal of Traumatic Stress, 16(2),* 175–186.

Coid, J., Yang, M., Ullrich, S., Roberts, A., & Hare, R. D. (2009). Prevalence and correlates of psychopathic traits in the household population of Great Britain. *International Journal of Law and Psychiatry, 32,* 65–73.

Cooke, D. J., Kosson, D. S., & Michie, C. (2001). Psychopathy and ethnicity: Structural, item, and test generalizability of the Psychopathy Checklist—Revised (PCL-R) in Caucasian and African American participants. *Psychological Assessment, 13,* 531–542.

Cooke, D. J., & Michie, C. (2001). Refining the construct of psychopathy: Towards a hierarchical model. *Psychological Assessment, 13*(2), 171–188.

Cooke, J. D., Michie, C., & Hart, S. D. (2006). Facets of clinical psychopathy. In C. J. Patrick (Ed.), *Handbook of psychopathy* (pp. 91–106). New York: Guilford.

Cooke, D. J., Michie, C., Hart, S. D., & Clark, D. (2005). Searching for the pan-cultural core of psychopathic personality disorder. *Personality and Individual Differences, 39*, 283–295.

Cooke, D. J., Michie, C., Hart, S. D., & Hare, R. D. (1999). Evaluating the Screening Version of the Hare Psychopathy Checklist—Revised (PCL:SV): An item response theory analysis. *Psychological Assessment, 11*, 3–13.

Cornell, D. G., Warren, J., Hawk, G., Stafford, D., Oram, G., & Pine, D. (1996). Psychopathy in instrumental and reactive violent offenders. *Journal of Consulting and Clinical Psychology, 64*, 783–790.

Das, J., de Ruiter, C., Doreleijers, T., & Hillege, S. (2009). Reliability and construct validity of the Dutch Psychopathy Checklist: Youth Version: Findings from a sample of male adolescents in a juvenile justice treatment institution. *Assessment, 16*, 88–102.

DeMatteo, D., & Edens, J. F. (2006). The role and relevance of the Psychopathy Checklist-Revised in court: A case law survey of U.S. courts (1991–2004). *Psychology, Public Policy, and Law, 12*, 214–241.

DeMatteo, D., Heilbrun, K., & Marczyk, G. (2006). An empirical investigation of psychopathy in a noninstitutionalized and noncriminal sample. *Behavioral Sciences and the Law, 24*, 133–146.

De Oliveira-Souza, R., Ignácio, F. A., Moll, J., & Hare, R. D. (2008). Psychopathy in a civil psychiatric outpatient sample. *Criminal Justice and Behavior, 35*, 427–437.

Dolan, M., & Doyle M. (2000). Violence risk prediction: Clinical and actuarial measures and the role of the Psychopathy Checklist. *British Journal of Psychiatry, 177*, 303–311.

Dolan, M., & Fullam, R. C. (2009). Psychopathy and functional magnetic resonance imaging blood oxygenation level-dependent responses to emotional faces in violent patients with schizophrenia. *Biological Psychiatry, 66*, 570–577.

Dolan, M., & Völlm, B. (2009). Antisocial personality disorder and psychopathy in women: A literature review on the reliability and validity of assessment instruments. *International Journal of Law and Psychiatry, 32*, 2–9.

Doren, D. M., & Yates, P. M. (2008). Effectiveness of sex offender treatment for psychopathic sexual offenders. *International Journal of Offender Therapy and Comparative Criminology, 52*, 234–245.

Douglas, K. S., Strand, S., Belfrage, H., Fransson, G., & Levander, S. (2005). Reliability and validity evaluation of the Psychopathy Checklist: Screening Version (PCL:SV) in Swedish correctional and forensic psychiatric samples. *Assessment, 12*, 145–161.

Douglas, K. S., & Webster, C. D. (1999). The HCR-20 violence risk assessment scheme: Concurrent validity in a sample of incarcerated offenders. *Criminal Justice and Behavior, 26*, 3–19.

Dowden, C., & Andrews, D. A. (2000). Effective correctional treatment and violent reoffending: A meta-analysis. *Canadian Journal of Criminology, 4*, 449–467.

Edens, J. F. (2001). Misuses of the Hare Psychopathy Checklist-Revised in court: Two case examples. *Journal of Interpersonal Violence, 16*, 1082–1093.

Edens, J. F., Boccaccini, M. T., & Johnson, D. W. (2010). Inter-rater reliability of the PCL-R total and factor scores among psychopathic sex offenders: Are personality features more prone to disagreement than behavioral features? *Behavioral Sciences and Law, 28*, 106–119.

Edens, J. F., & Cahill, M. A. (2007). Psychopathy in adolescence and criminal recidivism in young adulthood: Longitudinal results from a multiethnic sample of youthful offenders. *Assessment, 14*, 57–64.

Edens, J. F., Campbell, J. S., & Weir, J. M. (2007). Youth psychopathy and criminal recidivism: A meta-analysis of the Psychopathy Checklist Measures. *Law and Human Behavior, 31*, 53–75.

Edens, J. F., Desforges, D. M., Fernandez, K., & Palac, C. A. (2004). Effects of psychopathy and violence risk testimony on mock juror perceptions of dangerousness in a capital murder trial. *Psychology, Crime and Law, 10*, 393–412.

Edens, J. F., Guy, L. S., & Fernandez, K. (2003). Psychopathic traits predict attitudes toward a juvenile capital murderer. *Behavioral Sciences and the Law, 21*, 807–828.

Edens, J. F., Marcus, D. K., Lilienfeld, S. O., & Poythress, N. G. (2006). Psychopathic, not psychopath: Taxometric evidence for the dimensional structure of psychopathy. *Journal of Abnormal Psychology, 115*, 131–144.

Edens, J. F., Skeem, J. L., Cruise, K. R., & Cauffman, E. (2001). Assessment of "juvenile psychopathy" and its association with violence: A critical review. *Behavioral Sciences and the Law, 19*, 53–80.

Fergusson, D. M., & Lynskey, M. T. (1996). Adolescent resiliency to family adversity. *Journal of Child Psychology and Psychiatry, and Allied Disciplines, 37*(3), 281–292.

Fergusson, D. M., Lynskye, M. T., & Horwood, L. J. (1996). Childhood sexual abuse and psychiatric disorder in young adulthood: I. Prevalence of sexual abuse and factors associated with sexual abuse. *Journal of the American Academy of Child & Adolescent Psychiatry, 35*, 1355–1364.

Flight, J. I., & Forth, A. E. (2007). Instrumentally violent youths: The roles of psychopathic traits, empathy, and attachment. *Criminal Justice and Behavior, 34*, 739–751.

Forsman, M., Lichtenstein, P., Andershed, H., & Larsson, H. (2008). Genetic effects explain the stability of psychopathic personality from mid- to late adolescence. *Journal of Abnormal Psychology, 117*, 606–617.

Forth, A., & Book, A. S. (2007). Psychopathy in youth: A valid construct. In H. Hervé, & J. C. Yuille (Eds.), *The psychopath: Theory, research, and practice* (pp. 369–387). Mahwah, NJ: Lawrence Erlbaum.

Forth, A., & Book, A. S. (2010). Psychopathic traits in children and adolescents: The relationship with antisocial behaviors and aggression. In R. Salekin & D. Lynam (Eds.), *Handbook of youth psychopathy* (pp. 251–283). New York: Guilford.

Forth, A. E., Brown, S. L., Hart, S. D., & Hare, R. D. (1996). The assessment of psychopathy in male and female noncriminals: Reliability and validity. *Personality and Individual Differences, 20*, 531–543.

Forth, A. E., Kosson, D., & Hare, R. D. (2003). *The Hare Psychopathy Checklist: Youth Version*. Toronto, ON: Multi-Health Systems.

Frick, P. J. (2002). Juvenile psychopathy from a developmental perspective: Implications for construct development and use in forensic assessments. *Law and Human Behavior, 26*, 247–253.

Frick, P. J. (2009). Extending the construct of psychopathy to youth: Implications for understanding, diagnosing, and treating antisocial children and adolescents. *Canadian Journal of Psychiatry, 54*, 803–812.

Frick, P. J., & Hare, R. D. (2001). *The Antisocial Process Screening Device*. Toronto, ON: Multi-Health Systems.

Frick, P. J., Kimonis, E. R., Dandreaux, D. M., & Farrell, J. M. (2003). The 4 year stability of psychopathic traits in non-referred youth. *Behavioral Sciences and the Law, 21*, 713–736.

Frick, P. J., & Marsee, M. A. (2006). Psychopathy and developmental pathways to antisocial behavior in youth. In C. J. Patrick (Ed.), *Handbook of psychopathy* (pp. 353–374). New York: Guilford.

Furr, K. D. (1993). Prediction of sexual or violent recidivism among sexual offenders: A comparison of prediction instruments. *Annals of Sex Research, 6*, 271–286.

Gacono, C. B., & Hutton, H. E. (1994). Suggestions for the clinical and forensic use of the Hare Psychopathy Checklist—Revised (PCL-R). *International Journal of Law and Psychiatry, 17*, 303–317.

Glenn, A. L. & Raine, A. (2008). The neurobiology of psychopathy. *Psychiatric Clinics of North America, 31*, 463–475.

Glenn, A. L., Raine, A, & Schug, R. A. (2009). The neural correlates of moral decision-making in psychopathy. *Molecular Psychiatry, 14*, 5–6.

Glover, A. A. J., Nicholson, D. E., Hemmati, T., Bernfeld, G. A., & Quinsey, V. L. (2002). A comparison of predictors of general and violent recidivism among high-risk federal offenders. *Criminal Justice and Offenders, 29*, 235–249.

Grann, M. (2000). The PCL-R and gender. *European Journal of Psychological Assessment, 16*, 147–149.

Grann, M., Långström, N., Tengström, A., & Kullgren, G. (1999). Psychopathy (PCL-R) predicts violent recidivism among criminal offenders with personality disorders in Sweden. *Law and Human Behavior, 23*, 205–217.

Gray, N. S., Snowden, R. J., MacCulloch, S., Phillips, H., Taylor, J., & MacCulloch, M. J. (2004). Brief report: Relative efficacy of criminological, clinical, and personality measures of future risk of offending in mentally disordered offenders: A comparative study of HCR-20, PCL:SV, and OGRS. *Journal of Consulting and Clinical Psychology, 72*, 523–530.

Gretton, H., Hare, R. D., & Catchpole, R. (2004). Psychopathy and offending from adolescence to adulthood: A ten-year follow-up. *Journal of Consulting and Clinical Psychology, 72*, 636–645.

Gretton, H. M., McBride, M., Hare, R. D., O'Shaughnessy, R., & Kumka, G. (2001). Psychopathy and recidivism in adolescent sex offenders. *Criminal Justice and Behavior, 28*, 427–449.

Guay, J. P., Ruscio, J., Knight, R. A., & Hare, R. D. (2007). A taxometric analysis of the latent structure of psychopathy: Evidence for dimensionality. *Journal of Abnormal Psychology, 116*, 701–716.

Guy, L. S., & Douglas, K. S. (2006). Examining the utility of the PCL:SV as a screening measure using competing factor models of psychopathy. *Psychological Assessment, 18*, 225–230.

Hanson, R. K., Bourgon, G., Helmus, L., & Hodgson, S. (2009). The principles of effective treatment also apply to sexual offenders: A meta-analysis. *Criminal Justice and Behavior, 36*, 865–891.

Hare, R. D. (1980). A research scale for the assessment of psychopathy in criminal populations. *Personality and Individual Differences, 1*, 111–119.

Hare, R. D. (1991). *The Hare Psychopathy Checklist—Revised*. Toronto, ON: Multi-Health Systems.

Hare, R. D. (1996). Psychopathy: A clinical construct whose time has come. *Criminal Justice and Behavior, 23*, 25–54.

Hare, R. D. (1998). The Hare PCL-R: Some issues concerning its use and misuse. *Legal and Criminological Psychology, 3*, 101–123.

Hare, R. D. (2003). *The Hare Psychopathy Checklist—Revised*, (2nd ed.). Toronto, ON: Multi-Health Systems.

Hare, R. D., Clark, D., Grann, M., & Thornton, D. (2000). Psychopathy and the predictive validity of the PCL-R: An international perspective. *Behavioral Sciences & the Law, 18*, 623–645.

Hare, R. D., Harpur, T. J., Hakstian, A. R., Forth, A. E., Hart, S. D., & Newman, J. P. (1990). The Revised Psychopathy Checklist: Reliability and factor structure. *Psychological Assessment: A Journal of Consulting and Clinical Psychology, 2*, 338–341.

Hare, R. D., McPherson, L. M., & Forth, A. (1988). Male psychopaths and their criminal careers. *Journal of Consulting and Clinical Psychology, 56*(5), 710–714.

Hare, R. D., & Neumann, C. S. (2006). The PCL-R assessment of psychopathy: Development, structural properties, and new directions. In C. J. Patrick (Ed.), *Handbook of psychopathy* (pp. 58–88). New York: Guilford.

Hare, R. D., & Neumann, C. S. (2008). Psychopathy as a clinical and empirical construct. *Annual Review of Clinical Psychology, 4*, 217–246.

Hare, R. D., & Neumann, C. S. (2009). Psychopathy: Assessment and forensic implications. *Canadian Journal of Psychiatry, 54*, 791–802.

Harpur, T. J., Hakstian, R., & Hare, R. D. (1988). Factor structure of the Psychopathy Checklist. *Journal of Consulting and Clinical Psychology, 56*, 741–747.

Harpur, T. J., Hare, R. D., & Hakstian, R. (1989). Two-factor conceptualization of psychopathy: Construct validity and assessment implications. *Journal of Consulting and Clinical Psychology, 1*, 6–17.

Harris, G. T., & Rice, M. E. (2006). The treatment of psychopathy. In C. J. Patrick (Ed.), *Handbook of psychopathy* (pp. 555–572). New York: Guilford.

Hart, S. D., Cox, D. N., & Hare, R. D. (1995). *Manual for the Psychopathy Checklist: Screening Version (PCL:SV)*. Toronto, ON: Multi-Health Systems.

Hemphill, J. F., Hare, R. D., & Wong, S. (1998). Psychopathy and recidivism: A review. *Legal and Criminological Psychology, 3*, 139–170.

Hildebrand, M., de Ruiter, C., & de Vogel, V. (2004). Psychopathy and sexual deviance in treated rapists: Association with sexual and nonsexual recidivism. *Sexual abuse: Journal of Research and Treatment, 16*, 1–24.

Hill, C. D., Neumann, C. S., & Rogers, R. (2004). Confirmatory factor analysis of the Psychopathy Checklist: Screening Version in offenders with Axis-I disorders. *Psychological Assessment, 16*, 90–95.

Hill, C. D., Rogers, R, & Bickford, M. E. (1996). Predicting aggressive and socially disruptive behavior in a maximum security forensic psychiatric hospital. *Journal of Forensic Sciences, 41*, 56–59.

Hobson, J., Shine, J., & Roberts, R. (2000). How do psychopaths behave in a prison therapeutic community? *Psychology, Crime, and Law, 6*, 139–154.

Huchzermeier, C., Bruss, E., Geiger F., Kernbichler, A., &Aldenhoff, J. (2008). Predictive validity of the Psychopathy Checklist: Screening Version for intramural behaviour in violent offenders: A prospective study at a secure psychiatric hospital in Germany. *Canadian Journal of Psychiatry, 53*, 384–391.

Jones, S., & Cauffman, E. (2008). Juvenile psychopathy and judicial decision making: An empirical analysis of an ethical dilemma. *Behavioral Sciences and the Law, 26*, 151–165.

Jones, S., Cauffman, E., Miller, J. D., & Mulvey, E. (2006). Investigating different factor structures of the Psychopathy Checklist: Youth Version: Confirmatory factor analytic findings. *Psychological Assessment, 18*, 33–48.

Kiehl, K. A., Bates, A. T., Laurens, K. R., Hare, R. D., & Liddle, P. F. (2006). Brain potentials implicate temporal lobe abnormalities in criminal psychopaths. *Journal of Abnormal Psychology, 115*, 443–453.

Kiehl, K. A., Smith, A. M., Mendrek, A., Foster, B. B., Hare, R. D., & Liddle, P. F. (2004). Temporal lobe abnormalities in semantic processing by criminal psychopaths as revealed by functional magnetic resonance imaging. *Psychiatry Research: Neuroimaging, 130*, S297–312.

Kimonis, E. R., Frick, P. J., Skeem, J. L., Marsee, M. A., Cruise, K., Munoz, L. C., et al. (2008). Assessing callous-unemotional traits in adolescent offenders: Validation of the inventory of callous-unemotional traits. *International Journal of Law and Psychiatry, 31*(3), 241–252.

Kosson, D. S., Lorenz, R. A., & Newman, J. P. (2006). Effects of comorbid psychopathy on criminal offending and emotion processing in male offenders with antisocial personality disorder. *Journal of Abnormal Psychology, 115*, 798–806.

Kroner, D. G. & Mills, J. F. (2001). The accuracy of five risk appraisal instruments in predicting institutional misconduct and new convictions. *Criminal Justice and Behavior, 28*, 471–489.

Laakso, M. P., Vaurio, O., Koivisto, E., Savolainen, L., Eronen, M., & Aronen, H. J. (2001). Psychopathy and the posterior hippocampus. *Behavioural Brain Research, 118*, 187–193.

Langton, C. M., Barbaree, H. E., Harkins, L., & Peacock, E. J. (2006). Sex offenders' response to treatment and its association with recidivism as a function of psychopathy. *Sexual Abuse: A Journal of Research and Treatment, 18*, 99–120.

Leistico, A. R., Salekin, R. T., DeCoster, J., & Rogers, R. (2008). A large-scale meta-analysis relating the Hare measures of psychopathy to antisocial conduct. *Law and Human Behavior, 32*, 28–45.

Levenson, J. (2004). Sexual predator civil commitment: A comparison of selected and released offenders. *International Journal of Offender Therapy and Comparative Criminology, 48*, 638–648.

Lilienfeld, S. O., & Fowler, K. A. (2006). The self-report assessment of psychopathy: Problems, pitfalls, and promises. In C. Patrick (Ed.), *Handbook of psychopathy* (pp. 107–132). New York: Guilford.

Livesley, W. J. (2003). *Practical management of personality disorder*. New York: Guilford.

Livesley, W. J. (2010). Integrated treatment: Combining effective treatment principles and methods. In J. J. Magnavita (Ed.), *Evidence-based treatment of personality dysfunction: Principles, methods, and processes* (pp. 223–252). Washington, DC: American Psychological Association.

Lloyd, C. D., Clark, H. J., & Forth, A. E. (2010). Psychopathy, expert testimony, and indeterminate sentences: Exploring the relationship between Psychopathy Checklist-Revised testimony and trial outcome in Canada. *Legal and Criminological Psychology, 15*, 323–339.

Loeber, R., & Farrington, D. P. (2000). Young children who commit crime: Epidemiology, developmental origins, risk factors, early interventions, and policy implications. *Development and Psychopathology, 12*, 737–762.

Looman, J., Abracen, J., Maillet, G., & DiFazio, R. (2005). Phallometric nonresponding in sexual offenders. *Sexual Abuse: Journal of Research and Treatment, 10*, 325–336.

Looman, J, Abracen, J., Serin, R., & Marquis, P. (2005). Psychopathy, treatment change, and recidivism in high-risk, high-need sexual offenders. *Journal of Interpersonal Violence, 20*, 549–568.

Loucks, A. D., & Zamble, E. (2000). Predictors of criminal behavior and prison misconduct in serious female offenders. *Empirical and Applied Criminal Justice Review* [Online], *1*, 1–47.

Lykken, D. T. (2006). Psychopathic personality: The scope of the problem. In C. J. Patrick (Ed.), *Handbook of psychopathy* (pp. 3–13). New York: Guilford.

Lynam, D. R. (1997). Pursuing the psychopath: Capturing the fledgling psychopath in a nomological net. *Journal of Abnormal Psychology, 106*, 425–438.

Lynam, D. R., Derefinko, K. J., Caspi, A., Loeber, R., & Stouthamer-Loeber, M. (2007). The content validity of juvenile psychopathy: An empirical examination. *Psychological Assessment, 19*, 363–367.

Lynam, D. R., & Gudonis, L. (2005). Development of psychopathy. *Annual Review of Clinical Psychology, 1*, S381–407.

Lyon, D. R., & Ogloff, J. R. P. (2000). Legal and ethical issues in psychopathy assessment. In C. B. Gacono (Ed.), *The clinical and forensic assessment of psychopathy: A practitioner's guide* (pp. 139–173). Mahwah, NJ: Erlbaum.

Marshall, L. A., & Cooke, D. J. (1999). The childhood experiences of psychopaths: A retrospective study of familial and societal factors. *Journal of Personality Disorders, 13*(3), 211–225.

McDermott, P. A., Alterman, A. I., Cacciola, J. S., Rutherford, M. J., Newman, J. P., & Mulholland, E. M (2000). Generality of Psychopathy Checklist-Revised factors over prisoners and substance-dependent patients. *Journal of Consulting and Clinical Psychology, 68*, 181–186.

Millon, T. (1993). *The Millon Adolescent Clinical Inventory (MACI)*. Minneapolis, MN: NCS Assessments.

Millon, T., Simonsen, E., & Birket-Smith, M. (1998). Historical conceptions of psychopathy in the United States and Europe. In T. Millon, E. Simonsen, M. Birket-Smith, & R. D. Davis (Eds.), *Psychopathy: Antisocial, criminal and violent behavior* (pp. 3–31). New York: Guilford..

Moffitt, T. E., Arsenault, L., Jaffee, S. R., Kim-Cohen, J., Koenen, K. C., Odgers, C. L., et al. (2008). Research review: DSM-V conduct disorder: Research needs for an evidence base. *Journal of Child Psychology and Psychiatry, 49*, 3–33.

Moffitt, T. E., Caspi, A., Harrington, H., & Milne, B. J. (2002). Males on the life-course-persistent and adolescence-limited antisocial pathways: Follow-up at age 26 years. *Development and Psychopathology, 14*, 179–207.

Monahan, J., Steadman, H. J., Appelbaum, P. S., Robbins, P. C., Mulvey E. P., Silver, E., et al. (2000). Developing a clinically useful actuarial tool for assessing violence risk. *British Journal of Psychiatry, 176*, 312–319.

Murrie, D. C., Boccaccini, M. T., Johnson, J. T., & Janke, C. (2008). Does interrater (dis)agreement on Psychopathy Checklist score in sexually violent predator trials suggest partisan allegiance in forensic evaluations? *Law and Human Behavior, 32*, 352–362.

Murrie, D. C., Boccaccini, M. T., McCoy, W., & Cornell, D. G. (2007). Diagnostic labeling in juvenile court: How do descriptions of psychopathy and conduct disorder influence judges? *Journal of Clinical Child and Adolescent Psychology, 36*, 228–241.

Murrie, D. C., Boccaccini, M. T., Turner, D. B., Meeks, M., Woods, C., & Tussey, C. (2009). Rater (dis)agreement on risk assessment measures in sexually violent predator proceedings: Evidence of adversarial allegiance in forensic evaluation? *Psychology, Public Policy, and Law, 15*, 19–53.

Murrie, D. C., & Cornell, D. G. (2000). The Millon Adolescent Clinical Inventory and psychopathy. *Journal of Personality Assessment, 75*, 110–125.

Murrie, D. C., Cornell, D. G., Kaplan, S., McConville, D., & Levy-Elkon, A. (2004). Psychopathy scores and violence among juvenile offenders: A multi-measure study. *Behavioral Sciences & the Law, 22*, 49–67.

Murrie, D. C., Cornell, D. G., & McCoy, W. K. (2005). Psychopathy, conduct disorder, and stigma: Does diagnostic labeling influence juvenile probation officer recommendations? *Law and Human Behavior, 29*, 323–342.

Murrie, D. C., Marcus, D. K., Douglas, K. S., Lee, Z., Salekin, R. T., & Vincent, G. (2007). Youth with psychopathy features are not a discrete class: A taxometric analysis. *Journal of Child Psychology and Psychiatry, 48*, 714–723.

Neumann, C. S., & Hare, R. D. (2008). Psychopathic traits in a large community sample: Links to violence, alcohol use, and intelligence. *Journal of Consulting and Clinical Psychology, 76*, 893–899.

Neumann, C. S., Kosson, D. S., Forth, A. E., & Hare, R. D. (2006). Factor structure of the Hare Psychopathy Checklist: Youth Version in incarcerated adolescents. *Psychological Assessment, 18*, 142–154.

Nicholls, T. L., Ogloff, J. R. P., & Douglas, K. S. (2004). Assessing risk for violence among male and female civil psychiatric patients: The HCR-20, PCL:SV, and VSC. *Behavioral Sciences & the Law, 22*, 127–158.

Obradovic, J, Pardini, D. A., Long, J. D., & Loeber, R. (2007). Measuring interpersonal callousness in boys from childhood to adolescence: An examination of longitudinal invariance and temporal stability. *Journal of Clinical Child and Adolescent Psychology, 36*, 276–292.

Ogloff, J. (2006). Psychopathy/antisocial personality disorder conundrum. *Australian and New Zealand Journal of Psychiatry, 41*, 285–294.

Ogloff, J., Wong, S., & Greenwood, A. (1990). Treating criminal psychopaths in a therapeutic community program. *Behavioral Sciences and the Law, 8*, 181–190.

Olver, M. E., & Wong, S. C. P. (2009). Therapeutic responses of psychopathic sexual offenders: Treatment attrition, therapeutic change, and long-term recidivism. *Journal of Consulting and Clinical Psychology, 77*, 328–336.

De Oliveira-Souza, R., Moll, J., Ignácio, F. A., & Hare, R. D. (2008). Psychopathy in a civil psychiatric outpatient sample. *Criminal Justice and Behavior, 35*(4), 427–437.

Porter, S., Birt, A., & Boer, D. P. (2001). Investigation of the criminal and conditional release profiles of Canadian federal offenders as a function of psychopathy and age. *Law and Human Behavior, 25*, 647–661.

Porter, S., Fairweather, D., Drugge, J., Hervé, H., Birt, A., & Boer, D. P. (2000). Profiles of psychopathy in incarcerated sexual offenders. *Criminal Justice and Behavior, 27*, 216–233.

Quinton, D. & Rutter, M. (1988). Urbanism and child mental health. *Journal of Child Psychology and Psychiatry, 29*, 11–20.

Raine, A., Ishikawa, S. S., Arce, E., Lencz, T., Knuth, K. H., Bihrle, H., LaCasse, L., & Coletti, P. (2004). Hippocampal structural asymmetry in unsuccessful psychopaths. *Biological Psychiatry, 55(2)*, 185–191.

Raine, A. Lencz, T., Bihrle, S., LaCasse, L., & Coletti, P. (2000). Reduced prefrontal gray matter volume and reduced autonomic activity in antisocial personality disorder. *Archives of General Psychiatry, 57(2)*, 119–127.

Rice, M. E., & Harris, G. T. (1995). Violent recidivism: Assessing predictive validity. *Journal of Consulting and Clinical Psychology, 63(5)*, 737–748.

Rice, M. E., & Harris, G. T. (1997). Cross-validation and extension of the Violence Risk Appraisal Guide for child molesters and rapists. *Law and Human Behavior, 21*, 231–241.

Rice, M. E., Harris, G. T., & Cormier, C. A. (1992). An evaluation of a maximum security therapeutic community for psychopaths and other mentally disordered offenders. *Law and Human Behavior, 16*, 399–412.

Richards, H. J., Casey, J. O., & Lucente, S. W. (2003). Psychopathy and treatment response in incarcerated female substance abusers. *Criminal Justice and Behavior, 30*, 251–267.

Richards, H. J., Casey, J. O., Lucente, S. W., & Kafami, D. (2003). Differential association of Hare Psychopathy Checklist factor and facet Scores to HIV risk behaviors in incarcerated female substance abusers. *Individual Differences Research, 1*, 95–107.

Roberts, B. W., & DelVecchio, W. F. (2000). The rank-order consistency of personality traits from childhood to old age: A quantitative review of longitudinal studies. *Psychological Bulletin, 126*, 3–25.

Rockett, J. L., Murrie, D. C. & Boccaccini, M. (2007). Diagnostic labeling in juvenile justice settings: Do psychopathy and conduct disorder findings influence clinicians? *Psychological Services, 4*, 107–122.

Rogers, R, Duncan, J. C., Lynett, E., & Sewell, K. W. (1994). Prototypical analysis of antisocial personality disorder: DSM-IV and beyond. *Law and Human Behavior, 18*, 471–484.

Rutherford, M. J., Alterman, A. I., Gacciola, J. S., & McKay, J. R. (1996). Reliability and validity of the Revised Psychopathy Checklist in women methadone patients. *Assessment, 3*, 145–156.

Rutherford, M. J., Alterman, A. I., Gacciola, J. S., & McKay, J. R. (1997). Validity of the Psychopathy Checklist—Revised in male methadone patients. *Drug and Alcohol dependence, 44(2–3)*, 143–149.

Sadeh, N. & Verona, N. (2008). Psychopathy traits and selective impairments in attentional functioning. *Neuropsychology, 22*, 669–680.

Salekin, R. T., Neumann, C. S., Leistico, A. M., & Zalot, A. A. (2004). Psychopathy in youth and intelligence: An investigation of Cleckley's hypothesis. *Journal of Clinical Child and Adolescent Psychology, 33(4)*, 731–742.

Salekin, R., Rogers, R., & Sewell, K. W. (1996). A review and meta-analysis of the Psychopathy Checklist-Revised: Predictive validity of dangerousness. *Clinical Psychology: Science and Practice, 3*, 203–215.

Salekin, R., Rogers, R., Ustad, K. L., & Sewell, K. W. (1998). Psychopathy and recidivism in female inmates. *Law and Human Behavior, 22*, 109–128.

Salekin, R. T. (2006). Psychopathy in children and adolescents: Key issues in conceptualization and assessment. In C. J. Patrick (Ed.), *Handbook of psychopathy* (pp. 389–414). New York: Guilford.

Salekin, R. T., Brannen, D. N., Zalot, A. A., Leistico, A. M., & Neumann, C. S. (2006). Factor structure of psychopathy in youth: Testing the applicability of the new four-factor model. *Criminal Justice and Behavior, 33*, 135–157.

Salekin, R. T., Rosenbaum, J., & Lee Z. (2008). Child and adolescent psychopathy: Stability and change. *Psychiatry, Psychology, and Law, 15*, 224–236.

Salekin, R. T., Worley, C., & Grimes, R. D. (2010). Treatment of psychopathy: A review and brief introduction to the mental model approach for psychopathy. *Behavioral Sciences & the Law, 28*, 235–266

Salekin, R. T., Ziegler, T. A., Larrea, M. A., Anthony, V. L., & Bennett, A. D. (2003). Predicting dangerousness with two Millon Adolescent Clinical Inventory psychopathy scales: The importance of egocentric and callous traits. *Journal of Personality Assessment, 80*, 154–163.

Schmidt, F., McKinnon, L., Chattha, H. K., & Brownlee K. (2006). Concurrent and predictive validity of the Psychopathy Checklist: Youth version across gender and ethnicity. *Psychological Assessment, 18*, 393–401.

Seagrave, D., & Grisso, T. (2002). Adolescent development and the measurement of juvenile psychopathy. *Law and Human Behavior, 26,* 219–239.

Serin, R. C. (1996). Violent recidivism in criminal psychopaths. *Law and Human Behavior, 20,* 207–217.

Serin, R. C., & Amos, N. L. (1995). The role of psychopathy in the assessment of dangerousness. *International Journal of Law and Psychiatry, 18,* 231–238.

Serin, R. C., Mailloux, D. L., & Malcolm, P. B. (2001). Psychopathy, deviant sexual arousal, and recidivism among sexual offenders. *Journal of Interpersonal Violence, 16,* 234–246.

Seto, M. C., & Barbaree, H. E. (1999). Psychopathy, treatment behaviour, and sex offender recidivism. *Journal of Interpersonal Violence, 14,* 1235–1248.

Sevecke, K., Pukrop, R., Kosson, D. S., & Krischer, M. K. (2009). Factor structure of the Hare Psychopathy Checklist: Youth Version in German female and male detainees and community adolescents. *Psychological Assessment, 21,* 45–56.

Skeem, J. L., & Cauffman, E. (2003). View of the downward extension: Comparing the Youth Version of the Psychopathy Checklist with the Youth Psychopathic Traits Inventory. *Behavioral Sciences and the Law, 21,* 737–770.

Skeem, J. L., Edens, J. F., Camp, J. & Colwell, L. H. (2004). Are there ethnic differences in levels of psychopathy? A meta-analysis. *Law and Human Behavior, 28,* 505, 527.

Skeem, J. L., & Mulvey, E. P. (2001). Psychopathy and community violence among civil psychiatric patients: Results from the MacArthur Violence Risk Assessment Study. *Journal of Consulting and Clinical Psychology, 69,* 358–374.

Skeem, J. L., Mulvey, E. P., & Grisso, T. (2003). Applicability of traditional and revised models of psychopathy to the Psychopathy Checklist: Screening Version. *Psychological Assessment, 15,* 41–55.

Steinberg, L. (2002). The juvenile psychopath: Fads, fictions, and facts. In *National Institute of Justice perspectives on crime and justice: 2001 Lecture Series,* (Vol. V, pp. 35–64). Washington, DC: U.S. Department of Justice.

Sullivan, E. A., & Kosson, D. S. (2006). Ethnic and cultural variations in psychopathy. In C. J. Patrick (Ed.), *Handbook of psychopathy* (pp. 437–458). New York: Guilford.

Tengström, A., Grann, M., Långström, N., & Kullgren, G. (2000). Psychopathy (PCL-R) as a predictor of violent recidivsm among criminal offenders with schizophrenia. *Law and Human Behavior, 24,* 45–58.

Thornton, D., & Blud, L. (2007). The influence of psychopathic traits on response to treatment. In H. Hervé & J. C. Yuille (Eds.), *The psychopath: Theory, research, and practice* (pp. 505–539). Mahwah, NJ: Erlbaum.

Vanman, E. J., Mejia, V. Y., Dawson, M. E., Schell, A. M., & Raine, A. (2003). Modification of the startle reflex in a community sample: Do one or two dimensions of psychopathy underlie emotional processing? *Personality and Individual Differences, 35,* 2007–2021.

Vidal, S., & Skeem, J. L. (2007). Effect of psychopathy, abuse, and ethnicity on juvenile probation officers' decision-making and supervision strategies. *Law and Human Behavior, 31,* 479–498.

Viljoen, J. L., MacDougall, E. A. M., Gagnon, N. C., & Douglas, K. S. (2010). Psychopathy evidence in legal proceedings involving adolescent offenders. *Psychology, Public Policy, and Law, 16,* 254–283.

Viljoen, J. L., McLachlan, K., & Vincent, G. M. (2010). Assessing violence risk and psychopathy in juvenile and adult offenders: A survey of clinical practices. *Assessment, 17,* 377–395.

Vincent, G. M., & Kinscherff, R. (2008). The use of psychopathy in violence risk assessments of adolescent females. *Journal of Forensic Psychology Practice, 8,* 309–320.

Vincent, G. M., Odgers, C. L., McCormick, A. V., & Corrado, R. R. (2008). The PCL: YV and recidivism in male and female juveniles: A follow-up into young adulthood. *International Journal of Law and Psychiatry, 31,* 287–296.

Vitacco, M. J., Neumann, C. S., Caldwell, M. F., Leistico, A. M., & Van Rybroek, G. J. (2006). Testing factor models of the psychopathy checklist: Youth version and their association with instrumental aggression. *Journal of Personality Assessment, 87,* 74–83.

Vitacco, M. J., Neumann, C. S., & Jackson, R. L. (2005). Testing a four-factor model of psychopathy and its association with ethnicity, gender, intelligence, and violence. *Journal of Consulting and Clinical Psychology, 73,* 466–476.

Vitacco, M. J., Rogers, R., Neumann, C. S., Harrison, K. S., & Vincent, G. (2005). A Comparison of factor models on the PCL-R on mentally disordered offenders: The development of a four-factor model. *Criminal Justice and Behavior, 32*, 526–545.

Vitale, J. E., & Newman, J. P. (2001). Using the Psychopathy Checklist-Revised with female samples: Reliability, validity, and implications for clinical utility. *Clinical Psychology: Science, & Practice, 8*, 117–132.

Vitale, J. E., Smith, S. S., Brinkley, C. A., & Newman, J. P. (2002). The reliability and validity of the Psychopathy Checklist-Revised in a sample of female offenders. *Criminal Justice and Behavior, 29*, 202–231.

Walsh, T., & Walsh, Z. (2006). The evidentiary introduction of Psychopathy Checklist-Revised assessed psychopathy in U.S. Courts: Extent and appropriateness. *Law and Human Behavior, 30*, 493–507.

Walters, G. D. (2003). Predicting institutional adjustment and recidivism with the Psychopathy Checklist factor scores: A meta-analysis. *Law and Human Behavior, 27*, 541–558.

Walters, G. D., Gray, N. S., Jackson, R. L., Sewell, K. W., Rogers, R., Taylor, J., et al. (2007). A taxometric analysis of the Psychopathy Checklist: Screening Version (PCL:SV): Further evidence of dimensionality. *Psychological Assessment, 19*, 330–339.

Warren, J. I., Burnette, M., South, C. S., Chauhan, P., Bale, R., Friend, R., et al. (2003). Psychopathy in women: Structural modeling and co-morbidity. *International Journal of Law and Psychiatry, 26*, 223–242.

Webster, C. D., Douglas, K. S., Eaves, D., & Hart, S. D. (1997). *HCR-20 Assessing Risk for Violence* (Version 2). Burnaby, British Columbia: Mental Health, Law, and Policy Institute Simon Fraser University.

Weiler, B. L., & Widom, C. S. (1996). Psychopathy and violent behaviour in abused and neglected young adults. *Criminal Behaviour and Mental Health, 6*(3), 253–271.

Werner, E. E., & Smith, R. S. (1992). *Overcoming the odds: High risk children from birth to adulthood.* Ithaca, NY: Cornell University Press.

Williamson, S. E., Hare, R. D., & Wong, S. (1987). Violence: Criminal psychopaths and their victims. *Canadian Journal of Behavioral Science, 19*, 454–462.

Windle, M., & Dumenci, L. (1999). The factorial structure and construct validity of the Psychopathy Checklist-Revised (PCL-R) among alcoholic inpatients. *Structural Equation Modeling, 6*, 372–393.

Wong, S., & Burt, G. (2007). The heterogeneity of incarcerated psychopaths: Differences in risk, need, recidivism, and management approaches. In H. Hervé & J. C. Yuille (Eds.), *The psychopath: Theory, research, and practice* (pp. 461–484). Mahwah, NJ: Erlbaum.

Woodworth, M., & Porter, S. (2002). In cold blood: Characteristics of criminal homicides as a function of psychopathy. *Journal of Abnormal Psychology, 111*, 436–445.

Yang, Y., Raine, A., Narr, K. L., Coletti, P., & Toga, A. W. (2009). Localization of deformations within the amygdala in individuals with psychopathy. *Archives of General Psychiatry, 66*, 986–994.

Zinger, I., & Forth, A. (1998). Psychopathy and Canadian criminal proceedings: The potential for human rights abuses. *Canadian Journal of Criminology, 40*, 237–276.

CHAPTER 3

Alternatives to the Psychopathy Checklist—Revised

Katherine A. Fowler and Scott O. Lilienfeld

The Psychopathy Checklist-Revised (PCL-R; Hare, 1991a/2003) and its progeny, including the Psychopathy Checklist: Screening Version (Hart, Cox, & Hare, 1995) and Psychopathy Checklist: Youth Version (Forth, Kosson, & Hare, 2003), were major breakthroughs in the assessment and diagnosis of psychopathy. These measures were the first to provide systematic, reliable, and valid approaches to the detection of psychopathy, thereby facilitating meaningful research on this condition and advancing its scientific status as a meaningful construct. Accordingly, the PCL-R has increasingly come to be regarded as the "gold standard" in psychopathy research (e.g., Fulero, 1995). Nevertheless, genuine gold standards are probably unattainable in the domains of personality and psychopathology (Cronbach & Meehl, 1955; Faraone & Tsuang, 1994), as measures in these domains are necessarily fallible indicators of the constructs of interest.

Furthermore, there are pragmatic concerns regarding the PCL-R and its descendants, such as (a) the often lengthy nature of the interview (frequently 90 minutes or more), (b) need for extensive interviewer training, and (c) requirement for file or other corroborative information—the latter of which is often impractical in nonprison settings. These concerns have stimulated a number of alternatives to the PCL-R in psychopathy assessment.

Alternatives to the PCL-R comprise self-report, observer, and implicit measures of psychopathy, which differ from the PCL-R both methodologically and conceptually. Some rely on self-report, others on peer report, and still others on presumed implicit cognitive processes.

GOALS OF THE CHAPTER

In this chapter, we review conceptual and methodological alternatives to the PCL-R. We begin with self-report measures, followed by observer measures and finally implicit measures, and examine the advantages and disadvantages of each approach. Within each section, we survey the contemporary status of alternative psychopathy measures with an emphasis on their psychometric properties, research and clinical uses, and limitations.

We do not examine alternative measures of psychopathy in children or adolescents given that the childhood psychopathy construct and its controversies merit a substantial discussion beyond the scope of this chapter and are covered elsewhere in this volume. In addition, there is minimal overlap between adult and child psychopathy in the measures used.

We have also elected not to examine self-report measures of adult psychopathy that lack an adequate research base, such as the Minnesota Multiphasic Personality Inventory (MMPI)–based Sociopathy Scale (Spielberger, Kling, & O'Hagan, 1978), the Psychopathic States Inventory (Haertzen, Martin, Ross, & Neidert, 1980), Levenson's Psychopathy Scale (Levenson, 1988; not to be confused with the Levenson Primary and Secondary Psychopathy Scales, reviewed later), the Social Psychopathy Scale

(Edelmann & Vivian, 1988; Smith, 1985), or the Antisocial Personality Questionnaire (Blackburn & Fawcett, 1999).

SELF-REPORT MEASURES OF PSYCHOPATHY

The idea of assessing psychopathy by asking people to report their own responses to questions assessing the construct may seem paradoxical. After all, psychopathy is a condition marked by dishonesty, so how could psychopathic individuals be expected to provide useful self-reported responses? These noteworthy challenges aside (Lilienfeld, 1994, 1998), the self-report assessment of psychopathy has experienced a renaissance over the past two decades. Several advantages and disadvantages of self-report measures of psychopathy are worth noting. We will begin with the advantages.

Advantages of Self-Report Psychopathy Measures

Self-report measures may be of particular utility in the detection of subjective emotional states and traits. With respect to psychopathy, the relative *absence* of such states and traits, especially guilt, empathy, fear, and feelings of intimacy toward others, is ostensibly most relevant. Nevertheless, psychopaths may experience certain emotions, such as alienation and anger, more frequently than do nonpsychopaths.

Self-reports of personality converge moderately with reports from knowledgeable others ($r = .30–.50$; Kenrick & Funder, 1988). This substantial shared variance suggests at least some convergent validity for self-report measures; at the same time, the substantial amount of nonshared variance introduces the possibility that each information source possesses incremental validity (Meehl, 1959; Sechrest, 1963) above and beyond the other.

Another, more self-evident, advantage of using self-reports of psychopathy is economy, or ease of use. Self-report measures tend to be brief and easy to complete and require minimal training on the part of test administrators. In this respect, they stand in stark contrast to the PCL-R, which, as noted earlier, requires access to file information and extensive interviewer training.

An often unappreciated advantage of self-report measures is that they can detect response styles systematically (Lilienfeld & Fowler, 2006; Widiger & Frances, 1987). In this respect, they possess a key advantage over interview measures (the PCL-R included), few of which contain systematic response style indicators. Certain response styles, such as positive impression management and malingering, may be particularly problematic among psychopaths (Edens, 2004; Hart, Hare, & Harpur, 1992; Lilienfeld, 1994). Although such response styles may adversely affect the validity of psychopaths' responses, questionnaires can help to detect such response styles using validity scales (see Paulhus, 1991, for a review).

Finally, interrater reliability is not a consideration for self-report measures because these measures are completed by respondents and do not require "judgment calls" by interviewers or other observers. Many core features of psychopathy, such as lack of empathy and guilt, require considerable clinical inference on the part of observers and therefore are unlikely to achieve anywhere near perfect interrater reliability. Because validity is constrained by the square root of reliability (Meehl, 1986), this is an important consideration in interpreting the validity of interview and observer measures.

Disadvantages of Self-Report Psychopathy Measures

There are nonetheless several commonly discussed disadvantages of self-reports of psychopathy. The first is obvious: Psychopaths lie frequently and with minimal guilt or anxiety. Although such lying on questionnaires can sometimes be detected by response style indicators, many of these indicators, such as the MMPI-2 Lie Scale, tend to be insensitive to subtle or sophisticated forms of impression management (Greene, 2000; Kroger & Turnbull, 1975; Vincent, Linsz, & Greene, 1966).

Making matters more complicated, the nature of psychopaths' lying may depend largely on situational demands, and therefore cannot be predicted readily without knowledge of contextual variables. That is, if psychopaths are placed in a situation in which crafting a positive impression is desirable (e.g., applying for a job), they may attempt to make themselves appear healthy, whereas if they are placed in a situation in which crafting a negative impression is desirable (e.g., being evaluated for an insanity plea), they may attempt to make themselves appear disturbed.

Second, psychopaths often lack insight into the nature and extent of their psychological problems (Cleckley, 1941/1988), potentially resulting in *method–mode mismatches*: the use of a method, in this case self-report, that is inappropriate for the construct in question (Edens, Hart, Johnson, Johnson, & Olver, 2002). The failure of most psychopaths to perceive themselves as others do may constrain the usefulness of certain self-report items, especially those that require at least a modicum of accurate knowledge regarding the impact of one's behavior on others. At the same time, findings from a community sample revealed moderate to high levels of convergence (mean r = .64) between self-reported and informant-reported scores on three widely used psychopathy measures, raising the possibility that individuals with psychopathic traits can at least sometimes report veridically on their traits and behaviors (Miller, Jones, & Lynam, 2011).

One can conceptualize this disadvantage of self-reported psychopathy by means of the *Johari window* (named, curiously enough, after the first names of its developers, Joseph Luft and Harry Ingham), which schematically represents the four major "regions" of personality as perceived by both self and observers (Luft, 1969; see Figure 3.1). The Johari window consists of four cells: the region of personality known to both self and others (the "open" quadrant), the region of personality known to the self but not others (the "hidden" quadrant), the region of personality known to others but not the self (the "blind" quadrant), and finally and perhaps most interesting, the region of personality known to neither self nor others (the "unknown" quadrant). Observer reports

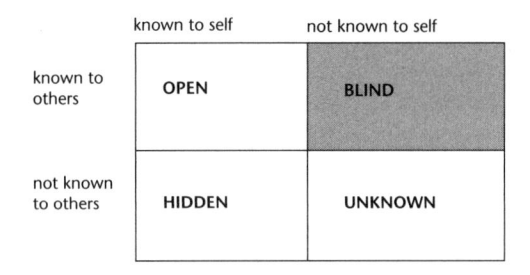

	known to self	not known to self
known to others	OPEN	BLIND
not known to others	HIDDEN	UNKNOWN

Figure 3.1 The Johari window conceptualizes personality as a multilayered construct, with quadrants representing combinations of self and other awareness (Luft, 1969).

are potentially of particular utility in assessing the blind quadrant, where others can report on attributes that psychopaths are either unable or unwilling to report.

Consequently, observers may be superior to psychopathic individuals when reporting on certain overt behaviors and their interpersonal consequences. For example, in a sample of more than 2000 military recruits, Oltmanns and Turkheimer (2009) found that peer reports of antisocial personality disorder (ASPD), a condition that overlaps moderately to highly with psychopathic personality; (Lilienfeld, 1994) were more accurate than self-reports in predicting early separation from service.

Third, it may be inherently paradoxical to ask individuals who have never experienced an emotion (or who have experienced only weak variants of it) to report on its presence or absence. As Kelly (1955) observed, a full understanding of a dimension requires an appreciation of both of its poles. For example, the experience of "cold" carries no subjective meaning unless one has experienced heat. Similarly, asking psychopaths to report on the absence of guilt may be fruitless given that they have had scant experience with its presence. Cleckley (1941/1988; see also Hare, 1993) likened psychopaths' experiences with emotion to the neurological syndrome of "semantic aphasia," a condition that ostensibly leads them to mislabel affective experiences. Psychopaths may erroneously learn to label certain emotions as "guilt" or "fear" even though they have never experienced them. They may learn to refer to "guilt," for example, when they experience negative affect after committing an antisocial act and receiving punishment for it, even though they are actually experiencing regret rather than remorse. From this perspective, psychopaths' reporting of many emotions may be inaccurate but not insincere.

Fourth, many self-report measures of psychopathology are heavily saturated with Negative Emotionality (NE), a pervasive higher-order personality dimension reflecting a disposition to experience negative affects of many kinds, including anxiety, irritability, hostility, and mistrust. Indeed, one of the great challenges in constructing self-report measures of psychopathology is to develop questionnaires that are not contaminated by NE (Finney, 1985; Tellegen, 1985). The substantial correlation of many self-report psychopathology measures with NE diminishes their discriminant validity because NE courses through many psychiatric conditions, including mood, anxiety, psychotic, eating, and somatoform disorders (Watson & Clark, 1984). Although one might expect measures of psychopathy to be largely independent of NE, many measures designed to detect psychopathy, such as the MMPI-II Psychopathic deviate scale, are substantially contaminated by this dimension (Lilienfeld, 1994). This is especially true of self-report measures that detect the antisocial lifestyle and impulsive behaviors associated with psychopathy (Harpur, Hare, & Hakstian, 1989).

Misconceptions and Misunderstandings Regarding the Self-Report Assessment of Psychopathy

The potential disadvantages of self-report psychopathy measures raise important questions regarding their validity. Nonetheless, we would be remiss not to address three widespread misconceptions that have led some authors to discount prematurely the potential value of self-reports in the assessment of this condition.

(1) *Requirement of veridical responding.* The first misconception is that the validity of self-report measures validity hinges on the assumption of veridical (truthful) responding (Lilienfeld, 1994). This assumption, if accurate, would be potentially problematic given psychopaths' dishonesty and lack of insight. But as Meehl (1945) noted, responses to self-report items can offer diagnostically helpful information regarding respondents' apperceptions of themselves and the world, regardless of their factual accuracy. Take the item, "I often get blamed for things that aren't my fault," which appears on the Psychopathic Personality Inventory (PPI; Lilienfeld & Andrews, 1996; see also Lilienfeld & Widows, 2005). A "True" response to this item is a valid indicator of psychopathy, even though it is unlikely to be factually accurate, because most psychopaths are probably not blamed nearly enough for things that go wrong in their lives! Nevertheless, this item provides useful information regarding psychopaths' well-known propensity to externalize blame (Hare, 1991a) and to perceive others as malevolent (Millon, 1981).

(2) *Propensity toward positive impression management.* A second misconception is that psychopaths consistently engage in positive impression management on self-report measures. In fact, self-report measures of psychopathy tend to correlate negatively with indices of social desirability and positive impression management (e.g., Hare, 1982; Lilienfeld & Andrews, 1996; Lilienfeld & Widows, 2005; Ray & Ray, 1982). Indeed, a recent meta-analysis of the PPI and the Levenson Primary and Secondary Psychopathy Scales (Levenson, Kiehl, & Fitzpatrick, 1995) revealed a slight average negative correlation ($r = .-11$) between total psychopathy scores and measures of positive impression management (Ray, Hall, Rivera-Hudson, Poythress, Lilienfeld, & Morano, 2013). Although this finding may appear puzzling, we might conjecture that psychopaths possess a different conception of what is socially undesirable, such as antisocial behaviors, recklessness, hostility, and poor impulse control, compared with the average person (Lilienfeld, 1994). It should also be borne in mind that response style measures, such as self-report lie scales, are not entirely independent of genuine trait variance (see Chan, 2009, for a discussion).

As a consequence, extreme (either high or low) scores on these scales are probably heterogeneous in origin, reflecting a mix of substantive (trait) and stylistic (response set and response style) variance.

(3) *Aptitude for malingering.* A third misconception is that psychopaths are particularly adept at manipulating their responses to self-report measures, rendering their responses even more untrustworthy than those of nonpsychopathic dissimulators. There is no empirical support for this claim, which has not been extensively researched. Edens, Buffington, and Tomicic (2000) asked 143 college students to take the PPI (as noted earlier, a self-report measure of psychopathy) under two conditions: (a) honestly or (b) with instructions to malinger psychosis. They found that PPI scores were not significantly related to malingering success, but significantly and positively correlated with willingness to malinger, as well as self-perceived ability to malinger a mental disorder. Thus, although psychopaths may be more inclined than nonpsychopaths to malinger on psychological tests when it is in their interests (Rogers et al., 2002), there is no evidence that they are especially adept at doing so (see also Poythress, Edens, & Watkins, 2002).

Longstanding Empirical Problems in the Self-Report Assessment of Psychopathy

Until fairly recently, the self-report assessment of psychopathy was regarded by many as a deeply troubled endeavor (e.g., Hare, 1985; Hart et al., 1992; Lilienfeld & Fowler, 2006), largely due to three empirical problems. As we will see, these problems persist even to the present day, although there has been progress toward their resolution.

(1) *Low correlations among psychopathy questionnaires.* Early studies revealed low correlations among questionnaires designed to assess psychopathy, suggesting that measures are assessing only slightly overlapping aspects of the same construct.

In a large sample of adult male inmates, Hare (1985) found low or at best moderate correlations among the most frequently used self-report psychopathy instruments at the time, including the MMPI Pd scale and Hypomania (Ma) scales; Gough's (1960) California Psychological Inventory (CPI) Socialization (So) Scale (see Kosson, Steuerwald, Newman, & Widom, 1994), which is often scored in reverse as a measure of psychopathy; and Hare's Self-Report Psychopathy Scale. The absolute value of the correlations ranged from .14 to .53, and some measures previously viewed as virtually interchangeable shared relatively little variance; for example, the MMPI Pd scale and So scale correlated only $r = -.34$.

(2) *Method covariance.* Although correlations among self-report measures of psychopathy are often low or modest, even these correlations may partly reflect method variance arising from the shared use of a questionnaire format.

In the aforementioned study, Hare (1985) conducted a principal components analysis of self-report and clinical–behavioral measures administered to prison inmates. This analysis yielded a two-component solution that appeared to reflect method variance rather than content variance. Specifically, the first component was marked by high loadings on the clinical–behavioral measures, whereas the second was marked by high loadings on the self-report measures. The content of the scales appeared to exert little impact on the pattern of intercorrelations. For example, even though the PCL and *Diagnostic and Statistical Manual of Mental Disorders—Third Edition (DSM–III)* criteria for ASPD ostensibly assess different constructs (psychopathy vs. ASPD, respectively; see, Hare, 1991a; Lilienfeld, 1994), they loaded more highly with each other than with self-report measures ostensibly assessing the same construct (see also Widom & Newman, 1985).

(3) *Nonspecific measures of behavioral deviance.* Another shortcoming of many older self-report psychopathy measures is that they appear primarily to be nonspecific measures of behavioral deviance (global antisocial

and criminal behavior) rather than of the core affective and interpersonal features of psychopathy.

Harpur et al. (1989) examined the correlations of the two major PCL factors with several self-report indices relevant to psychopathy, including the MMPI Pd and Ma scales, the So scale, the Eysenck Personality Questionnaire Psychoticism Scale (Eysenck & Eysenck, 1975), the Sensation Seeking Scale (SSS; Zuckerman, Kolin, Price, & Zoob, 1964), and the Self-Report Psychopathy Scale (SRP; Hare, 1985). The correlations of these questionnaires with PCL Factor 2, which assesses an antisocial and impulsive lifestyle, were moderately high and were generally in the $r = .3$ to $.5$ range. In contrast, the correlations of these measures with PCL Factor 1, which assesses the core affective and interpersonal features of psychopathy, were negligible to low, and were generally in the $r = .05$ to $.15$ range. Perhaps most surprisingly, two of the most frequently administered self-report measures of psychopathy (Hare & Cox, 1978), the MMPI Pd scale and the So scale, correlated with PCL Factor 1 at only $r = .05$ and $-.06$, respectively. Hart, Forth, and Hare (1991) and Edens et al. (2000) examined the Antisocial and Aggressive/Sadistic scales of the Millon Clinical Multiaxial Inventory-II (MCMI-II; Millon, 1987), and the Antisocial scale of the Personality Assessment Inventory (PAI; Morey, 1991) respectively, and obtained similar correlations with PCL/PCL-R factor scores.

These findings suggest that several widely used self-report measures of psychopathy, including the MMPI Pd scale, are largely unrelated to the core personality features of this condition (see also Hawk & Peterson, 1973; Lovering & Douglas, 2004). Instead, these measures are probably markers of nonspecific behavioral deviance, which do not distinguish psychopathy from a variety of other conditions often associated with antisocial and criminal behavior (Lykken, 1995).

The shortcomings of extant psychopathy questionnaires have led several investigators to develop new self-report measures of psychopathy over the past two decades. The comparative merit of these measures is an active topic of debate among researchers. There are currently three classes of self-report psychopathy measures: (a) those that measure patterns of normal personality and measure match to a profile considered consistent with psychopathy, (b) those that measure general patterns of pathological personality ("omnibus measures") and include subscales developed to assess psychopathy, and (c) those developed to assess psychopathy directly and specifically.

First, we briefly address measures of normal and pathological personality that contain "profiles" associated with psychopathy. We then address self-report psychopathy measures in greater detail.

Measures of Normal and Pathological Personality with Psychopathy "Profiles"

The lion's share of evidence indicates that psychopathy is best conceptualized dimensionally (Marcus, John, & Edens, 2004), and diagnostic criteria for psychopathy are replete with personality traits (e.g., poor impulse control, egocentricity) that are presumably distributed continuously. Therefore, several researchers have suggested examining dimensionally conceptualized normal personality and broad range personality pathology as novel ways of assessing psychopathy.

Research on normal-range structural models of personality and psychopathy has focused primarily on configurations derived from the Five-Factor Model (FFM; Costa & McCrae, 1990) of personality. The FFM encompasses five broad domains relevant to normal personality: Extraversion (E), Agreeableness (A), Conscientiousness (C), Neuroticism (N), and Openness (O). Derefinko and Lynam (2006) conducted a meta-analysis of 11 studies that examined the FFM in relation to psychopathy. Criterion measures included the PCL-R, PPI, and Levenson Primary and Secondary Scales (LPSP), to be discussed later. They found that all FFM domains bore significant relations to psychopathy, although to differing degrees. The weighted mean effect sizes (WMESs) for Neuroticism (.16), Extraversion (−.05), and Openness (−.09) were smaller, while Agreeableness (−.54) and Conscientiousness (−.38) were larger. Examining relations with Factor 1 and Factor 2 separately, the primary differences were that Factor 2 (vs. Factor 1) was more strongly negatively related to Agreeableness and Conscientiousness (WMES = −.43, .46 vs. −.39, −.22), and was moderately positively related to Neuroticism (WMES = .35 vs. −.01). Lynam and colleagues emphasized the importance of low Agreeableness (or as they also call it, high Antagonism) in this model. They suggested using the NEO-PI (Costa & McCrae, 1992) as an assessment measure of psychopathy, as it is a widely used and psychometrically sound measure of the FFM. However, they also noted that further research should examine the utility of this approach in real-world settings (Lynam & Derefinko, 2006).

There is a longer history of including scales that purportedly measure psychopathy in omnibus measures of psychopathology. As mentioned previously, "Psychopathic Deviate" (Pd, Scale 4) is among the original MMPI clinical scales, but it converges poorly with clinical conceptualizations of psychopathy and bears weak associations with measures of the core affective and interpersonal features of psychopathy. Recent attempts to create more valid MMPI psychopathy scales have yielded improvements. Sellbom, Ben-Porath, and Stafford (2007) compared the MMPI-2 Pd scale with other selected MMPI-2 scales, namely Disconstraint (DISC), Antisocial Practices (ASP), and Restructured Clinical Scale 4 (RC4). They found that RC4 performed the best, correlating $r = .50$ with Psychopathy Checklist: Short Version (PCL:SV; Hart et al., 1995) total scores. RC4 was a significantly better predictor of juvenile and adult antisocial behavior than the other scales, but it did not outperform ASP and DISC in predicting the interpersonal and affective features of psychopathy ($r = .29-.36$ with PCL:SV Factor 1). All scales correlated more highly with PCL:SV Factor 2 scores ($r = .41-.62$), indicating that they still capture behavioral deviance to a greater degree than personality features of psychopathy (however, Sellbom, Ben-Porath, Lilienfeld, Patrick, & Graham, 2005 discussed more complex MMPI-2 scale configurations that may better represent Factor 1 features).

Noting the success of generating psychopathy-relevant profiles using the FFM, Pryor, Miller, and Gaughan (2009) adopted a similar approach using the Schedule for Nonadaptive and Adaptive Personality (SNAP, Clark, 1993) and Dimensional Assessment of Personality Pathology-Basic Questionnaire (DAPP-BQ; Livesley, 1990), two measures with scales assessing higher-order temperament and lower-order traits associated with personality pathology. Comparing the two measures and examining their relation with two psychopathy self-report measures (PPI and LPSP), they found that both the SNAP and DAPP-BQ demonstrated good predictive validity, each accounting for a substantial amount of variance in psychopathy scores (57%–77%).

In sum, dimensional models of normal and pathological personality have yielded configurations relevant to psychopathy to varying degrees. This work is timely owing to the ongoing debate surrounding the nature of personality disorders as hybrids of personality and problematic behaviors (Pryor, Miller, & Gaughan, 2009). Nevertheless, from a pragmatic standpoint, the utility of these measures in the assessment of psychopathy remains unresolved. For example, the degree to which they relate to external psychopathy-relevant criteria, such as recidivism, or demonstrate clinical utility over and above self- or other-reported psychopathy, requires further research.

Psychopathy-Specific Measures

In this section, we review three self-report measures of psychopathy, the Levenson Primary and Secondary Psychopathy Scales (LPSP; Levenson et al., 1995); the Self-Report Psychopathy Scale (SRP) and its revisions, the SRP-II and SRP-III (Hare, 1985, 1991b; Paulhus, Hemphill, & Hare, 2006); and the Psychopathic Personality Inventory and its revision, the PPI-R (Lilienfeld, 1990; Lilienfeld & Widows, 2005). We focus on these measures because they were designed to address the shortcomings of previously developed psychopathy self-report measures (e.g., MMPI-Pd, CPI-So) and have been examined in numerous published studies.

Levenson Primary and Secondary Psychopathy Scales

Construction and Format

The LPSP was developed by Levenson et al. (1995) to assess self-reported psychopathy in noninstitutional samples. It consists of 26 items in a 1–4 Likert-type format. The Primary and Secondary scales of the LPSP were rationally constructed (i.e., logically extended from what is known about a construct) based on Karpman's (1948) classic clinical writings, which describe two manifestations of psychopathic features: one "primary" (Cleckley) psychopathy, and one "secondary" or "pseudopsychopathy" (Lykken, 1995). In contrast with primary psychopathy, secondary psychopathy is presumably characterized by elevated neuroticism/anxiety accompanied by impulsivity but intact interpersonal functioning. A sample item from the Primary psychopathy scale is "Looking out for myself is my top priority," whereas a sample item from the Secondary psychopathy scale is "I am often bored."

Psychometric Properties and Validity

Across several studies, in large samples of undergraduates ($N = 487–1,154$) and male prisoners ($N = 70–549$), the LPSP Primary and Secondary psychopathy scales demonstrated acceptable-to-good internal consistency (Levenson et al., 1995; Lynam, Whiteside, & Jones, 1999; Brinkley, Schmitt, Smith, & Newman, 2001). The Primary scale has consistently demonstrated higher internal consistency (Cronbach's alpha = .82–.84) than the Secondary scale (Cronbach's alpha = .63–.68). Men have scored higher than women, in keeping with other findings in the psychopathy literature (Cale & Lilienfeld, 2002; Lykken, 1995).

Levenson et al. intended the LPSP to provide indices of PCL-R Factors 1 (Interpersonal and affective deficits) and 2 (Antisocial behavior) and conducted exploratory factor analyses demonstrating that the LPSP yields a two-factor structure parallel to that of the PCL-R. Others conducted confirmatory factor analyses replicating this finding (Brinkley et al., 2001; Lynam et al., 1999). In a sample of 270 Caucasian and 279 African American prisoners, Brinkley et al. (2001) found that although a two-factor structure emerged, the pattern of item loadings differed somewhat for the two groups.

Several researchers have examined the examined the construct validity of the LPSP its relation to measures of anxiety, personality, and laboratory correlates of psychopathy. Levenson et al. suggested that the primary and secondary psychopathy scales could be differentiated on the basis of low (high scorers on the Primary scale) vs. high (high scorers on the Secondary scale) trait anxiety. They found that both scales correlated positively and significantly with a self-report measure of trait anxiety, although the Primary scale correlation was weak ($r = .09$). Lynam et al. (1999) examined correlations between the LPSP and Big Five dimensions, and found that the Primary scale was negatively related to Agreeableness, whereas the Secondary scale was negatively related to Agreeableness and Conscientiousness and positively related to Neuroticism (also consistent with the view that secondary psychopathy is associated with elevated anxiety).

Both scales correlated positively and significantly with a measures of antisocial behavior (Levenson et al., 1995; Lynam et al., 1999) and alcohol and drug use (Lynam et al., 1999), with most correlations in the $r = .20–.30$ range. Converging with these findings, Levenson et al. found that both LPSP scales correlated positively with the Boredom Susceptibility and Disinhibition subscales of the SSS (Zuckerman, 1989), further supporting the construct validity of the LPSP given the centrality of risk-taking to psychopathy. In a large sample of male prisoners ($N = 549$), Brinkley et al. (2001) found that only the Secondary scale correlated positively and significantly with alcohol use.

In a sample of 70 male prisoners, Lynam et al. (1999) examined the relation of the LPSP to two laboratory measures relevant to psychopathy, the go-no task and the Q task (see Newman & Wallace, 1993). Both measures index deficits in response modulation (typically defined as the ability to switch attention to extraneous stimuli when focused on a dominant response set), a construct considered central to psychopathy by Newman and his colleagues (e.g., Newman & Kosson, 1986). The LPSP scales were significantly (marginally for the Secondary scale) related to passive avoidance (commission) errors on the go-no task (see also Brinkley et al., 2001) and marginally significantly related to (low) response interference on the Q task. The absolute values of the correlations ranged from $r = .18$ to $.23$, and there was no significant difference in the correlates of the two LPSP scales.

Other recent work has examined the relation of the LPSP scales to a variety of personality traits, both in the normal and abnormal domains. In a male college sample, Falkenbach, Falki, Poythress, and Manchak (2007) found that both LPSP scales were positively associated with measures of self-reported anger, hostility, and verbal and physical aggression. In contrast to Levenson et al. (1995), both scales were moderately related to trait anxiety, with the Secondary scale correlation being higher than the Primary scale correlation.

Summary of Advantages and Disadvantages
The LPSP holds promise as a self-report measure of psychopathy. Its internal consistency is acceptable to good, and it exhibits a two-factor structure similar to the PCL-R. In addition, it demonstrates theoretically meaningful relations with self-report measures of sensation seeking and antisocial behavior, as well as with laboratory measures of response modulation.

Nonetheless, several investigators have reported findings that raise questions concerning the discriminant validity of the LPSP scales. Brinkley et al. found that while the Primary scale correlated significantly more highly with PCL-R Factor 1, the Secondary scale correlated equally with PCL-R Factor 1 and Factor 2. Lilienfeld, Skeem, and Poythress (2004) similarly reported that the LPSP Primary Scale correlated highly ($r = .62$) with the Factor 2 scale of the Psychopathic Personality Inventory (Lilienfeld, 1990; see "Psychopathic Personality Inventory"), and Wilson, Frick, and Clements (1999) and Lilienfeld and Hess (2001) similarly found that the LPSP Primary Scale was correlated highly with other Factor 2 self-report psychopathy measures.

The Self-Report Psychopathy Scale

Construction and Format
The SRP was constructed by Hare and colleagues using a combination of approaches. Hare (1985) initially identified 75 rationally constructed items that distinguished high from low PCL scorers. This preliminary item pool was refined by selecting 29 items that exhibited high correlations with the PCL total score. The original version of the SRP correlated only modestly with the PCL (Hare, 1985) and did not provide adequate content coverage of several traits traditionally deemed central to psychopathy, including superficial charm, callousness, and dishonesty.

These concerns were addressed in a revised version of the SRP, the SRP-II (Hare, 1991b; Hare, Hemphill, & Paulhus, 2002). Like the PCL-R, this revised measure, called the SRP-II, contains two factors, with the first factor assessing primarily the core interpersonal and affective features of psychopathy and the second factor primarily assessing antisocial and impulsive lifestyle. SRP-II items include "I can read people like a book" (Factor 1) and "I have often done something dangerous just for the thrill of it" (Factor 2). The SRP-II has since been revised, and is now in its third version (SRP-III; Paulhus, Hemphill, & Hare, 2006).

Psychometric Properties
There has been somewhat less published research on the reliability and validity of the SRP its progeny than other self-report psychopathy measures. In a prison sample, Hare (1985a) found that the SRP was internally consistent (Cronbach's alpha = .80; see Paulhus & Williams, 2002, for similar findings on the SRP-III). In an undergraduate sample, Lilienfeld and Penna (2001) reported that that although the SRP-II total score was internally consistent (Cronbach's alpha = .91), the internal consistency of SRP-II Factor 1 was marginal (Cronbach's alpha = .59), and the internal consistency of SRP-II Factor 2 was adequate (Cronbach's alpha = .72). In another undergraduate sample, Zagon and Jackson (1994) reported that males scored significantly higher than females on both the SRP total score and its factors. They also found that the two SRP-II factors correlated significantly but moderately ($r = .37$), roughly paralleling findings for the two factors of the PCL-R (Hare, 1991a).

Williams, Nathanson, and Paulhus (2002) conducted an oblique factor analysis of an abbreviated 31-item version of the SRP-II in an undergraduate sample. They found both two- (Cold Affect and Antisocial Behavior) and three-factor (Deficient Affect, Interpersonal Callousness, and Antisocial Behavior) solutions to be interpretable, although they did not test the model fit of these two solutions against each other. Because they used an abbreviated version of the SRP-II, the extent to which their findings apply to the full SRP-II is unclear.

Several researchers have examined the convergent validity of the SRP and its revised versions. Hare et al. found that the SRP correlated $r = .26$ with the MMPI Pd scale, $r = -.53$ with the CPI So scale, $r = .35$ with clinician ratings of the *DSM–III* criteria for ASPD, and $r = .38$ with the PCL. In the *DSM–IV* field trials, Widiger et al. (1996) administered the SRP-II, along with various measures of psychopathy and ASPD, to more than 400 men recruited from various clinical (prison, psychiatric) sites. The SRP-II correlated at an average of $r = .35$ with *DSM–III–R* diagnoses of ASPD and at an average of $r = .38$ with a 10-item abbreviated version of the PCL-R (Zagon & Jackson, 1994).

Zagon and Jackson (1994) reported that the SRP-II correlated significantly and moderately ($r = .62$) with a self-report measure of narcissism, the Narcissistic Personality Inventory (NPI; Raskin & Hall, 1979). This correlation is consistent with the egocentricity traditionally considered central to psychopathy (Cleckley, 1941/1988; Hare, 1991b). In addition, the SRP-II correlated significantly and negatively ($r = -.30$ in both cases) with a self-report measure of trait anxiety, the trait form of the State-Trait Anxiety Inventory (Spielberger, Gorsuch, Lushene, Vagg, & Jacobs, 1983) and a self-report measure of empathy, the Interpersonal Reactivity Index (Davis, 1983). Trait anxiety was selectively related to SRP-II Factor 1, whereas narcissism and empathy were related to both SRP-II factors.

Williams et al. (2002) found that the SRP-II total score correlated significantly ($r = .38$) with self-reported delinquency (including violent crime, cheating, and bullying), even after three SRP-II items tapping delinquent acts were removed from computation of the total score.

Paulhus and Williams (2002) found that the SRP-III correlated significantly with the NPI ($r = .50$) and with a questionnaire measure of Machiavellianism, the MACH-IV ($r = .31$) (Christie & Geis, 1970). The former finding replicates that of Zagon and Jackson (1994) regarding the SRP-II. Paulhus and Williams also reported significant correlations between the SRP-III and all five dimensions of the five-factor model as measured by the Big Five Inventory (John & Srivastava, 1999). Specifically, they found that the SRP-III correlated positively with Extraversion and Openness to experience ($r = .34$ and $.24$, respectively) and negatively with Agreeableness, Conscientiousness, and Neuroticism ($r = -.25$, $-.24$, and $-.34$, respectively). These findings are broadly consistent with those reported for the screening version of the PCL-R (Hart & Hare, 1994). They also found a small but significant correlation ($r = .14$) between SRP-III scores and participants' propensity to overestimate their own intelligence.

Finally, in a large sample of undergraduates, Derefinko and Lynam (2006) examined the association between the SRP and FFM scores. Broadly corroborating the findings of Paulhus and Williams (2002), SRP total scores were nonsignificantly related to FFM Extraversion and Neuroticism, weakly but significantly associated with FFM Openness to Experience ($r = -11$), and markedly negatively associated with FFM Agreeableness ($r = -.62$) and Conscientiousness ($r = -.34$).

Summary of Advantages and Disadvantages

The SRP and its revised editions have shown promising internal consistency and construct validity in various samples, although questions remain regarding the internal consistency of Factor 1. The SRP total score correlates highly with other self-report measures relevant to psychopathy, and exhibits meaningful convergent relations with self-report measures of traits ostensibly related to psychopathy, including narcissism, empathy, (reversed) agreeableness, and (reversed) conscientiousness.

There is little published research on the differential correlates of the two SRP factors or the relation between the SRP and laboratory measures that ostensibly tap the core deficits of psychopathy.

The Psychopathic Personality Inventory and Psychopathic Personality Inventory-Revised

Construction and Format

The PPI was developed by Lilienfeld (1990) to detect psychopathic traits in community samples. It consists of 187 items in a 4-point Likert-type format and comprises 8 subscales that assess lower-order facets of psychopathy and that yield a total score ostensibly representing global psychopathy. The PPI also has psychopathy-relevant validity scales assessing positive impression management (now called the Virtuous Responding scale), malingering (Deviant Responding scale), and careless or random responding (Variable Response Inconsistency scale).

Using an exploratory approach to test construction (Loevinger, 1957; Tellegen & Waller, 2008), Lilienfeld (1990) wrote a large number of items based on a broad array of characteristics deemed potentially relevant to psychopathy by diverse authors (e.g., Albert, Brigante, & Chase, 1959; Gray & Hutchinson, 1964; Hare, 1991a). He submitted responses to these items to successive factor analyses in undergraduate samples. He revised scales and items based on the results of each factor analysis, across three rounds of factor analysis involving 1,156 participants.

Factor analyses of the PPI item pool yielded eight replicable factors: Machiavellian Egocentricity (a ruthless willingness to manipulate and take advantage of others, e.g., "I sometimes try to get others to 'bend the rules' for me if I can't change them any other way"), Social Potency (interpersonal impact and skill at influencing others, e.g., "Even when others are upset with me, I can usually win them over with my charm"), Fearlessness (a willingness to take physical risks and an absence of anticipatory anxiety, e.g., "Making a parachute jump would really frighten me," keyed in the false direction), Coldheartedness (callousness, guiltlessness, and absence of empathy, e.g., "I have had 'crushes' on people that were so intense that they were painful," keyed in the false direction), Impulsive Nonconformity (a flagrant disregard for tradition, e.g., "I sometimes question authority figures 'just for the hell of it'"), Blame Externalization (attribution of responsibility for one's mistakes to others, e.g., "When I'm in a group of people who do something wrong, somehow it seems like I'm usually the one who ends up getting blamed"), Carefree Nonplanfulness (an insouciant attitude toward the future, e.g., "I weigh the pros and cons of major decisions carefully before making them," keyed in the false direction), and Stress Immunity (absence of tension in anxiety-provoking situations, e.g., "I can remain calm in situations that would make many other people panic").

Factor analyses by Benning, Patrick, Hicks, Blonigen, and Krueger (2003) demonstrated that the PPI also conforms roughly to a two-factor structure conceptually similar to that of PCL-R (but see Neumann, Malterer, & Newman, 2008, for analyses calling this factor solution into question). In these factor analyses, Social Potency, Fearlessness, and Stress Immunity loaded on Factor I, whereas Machiavellian Egocentricity, Impulsive Nonconformity, Blame Externalization, and Carefree Nonplanfulness loaded on Factor 2. Coldheartedness, which assesses many of the deficits traditionally regarded as central to psychopathy, particularly guiltlessness and lovelessness (McCord & McCord, 1964), did not load substantially on either factor. In sharp contrast to the two PCL-R factors, which are moderately correlated, the two PPI factors are essentially orthogonal. Therefore, PPI-assessed psychopathy seems to mark a condition marked by a configuration of largely unrelated attributes rather than a classic psychopathological syndrome, which is traditionally marked by covarying signs and symptoms (see Kazdin, 1983; Lilienfeld, Waldman, & Israel, 1994).

The PPI was recently revised to the PPI-R (Lilienfeld & Widows, 2005) to lower its reading level, eliminate psychometrically problematic or culturally specific items, and develop adequate norms for college/community and prison samples. Two PPI content subscales were renamed: "Impulsive Nonconformity" was renamed "Rebellious Nonconformity," and "Social Potency" was renamed "Social Influence." Because the factor structure and external correlates of the PPI and PPI-R appear to be extremely similar (Lilienfeld & Widows, 2005), we will in most cases review both measures interchangeably.

Psychometric Properties

Across four undergraduate samples, Lilienfeld and Andrews (1996) found good internal consistency for PPI total (Cronbach's alpha = .90–.93), and subscale scores (Cronbach's alpha = .70–.90). In addition, they reported PPI total score test–retest reliability of $r = .95$ over a mean 26-day interval. Test–retest reliabilities of the PPI subscales ranged from $r = .82$ to .94 (see Chapman, Gremore, & Farmer, 2003, for comparable data). PPI-R total score internal consistencies (alphas) in college/community ($N = 985$) and offender ($N = 154$) samples were .92 and .84, respectively; PPI-R subscales in these two samples ranged from .78 to .92 (college/community sample) and .71 to .84 (offender sample; Lilienfeld & Widows, 2005).

Although most of the PPI subscales correlate positively, some (e.g., Blame Externalization and Stress Immunity) correlate slightly to moderately negatively (Lilienfeld & Andrews, 1996; see also Chapman et al., 2003). The same basic pattern holds for the subscales of the PPI-R (Lilienfeld & Widows, 2005).

Lilienfeld and Andrews (1996) found that PPI total scores were significantly higher in men than in women, as were scores on six PPI subscales (Machiavellian Egocentricity, Fearlessness, Coldheartedness, Impulsive Nonconformity, Blame Externalization, and Stress Immunity). They further reported that the PPI displayed adequate convergent validity with self-report measures of psychopathy and antisocial behavior, including the CPI So scale ($r = -.59$); MMPI-2 ASP content scale ($r = .56$ and .58 in two samples); the Personality Diagnostic Questionnaire-Revised ASPD scale ($r = .58$ and .43 in two samples); and several Multidimensional Personality Questionnaire (MPQ; Tellegen, in press) scales, including Social Potency ($r = .39$), Aggression ($r = .38$), Harm Avoidance ($r = -.55$), Control vs. Impulsiveness ($r = -.27$), and Traditionalism ($r = -.20$). In addition, the PPI displayed convergent validity with the Minnesota Temperament Inventory (MTI), a measure of peer-rated Cleckley psychopathy to be discussed later ($r = .45$), interview-rated Cleckley psychopathy ($r = .60$), and with ASPD ($r = .59$) and Narcissistic Personality Disorder ($r = .35$) as measured by the Structured Clinical Interview for *DSM–III–R* (Spitzer, Williams, & Gibbon, 1987).

With respect to PPI Factor 1 and Factor 2, Lilienfeld, Skeem, and Poythress (2004) found that, in contrast to the LPSP primary and secondary psychopathy scales, the two PPI factors displayed a relatively clear convergent/discriminant pattern of relations with other measures. PPI Factor 1 correlated more highly with PCL-R Factor 1 than with PCL-R Factor 2 (although this difference did not reach significance) and vice versa for PPI Factor 2. In addition, PPI Factor 1 was virtually unrelated to self-reported features of *DSM–IV* ASPD ($r = .06$) whereas PPI Factor 2 was highly related to these features ($r = .61$).

In the same undergraduate sample described earlier, Derefinko and Lynam (2006) reported a similar pattern of correlates for PPI total scores as they had found for SRP total scores. Specifically, PPI total scores were nonsignificantly associated with FFM Extraversion and Neuroticism, but positively and significantly associated with FFM Openness to Experience ($r = .11$), Agreeableness ($r = -.52$), and Conscientiousness ($r = -.43$).

Although the PPI was designed for noncriminal samples, several investigators have examined its correlates in prisoners. In a sample of 100 male inmates, Sandoval, Hancock, Poythress, Edens, and Lilienfeld (2000) found that the PPI correlated $r = -.45$ with the Questionnaire Measure of Emotional Empathy (Mehrabian & Epstein, 1972) and $r = .60$ with the Buss and Perry (1992) Aggression Questionnaire. Both correlations are consistent with the clinical portrait of psychopaths as callous and as having low frustration tolerance (Hare, 1991a).

In a sample of 60 male inmates, Edens, Poythress, and Watkins (2001) found that the PPI exhibited a theoretically meaningful pattern of relations with the scales of the Personality Assessment Inventory (PAI; Morey, 1991). For example, the PPI correlated significantly with the PAI Antisocial ($r = .68$), Aggression ($r = .57$), and Dominance ($r = .38$) scales. Edens et al. also found that the PPI correlated significantly with physical ($r = .26$) and nonaggressive ($r = .37$) disciplinary infractions (see Edens, Poythress, & Lilienfeld, 1999, for additional data on the PPI and disciplinary infractions).

In a sample of 50 inmates, Poythress, Edens, and Lilienfeld (1998) found that the PPI and PCL-R correlated at $r = .54$. Moreover, the PPI correlated $r = .54$ with PCL-R Factor 1 and $r = .40$ with PCL-R Factor 2, indicating that the PPI may be the first self-report measure of psychopathy to be associated substantially with the core interpersonal and affective features of psychopathy. Further, partial correlation analyses between the PPI and each PCL-R factor controlling for the variance shared with the other PCL-R factor revealed that the PPI was selectively associated with PCL-R Factor I (partial $r = .40$, $p < .01$) rather than with PCL-R Factor 2 (partial $r = .14$, ns). Chapman et al. (2003) examined the psychometric properties of the PPI in 153 female inmates. They found that the PPI total score correlated $r = -.60$ with CPI So and $r = .81$ with the PAI ANT scale. Interestingly, Chapman et al. also reported that PPI total scores in their sample did not differ significantly from those obtained from female undergraduates in a previous study (Hamburger, Lilienfeld, & Hogben, 1996). This finding raises concerns regarding the PPI's criterion-related validity, although differences in age, education, social class, and other potential covariates between the two samples render this finding difficult to interpret. Moreover, Chapman et al. did not separate the two PPI factors in their sample; in subsequent work using the PPI-R, prisoners scored higher than undergraduates on PPI Factor 2, but lower on PPI Factor 1, probably reflecting the latter dimension's emphasis on adaptive functioning (Lilienfeld & Widows, 2005).

Indeed, the two PPI factors exhibit strikingly divergent, in some cases even opposing, correlates. In a sample of 353 adult community males, Benning et al. (2003) found that PPI Factor 1 was positively correlated with educational level, high school class rank, and adult antisocial behavior. In contrast, PPI Factor 2 correlated negatively with educational achievement, income, verbal intelligence, and age at first substance use and correlated positively with both child and adult antisocial behaviors. Benning et al. suggested that PPI Factor 1 may reflect emotional resilience, whereas PPI Factor 2 may reflect a broad predisposition toward externalizing behavior. This possibility is consistent with preliminary analyses from prison and substance abuse samples indicating that PPI Factor 2 is positively associated with suicide ideation/attempts, whereas PPI Factor 1 is negatively associated with these variables (Douglas et al., 2008). One interpretation of this divergence is that features of psychopathy linked to high interpersonal influence and low anticipatory anxiety exert a protective influence against suicidal thinking and behavior.

Summary of Advantages and Disadvantages

The PPI and its successor, the PPI-R, hold considerable potential as self-report measures of psychopathy. Both its total score and subscales are internally consistent and temporally stable. The PPI total score displays good convergent and discriminant validity with self-report, interview, and observer-rated measures of psychopathy, antisocial behavior, DSM personality disorders, and normal-range personality traits. In addition, the PPI correlates moderately to highly with PCL-R Factor 1, suggesting that it assesses at least some of the core affective and interpersonal deficits of psychopathy. Nevertheless, the PPI also correlates moderately to highly with PCL-R Factor 2, indicating that it is not selective to these deficits. Finally, the PPI's two factors appear to exhibit promising convergent and discriminant validity with measures of psychopathy and antisocial behavior, as well as strikingly different demographic and personality correlates.

Still, several questions regarding the PPI's validity remain. Relatively little published research has examined the PPI's relation to laboratory indices relevant to psychopathy, such as go/no-go tasks and neuropsychological measures of prefrontal lobe dysfunction (but see Sellbom & Verona, 2007, for data on the PPI and prefrontal tasks). Nevertheless, neuroimaging findings lend at least some support to the construct validity of the PPI. Using functional magnetic resonance imaging (fMRI), Gordon, Baird, and End (2004) found that high scorers on PPI Factor 1 showed heightened dorsolateral prefrontal activity in response to emotional faces compared with low scorers, suggesting a more cognitive and perhaps "cool" approach to this affectively laden task. In addition, also using fMRI, Harenski, Kim, and

Hamann (2009) reported that total PPI scores were negatively associated with medial prefrontal activation in response to photographs depicting moral violations (e.g., a person abusing another person), suggesting less emotional processing of ethical stimuli in psychopathic participants. More evidence along these lines should help fill out the nomological network linking the PPI to measures of psychobiological deficits ostensibly underpinning psychopathy. Finally, the virtual orthogonality of the two PPI factors suggests that the PPI does not assess a unitary construct. It is unclear whether this finding calls into question the PPI's structural validity or whether it points toward the need to reconceptualize the construct of psychopathy as a constellation of two or more largely unrelated attributes rather than a classical syndrome (Fowles & Dindo, 2009; Lilienfeld & Fowler, 2006).

OBSERVER MEASURES

Psychopaths' noted lack of insight into their disorder has sparked interest in the possibility that observers' reports offer a valuable contribution to the assessment of psychopathy. Observers may possess access to important information regarding psychopathic traits that is missed by self-report measures or interviews. In this way, they may be able to fill in the "blind spots" ostensibly generated by psychopathic individuals' lack of awareness of their own shortcomings (Grove & Tellegen, 1991).

The Interpersonal Measure of Psychopathy

The Interpersonal Measure of Psychopathy (IM-P; Kosson, Steuerwald, Forth, & Kirkhart, 1997) consists of 21 items that observers use to rate a variety of interpersonal interactions and nonverbal behaviors relevant to psychopathy, typically based on the PCL-R or other interviews. Kosson et al. based the IM-P on the notion that assessing overt behaviors in the context of a specific situation (in this case, in the course of an interview) could reduce subjectivity in judging interpersonal aspects of psychopathy. They used three major sources to generate items: the relevant theoretical and empirical literature, clinical impressions formed during inmate interviews, and a survey asking psychopathy researchers to characterize their interpersonal interactions with psychopaths.

The 21 IM-P items are scored on a 1- to 4-point scale (with 1 describing the individual *not at all* and 4 describing the individual *perfectly*). Items such as "Interrupts," "Makes personal comments," and "Expressed narcissism" are scored based on verbal and nonverbal interviewee behaviors. Kosson et al. and Zolondek, Lilienfeld, Patrick, and Fowler (2006) reported good internal consistency for the IM-P in a sample of (Cronbach's alpha = .89, .91), and good-to-moderate interrater reliability in prison (ICCs = .77 and .83) and undergraduate (ICC = .60) samples.

Kosson et al. (1997) found that IM-P scores correlated preferentially with Factor 1 of the PCL-R (r = .33–.62 with Factor 1, vs. r = 15–.31 with Factor 2). The results of regression analyses indicated that above and beyond the variance predicted by PCL-R factor scores, IM-P scores uniquely predicted interviewer-reported feelings of trepidation, as well as a desire to avoid conflict, observer ratings of interpersonal dominance, and a history of adult fights (Zolondek et al., however, did not find a significant association with adult fights). After controlling for variance explained by PCL-R factor scores, IM-P scores were negatively associated with nonviolent disciplinary infractions, criminal and drug use versatility, and adult symptoms of ASPD (but see Zolondek et al. for different findings). Taken together, these data suggest that IM-P scores are less of a measure of behavioral deviance than are the PCL-R factors, suggesting that they may be helpful in detecting "successful," subclinical, or noncriminal individuals with psychopathic features (Hall & Benning, 2006).

Although Kosson et al. suggested that the IM-P be used as an adjunct to a PCL-R interview, IM-P ratings could be derived from other interview formats as well, as the ratings are not dependent on the PCL-R per se. Zolondek et al. suggested that a greater range of behaviors could be detected if the IM-P were administered while observing inmates engaged in everyday social interactions. Further research should examine the potential incremental validity of the IM-P across different contexts.

The Psychopathy Q-Sort

Reise and Oliver (1994) developed a Q-sort measure (the Psychopathy Q-Sort[PQS]) to permit the assessment of psychopathy by observers. They asked seven judges with psychopathy expertise to sort the 100 items of the California Q-set (CAQ; Block, 1961) according to their conceptualization of a prototypical psychopath. The CAQ consists of 100 descriptive statements about personality sorted into a forced quasi-normal distribution of nine categories by either an observer or self, from *most uncharacteristic* to *most characteristic*. This distribution eliminates certain response biases, such as extreme item ratings. Sample CAQ statements include "Is critical, skeptical, not easily impressed," "Is a talkative individual," and "Initiates humor." To obtain PQS scores, the rater's sorted responses are correlated with the expert-generated psychopathy prototype, and reflect the degree of match ($r = .00–1.0$).

PQS prototype reliability aggregated over seven judges was .90. It correlated $r = .51$ ($p < .01$) with a CAQ narcissism prototype vs. $r = .16$ (*ns*) with a CAQ hysteria prototype (Reise & Oliver, 1994), demonstrating discriminant validity. In a community sample ($N = 350$), Reise and Wink (1995) found that the PQS was positively associated with features of Cluster B personality disorders (e.g., antisocial, borderline), negligibly or negatively associated with features of other personality disorders, and negatively associated with CPI (So) Scale scores, further demonstrating convergent and discriminant validity.

Fowler and Lilienfeld (2007) reported significant correlations between self-reported PQS scores and several self-report measures of psychopathy, including the LPSP total ($r = .57$), Primary ($r = .55$), and Secondary ($r = .50$) scales; PPI total ($r = .67$) and subscale scores (Machiavellian Egocentricity, Social Potency, Coldheartedness, Fearlessness, Impulsive Nonconformity, Stress Immunity; $r = .31–.48$); and a measure of ASPD ($r = .50$). Peer-reported PQS scores displayed significant, albeit relatively modest, correlations with PPI total ($r = .38$) and Fearlessness ($r = .38$) with ASPD scores ($r = .36$).

Other Observer Measures of Psychopathy

The Psychopathy-Scan

The Psychopathy-Scan (P-Scan; Hare & Herve, 1990) is a psychopathy scale designed for use by nonclinical raters. It consists of three scales (interpersonal, affective, behavioral) comprising 30 items each. The three-scale structure parallels the three-factor model of psychopathy proposed by Cooke and Michie (2001), which largely divides Factor 1 into separable interpersonal and affective facets. Items are rated from 0 (*does not apply*) to 2 (*definitely applies*).

There is little extant research on the validity of the P-Scan. Elwood, Poythress, and Douglas (2004) reported on the psychometric properties of the P-Scan in a large sample of undergraduates. They reported excellent internal consistency for the three scales (Cronbach's alpha = .90–.92; mean interitem correlations $r = .22–.24$). They found modest evidence for convergent validity with LPSP scores ($r = .21–.33$), and external validity ($r = .22–.24$ with participants' scores on a self-report measure of antisocial activity). Elwood et al. concluded that despite the high internal consistency of the P-Scan, the modest concurrent and predictive validity indicated by their findings underscores the need to investigate the appropriateness of using the P-Scan in applied settings.

The B-Scan

The B-Scan (Babiak & Hare, in press) is designed to permit supervisors and co-workers to assess psychopathic features in individuals in work settings. It consists of four scales (personal style, emotional style, organizational maturity, and antisocial tendencies), intended to parallel Hare's (2003) four-factor model of psychopathy (Mathieu, Hare, Jones, Babiak, & Neumann, in press). This measure is currently undergoing validation research, but has not yet been examined in peer-reviewed studies.

Minnesota Temperament Inventory

The Minnesota Temperament Inventory (MTI; Taylor, Loney, Bobadilla, Iacono, & McGue, 2003) is a 19-item self-report questionnaire created explicitly to capture Cleckley's (1941; 1988) classic conceptualization of psychopathy (Harkness, 1992; Lilienfeld & Andrews, 1996). The MTI was initially designed for observer report, although it can be adapted for self-report (see data on the construct validity of the MTI used as a self-report measure). Item scores range from 1 ("not at all true of me") to 4 ("very true of me").

As noted earlier, Lilienfeld and Andrews (1996) reported that in college students, MTI scores (as completed by peers nominated by students) correlated $r = .45$ with PPI total scores; Cale and Lilienfeld (2002) found a similar, although somewhat lower correlation ($r = .33$) in a sample of 75 theatre actors. The validity of the MTI as an observer measure of psychopathy requires further investigation.

IMPLICIT MEASURES

Psychopaths' lack of insight into their disorder (Cleckley, 1941/1988) has led some researchers to develop implicit measures to detect the core traits of the condition. Such measures, which fall into the "unknown," quadrant of the Johari window (Luft, 1969), do not depend on conscious knowledge of one's personality attributes. In this respect, they may circumvent some of the shortcomings of self-report and interview-based measures of psychopathy, many of which assume at least a modicum of self-understanding.

Projective Devices

The best known implicit measures of personality and psychopathology are projective devices, most of which present respondents with ambiguous stimuli and ask them to interpret or "make sense" of them (Lilienfeld, Wood, & Garb, 2000). In turn, the best known projective device is the Rorschach Inkblot Test, which consists of 10 symmetrical inkblots, five in black-and-white and five containing color. A number of authors, most recently Gacono and Meloy (1994), have argued that several Rorschach indicators, such as low numbers of texture responses (ostensibly reflecting a low need for intimacy) and shading responses (ostensibly reflecting minimal anxiety), and high numbers of reflection responses (ostensibly reflecting narcissism), are highly diagnostic of psychopathy. Indeed, Gacono and Meloy (2004) deemed the Rorschach "ideally suited" for detecting psychopathy, maintaining that "we have validated the use of the Rorschach as a sensitive instrument to discriminate between psychopathic and nonpsychopathic subjects" (pp. 236–237).

A recent meta-analysis (Wood et al., 2010) examined these claims. The authors pooled data across 22 studies ($N = 780$ inmates) that examined 37 Rorschach variables purportedly linked to psychopathy. The mean correlation (r) between these variables and scores on various versions of the PCL-R was only .06 (median = .07), with a range of $-.11$ to .24. Thirty-two of the 37 mean r values were nonsignificant. Interestingly, the few significant findings, all which were weak in magnitude, stemmed largely from highly face valid Rorschach "content indicators" (e.g., number of aggressive responses, number of cooperative responses), which are markedly deemphasized in recent scoring systems for the Rorschach (Wood, Nezworski, Lilienfeld, & Garb, 2003). These results provide feeble support for the use of the Rorschach to detect psychopathy in forensic settings.

Conditional Reasoning Tests for Aggression, and berrant Self-Promotion

James (1998) and colleagues (e.g., Bing et al., 2007) developed a series of inductive reasoning tasks, which they termed "conditional reasoning tests" (CRTs), designed to circumvent self-enhancement biases that may diminish the accuracy of self-report measures of personality. Two CRTs are particularly relevant to psychopathy: one pertaining to antisocial/aggressive behavior and one pertaining to a narcissistic personality configuration combined with antisocial behavior, which they call "Aberrant self-promotion."

The Aggression CRT (CRT-A) was designed to detect "justification mechanisms" that excuse antisocial behaviors. In developing the CRT for aggression, James proposed six primary implicit biases that serve as justification mechanisms for aggressive acts: hostile attribution bias (a tendency to perceive hostility and danger in the behavior of others), potency bias (a focus on dominance and submission in social interactions), retribution bias (a tendency to view retaliation as more logical than reconciliation), victimization by powerful others (a tendency to view powerful others as exploitative), derogation of target bias (an unconscious tendency to view targets of aggression as evil and immoral), and social discounting bias (a tendency to frame to social norms as repressive and unduly restrictive of one's choices). Individuals are presented with a series of 22 problems, with four possible solutions to choose from. Each set of solutions contains one possible prosocial alternative, one answer representative of a justification mechanism, and two illogical choices.

James and LeBreton (2010) summarized reliability and validity findings for the CRT-A across a number of large community samples ($N = 60–225$). They found adequate test-retest reliability ($r = .82$) and internal consistency (Cronbach's alpha = .76) in large samples of students and students/employees. They further reported the results of a large meta-analysis that showed moderate-to-high correlations ($r = .22–.64$) between CRT-A scores and a number of pragmatically important criterion measures, including theft, absenteeism from work, attrition, and sports fouls. Clearly, aggression and justification for aggression are relevant to psychopathy, as are several of the behavioral criterion measures used in the validation of the CRT-A. Further, implicit biases (e.g., regarding hostility and dominance) that justify aggression are relevant to psychopathic individuals' tendency to externalize blame for antisocial actions. In addition, as LeBreton, Binning, and Adorno (2006) observed, the CRT-A might be useful for detecting subclinical psychopathy, perhaps especially in vocational settings. Nevertheless, as James and colleagues acknowledged, there are no data at present on the reliability or validity of the CRT-A with psychopaths.

Similarly, Gustafson and colleagues (1999, 2000a, b; Gustafson & Ritzer, 1995) have written about a leadership style termed "aberrant self-promotion." According to Gustafson, aberrant self-promoters (ASPs) differ from psychopaths in the degree of severity of antisocial behavior in that they may engage in behaviors that violate social norms. Like psychopaths, ASPs are posited to exhibit such personality features as superficial charm, grandiosity, manipulativeness, and lack of guilt and empathy. O'Shea, Gustafson, Hense, Hawes, and Lowe (2004) developed a CRT for Aberrant Self Promotion based on the hypothesis that ASPs' antisocial behavior arises from a justification mechanism of self-perceived superiority over others, including a belief that they have a "special destiny" (Gustafson, 1999, 2000a, b). Although this measure could prove to be a useful tool for detecting subclinical psychopathy in organizational settings, there are scant validation data on this instrument to date.

DISCUSSION AND FUTURE DIRECTIONS

The past 15 years have witnessed significant advances in the development of alternatives to the PCL-R. Revisiting the Johari window demonstrates the unique potential contributions of each of the classes of measures reviewed here (self, observer, and implicit reports). Each type can be thought of as covering a different quadrant of the window: Self-report measures capture the region of personality that is known to self but not others; observer measures capture the region known to others but not to the self ("blind spots"); and implicit measures capture the region that is unknown to the self and to others. Used in conjunction with one another, therefore, self, observer, and implicit measures could capture different aspects of psychopathy and generate particularly powerful assessments. Nevertheless, further research is needed to ascertain the extent to which this approach yields incremental validity above and beyond using one information source in isolation.

Psychometric advances over the past 15 years have yielded viable, well-validated alternatives to PCL-R assessment of psychopathy, especially in the domain of self-report. Previous pessimistic conclusions regarding the low correlations among self-report psychopathy measures (e.g., Hundleby & Ross, 1977) must now be revised in light of evidence supporting newer measures. The convergent validities among the LPSP, SRP, and PPI and their convergence with measures of normal-range personality traits and *DSM–IV* personality disorder features support the use of self-report measures of psychopathy. In addition, the PPI correlates moderately to highly with PCL-R Factor 1 (Poythress et al., 1998), suggesting that at least some self-report psychopathy measures can adequately assess the core interpersonal and affective features of psychopathy.

Observer and implicit measures of psychopathy have the potential to circumvent the notable lack of insight observed in psychopaths, but few have been developed and well validated. Their contribution to the assessment of psychopathy should not be overlooked, and we hope that the measures showing initial promise will be investigated further in coming years.

In conclusion, there are compelling indications that several current alternative measures to the PCL-R are viable. The pragmatic needs of workers in institutional, community, or clinical settings may point to specific advantages of using one or more alternative measures of psychopathy, and we hope that this chapter provides guidance when evaluating the options.

REFERENCES

Albert, R. S., Brigante, T. R., & Chase, M. (1959). The psychopathic personality: A content analysis of the concept. *Journal of General Psychology, 60,* 17–28.

Babiak, P., & Hare, R. D. (in press). *The B-Scan 360 manual.* Toronto, Ontario, Canada: Multi-Health Systems.

Benning S. D., Patrick, C. J., Hicks, B. M., Blonigen, D. M., & Krueger, R. F. (2003). Factor structure of the Psychopathic Personality Inventory: Validity and implications for clinical assessment. *Psychological Assessment, 15,* 340–350.

Bing, M. N., Stewart, S. M., Davison, H. K., Green, P. D., McIntyre, M. D., & James, L. R. (2007). An integrative typology of personality assessment for aggression: Implications for predicting counterproductive workplace behavior. *Journal of Applied Psychology, 92,* 722–744.

Blackburn, R., & Fawcett, D. (1999). The antisocial personality questionnaire: An inventory for assessing personality deviation in offender populations. *European Journal of Psychological Assessment, 15,* 14–24.

Block, J. (1961). *The Q-Sort method in personality assessment and psychiatric research.* Springfield, IL: Charles C Thomas.

Brinkley, C. A., Schmitt, W. A., Smith, S. S., & Newman, J. P. (2001). Construct validation of a self-report psychopathy scale: Does Levenson's self-report psychopathy scale measure the same constructs as Hare's psychopathy checklist-revised? *Personality and Individual Differences, 31,* 1021–1038.

Buss, A. H., & Perry, M. (1992). The aggression questionnaire. *Journal of Personality and Social Psychology, 63,* 452–459.

Cale, E. M., & Lilienfeld, S. O. (2002). Sex differences in psychopathy and antisocial personality disorder: A review and integration. *Clinical Psychology Review, 22*(8), 1179–1207.

Chan, D. (2009). So why ask me? Are self-report data really that bad? In: C. E. Lance & R. J. Vandenberg (Eds.), *Statistical and methodological myths and urban legends: Doctrine, verity, and fable in the organizational and social sciences* (pp. 309–335). New York: Routledge.

Chapman, A. L., Gremore, T. M., & Farmer, R. F. (2003). Psychometric analysis of the Psychopathic Personality Inventory (PPI) with female inmates. *Journal of Personality Assessment, 80,* 164–172.

Christie, R., & Geis, F. L. (1970). *Studies in Machiavellianism.* London: Academic Press.

Clark, L. A. (1993). *Schedule for Nonadaptive and Adaptive Personality (SNAP).* Minneapolis: University of Minnesota Press.

Cleckley, H. (1941/1988). *The mask of sanity.* St. Louis, MO: Mosby.

Cooke, D. J., & Michie, C. (2001). Refining the construct of psychopathy: Towards a hierarchical model. *Psychological Assessment, 13,* 171–188.

Costa P. T., Jr., & McCrae, R. R. (1990). Personality disorders and the five-factor model of personality. *Journal of Personality Disorders, 4,* 362–371.

Costa P. T., Jr., & McCrae, R. R. (1992). *Revised NEO Personality Inventory (NEO PIR) Professional Manual.* Odessa, FL: PAR.

Cronbach, L. J., & Meehl, P. E. (1955). Construct validity in psychological tests. *Psychological Bulletin, 52,* 281–302.

Davis, M. H. (1983). Measuring individual differences in empathy: Evidence for a multidimensional approach. *Journal of Personality and Social Psychology, 44,* 113–126.

Derefinko, K. J., & Lynam, D. R. (2006). Convergence and divergence among self-report psychopathy measures: A personality-based approach. *Journal of Personality Disorders, 20,* 261–280.

Douglas, K. S., Lilienfeld, S. O., Skeem, J. L., Poythress, N. G., Edens, J. F., & Patrick, C. J. (2008). Relation of antisocial and psychopathic traits to suicide-related behavior among offenders. *Law and Human Behavior, 32*(6), 511–525.

Edelmann, R. J., & Vivian, S. E. (1988). Further analysis of the social psychopathy scale. *Personality and Individual Differences, 9,* 581–587.

Edens, J. F. (2004). Effect of response distortion on the assessment of divergent facets of psychopathy. *Assessment, 11,* 109–112.

Edens, J. F., Buffington, J. K., & Tomicic, T. L. (2000). An investigation of the relationship between psychopathic traits and malingering on the Psychopathic Personality Inventory. *Assessment, 7,* 281–296.

Edens, J. F., Hart, S. D., Johnson, D. W., Johnson, J. K., & Olver, M. E. (2000). Use of the Personality Assessment Inventory to assess psychopathy in offender populations. *Psychological Assessment, 12,* 132–139.

Edens, J. F., Poythress, N. G., & Lilienfeld, S. O. (1999). Identifying inmates at risk for disciplinary infractions: A comparison of two measures of psychopathy. *Behavioral Sciences and the Law, 17,* 435–443.

Edens, J. F., Poythress, N. G., & Watkins, M. M. (2001). Further validation of the psychopathic personality inventory among offenders: Personality and behavioral correlates. *Journal of Personality Disorders, 15,* 403–415.

Elwood, C. E., Poythress, N. G., & Douglas, K. S. (2004). Evaluation of the Hare P-SCAN in a non-clinical population. *Personality and Individual Differences, 36,* 833–843.

Eysenck, S. B. G., & Eysenck, H. J. (1975). *Manual of the Eysenck Personality Questionnaire.* London: University of London Press.

Falkenbach, D., Poythress, N., Falki, M., & Manchak, S. (2007). Reliability and validity of two self-report measures of psychopathy. *Assessment, 14,* 341–350.

Faraone, S. V., & Tsuang, M. T. (1994). Measuring diagnostic accuracy in the absence of a "gold standard." *American Journal of Psychiatry, 151,* 650–657.

Finney, J. C. (1985). Anxiety: Its measurement by objective personality tests and self-report. In: A.H. Tuma & J.D. Maser (Eds.), *Anxiety and the anxiety disorders* (pp. 645–673). Hillsdale, NJ: Erlbaum.

Forth, A. E., Kosson, D. S., & Hare, R. D. (2003). *The Hare PCL:YV.* Toronto, Ontario, Canada: Multi-Health Systems.

Fowler, K. A., & Lilienfeld, S. O. (2007). The psychopathy Q-Sort: Construct validity evidence in a nonclinical sample. *Assessment, 14,* 75–79.

Fowles, D. C., & Dindo, L. (2009). Temperament and psychopathy: A dual-pathway model. *Current Directions in Psychological Science, 18,* 179–183.

Fulero, S. (1995). Review of the Hare Psychopathy Checklist-Revised. In: J. C. Conoley, J. C. Impara, & L. L.. Murphy (Eds.), *Twelfth mental measurements yearbook* (pp. 453–454). Lincoln, NE: Buros Institute.

Gacono, C., & Meloy, R. (1994). *The Rorschach assessment of aggressive and psychopathic personalities.* Hillsdale, NJ: Erlbaum.

Golden, R. R., & Meehl, P. E. (1979). Detection of the schizoid taxon with MMPI indicators. *Journal of Abnormal Psychology, 88,* 217–233.

Gordon, H. L., Baird, A. A., & End, A. (2004). Functional differences among those high and low on a trait measure of psychopathy. *Biological Psychiatry, 56,* 516–521.

Gough, H. G. (1960). Theory and method of socialization. *Journal of Consulting and Clinical Psychology, 24,* 23–30.

Gray, K. G., & Hutchinson, H. C. (1964). The psychopathic personality: A survey of Canadian psychiatrists' opinions. *Canadian Psychiatric Association Journal, 9,* 452–461.

Greene, R. L. (2000). *The MMPI-2: An interpretive manual* (2nd ed.). Boston: Allyn & Bacon.

Grove, W. M., & Tellegen, A. (1991). Problems in the classification of personality disorders. *Journal of Personality Disorders, 5,* 31–42.

Gustafson, S. B. (1999, April). Out of their own mouths: A conditional reasoning instrument for identifying aberrant self-promoters. In: P. Babiak (Chair), *Liars of the dark side: Can personality interfere with personality measurement?* Paper presented at the Thirteenth Annual Conference of the Society for Industrial and Organizational Psychology, Atlanta, GA.

Gustafson, S. B. (2000a, April). Out of their own mouths II: Continuing support for the validity of a conditional reasoning instrument for identifying aberrant self-promoters. In: S. Gustafson (Chair), *Personality in the shadows: A continuum of destructiveness.* Paper presented at the

Fourteenth Annual Conference of the Society for Industrial and Organizational Psychology, New Orleans, LA.

Gustafson, S. B. (2000b). Personality and organizational destructiveness: Fact, fiction, and fable. In: L. R. Bergman, R. B. Cairns, L. G. Nilsson, & L. Nystedt (Eds.), *Developmental science and the holistic approach.* (pp. 299–313). Mahwah, NJ: Erlbaum.

Gustafson, S. B., & Ritzer, D. R. (1995). The dark side of normal: A psychopathy-linked pattern called aberrant self-promotion. *European Journal of Personality, 9,* 147–83.

Haertzen, C. A., Martin, W. R., Ross, F. E., & Neidert, G. L. (1980). Psychopathic states inventory (PSI): Development of a short test for measuring psychopathic states. *International Journal of the Addictions, 15,* 137–146.

Hall, J. R., & Benning, S. D. (2006). The "successful" psychopath: Adaptive and subclinical manifestations of psychopathy in the general population. In: C. J. Patrick (Ed.), *Handbook of Psychopathy.* (pp. 459–481). New York: Guilford.

Hamburger, M. E., Lilienfeld, S. O., & Hogben, M. (1996). Psychopathy, gender, and gender roles: Implications for antisocial and histrionic personality disorders. *Journal of Personality Disorders, 10,* 41–55.

Hare, R. D. (1982). Psychopathy and the personality dimensions of psychoticism, extraversion, and neuroticism. *Personality and Individual Differences, 3,* 35–42.

Hare, R. D. (1985). A comparison of procedures for the assessment of psychopathy. *Journal of Consulting and Clinical Psychology, 53,* 7–16.

Hare, R. D. (1991a). *The Hare Psychopathy Checklist—Revised.* Toronto, Ontario, Canada: Multi-Health Systems.

Hare, R. D. (1991b). *The Self-Report Psychopathy Scale—II.* Vancouver, Canada: Unpublished test, University of British Columbia.

Hare, R. D. (1993). *Without conscience: The disturbing world of psychopaths among us.* New York: Simon & Schuster (Pocket Books).

Hare, R. D. (2003). *Manual for the Psychopathy Checklist-Revised (PCL-R).* Toronto, Ontario, Canada: Multi-Health Systems.

Hare, R. D., & Cox, D. N. (1978). Clinical and empirical conceptions of psychopathy, and the selection of subjects for research. In: R.D. Hare & D. Schalling (Eds.), *Psychopathic behaviour. Approaches to research* (pp. 1–21). Chichester, UK: Wiley.

Hare, R. D., Hemphill, J. F., & Paulhus, D. (2002). *The Self-Report Psychopathy Scale – II (SRP-II).* Manual in preparation.

Hare, R. D., & Herve, H. F. (1999). *Hare P-Scan.* Toronto, Ontario, Canada: Multi-Health Systems.

Harenski, C. L., Kim, S. H., & Hamann, S. (2009). Neuroticism and psychopathy predict brain activation during moral and nonmoral emotion regulation. *Cognitive, Affective and Behavioral Neuroscience, 9,* 1–15.

Harkness, A. R. (1992). Fundamental topics in the personality disorders: Candidate trait dimensions from lower regions of the hierarchy. *Psychological Assessment, 4,* 251–259.

Hart, S. D., Cox, D. N., & Hare, R. D. (1995). *Manual for the Psychopathy Checklist: Screening Version (PCL:SV).* Toronto, Ontario, Canada: Multi-Health Systems.

Harpur, T. J., Hare, R. D., & Hakstian, A. R. (1989). Two-factor conceptualization of psychopathy: Construct validity and assessment implications. *Psychological Assessment, 1,* 6–17.

Hart, S. D., Forth, A. E., & Hare, R. D. (1991). The MCMI-II and psychopathy. *Journal of Personality Disorders, 5,* 318–327.

Hart, S. D., & Hare, R. D. (1994). Psychopathy and the Big Five: Correlations between observers' ratings of normal and pathological personality. *Journal of Personality Disorders, 8,* 32–40.

Hart, S. D., Hare, R. D., & Harpur, T. J. (1992). The Psychopathy Checklist: Overview for researchers and clinicians. In: J. Rosen & P. McReynolds (Eds.), *Advances in psychological assessment,* Vol. 7 (pp. 103–130). New York: Plenum Press.

Hawk, S. S., & Peterson, R. A. (1973). Do MMPI psychopathic deviancy scores reflect psychopathic deviancy or just deviancy? *Journal of Personality Assessment, 38,* 362–368.

Hundleby, J. D., & Ross, B. E. (1977). A comparison of questionnaire measures of psychopathy. *Journal of Consulting and Clinical Psychology, 45,* 702–703.

James, L. R. (1998). Measurement of personality via conditional reasoning. *Organizational Research Methods, 1,* 131–163.

James, L. R., & LeBreton, J. M. (2010). Assessing aggression using conditional reasoning. *Current Directions in Psychological Science, 19,* 30–35.

James, L. R., & McIntyre, M. D. (2000). *Conditional Reasoning Test of Aggression test manual.* San Antonio, TX: Psychological Corporation.

John, O. P., & Srivastava, S. (1999). The Big Five trait taxonomy: History, measurement, and theoretical perspectives. In: L. A. Pervin & O. P. John (Eds.), *Handbook of personality: Theory and research* (2nd ed., pp. 102–138). New York: Guilford.

Karpman, B. (1948). The myth of psychopathic personality. *American Journal of Psychiatry, 103*, 523–534.

Kazdin, A. E. (1983). Psychiatric diagnosis, dimensions of dysfunction, and child behavior therapy. *Behavior Therapy, 14*, 73–99.

Kelly, G. A. (1955). *The psychology of personal constructs*, Vols. 1 and 2. New York: Norton.

Kenrick, D. T., & Funder, D. C. (1988). Profiting from controversy: Lessons from the person-situation debate. *American Psychologist, 43*, 23–34.

Kosson, D. S., Forth, A. E., Steuerwald, B. L., & Kirkhart, K. J. (1997). A new method for assessing the interpersonal behavior of psychopathic individuals: Preliminary validation studies. *Psychological Assessment, 9*, 89–101.

Kosson, D. S., Steuerwald, B. L., Newman, J. P., & Widom, C. S. (1994). The relation between socialization and antisocial behavior, substance use, and family conflict in college students. *Journal of Personality Assessment, 63*, 473–488.

Kroger, R. O., & Turnbull, W. (1975). Invalidity of validity scales: The case of the MMPI. *Journal of Consulting and Clinical Psychology, 43*, 48–55.

LeBreton, J. M., Binning, J. F., & Adorno, A. J. (2006). Subclinical psychopaths. In J. C. Thomas & D. Segal (Eds.), *Comprehensive handbook of personality and psychopathology, Vol. I, Personality and everyday functioning* (pp. 388–412). New York: John Wiley and Sons, Inc.

Levenson, M. R. (1988). Is there a risk-taking personality? Unpublished manuscript, Boston Veterans Administration Outpatient Clinic.

Levenson, M. R., Kiehl, K. A., & Fitzpatrick, C. M. (1995). Assessing psychopathic attributes in a noninstitutionalized population. *Journal of Personality and Social Psychology, 68*, 151–158.

Lilienfeld, S. O. (1990). *Development and preliminary validation of a self-report measure of psychopathic personality.* Doctoral dissertation, University of Minnesota, Minneapolis.

Lilienfeld, S. O. (1994). Conceptual problems in the assessment of psychopathy. *Clinical Psychology Review, 14*, 17–38.

Lilienfeld, S. O. (1998). Methodological advances and developments in the assessment of psychopathy. *Behaviour Research and Therapy, 36*, 99–125.

Lilienfeld, S. O., & Andrews, B. P. (1996). Development and preliminary validation of a self report measure of psychopathic personality traits in noncriminal populations. *Journal of Personality Assessment, 66*, 488–524.

Lilienfeld, S. O., & Fowler, K. A. (2006). The self-report assessment of psychopathy: Problems, pitfalls, and promises. In: C. J. Patrick (Ed.), *The handbook of psychopathy* (pp. 107–132). New York: Guilford.

Lilienfeld, S. O., & Hess, T. (2001). Psychopathic personality traits and somatization: Sex differences and the mediating role of negative emotionality. *Journal of Psychopathology and Behavioral Assessment, 23*, 11–24.

Lilienfeld, S. O., & Penna, S. (2001). Anxiety sensitivity: Relations to psychopathy, DSM-IV personality disorders, and personality traits. *Journal of Anxiety Disorders, 15*, 367–393.

Lilienfeld, S. O., Skeem, J., & Poythress, N. G. (2004, March). *Psychometric properties of self-report psychopathy measures.* In: N. Poythress (Chair), Contemporary issues in psychopathy research. Symposium at the Annual Convention of the American Psychology-Law Society, Scottsdale, AZ.

Lilienfeld, S. O., Waldman, I. D., & Israel, A. C. (1994). A critical note on the use of the term and concept of "comorbidity" in psychopathology research. *Clinical Psychology: Science and Practice, 1*, 71–83.

Lilienfeld, S. O., & Widows, M. R. (2005). *Psychological assessment inventory-revised (PPI-R).* Lutz, FL: Psychological Assessment Resources.

Lilienfeld, S. O., Wood, J. M., & Garb, H. N. (2000). The scientific status of projective techniques. *Psychological Science in the Public Interest, 1*, 27–66.

Livesley, W. J. (1990). *Dimensional assessment of personality pathology-basic questionnaire.* University of British Columbia.

Loevinger, J. (1957). Objective tests as instruments of psychological theory. *Psychological Reports, 9*, 635–694.

Lovering, A., & Douglas, K. S. (2004, March). *Comparative analysis of multiple self-report measures' association with the construct of psychopathy among criminal offenders.* Poster presented at the Annual Meeting of the American Psychology-Law Society, Scottsdale, AZ.

Luft, J. (1969). *Of human interaction.* Palo Alto, CA: National Press.

Lykken, D. T. (1995). *The antisocial personalities.* Mahwah, NJ: Erlbaum.

Lynam, D. R. & Derefinko, K. (2006). Psychopathy and personality. In C. J. Patrick (Ed.), *Handbook of psychopathy* (pp. 133–155). New York: Guilford.

Lynam, D. R., Whiteside, S., & Jones, S. (1999). Self-reported psychopathy: A validation study. *Journal of Personality Assessment, 73*, 110–132.

Marcus, D. K., John, S. L., & Edens, J. F. (2004). A taxometric analysis of psychopathic personality. *Journal of Abnormal Psychology, 113*, 626–635.

Mathieu, C., Hare, R. D., Jones, D. N., Babiak, P., & Neumann, C. S. (in press). Factor structure of the B-Scan 360: A measure of corporate psychopath. *Psychological Assessment*.

McCord, W., & McCord, J. (1964). *The psychopath: An essay on the criminal mind*. Princeton, NJ: Van Nostrand.

Meehl, P. E. (1945). The dynamics of "structured" personality tests. *Journal of Clinical Psychology, 1*, 296–303.

Meehl, P. E. (1959). Some ruminations on the validation of clinical procedures. *Canadian Journal of Psychology, 13*, 102–128.

Meehl, P. E. (1986). Diagnostic taxa as open concepts: Meta-theoretical and statistical questions about the reliability and construct validity in the grand strategy of nosological questions. In: T. Millon & G. L. Klerman (eds.), *Contemporary directions in psychopathology: Toward the DSM-IV* (pp. 215–231). New York: Guilford.

Mehrabian, A., & Epstein, N. (1972). A measure of emotional empathy. *Journal of Personality, 40*, 525–543.

Miller, J. D., Jones, S. E., & Lynam, D. R. (2011). Psychopathic traits from the perspective of self and informant reports: Is there evidence for a lack of insight? *Journal of Abnormal Psychology, 120*, 758–764.

Millon, T. (1981). *Disorders of personality, DSM-III: Axis II*. New York: Wiley.

Millon, T. (1987). *Manual for the MCMI-II*. Minneapolis, MN: National Computer Systems.

Morey, L. (1991). *Personality Assessment Inventory: Professional manual*. Tampa, FL: Psychological Assessment Resources.

Neumann, C. S., Malterer, M. B., & Newman, J. P. (2008). Factor structure of the Psychopathic Personality Inventory (PPI): Findings from a large incarcerated sample. *Psychological Assessment, 20*, 169–174.

Newman, J. P., & Kosson, D. S. (1986). Passive avoidance learning in psychopathic and nonpsychopathic offenders. *Journal of Abnormal Psychology, 95*, 257–63.

Newman, J. P., & Wallace, J. F. (1993). Diverse pathways to deficient self-regulation: Implications for disinhibitory psychopathology in children. *Clinical Psychology Review, 13*, 699–720.

Oltmanns, T. F., & Turkheimer, E. (2009). Person perception and personality pathology. *Current Directions in Psychological Science, 18*, 32–36.

O'Shea, P. G., Gustafson, S. B., Hense, R., Hawes, S. R., & Lowe, J. (2004, April). The conditional reasoning item development process: Pitfalls, successes, and lesson learned. In: S. B. Gustafson (Chair), *Making conditional reasoning tests work: Reports from the frontier*. Symposium conducted at the annual meeting of the Society for Industrial and Organizational Psychology, Chicago, IL.

Paulhus, D. L. (1991). Measurement and control of response bias. In: J. P. Robinson & P. R. Shaver (Eds.), *Measures of personality and social psychological attitudes* (pp. 17–59). San Diego: Academic Press.

Paulhus, D. L., Hemphill, J. F., & Hare, R. D. (2006). *Scoring manual for the Hare Self-Report Psychopathy Scale-III*. Toronto, Ontario, Canada: Multi-Health Systems.

Paulhus, D. L., & Williams, K. M. (2002). The dark triad of personality: Narcissism, Machiavellianism, and psychopathy. *Journal of Research in Personality, 36*, 556–563.

Poythress, N. G., Edens, J. F., & Lilienfeld, S. O. (1998). Criterion-related validity of the Psychopathic Personality Inventory in a prison sample. *Psychological Assessment, 10*, 426–430.

Poythress, N. G., Edens, J. F., & Watkins, M. M. (2002). The relationship between psychopathic personality features and malingering symptoms of major mental illness. *Law and Human Behavior, 25*, 567–582.

Pryor, L. R., Miller, J. D., & Gaughan, E. T. (2009). Testing two alternative pathological personality measures in the assessment of psychopathy: An examination of the snap and dapp-bq. *Journal of Personality Disorders, 23*, 85–100.

Raskin, R. R., & Hall, C. S. (1979). Narcissistic Personality Inventory. *Psychological Reports, 45*, 590.

Ray, J. V., Hall, J., Rivera-Hudson, N., Poythress, N. G., Lilienfeld, S. O., & Morano, M. (2012). The relation between self-reported psychopathic traits and distorted response styles: A meta-analytic review. *Personality Disorders: Theory, Research, and Treatment, 4*, 1–14.

Ray, J. J., & Ray, J. A. B. (1982). Some apparent advantages of sub-clinical psychopathy. *Journal of Social Psychology, 117,* 135–142.

Reise, S. P., & Oliver, C. J. (1994). Development of a California Q-set indicator of primary psychopathy. *Journal of Personality Assessment, 62,* 130–144.

Reise, S. P., & Wink, P. (1995). Psychological implications of the Psychopathy Q-Sort. *Journal of Personality Assessment, 65,* 300–312.

Rogers, R., Vitacco, M. J., Jackson, R. L., Martin, M., Collins, M., & Sewell, K. W. (2002). Faking psychopathy: An examination of response styles with antisocial youth. *Journal of Personality Assessment, 78,* 31–46.

Sandoval, A. R., Hancock, D., Poythress, N. G., Edens, J. F., & Lilienfeld, S. O. (2000). Construct validity of the Psychopathic Personality Inventory in a correctional sample. *Journal of Personality Assessment, 74,* 262–281.

Sechrest, L. (1963). Incremental validity: A recommendation. *Educational and Psychological Measurement, 23,* 153–158.

Sellbom, M., Ben-Porath, Y. S., Lilienfeld, S. O., Patrick, C. J., & Graham, J. R. (2005). Assessing psychopathic personality traits with the MMPI-2. *Journal of Personality Assessment, 85,* 334–343.

Sellbom, M., Ben-Porath, Y. S., & Stafford, K. P. (2007). A Comparison of MMPI-2 measures of psychopathic deviance in a forensic setting. *Psychological Assessment, 19,* 430–436.

Sellbom, M., & Verona, E. (2007). Neuropsychological correlates of psychopathic traits in a non-incarcerated sample. *Journal of Research in Personality, 41,* 276–294.

Smith, R. J. (1985). The concept and measurement of social psychopathy. *Journal of Research in Personality, 19,* 219–231.

Spielberger, C. D., Gorsuch, R. L., Lushene, R. E., Vagg, P. R., & Jacobs, G. A. (1983). *Manual for the State-Trait Anxiety Inventory*. Palo Alto, CA: Consult. Psychologists Press.

Spielberger, C. D., Kling, J. K., & O'Hagan, S. E. (1978). Dimensions of psychopathic personality: Antisocial behavior and anxiety. In: R. D. Hare & D. Schalling (Eds.), *Psychopathic behaviour: Approaches to research* (pp. 23–46). Chichester, UK: Wiley.

Spitzer, R. L., Williams, J. B. W., & Gibbon, M. (1987). *Structured clinical interview for DSM-III-R Axis II (SCID-II)*. Washington, DC: American Psychiatric Association Press.

Taylor, J., Loney, B. R., Bobadilla, L., Iacono, W. G., & McGue, M. (2003). Genetic and environmental influences on psychopathy trait dimensions in a community sample of male twins. *Journal of Abnormal Child Psychology, 31,* 633–645.

Tellegen, A. (1985). Structure of mood and personality and their relevance to assessing anxiety, with an emphasis on self-report. In: A. H. Tuma & J. D. Maser (Eds.), *Anxiety and the anxiety disorders* (pp. 681–706). Hillsdale, NJ: Erlbaum.

Tellegen, A., & Waller, N. (2008). Exploring personality through test construction: Development of the Multidimensional Personality Questionnaire. In: S. R. Briggs, & J. M. Cheek (Eds.), *Personality measures: Development and evaluation* (Vol. I, pp. 261–292). Greenwich, CT: JAI Press.

Vincent, N. M. P., Linsz, N. L., & Greene, M. I. (1966). The L scale of the MMPI as an index of falsification. *Journal of Clinical Psychology, 22,* 214–215.

Watson, D., & Clark, L. A. (1984). Negative affectivity: The disposition to experience aversive emotional affects. *Psychological Bulletin, 55,* 465–490.

Widiger, T. A., Cadoret, R., Hare, R., Robins, L., Rutherford, M., Zanarini, M., et al. (1996). DSM—IV antisocial personality disorder field trial. *Journal of Abnormal Psychology, 105,* 3–16.

Widiger, T. A., & Frances, A. (1987). Interviews and inventories for measurement of personality disorders. *Clinical Psychology Review, 7,* 49–75.

Widom, C. S., & Newman, J. P. (1985). Characteristics of noninstitutionalized psychopaths. In: J. Gunn & D. Farrington (Eds.), *Current research in forensic psychiatry and psychology*, Vol. 2 (pp. 57–80). New York: Wiley.

Williams, K., Nathanson, C., & Paulhus, D. (2002). *Factor structure of the self-report psychopathy scale: Two and three factor solutions*. Paper presented at the Annual Meeting of the Canadian Psychological Association, Vancouver, British Columbia.

Wilson, D. L., Frick, P. J., & Clements, C. B. (1999). Gender, somatization, and psychopathic traits in a college sample. *Journal of Psychopathology and Behavioral Assessment, 21,* 221–235.

Wood, J. M., Lilienfeld, S. O., Nezworski, M. T., Garb, H. N., Allen, K. H., & Wildermuth, J. L (2010). The validity of the Rorschach Inkblot Test for discriminating psychopaths from non-psychopaths in forensic populations: A meta-analysis. *Psychological Assessment, 22,* 336–349.

Wood, J. M., Nezworski, M. T., Lilienfeld, S. O., & Garb, H. N. (2003). *What's wrong with the Rorschach? Science confronts the controversial inkblot test.* New York: Jossey-Bass.

Zagon, I., & Jackson, H. (1994). Construct validity of a psychopathy measure. *Personality and Individual Differences, 17,* 125–135.

Zolondek, S., Lilienfeld, S. O., Patrick, C. J., & Fowler, K. A. (2006). The interpersonal measure of psychopathy: Construct and incremental validity in male prisoners. *Assessment, 13,* 470–482.

Zuckerman, M. (1989). Personality in the third dimension: A psychobiological approach. *Personality and Individual Differences, 10,* 391–418.

Zuckerman, M., Kolin, E. A., Price, L., & Zoob, I. (1964). Development of a Sensation-Seeking Scale. *Journal of Consulting Psychology, 28,* 477–482.

PART TWO
DEVELOPMENTAL PERSPECTIVES ON PSYCHOPATHY

CHAPTER 4

Developmental Conceptualizations of Psychopathic Features

Dustin A. Pardini and Amy L. Byrd

Over the past several decades, a growing number of studies have attempted to identify features of adult psychopathy in children and adolescents. These studies initially focused on indexing developmentally appropriate manifestations of psychopathic characteristics in youth, as well as validating their importance for understanding the development of serious and persistent antisocial behavior (Frick, O'Brien, Wootton, & Mcburnett, 1994; Lynam, 1997). An increased emphasis has now been placed on examining the developmental continuity of psychopathic features across the lifespan, including delineating the etiological factors that lead to the early emergence and change in these features over time (Pardini & Loeber, 2007). The knowledge gained from these studies has provided unique insights that are applicable to developing effective interventions designed to prevent at-risk youth from developing a solidified psychopathic personality in adulthood. This chapter provides an overview of contemporary conceptualizations of psychopathic characteristics in childhood and early adolescence, discusses evidence for stability and change in these features over time, examines the potential etiological underpinnings of these characteristics, and reviews findings regarding the impact of early interventions on psychopathic features. Implications for the juvenile justice system are addressed within the context of limitations in the existing literature.

ASSESSING PSYCHOPATHIC FEATURES IN YOUTH

Attempts to apply the concept of adult psychopathy to youth began by delineating those symptoms that could be reliably assessed in a developmentally appropriate manner (Frick et al., 1994). Certain psychopathic features such as maintaining a parasitic lifestyle and sexual promiscuity were by and large excluded because they were not particularly meaningful in young children. There was also a recognition that psychopathic features in youth should be at least somewhat distinct from existing conceptualizations of attention-deficit/hyperactivity disorder (ADHD), oppositional defiant disorder (ODD), and conduct disorder (CD) outlined in the *Diagnostic and Statistical Manual of Mental Disorders*-4th Edition (*DSM-IV*; American Psychiatric Association, 2000). This led to an increased emphasis on identifying the interpersonal (e.g., grandiosity, superficial charm, inflated sense of self-worth, manipulativeness, deceitfulness) and affective (e.g., lack of empathy, deficient guilt/remorse, shallow affect) components of psychopathy in youth (Dadds, Fraser, Frost, & Hawes, 2005; Frick, Bodin, & Barry, 2000; Lynam, 1997; Pardini, Obradović, & Loeber, 2006). Although some of these features were included in the *DSM-III-R* formulation of CD as part of an undersocialized–aggressive subtype, they were eliminated during the transition to *DSM-IV* due to limited evidence for their clinical utility (Pardini, Frick, & Moffitt, 2010). Because early manifestations of the impulsive/unstable lifestyle and antisocial facets of psychopathy in youth largely overlap with symptoms of ADHD, ODD, and CD (Dadds et al., 2005; Frick, Bodin, et al., 2000), the remainder of this chapter focuses solely on developmental conceptualizations of the interpersonal and affective features of psychopathy, with a particular emphasis on manifestations in childhood and early adolescence.

Methods for assessing psychopathic features in children have largely relied on parent and teacher ratings scales. The Antisocial Processes Screening Device (APSD; Frick & Hare, 2001) was the first measure developed explicitly to index childhood features of adult psychopathy as delineated in the Hare Psychopathy Checklist-Revised (PCL-R; Hare, 1991, 2003). Since its original development, a modified version of the APSD has been created for use with preschool-age children as young as age 3 (Dadds et al., 2005). Although several other ad hoc parent and teacher rating scales have been created to measure psychopathic features using subsets of items from existing measures (Lynam, 1997; Pardini et al., 2006; Willoughby, Waschbusch, Moore, & Propper, 2011), none of them have been researched as extensively as the APSD. Once children begin to enter adolescence, psychopathic features are typically assessed using a diverse array of self-report measures (Benning, Patrick, Salekin, & Leistico, 2005) and the Hare Psychopathy Checklist-Youth Version (PCL-YV; Forth, Kosson, & Hare, 2003), which consists of a semistructured interview similar to the PCL-R for adults.

Early studies validating measures of psychopathy in youth have consistently found that the affective features of the disorder, commonly referred to as callous-unemotional (CU) traits, can be reliably distinguished from externalizing disorder symptoms. Specifically, factor analytic studies indicate that parent and teacher ratings of CU traits are distinct from behaviors more consistent with ADHD, ODD, and CD in both clinic- and community-based samples of children (Dadds et al., 2005; Fite, Greening, Stoppelbein, & Fabiano, 2009; Frick, Bodin, et al., 2000). These findings have been extended to a sample of preschool-aged children as young as age 3 (Dadds et al., 2005), as well as a large community sample of young girls 5 to 12 years of age (Loeber et al., 2009). Longitudinal studies have also supported the clinical utility of CU traits in predicting several maladaptive outcomes even after controlling for co-occurring conduct problems. These outcomes include future aggression/bullying (Pardini, Lochman, & Powell, 2007; Pardini, Stephanie, Hipwell, Stouthamer-Loeber, & Loeber, 2012), CD symptoms (Pardini et al., 2012), juvenile arrests and delinquency (Frick, Cornell, Barry, Bodin, & Dane, 2003; McMahon, Witkiewitz, Kotler, & Conduct Problems Prevention Research Group, 2010; Pardini & Fite, 2010), overall clinical impairment (Keenan, Wroblewski, Hipwell, Loeber, & Stouthamer-Loeber, 2010), and antisocial personality disorder (McMahon et al., 2010). In contrast, youth with CU traits do not appear to be at significant risk for developing internalizing problems (Frick, Lilienfeld, Ellis, Loney, & Silverthorn, 1999; Pardini et al., 2007; Pardini et al., 2012), with some longitudinal evidence suggesting that CU traits may actually buffer conduct problem youth from developing anxiety problems over time (Pardini & Fite, 2010).

Evidence supporting the uniqueness of the interpersonal features of psychopathy in children relative to externalizing disorders has been somewhat mixed. Using the APSD in a community sample, Frick and colleagues (2000) identified a facet of psychopathy in children distinct from CU traits and impulsive behaviors that they labeled "narcissism." The narcissism factor included behaviors such as bragging, conning and manipulating others, and having an inflated sense self-worth, consistent with the interpersonal features of adult psychopathy. Pardini and colleagues (2006) also found that features largely consistent with the interpersonal component of adult psychopathy (e.g., being a smooth talker, manipulating others, exaggerating) could be distinguished from ADHD, ODD, and CD behaviors in a large community sample of boys aged 7 to 13 years. However, these features seem to overlap more substantially with aggression and ODD/CD symptoms than with CU traits (Frick, Bodin, et al., 2000; Munoz & Frick, 2007).

In fact, at least two factor analytic studies with clinically disordered samples found that items on the narcissism scale of the APSD could not be reliably distinguished from conduct problems and impulsive behaviors (Dadds et al., 2005; Fite et al., 2009). Evidence supporting the uniqueness of the interpersonal facet of psychopathy using self-report measures and interviewer ratings has been more consistent in older adolescents (Andershed, Kerr, Stattin, & Levander, 2002; Jones, Cauffman, Miller, & Mulvey, 2006; Poythress, Dembo, Wareham, & Greenbaum, 2006). Although it is possible that complex interpersonal behaviors such as manipulating others and being superficially charming may become more distinct from childhood to adolescence, some evidence suggest that these behaviors are indexing a similar underlying construct across this developmental span (Obradović, Pardini, Long, & Loeber, 2007).

Despite these complexities, the interpersonal features of psychopathy seem to provide unique information about the developmental course of antisocial behavior in youth. Cross-sectional studies have found that features of narcissism are associated with increased conduct problem severity (Barry, Frick, & Killian, 2003), self-reported delinquency (Barry, Grafeman, Adler, & Pickard, 2007; Barry, Pickard, & Ansel, 2009; Lau, Marsee, Kunimatsu, & Fassnacht, 2011), and aggressive behaviors (Barry et al., 2009; Barry, Thompson, et al., 2007; Fite et al., 2009; Lau et al., 2011; Washburn, McMahon, King, Reinecke, & Silver, 2004) in youth. In laboratory settings, narcissistic children have also been shown to exhibit increased levels of aggression after receiving critical feedback from their peers (Thomaes, Bushman, Stegge, & Olthof, 2008). Although still relatively rare, longitudinal studies have begun supporting the predictive utility of the interpersonal features of psychopathy in children and young adolescents. Features of narcissism have been shown to predict self-reported delinquency and police contacts across a 3-year follow-up in children, even after controlling for co-occurring conduct problems, CU traits, impulsivity, and parenting practices (Barry, Frick, Adler, & Grafeman, 2007). Boys exhibiting many of the interpersonal features of psychopathy in early adolescence also tend to exhibit persistent delinquency and higher levels of antisocial personality and psychopathy in adulthood (Burke, Loeber, & Lahey, 2007; Byrd, Loeber, & Pardini, 2012; Loeber, Pardini, Stouthamer-Loeber, & Raine, 2007; Pardini et al., 2006).

STABILITY AND CHANGE IN CHILDHOOD PSYCHOPATHIC FEATURES

An area of continued debate is whether psychopathic features in youth should be viewed as malignant personality traits (Johnstone & Cooke, 2004; Seagrave & Grisso, 2002). There are now a fairly substantial number of longitudinal studies examining whether the affective and interpersonal features of psychopathy are relatively stable from childhood through adolescence. In fact, more longitudinal studies have documented the stability of psychopathic features in children than adults. Although exceptions do exist, most studies have found that parent-report instruments assessing features of psychopathy in youth are as stable as measures of adult personality traits. For example, Loeber and colleagues (2009) found high 1-year stability estimates for parent-reported CU traits in young girls across ages 5 to 11 (intraclass correlation coefficients [ICCs] ranging from .71 to .76). Similar levels of stability have been reported across a 2-year span for parent-reported CU traits (ICCs = .76) and narcissism (ICC = .89) in aggressive children (Barry, Barry, Deming, & Lochman, 2008). Frick and colleagues also found strong stability for parent-report of both CU traits (ICC = .71) and narcissism (ICC = .77) across a 4-year span from childhood to early adolescence in a high-risk sample (Frick, Kimonis, Dandreaux, & Farell, 2003). In one of the most extensive stability studies conducted to date, Obradović and colleagues (2007) reported high year-to-year stability estimates in boys from age 8 to 16 ($r = .77–.84$) using a parent-report scale that primarily assessed the interpersonal features of psychopathy. More importantly, moderate levels of stability were found across the entire 8-year span ($r = .50$). These findings indicate that parent-reported ratings of youth psychopathic features are at least as stable as interviewer ratings of adult psychopathy (Rutherford, Cacciola, Alterman, McKay, & Cook, 1999).

In contrast, teacher-report measures of psychopathic features in youth are far less stable and more variable across studies. For example, Barry and colleagues (2007) found high levels of stability across a 2-year span for teacher-reported CU traits (ICC = .62), but stability of narcissistic features was relatively poor (ICC = .33). Although another study found that teacher-report of the interpersonal features of psychopathy in boys were relatively stable across year-to-year assessments from age 8 to 16 ($r = .52–.65$), the 8-year stability of teacher ratings was relatively low ($r = .27$; Obradović et al., 2007). The attenuation in stability in teacher versus parent-report measures is not surprising given that teachers change from year to year and the amount of time teachers spend interacting with any one student decreases from elementary to high school. However, this discrepancy speaks to a larger issue about how to combine information across multiple informants, particularly because the correlation between parent and teacher ratings of psychopathic features is typically low to moderate (Barry et al., 2008; Barry, Thompson, et al., 2007; Frick, Bodin, et al., 2000). Although the issue of low concordance across informants is common to most mental health problems, there is an ongoing debate about whether or not to require psychopathic features in youth to be endorsed by multiple informants across different settings to be considered clinically significant (Frick & Moffitt, 2010).

Although psychopathic features appear somewhat stable in children, there are subgroups of youth who experience significant changes in these behaviors over time. One recent study found substantial instability in CU traits from ages 7 to 12 in a large twin sample (Viding, Fontaine, Oliver, & Plomin, 2009), with trajectory group analyses indicating that approximately 80% of children with initially high levels of CU traits had precipitous decreases in these features over time. Although youth with decreasing CU traits continued to have significant conduct problems, hyperactivity, and emotional problems at age 12, these problems were less severe than those found in youth with chronically high CU traits. A group of youth who exhibited significant increases in CU traits from age 7 to 12 (9.6% of the total sample) was also identified, suggesting that CU traits continue to emerge into late childhood. Similarly, Pardini and Loeber (2008) found that significant within-individual change in the interpersonal features of psychopathy occurred from ages 14 to 18, with decreases across the period predicting lower levels of antisocial personality disorder symptoms in early adulthood (i.e., age 25). In sum, meaningful changes in psychopathic features do appear to occur across childhood and adolescence.

DEVELOPMENTAL ORIGINS OF AFFECTIVE PSYCHOPATHIC FEATURES

To promote effective prevention efforts, recent studies have attempted to identify factors that lead to the early emergence and subsequent change in the affective features of psychopathy over time. Work in this area is embedded within a larger literature investigating conscience development in normal children (Kochanska & Aksan, 2006). The term "conscience" is often used to reference a broad-based construct that involves the cognitive and emotional internalization of societal rules that guide prosocial behaviors (Kochanska, 1991). The affective components of conscience include experiencing affective discomfort following wrongdoing (i.e., guilt) and having empathetic concern for the well-being of others (Kochanska, Forman, Aksan, & Dunbar, 2005; Zahn-Waxler & Kochanska, 1990), which are core characteristics of CU traits (Frick, Barry, & Bodin, 2000). Developmental research suggests that children begin to exhibit sophisticated emotional reactions to wrong-doing and others' distress between the ages of 2 to 3 (Kochanska, 1993). Longitudinal studies have also found that infants and toddlers with low levels of temperamental fearfulness have particular difficulties developing the affective components of a conscience, including concern for the suffering of others (Fox, Young, & Zahn-Waxler, 1999; Kochanska, 1995; Rothbart, Ahadi, & Hershey, 1994) and guilt following transgressions (Kochanska, Gross, Lin, & Nichols, 2002; Rothbart et al., 1994).

Similar findings have emerged in a diverse array of studies explicitly assessing the affective features of psychopathy in youth. Cross-sectional studies have found that low levels of anxiety and fear are associated with increased CU traits in a diverse array of samples (Frick et al., 1999; Hipwell et al., 2007; Lynam et al., 2005; Pardini, Lochman, & Frick, 2003). These findings are particularly robust after controlling for co-occurring conduct problems, which tend to be associated with increased levels of anxiety and emotional dysregulation (Frick et al., 1999; Hipwell et al., 2007; Pardini et al., 2007). However, longitudinal studies have provided somewhat mixed results. Frick, Kimonis, and colleagues (2003) found no evidence that fearfulness predicted changes in CU traits over a 4-year period in children. Although one study found that aggressive children with low anxiety exhibited increases in CU traits across a 1-year follow-up, this was true only for children who did not have a warm and involved relationship with their parents (Pardini et al., 2007).

Some have speculated that children with low temperamental fear may be at risk for developing CU traits because they are relatively insensitive to the socializing influences of punishment (Dadds & Salmon, 2003; Pardini, 2006). Children who experience low anxious arousal when being disciplined for misbehavior may be less likely to encode parental messages about the acceptability of behaviors, which some have argued is essential to the internalization of moral beliefs and emotions (Kochanska, 1997). Through repeated disciplinary interactions, typically developing children likely become conditioned to experience increases in negative arousal when contemplating or engaging in misconduct even in the absence of an authority figure, which is a key feature of guilt and remorse. In support of this conceptual model, children with high levels of CU traits tend to display less sensitivity to punishment while engaged in a goal-directed card-playing task (Blair, Colledge, & Mitchell, 2001), regardless of whether or not they have significant conduct problems (Frick, Cornell, Bodin, et al., 2003; O'Brien & Frick, 1996). Both adolescents and children with CU traits also report lower levels of concern about being punished for aggressive behavior (Jones, Happe, Gilbert, Burnett, & Viding, 2010; Pardini & Byrd, 2012; Pardini et al., 2003), with the association between low fearfulness and CU traits in incarcerated adolescents being mediated by a lack of concern about being punished (Pardini, 2006). Although these results are promising, longitudinal studies examining this proposed developmental model of CU traits are still needed.

A distinct, yet related, set of studies have postulated that relatively fearless children have an impaired ability to recognize distress cues in others, which places them at risk for developing CU traits (Marsh & Blair, 2008). According to the violence inhibition model (VIM), humans possess a basic neural system that responds to cues of distress in others (particularly fearful faces) by initiating increased attention, behavioral freezing, and aversive arousal (Blair, 2001). As a result, normally developing children learn to avoid initiating violent behavior because the fearful distress it produces in the victim is repeatedly paired with aversive arousal in the perpetrator. Children with CU traits are believed to have subtle neurological impairments in limbic brain regions (particularly the amygdala) that impair their ability to recognize, and become aroused by fearful distress cues in others (Blair, 2005). This developmental model is supported by studies indicating that children and adolescents exhibiting antisocial behavior have difficulties recognizing fearful distress cues in others (Marsh & Blair, 2008), with some evidence suggesting this deficit may be unique to youth with CU traits (Dadds et al., 2006; Munoz, 2009). More recently, neuroimaging studies have found that children and adolescents with high levels of both conduct problems and CU traits exhibit reduced responding to others' fearful distress in a cortic-limbic network that includes the amygdala relative to healthy children (Jones, Laurens, Herba, Barker, & Viding, 2009; Marsh et al., 2008). Because these functional magnetic resonance imaging (fMRI) studies did not include a control group with high levels of conduct problems without CU traits, it remains unclear whether these deficits are uniquely related to the affective features of psychopathy in youth. There is also evidence suggesting that the emotion processing deficits associated with CU traits are not circumscribed to fearful stimuli (Munoz, 2009) and may be overcome by simply having youth attend to salient facial features such as the eyes (Dadds, El Masry, Wimalaweera, & Guastella, 2008). Moreover, there are no known longitudinal studies linking emotional deficits in processing fearful distress in others to the early emergence of CU traits or the stability of features over time.

Although investigators often emphasize the importance of individual-level factors in the development of the affective features of psychopathy, early parenting practices that facilitate positive parent–child attachment also appear to be important. Within typically developing infants, exposure to high levels of parental warmth and responsiveness have been associated with the development of empathic responding (Kiang, Moreno, & Robinson, 2004) and feeling guilt (Kochanska et al., 2005) in childhood. Positive parental reinforcement and involvement have also been associated with decreases in CU traits across a 4-year period from childhood to adolescents (Frick, Kimonis, et al., 2003). Moreover, there is evidence suggesting that a warm/involved parent–child relationship may protect aggressive children with low anxiety from developing CU traits (Pardini et al., 2007) and buffer children with high CU traits from developing serious conduct problems (Kroneman, Hipwell, Loeber, Koot, & Pardini, 2011; Pasalich, Dadds, Hawes, & Brennan, 2011).

Children who are exposed to excessive amounts of harsh parental discipline (e.g., physical punishment, verbal abuse) may also be prone to developing CU traits over time. The use of harsh discipline may produce excessive arousal that interferes with the child's ability to encode parental messages about appropriate conduct and desensitize children to victim suffering (Gershoff, 2002; Kochanska, 1997). Longitudinal research indicates that typically developing children who are exposed to harsh forms of discipline display less guilt following transgressions (Kochanska et al., 2002) and are less concerned about the feelings of others (Hastings, Zahn-Waxler, Robinson, Usher, & Bridges, 2000). More recent longitudinal evidence also suggests that children who are exposed to negative parenting practices (including corporal punishment) exhibit increases in CU traits over time (Frick, Kimonis, et al., 2003; Pardini et al., 2007). Although one recent study found that harsh parenting did not distinguish conduct problem children with chronic CU traits from those with declining CU traits from age 7 to 12, the harsh parenting scale used consisted of only two items, which may have limited the ability to detect significant effects (Fontaine, McCrory, Boivin, Moffitt, & Viding, 2011).

DEVELOPMENTAL ORIGINS OF INTERPERSONAL PSYCHOPATHIC FEATURES

In contrast to studies on the affective features of psychopathy, research on the developmental origins of the interpersonal characteristics of psychopathy in children remains sparse. However, developmental studies have examined the early emergence of lying and deceitful behaviors in children, which are core features of the interpersonal facet of adult psychopathy (Cooke & Michie, 2001). Children have a basic understanding that intentionally lying to cover up misdeeds is morally wrong around the age of 3 (Talwar, Lee, Bala, & Lindsay, 2002). Despite this fact, most preschool-aged children will lie to conceal rule-breaking behavior in laboratory settings, with this type of lying increasing substantially between the ages of 3 and 7 (Talwar & Lee, 2002). What remains unclear is at what point lying behavior becomes less normative and develops into more sophisticated forms of deception, such as conning others for personal gain. Preschool and elementary school–age children are often unable to maintain a lie convincingly upon questioning (Talwar, Gordon, & Lee, 2007), making it unlikely that they are able to con or manipulate others effectively. Although no known empirical studies have examined what factors contribute to continuity and change in lying behaviors across early childhood, Pardini and Loeber (2008) found that poor parent–child communication was associated with chronically high levels of largely deceptive and manipulative behaviors from early to late adolescence. However, the parenting and peer factors examined in this study were unrelated to changes in these features over time.

A distinct literature relevant for understanding the development of the interpersonal features of psychopathy has focused on the early emergence of maladaptive narcissism. Considered a hallmark feature adult psychopathy (Blackburn, 2007; Cleckley, 1976; Hare, Hart, & Harpur, 1991), maladaptive narcissism is characterized by an inflated sense of self-worth, entitlement, arrogance, a need for excessive admiration, interpersonal exploitative behaviors, and a desire to be viewed as superior by others (Raskin, Novacek, & Hogan, 1991). Narcissistic features in youth are often assessed using the Narcissistic Personality Inventory for adults (Washburn et al., 2004) or a modified child version of this measure (Barry et al., 2003), though at least one promising alternative has been developed (Thomaes, Stegge, Bushman, Olthof, & Denissen, 2008). These self-report measures have been validated for use in children as young as 10 years of age (Barry et al., 2003; Thomaes, Bushman, et al., 2008), with studies indicating that self-reported narcissism in children is distinct from self-esteem and positively associated with antisocial and aggressive behavior (Barry et al., 2003; Lau et al., 2011; Thomaes, Stegge, et al., 2008). Similar to narcissistic adults, children who endorse high levels of narcissistic features tend to react aggressively when shamed or criticized by others (Barry et al., 2003; Thomaes, Bushman, et al., 2008; Thomaes, Stegge, et al., 2008). However, some contradictory findings have been reported. Specifically, one study found that the association between narcissism and aggression occurs only for children with low self-esteem (Barry et al., 2003), while another reported the association held only for children with high self-esteem (Thomaes, Bushman, et al., 2008).

Although there has been considerable theoretical conjecture about the developmental origins of narcissism, studies in this area remain scarce (Thomaes, Bushman, De Castro, & Stegge, 2009). It is believed that environmental factors are likely to play an important role, as evidence from twin studies suggests that approximately 40% of the variability in adult narcissistic features can be attributed to nonshared environmental factors (Vernon, Villani, Vickers, & Harris, 2008). The relatively few studies of the developmental precursors to narcissism have focused on parent–child relationship quality as a potentially important etiological factor. Older adolescents and adults with narcissistic features tend to describe their parents as being both cold and controlling, as well as indulgent and permissive (Otway & Vignoles, 2006). This vacillation between extremes has been postulated to underlie the need for constant admiration and hypersensitivity to critical feedback that is a hallmark of the adult narcissistic personality. However, studies with younger children and early adolescents have failed to directly support this theoretical model. One longitudinal study found that negative parenting practices (e.g., inconsistent punishment, corporal punishment), but not positive parenting factors (e.g., involvement, reinforcement) predicted increases in the narcissism factor of the APSD in youth across a 4-year period (Frick, Kimonis, et al., 2003). A more recent study found that preschool children who were rated as high in attention-seeking, hostility, and impulsivity exhibited higher levels of maladaptive narcissism in young adulthood, but only when exposed to a maternal caregiver who had a more authoritarian (e.g., overly strict, controlling) than authoritative (e.g., respects child's feelings and interests) parenting style (Carlson & Gjerde, 2009; Cramer, 2011). Because this study focused on a small normative sample, it is unclear whether the findings are relevant to understanding the development of the pathological aspects of narcissism found in adults with elevated psychopathic features.

EARLY INTERVENTIONS AND CHILDHOOD PSYCHOPATHIC FEATURES

There has been some speculation that existing treatments may not be effective for children and young adolescents with co-occurring conduct problems and high levels of psychopathic traits. In contrast to this pessimistic view, empirical studies have produced largely positive results regarding the treatment of youth with psychopathic features. For example, one study found that a parent-focused cognitive–behavioral treatment for young children with ODD resulted in positive changes in parent-reported CU features from pretreatment to 6-month follow-up ($d = .57$), and approximately half of those youth with high pre-treatment CU traits exhibited substantive reductions in these features over time (Hawes & Dadds, 2007). However, the authors did find that children with higher levels of CU traits before treatment had poorer behavior outcomes at the posttreatment assessment. Another recent investigation looked at the impact of a multifaceted intervention (e.g., medication management, parent management training, family therapy, teacher consultation) on psychopathic features in 6- to 11-year-old children with ODD/CD. Significant reductions in teacher-reported CU traits and narcissistic features were found from pre- to posttreatment ($d = .44$ and $.47$, respectively) and these treatment gains were sustained across a 3-year follow-up (Kolko et al., 2009). Moreover, children with high levels of CU traits and narcissism at pretreatment did not evidence poorer treatment outcomes, contrary to the notion that youth with psychopathic traits are unresponsive to traditional treatment methods (Kolko & Pardini, 2010). In contrast, a more recent study found that pretreatment CU traits were associated with fewer improvements in social skills and problem-solving abilities after an 8-week summer program for children with co-occurring CD/ADHD (Haas et al., 2011).

A limiting factor in these early treatment studies is the lack of a randomized comparison group. As a result, it is difficult to determine whether the improvements in psychopathic features observed were due to naturally occurring declines in these characteristics, a statistical artifact such as regression to the mean, and/or the effects of treatment. A recent randomized treatment trial addressed this limitation by examining whether a manualized parenting intervention produced greater reductions in childhood psychopathic features relative to treatment as usual in the community (McDonald, Dodson, Rosenfield, & Jouriles, 2011). Participants were 66 mothers in a domestic violence shelter with a child between the ages of 4 and 9 who met criteria for ODD or CD. Families were randomly assigned to receive services as usual in the community or a manualized intervention (i.e., Project Support) lasting up to 8 months that consisted of a combination of child management skills training and therapeutic emotional support for the mothers. Most of the families in the control condition received either no services during the course of the study (67.6%) or fewer than four therapy sessions (33.3%). Findings indicated that Project Support produced significant reductions in the interpersonal and affective features of psychopathy during the intervention period relative to the control condition, and these improvements were sustained across a 3-month follow-up. In contrast, children in the comparison condition experienced a significant increase in psychopathic features across the follow-up period. Improvements in psychopathic features were mediated in part by reductions in harsh and inconsistent parenting. Although this study suffered from limitations such as the use of a small sample and a failure to include positive parenting characteristics as potential treatment mediators, it suggests that interventions can produce reductions in childhood psychopathic features by facilitating improvements in maladaptive parenting behaviors.

There are several other empirically supported interventions for young children with conduct problems that may also be effective in both reducing psychopathic features and preventing their development. Although the impact of these interventions on psychopathic features has not been directly evaluated, they target parent (e.g., parental warmth) and child (e.g., empathetic concern) factors that may drive the development of these characteristics over time. Comprehensive reviews of these treatments can be found elsewhere (Nock, 2003; Pardini, 2008), but we highlight two that seem particularly promising for preschool-aged children based on prior studies: Parent–Child Interaction Therapy (PCIT) and The Incredible Years Program. PCIT is a parent-focused intervention that is implemented in the context of naturalistic play settings between the parent and child in which a therapist provides guidance behind a one-way mirror through the use of a hidden ear device worn by the parent (Eyberg, Boggs, & Algina, 1995; Herschell, Calzada, Eyberg, & McNeil, 2002;

Hood & Eyberg, 2003). This technique provides the parent with the unique opportunity to practice and master skills in vivo rather than passively learning through didactic interactions with the therapist. Sessions focus on strengthening the parent–child relationship through child-directed play, reinforcing positive behaviors, and decreasing maladaptive behaviors through the use of consistently implemented behavioral management techniques. A review of 17 studies that used PCIT found strong evidence of statistically and clinically significant improvements in children's behavioral functioning (Gallagher, 2003). PCIT's focus on increasing positive parent–child bonding may be particularly useful for reducing CU traits as it may help to facilitate the development of emotional connectedness with others.

The Incredible Years program also has promise as a treatment for psychopathic features in youth (Webster-Stratton, Reid, & Hammond, 2004). The program is designed for children ages 4 to 7 who have conduct problems significant enough to warrant a diagnosis of either ODD or CD, and the full intervention includes both a parent and child component (Beauchaine, Webster-Stratton, & Reid, 2005; Webster-Stratton & Hammond, 1997; Webster-Stratton et al., 2004). The child component consists of groups of six or seven children who attend weekly 2-hour sessions for approximately 17 weeks. Videotaped vignettes and life-size puppets are used to promote emotional empathy, perspective-taking skills, conflict resolution skills, anger regulation, and friendship building. The Parent Training intervention consists of approximately 22 sessions that use videotapes to model appropriate ways to deal with problematic parent–child interactions. Sessions address topics such as promoting effective play techniques, limit setting, handling misbehavior, and the communication of emotions. The parenting component of the intervention has been shown to reduce harsh and ineffective discipline, increase nurturing and supportive parenting, and reduce children's behavior problems (Beauchaine et al., 2005; Reid, Webster-Stratton, & Hammond, 2003). Moreover, reductions in negative parenting and improvements in positive parenting have been shown to mediate (at least in part) reductions in children's behavior problems (Beauchaine et al., 2005; Gardner, Hutchings, Bywater, & Whitaker, 2010). The child intervention has also been shown to produce significant reductions in the amount of conduct problems children exhibit at home and school, as well as increases in social problem-solving skills in comparison to waitlist control conditions (Beauchaine et al., 2005; Reid et al., 2003; Webster-Stratton & Hammond, 1997; Webster-Stratton et al., 2004). It is possible that facets of the child component that focus on developing emotional empathy and friendship building, as well as parenting sessions that reinforce consistent nonharsh discipline and communication of emotions, may help to reduce psychopathic features over time.

LEGAL AND CLINICAL IMPLICATIONS

Although a considerable amount of research has been generated over the past decade on psychopathic features in youth, incorporating these features into juvenile justice decision making remains controversial. This is due in part to unresolved measurement issues associated with the assessment of psychopathic features in youth. For example, there is currently no consensus regarding what constitutes clinically significant levels of psychopathic traits in children and adolescents. Evidence suggests that it is best to view most mental health problems in children and adults as falling on a continuum rather than attempting to place people artificially into categories (Brown & Barlow, 2005; Fergusson, Boden, & Horwood, 2010). This is also relevant for conceptualizations of psychopathic features, as there is no convincing evidence indicating that a latent categorical taxon underlies psychopathic characteristics in youth (Murrie et al., 2007; Vasey, Kotov, Frick, & Loney, 2005). Other measurement issues include relatively modest correlations across divergent measures of the interpersonal and affective features of psychopathy (Lee, Vincent, Hart, & Corrado, 2003; Poythress et al., 2006; Salekin, Neumann, Leistico, DiCicco, & Duros, 2004), relatively low agreement across multiple informants (Barry et al., 2008; Viding et al., 2009), and some lingering problems with measurement unreliability, particularly for the core affective dimension of psychopathy (Kotler & McMahon, 2005; McDonald et al., 2011; Viding et al., 2009). These issues have made it difficult to come up with specific recommendations on how to best assess psychopathic features from childhood through adolescence and use this information in clinical and legal decision making.

Attempts to incorporate psychopathic features into legal decision making should also recognize that many youth with elevated psychopathic traits will not continue to exhibit the core features of psychopathy into adulthood. Only two published longitudinal studies have investigated the association between early psychopathic features and scores on the PCL-R in early adulthood (Burke et al., 2007; Lynam, Caspi, Moffitt, Loeber, & Stouthamer-Loeber, 2007). Both of these studies found that psychopathic features in youth were related primarily to the impulsive lifestyle and antisocial behavior components of adult psychopathy. Neither study found that the features of psychopathy in youth uniquely predicted the interpersonal/affective facet of adult psychopathy after controlling for the impulsive/antisocial dimension of the disorder. Therefore, although early psychopathic features in youth may be associated with antisocial outcomes in adulthood, evidence suggests that they may tell us little about who will exhibit the core interpersonal and affective features of psychopathy into adulthood. Making reference to youth as "psychopaths" may be misconstrued as indicating that these features are strongly linked to manifestations of adult psychopathy, which can have profound negative consequences given the stigma associated with this largely pejorative label.

It is also important to emphasize that the effect sizes reported between the interpersonal/affective psychopathic features and future offending are generally small in magnitude after controlling for prior antisocial behavior (Corrado, Vincent, Hart, & Cohen, 2004; Pardini & Fite, 2010). In many cases, the strength of the associations reported is comparable to other risk factors, such as CD symptom severity, peer delinquency, and poor school motivation/achievement (Burke et al., 2007; Loeber et al., 2005). In addition, the clinical utility of assessing psychopathic traits to measure offending risk within the juvenile justice setting outside of a research context is unknown. It may be more difficult to collect accurate information about psychopathic features from parents and youth when there is a possibility that such a disclosure would result in adverse legal consequences. Even if these youth are at greater risk for offending, it is important to remember that comprehensive empirically supported treatments do seem to reduce psychopathic features in youth when they are administered with fidelity (Hawes & Dadds, 2007; Kolko et al., 2009; Kolko & Pardini, 2010; McDonald et al., 2011).

With these limitations in mind, efforts are currently underway to incorporate the affective features of psychopathy into existing definitions of CD in order to inform clinical decision making. Specifically, the workgroup on Disruptive Behavior Disorders for the 5th Edition of the *Diagnostic and Statistical Manual of Mental Disorders* (*DSM–5*) has recently proposed including a CU subtype to the diagnosis of CD based upon extensive research using the APSD (Frick & Moffitt, 2010; Pardini et al., 2010). To meet criteria for the subtype, youth with CD would have to exhibit two out of four of the following characteristics for at least 12 months and in more than one relationship or setting: (1) lack of remorse or guilt; (2) callous/lack of empathy; (3) unconcerned about performance; and (4) shallow or deficient affect. Recent studies have begun providing some empirical support for the clinical utility of this proposed subtyping scheme. Cross-sectional evidence indicates that the proposed CU specifier identifies a subgroup of children and adolescents with CD that have particularly high levels of aggression and cruelty toward others (Kahn, Frick, Youngstrom, Findling, & Youngstrom, 2011). In terms of predictive utility, a recent longitudinal study found that girls 6 to 8 years of age who met criteria for the CU subtype of CD exhibited lower anxiety problems, but more severe aggression/bullying, CD symptoms, academic problems, and global impairment at a 6-year follow-up than girls with CD alone (Pardini et al., 2012). Adolescents who meet criteria for the CU subtype of CD also appear to be at particularly high risk for exhibiting criminal behavior into adulthood (McMahon et al., 2010).

In contrast, one study found that children with the CU subtype of CD exhibited similar levels of antisocial behaviors to those with CD alone across a 3-year follow-up after completing a comprehensive multimodal intervention (Kolko & Pardini, 2010). Despite these advances, issues about how to incorporate information on CU traits across multiple informants and the optimal methods for assessing these features still need to be resolved, most likely as part of the *DSM-5* field trials. If the CU subtype of CD is eventually incorporated into *DSM-5*, then features of psychopathy will likely become part of the juvenile justice decision making process as youth who are diagnosed with this disorder begin appearing in court.

In conclusion, it is clear that youth with psychopathic features should be targeted for interventions early in life, before they become ensnared in a criminal lifestyle that limits opportunities for positive change. Despite noted clinical pessimism, recent empirical work outlined in this chapter has demonstrated promising results with regard to intervention, suggesting that these youth are by no means untreatable. Research has also shown considerable within-individual change in these characteristics across development and has documented several factors (e.g., warm parent–child relationship, consistent nonharsh discipline) that may help to prevent the development of psychopathic features in youth and/or aid in the reduction of these characteristics over time. Thus, although these features delineate a subgroup of youth at risk for exhibiting severe and persistent criminal behavior, these youth should not be considered a "lost cause."

REFERENCES

American Psychiatric Association. (2000). *Diagnostic and statistical manual of mental disorders* (4th ed. text revision): Washington, DC: American Psychiatric Association.

Andershed, H., Kerr, M., Stattin, H., & Levander, S. (2002). Psychopathic traits in non-referred youths: A new assessment tool. In E. Blaauw & L. Sheridan (Eds.), *Psychopaths: Current international perspectives* (pp. 131–158). The Hague: Elsevier.

Barry, C., Frick, P., Adler, K., & Grafeman, S. (2007). The predictive utility of narcissism among children and adolescents: Evidence for a distinction between adaptive and maladaptive narcissism. *Journal of Child and Family Studies*, 16(4), 508–521.

Barry, C. T., Frick, P. J., & Killian, A. L. (2003). The relation of narcissism and self-esteem to conduct problems in children: A preliminary investigation. *Journal of Clinical Child and Adolescent Psychology*, 32(1), 139–152.

Barry, C. T., Grafeman, S. J., Adler, K. K., & Pickard, J. D. (2007). The relations among narcissism, self-esteem, and delinquency in a sample of at-risk adolescents. *Journal of Adolescence*, 30(6), 933–942.

Barry, C. T., Pickard, J. D., & Ansel, L. L. (2009). The associations of adolescent invulnerability and narcissism with problem behaviors. *Personality and Individual Differences*, 47(6), 577–582.

Barry, T. D., Barry, C. T., Deming, A. M., & Lochman, J. E. (2008). Stability of psychopathic characteristics in childhood—The influence of social relationships. *Criminal Justice and Behavior*, 35(2), 244–262.

Barry, T. D., Thompson, A., Barry, C. T., Lochman, J. E., Adler, K., & Hill, K. (2007). The importance of narcissism in predicting proactive and reactive aggression in moderately to highly aggressive children. *Aggressive Behavior*, 33(3), 185–197.

Beauchaine, T. P., Webster-Stratton, C., & Reid, M. J. (2005). Mediators, moderators, and predictors of 1-year outcomes among children treated for early-onset conduct problems: A latent growth curve analysis. *Journal of Consulting and Clinical Psychology*, 73(3), 371–388.

Benning, S. D., Patrick, C. J., Salekin, R. T., & Leistico, A. M. (2005). Convergent and discriminant validity of psychopathy factors assessed via self-report: A comparison of three instruments. *Assessment*, 12(3), 270–289.

Blackburn, R. (2007). Personality disorder and antisocial deviance: Comments on the debate on the structure of the psychopathy checklist-revised. *Journal of Personality Disorders*, 21(2), 142–159.

Blair, R. J. (2001). Neurocognitive models of aggression, the antisocial personality disorders, and psychopathy. *Journal of Neurology, Neurosurgery, and Psychiatry*, 71(6), 727–731.

Blair, R. J. (2005). Applying a cognitive neuroscience perspective to the disorder of psychopathy. *Development and Psychopathology*, 17(3), 865–891.

Blair, R. J. R., Colledge, E., & Mitchell, D. G. V. (2001). Somatic makers and response reversal:

Is there orbitofrontal cortex dysfunction in boys with psychopathic tendencies? *Journal of Abnormal Child Psychology, 29,* 499–511.

Brown, T. A., & Barlow, D. H. (2005). Dimensional versus categorical classification of mental disorders in the fifth edition of the Diagnostic and Statistical Manual of Mental Disorders and beyond: Comment on the special section. *Journal of Abnormal Psychology, 114*(4), 551–556.

Burke, J. D., Loeber, R., & Lahey, B. B. (2007). Adolescent conduct disorder and interpersonal callousness as predictors of psychopathy in young adults. *Journal of Clinical Child and Adolescent Psychology, 36*(3), 334–346.

Byrd, A. L., Loeber, R., & Pardini, D. A. (2012). Understanding desisting and persisting forms of delinquency: The unique contributions of disruptive behavior disorders and interpersonal callousness. *Journal of Child Psychology and Psychiatry, 53,* 371–380.

Carlson, K. S., & Gjerde, P. F. (2009). Preschool personality antecedents of narcissism in adolescence and young adulthood: A 20-year longitudinal study. *Journal of Research in Personality, 43*(4), 570–578.

Cleckley, H. (1976). *The mask of sanity* (5th ed.). St. Louis, MO: Mosby.

Cooke, D. J., & Michie, C. (2001). Refining the construct of psychopathy: Towards a hierarchical model. *Psychological Assessment, 13*(2), 171–188.

Corrado, R. R., Vincent, G. M., Hart, S. D., & Cohen, I. M. (2004). Predictive validity of the Psychopathy Checklist: Youth Version for general and violent recidivism. *Behavioral Science and the Law, 22*(1), 5–22.

Cramer, P. (2011). Young adult narcissism: A 20 year longitudinal study of the contribution of parenting styles, preschool precursors of narcissism, and denial. *Journal of Research in Personality, 45*(1), 19–28.

Dadds, M. R., El Masry, Y., Wimalaweera, S., & Guastella, A. J. (2008). Reduced eye gaze explains "fear blindness" in childhood psychopathic traits. *Journal of the American Academy of Child and Adolescent Psychiatry, 47*(4), 455–463.

Dadds, M. R., Fraser, J., Frost, A., & Hawes, D. J. (2005). Disentangling the underlying dimensions of psychopathy and conduct problems in childhood: A community study. *Journal of Consulting and Clinical Psychology, 73*(3), 400–410.

Dadds, M. R., Perry, Y., Hawes, D. J., Merz, S., Riddell, A. C., Haines, D. J., et al. (2006). Attention to the eyes and fear-recognition deficits in child psychopathy. *British Journal of Psychiatry, 189,* 280–281.

Dadds, M. R., & Salmon, K. (2003). Punishment insensitivity and parenting: temperament and learning as interacting risks for antisocial behavior. *Clinical Child and Family Psychology Review, 6*(2), 69–86.

Eyberg, S. M., Boggs, S. R., & Algina, J. (1995). Parent-Child Interaction Therapy: A psychosocial model for the treatment of young children with conduct problem behavior and their families. *Psychopharmacology Bulletin, 31*(1), 83–91.

Fergusson, D. M., Boden, J. M., & Horwood, L. J. (2010). Classification of behavior disorders in adolescence: Scaling methods, predictive validity, and gender differences. *Journal of Abnormal Psychology, 119*(4), 699–712.

Fite, P. J., Greening, L., Stoppelbein, L., & Fabiano, G. A. (2009). Confirmatory factor analysis of the antisocial process screening device with a clinical inpatient population. *Assessment, 16*(1), 103–114.

Fontaine, N. M. G., McCrory, E. J. P., Boivin, M., Moffitt, T. E., & Viding, E. (2011). Predictors and outcomes of joint trajectories of callous–unemotional traits and conduct problems in childhood. *Journal of Abnormal Psychology, 120*(3), 730–742.

Forth, A., Kosson, D., & Hare, R. (2003). *The Hare Psychopathy Checklist: Youth Version, technical manual.* New York: Multi-Health Systems.

Fox, N. A., Young, S. K., & Zahn-Waxler, C. (1999). The relations between temperament and empathy in 2-year-olds. *Developmental Psychology, 35*(5), 1189–1197.

Frick, P. J., Barry, C. T., & Bodin, S. D. (2000). Applying the concept of psychopathy to children: Implications for the assessment of antisocial youth. In C.B. Gacono (Ed.), *The clinical and forensic assessment of psychopathy* (pp. 3–24). Mahwah, NJ: Erlbaum.

Frick, P. J., Bodin, S. D., & Barry, C. T. (2000). Psychopathic traits and conduct problems in community and clinic-referred samples of children: Further development of the psychopathy screening device. *Psychological Assessment, 12*(4), 382–393.

Frick, P. J., Cornell, A. H., Barry, C. T., Bodin, S. D., & Dane, H. E. (2003). Callous-unemotional traits and conduct problems in the prediction of conduct problem severity, aggression, and self-report of delinquency. *Journal of Abnormal Child Psychology, 31*(4), 457–470.

Frick, P. J., Cornell, A. H., Bodin, S. D., Dane, H. E., Barry, C. T., & Loney, B. R. (2003). Callous-unemotional traits and developmental pathways to severe conduct problems. *Developmental Psychology, 39*(2), 246–260.

Frick, P. J., & Hare, R. D. (2001). *Antisocial Process Screening Device technical manual*. Toronto, Ontario, Canada: Multi-Health Systems.

Frick, P. J., Kimonis, E. R., Dandreaux, D. M., & Farell, J. M. (2003). The 4 year stability of psychopathic traits in non-referred youth. *Behavioral Sciences and the Law, 21*(6), 713–736.

Frick, P. J., Lilienfeld, S. O., Ellis, M., Loney, B., & Silverthorn, P. (1999). The association between anxiety and psychopathy dimensions in children. *Journal of Abnormal Child Psychology, 27*(5), 383–392.

Frick, P. J., & Moffitt, T. E. (2010). A proposal to the DSM-V Childhood Disorders and the ADHD and Disruptive Behavior Disorders Work Groups to include a specifier to the diagnosis of Conduct Disorder based on the presence of callous-unemotional traits. Retrieved from: http://www.dsm5.org/Proposed%20Revision%20Attachments/Proposal%20for%20Callous%20and%20Unemotional%20Specifier%20of%20Conduct%20Disorder.pdf. Accessed July 23, 2010.

Frick, P. J., Obrien, B. S., Wootton, J. M., & Mcburnett, K. (1994). Psychopathy and conduct problems in children. *Journal of Abnormal Psychology, 103*(4), 700–707.

Gallagher, N. (2003). Effects of Parent-Child Interaction Therapy on young children with disruptive behavior disorders. *Bridges, 1*(4), 1–17.

Gardner, F., Hutchings, J., Bywater, T., & Whitaker, C. (2010). Who benefits and how does it work? Moderators and mediators of outcome in an effectiveness trial of a parenting intervention. *Journal of Clinical Child & Adolescent Psychology, 39*, 568–280.

Gershoff, E. T. (2002). Corporal punishment by parents and associated child behaviors and experiences: A meta-analytic and theoretical review. *Psychological Bulletin, 128*(4), 539–579.

Haas, S. M., Waschbusch, D. A., Pelham, W. E., King, S., Andrade, B. F., & Carrey, N. J. (2011). Treatment response in CP/ADHD children with callous/unemotional traits. *Journal of Abnormal Child Psychology, 39*(4), 541–552.

Hare, R. D. (1991). *The Hare Psychopathy Checklist Revised*. Toronto, Ontario, Canada: Multi-Health Systems.

Hare, R. D. (2003). *Hare Psychopathy Checklist-Revised (PCL-R)* (2nd ed.). North Toawanda, NY: Multi-Health Systems.

Hare, R. D., Hart, S. D., & Harpur, T. J. (1991). Psychopathy and the DSM-IV criteria for antisocial personality disorder. *Journal of Abnormal Psychology, 100*, 391–398.

Hastings, P. D., Zahn-Waxler, C., Robinson, J., Usher, B., & Bridges, D. (2000). The development of concern for others in children with behavior problems. *Developmental Psychology, 36*(5), 531–546.

Hawes, D. J., & Dadds, M. R. (2007). Stability and malleability of callous-unemotional traits during treatment for childhood conduct problems. *Journal of Clinical Child and Adolescent Psychology, 36*(3), 347–355.

Herschell, A. D., Calzada, E. J., Eyberg, S. M., & McNeil, C. B. (2002). Parent-Child Interaction Therapy: New directions in research. *Cognitive and Behavioral Practice, 9*, 9–16.

Hipwell, A. E., Pardini, D. A., Loeber, R., Sembower, M., Keenan, K., & Stouthamer-Loeber, M. (2007). Callous-unemotional behaviors in young girls: Shared and unique effects relative to conduct problems. *Journal of Clinical Child and Adolescent Psychology, 36*(3), 293–304.

Hood, K. K., & Eyberg, S. M. (2003). Outcomes of parent-child interaction therapy: Mothers' reports of maintenance three to six years after treatment. *Journal of Clinical and Child and Adolescent Psychology, 32*(3), 419–429.

Johnstone, L., & Cooke, D. J. (2004). Psychopathic-like traits in childhood: Conceptual and measurement concerns. *Behavioral Sciences and the Law, 22*(1), 103–125.

Jones, A. P., Happe, F. G. E., Gilbert, F., Burnett, S., & Viding, E. (2010). Feeling, caring, knowing: different types of empathy deficit in boys with psychopathic tendencies and autism spectrum disorder. *Journal of Child Psychology and Psychiatry, 51*(11), 1188–1197.

Jones, A. P., Laurens, K. R., Herba, C. M., Barker, G. J., & Viding, E. (2009). Amygdala hypoactivity to fearful faces in boys with conduct problems and callous-unemotional traits. *American Journal of Psychiatry, 166*(1), 95–102.

Jones, S., Cauffman, E., Miller, J. D., & Mulvey, E. (2006). Investigating different factor structures of the psychopathy checklist: Youth version: Confirmatory factor analytic findings. *Psychological Assessment, 18*(1), 33–48.

Kahn, R. E., Frick, P. J., Youngstrom, E., Findling, R. L., & Youngstrom, J. K. (2012). The effects of including a callous-unemotional specifier for the diagnosis of conduct disorder. *Journal of Child Psychology and Psychiatry, 53*, 271–282.

Keenan, K., Wroblewski, K., Hipwell, A., Loeber, R., & Stouthamer-Loeber, M. (2010). Age of onset, symptom threshold, and expansion of the nosology of conduct disorder for girls. *Journal of Abnormal Psychology, 119*(4), 689–698.

Kiang, L., Moreno, A. J., & Robinson, J. L. (2004). Maternal preconceptions about parenting predict child temperament, maternal sensitivity, and children's empathy. *Developmental Psychology, 40*(6), 1081–1092.

Kochanska, G. (1991). Socialization and temperament in the development of guilt and conscience. *Child Development, 62*, 1379–1392.

Kochanska, G. (1993). Toward a synthesis of parental socialization and child temperament in early development of conscience. *Child Development, 64*(2), 325–347.

Kochanska, G. (1995). Children's temperament, mother's discipline, and security of attachment: Multiple pathways to emerging internalization. *Child Development, 66*(3), 597–615.

Kochanska, G. (1997). Multiple pathways to conscience for children with different temperaments: From toddlerhood to age 5. *Developmental Psychology, 33*(2), 228–240.

Kochanska, G., & Aksan, N. (2006). Children's conscience and self-regulation. *Journal of Personality and Social Psychology, 74*(6), 1587–1617.

Kochanska, G., Forman, D. R., Aksan, N., & Dunbar, S. B. (2005). Pathways to conscience: Early mother-child mutually responsive orientation and children's moral emotion, conduct, and cognition. *Journal of Child Psychology and Psychiatry, 46*(1), 19–34.

Kochanska, G., Gross, J. N., Lin, M., & Nichols, K. E. (2002). Guilt in young children: Development, determinants, and relations with a broader system of standards. *Child Development, 73*, 461–482.

Kolko, D. J., Dorn, L. D., Bukstein, O. G., Pardini, D., Holden, E. A., & Hart, J. (2009). Community vs. clinic-based modular treatment of children with early-onset ODD or CD: A clinical trial with 3-year follow-up. *Journal of Abnormal Child Psychology, 37*(5), 591–609.

Kolko, D. J., & Pardini, D. A. (2010). ODD dimensions, ADHD, and callous-unemotional traits as predictors of treatment response in children with disruptive behavior disorders. *Journal of Abnormal Psychology, 119*(4), 713–725.

Kotler, J. S., & McMahon, R. J. (2005). Child psychopathy: theories, measurement, and relations with the development and persistence of conduct problems. *Clinical Child and Family Psychology Review, 8*(4), 291–325.

Kroneman, L. M., Hipwell, A. E., Loeber, R., Koot, H. M., & Pardini, D. A. (2011). Contextual risk factors as predictors of disruptive behavior disorder trajectories in girls: The moderating effect of callous-unemotional features. *Journal of Child Psychology and Psychiatry, 52*(2), 167–175.

Lau, K. S. L., Marsee, M. A., Kunimatsu, M. M., & Fassnacht, G. M. (2011). Examining associations between narcissism, behavior problems, and anxiety in non-referred adolescents. *Child & Youth Care Forum, 40*(3), 163–176.

Lee, Z., Vincent, G. M., Hart, S. D., & Corrado, R. R. (2003). The validity of the Antisocial Process Screening Device as a self-report measure of psychopathy in adolescent offenders. *Behavioral Sciences and the Law, 21*(6), 771–786.

Loeber, R., Pardini, D., Homish, D. L., Wei, E. H., Crawford, A. M., Farrington, D. P., et al. (2005). The prediction of violence and homicide in young men. *Journal of Consulting and Clinical Psychology, 73*(6), 1074–1088.

Loeber, R., Pardini, D. A., Hipwell, A., Stouthamer-Loeber, M., Keenan, K., & Sembower, M. A. (2009). Are there stable factors in preadolescent girls' externalizing behaviors? *Journal of Abnormal Child Psychology, 37*(6), 777–791.

Loeber, R., Pardini, D. A., Stouthamer-Loeber, M., & Raine, A. (2007). Do cognitive, physiological, and psychosocial risk and promotive factors predict desistance from delinquency in males? *Development and Psychopathology, 19*(3), 867–887.

Lynam, D. R. (1997). Pursuing the psychopath: capturing the fledgling psychopath in a nomological net. *Journal of Abnormal Psychology, 106*(3), 425–438.

Lynam, D. R., Caspi, A., Moffitt, T. E., Loeber, R., & Stouthamer-Loeber, M. (2007). Longitudinal evidence that psychopathy scores in early adolescence predict adult psychopathy. *Journal of Abnormal Psychology, 116*(1), 155–165.

Lynam, D. R., Caspi, A., Moffitt, T. E., Raine, A., Loeber, R., & Stouthamer-Loeber, M. (2005). Adolescent psychopathy and the big five: Results from two samples. *Journal of Abnormal Child Psychology, 33*(4), 431–443.

Marsh, A. A., & Blair, R. J. R. (2008). Deficits in facial affect recognition among antisocial populations:

A meta-analysis. *Neuroscience and Biobehavioral Reviews, 32*(3), 454–465.

Marsh, A. A., Finger, E. C., Mitchell, D. G., Reid, M. E., Sims, C., Kosson, D. S., et al. (2008). Reduced amygdala response to fearful expressions in children and adolescents with callous-unemotional traits and disruptive behavior disorders. *American Journal of Psychiatry, 165*(6), 712–720.

McDonald, R., Dodson, M. C., Rosenfield, D., & Jouriles, E. N. (2011). Effects of a parenting intervention on features of psychopathy in children. *Journal of Abnormal Child Psychology, 39*, 1013–1023.

McMahon, R. J., Witkiewitz, K., Kotler, J. S., & Conduct Problems Prevention Research Group. (2010). Predictive validity of callous-unemotional traits measured in early adolescence with respect to multiple antisocial outcomes. *Journal of Abnormal Psychology, 119*(4), 752–763.

Munoz, L. C. (2009). Callous-unemotional traits are related to combined deficits in recognizing afraid faces and body poses. *Journal of the American Academy of Child and Adolescent Psychiatry, 48*(5), 554–562.

Munoz, L. C., & Frick, P. J. (2007). The reliability, stability, and predictive utility of the self-report version of the Antisocial Process Screening Device. *Scandinavian Journal of Psychology, 48*(4), 299–312.

Murrie, D. C., Marcus, D. K., Douglas, K. S., Lee, Z., Salekin, R. T., & Vincent, G. (2007). Youth with psychopathy features are not a discrete class: A taxometric analysis. *Journal of Child Psychology and Psychiatry, 48*(7), 714–723.

Nock, M. K. (2003). Progress review of the psychological treatment of child conduct problems. *Clinical Psychology: Science and Practice, 10*(1), 1–28.

Obradović, J., Pardini, D. A., Long, J. D., & Loeber, R. (2007). Measuring interpersonal callousness in boys from childhood to adolescence: An examination of longitudinal invariance and temporal stability. *Journal of Clinical Child and Adolescent Psychology, 36*(3), 276–292.

O'Brien, B. S., & Frick, P. J. (1996). Reward dominance: Associations with anxiety, conduct problems, and psychopathy in children. *Journal of Abnormal Child Psychology, 24*(2), 223–240.

Otway, L. J., & Vignoles, V. L. (2006). Narcissism and childhood recollections: A quantitative test of psychoanalytic predictions. *Personality and Social Psychology Bulletin, 32*(1), 104–116.

Pardini, D., Obradović, J., & Loeber, R. (2006). Interpersonal callousness, hyperactivity/impulsivity, inattention, and conduct problems as precursors to delinquency persistence in boys: A comparison of three grade-based cohorts. *Journal of Clinical Child and Adolescent Psychology, 35*(1), 46–59.

Pardini, D. A. (2006). The callousness pathway to severe violent delinquency. *Aggressive Behavior, 32*(6), 590–598.

Pardini, D. A. (2008). Empirically supported treatments for conduct disorders in children and adolescents. In *Best practices in the behavioral management of health from preconception to adolescence* (pp. 290–321): Los Altos, CA: Institute for Brain Potential.

Pardini, D. A., & Byrd, A. L. (2012). Perceptions of aggressive conflicts and others' distress in children with callous-unemotional traits: "I'll show you who's boss, even if you suffer and I get in trouble." *Journal of Child Psychology and Psychiatry, 53*, 283–291.

Pardini, D. A., & Fite, P. J. (2010). Symptoms of conduct disorder, oppositional defiant disorder, attention-deficit/hyperactivity disorder, and callous-unemotional traits as unique predictors of psychosocial maladjustment in boys: Advancing an evidence base for DSMV. *Journal of the American Academy of Child and Adolescent Psychiatry, 49*(11), 1134–1144.

Pardini, D. A., Frick, P. J., & Moffitt, T. E. (2010). Building an evidence base for DSM-5 conceptualizations of oppositional defiant disorder and conduct disorder: Introduction to the special section. *Journal of Abnormal Psychology, 119*(4), 683–688.

Pardini, D. A., Lochman, J. E., & Frick, P. J. (2003). Callous/unemotional traits and social-cognitive processes in adjudicated youths. *Journal of the American Academy of Child and Adolescent Psychiatry, 42*(3), 364–371.

Pardini, D. A., Lochman, J. E., & Powell, N. (2007). The development of callous-unemotional traits and antisocial behavior in children: Are there shared and/or unique predictors? *Journal of Clinical Child and Adolescent Psychology, 36*(3), 319–333.

Pardini, D. A., & Loeber, R. (2007). Interpersonal and affective features of psychopathy in children and adolescents: Advancing a developmental perspective introduction to special section. *Journal of Clinical Child and Adolescent Psychology, 36*(3), 269–275.

Pardini, D. A., Stephanie, S., Hipwell, A., Stouthamer-Loeber, M., & Loeber, R. (2012).

The clinical utility of the proposed DSM-5 callous-unemotional subtype of conduct disorder in young girls. *Journal of the American Academy of Child and Adolescent Psychiatry, 51,* 62–73.

Pasalich, D. S., Dadds, M. R., Hawes, D. J., & Brennan, J. (2011). Do callous-unemotional traits moderate the relative importance of parental coercion versus warmth in child conduct problems? An observational study. *Journal of Child Psychology and Psychiatry, 52,* 1308–1315.

Poythress, N. G., Dembo, R., Wareham, J., & Greenbaum, P. E. (2006). Construct validity of the youth psychopathic traits inventory (YPI) and the antisocial process screening device (APSD) with justice-involved adolescents. *Criminal Justice and Behavior, 33*(1), 26–55.

Raskin, R., Novacek, J., & Hogan, R. (1991). Narcissism, self-esteem, and defensive self-enhancement. *Journal of Personality, 59*(1), 19–38.

Reid, M. J., Webster-Stratton, C., & Hammond, M. (2003). Follow-up of children who received the incredible years intervention for oppositional-defiant disorder: Maintenance and prediction of 2-year outcome. *Behavior Therapy, 34*(4), 471–491.

Rothbart, M. K., Ahadi, S. A., & Hershey, K. L. (1994). Temperament and social behavior in childhood. *Merrill-Palmer Quarterly, 40*(1), 21–39.

Rutherford, M., Cacciola, J. S., Alterman, A. I., McKay, J. R., & Cook, T. G. (1999). The 2-year test-retest reliability of the Psychopathy Checklist Revised in methadone patients. *Assessment, 6*(3), 285–292.

Salekin, R. T., Neumann, C. S., Leistico, A. M. R., DiCicco, T. M., & Duros, R. L. (2004). Psychopathy and comorbidity in a young offender sample: Taking a closer look at psychopathy's potential importance over disruptive behavior disorders. *Journal of Abnormal Psychology, 113*(3), 416–427.

Seagrave, D., & Grisso, T. (2002). Adolescent development and the measurement of juvenile psychopathy. *Law and Human Behavior, 26*(2), 219–239.

Talwar, V., Gordon, H. M., & Lee, K. (2007). Lying in the elementary school years: Verbal deception and its relation to second-order belief understanding. *Developmental Psychology, 43*(3), 804–810.

Talwar, V., & Lee, K. (2002). Development of lying to conceal a transgression: Children's control of expressive behavior during verbal deception. *International Journal of Behavioral Development, 26*(5), 436–444.

Talwar, V., Lee, K., Bala, N., & Lindsay, R. C. L. (2002). Children's conceptual knowledge of lying and its relation to their actual behaviors: Implications for court competence examinations. *Law and Human Behavior, 26*(4), 395–415.

Thomaes, S., Bushman, B. J., De Castro, B. O., & Stegge, H. (2009). What makes narcissists bloom? A framework for research on the etiology and development of narcissism. *Development and Psychopathology, 21*(4), 1233–1247.

Thomaes, S., Bushman, B. J., Stegge, H., & Olthof, T. (2008). Trumping shame by blasts of noise: Narcissism, self-esteem, shame, and aggression in young adolescents. *Child Development, 79*(6), 1792–1801.

Thomaes, S., Stegge, H., Bushman, B. J., Olthof, T., & Denissen, J. (2008). Development and validation of the childhood narcissism scale. *Journal of Personality Assessment, 90*(4), 382–391.

Vasey, M. W., Kotov, R., Frick, P. J., & Loney, B. R. (2005). The latent structure of psychopathy in youth: A taxometric investigation. *Journal of Abnormal Child Psychology, 33*(4), 411–429.

Vernon, P. A., Villani, V. C., Vickers, L. C., & Harris, J. A. (2008). A behavioral genetic investigation of the Dark Triad and the Big 5. *Personality and Individual Differences, 44*(2), 445–452.

Viding, E., Fontaine, N. M., Oliver, B. R., & Plomin, R. (2009). Negative parental discipline, conduct problems and callous-unemotional traits: monozygotic twin differences study. *British Journal of Psychiatry, 195*(5), 414–419.

Washburn, J. J., McMahon, S. D., King, C. A., Reinecke, M. A., & Silver, C. (2004). Narcissistic features in young adolescents: Relations to aggression and internalizing symptoms. *Journal of Youth and Adolescence, 33*(3), 247–260.

Webster-Stratton, C., & Hammond, M. (1997). Treating children with early-onset conduct problems: A comparison of child and parent training interventions. *Journal of Consulting and Clinical Psychology, 65*(1), 93–109.

Webster-Stratton, C., Reid, M. J., & Hammond, M. (2004). Treating children with early-onset conduct problems: Intervention outcomes for parent, child, and teacher training. *Journal of Clinical Child and Adolescent Psychology, 33*(1), 105–124.

Willoughby, M. T., Waschbusch, D. A., Moore, G. A., & Propper, C. B. (2011). Using the ASEBA to screen for callous unemotional traits in early childhood: Factor structure, temporal stability, and utility. *Journal of Psychopathology and Behavioral Assessment, 33*(1), 19–30.

Zahn-Waxler, C., & Kochanska, G. (1990). The origins of guilt. In: R. Thompson (Ed.), *Nebraska symposium on motivation* (Vol. 36, pp. 183–258). Lincoln, NE: University of Nebraska Press.

CHAPTER 5

Adolescent Psychopathy and the Law

Michael J. Vitacco and Randall T. Salekin

The application of psychopathy to legal proceedings involving both adults and adolescents has recently garnered greater attention (Vitacco, Lishner, & Neumann, 2012). This attention is exemplified by recent programming at the Society for the Scientific Study of Psychopathy, which hosted a special series of invited lectures on neuroscientific issues related to psychopathy, as well as various special issues in psychology and law journals focusing on the nature of psychopathic traits and their relationship to legal issues. Beyond academic endeavors, National Public Radio (NPR) hosted a three-part series entitled Neuroscience and Responsibility in which psychopathy was prominently featured. In 2011, NPR hosted a special series on the Psychopathy Checklist-Revised (PCL-R, Hare, 2003) and discussed with several prominent scholars various problems that can accompany labeling an individual as a "psychopath." Concerns regarding psychopathy are prominent in adolescents. The application of the construct of psychopathy to youth has been long considered controversial due to concerns about developmental inappropriateness (Seagrave & Grisso, 2002; Vincent & Grisso, 2005) and questions concerning the clinical utility of instruments designed to measure the construct in adolescents (Hart, Watt, & Vincent, 2002).

Yet, the assessment of psychopathy continues to be liberally used in pyscholegal evaluations involving adolescents. Its usefulness is firmly rooted in the prediction of violence and its potential incremental validity over other indicators of dysfunctional behavior (Salekin, Leistico, Neumann, DiCicco, & Duros, 2004). Juvenile forensic evaluators routinely use psychopathy assessments as part of assessing violence risk. A recent survey found 79% of clinicians involved in juvenile risk evaluations have used assessments of psychopathy at some point in their clinical practice to inform courts on violence risk. However, the vast majority of clinicians reported the use of psychopathy as a diagnostic label was inappropriate (Viljoen, McLachlan, & Vincent, 2010). This literature informs us that assessments of adolescent psychopathy are figuring prominently in risk assessment evaluations. It is imperative that mental health practitioners who assess adolescents with measures of psychopathy do so in a manner that is consistent with the highest ethical standards and in line with empirically based evidence.

COURT PROCEEDINGS AND PSYCHOPATHY

This chapter aims to unpack issues related to adolescent psychopathy and the law and provide a basis for both the appropriate and inappropriate assessment of psychopathy with this population. Although some clinicians and scholars have posited that psychopathy should never be assessed in adolescents, the position now appears more untenable given recent movement in the fields of forensic psychology and psychiatry. One way of understanding the clinical–legal applications of psychopathy is to review articles summarizing the use of psychopathy in legal proceedings.

Two articles have been especially informative regarding uses of psychopathy in the courtroom. Walsh and Walsh (2006) reported on nine different types of cases in which the Psychopathy Checklist (PCL, Hare, 1991) and the Psychopathy Checklist-Revised (Hare, 2003) were admitted in judicial proceedings. These cases encompassed the gamut of psycholegal issues and included civil commitment for individuals convicted of sex offenses, sentencing, termination of parental rights, and competency to stand trial, to name a few. DeMatteo and Edens (2006) discussed the use and relevance of the PCL-R in legal proceedings and focused on three separate examples outlining potential misuses of psychopathy in court. These included sexually violent persons commitment (e.g., civil commitment for individuals convicted of a sex offense), capital sentencing (e.g., used to support a death penalty sentence), and the insanity defense. Such issues underscore the difficulty with using psychopathy to assist the trier of fact in rendering an appropriate decision.

The use of psychopathy with adolescent offenders is more contentious than with adults. Again, understanding how psychopathy is applied in these specialized cases is enhanced by reviewing how psychopathy has been applied in legal proceedings with adolescents. Viljoen, MacDougall, Gagnon, and Douglas (2010) evaluated the use of psychopathy in adolescent cases throughout the United States and Canada. The results of this analysis are informative. Review of 111 cases (71 Canadian and 40 American) found the legal system has witnessed a dramatic increase in the number of cases in which psychopathy evidence was introduced over the last 70 years, with the greatest increase occurring over the last 20 years. Overall, there were 10 types of court cases in which psychopathy was introduced into evidence, with the majority of cases focusing on some issue related to violence risk assessment. However, there were some atypical uses related to cases involving competency to stand trial and criminal responsibility (e.g., not guilty by reason of insanity) in which the use of psychopathy appears more dubious and likely inappropriate.

The review by Viljoen et al. (2010) refocused attention on salient questions regarding adolescent psychopathy and (1) risk assessment; (2) diagnostic labeling; (3) issues related to the treatment of psychopathy, especially in adolescents; and (4) inappropriate uses of psychopathy in the courtroom. The next three sections of this chapter will confront each of these issues by providing recent empirical evidence and attempting to assist clinicians who decide that a psychopathy assessment is warranted. In each section, we provide a list of recommendations related to the topic, with the primary goal to develop recommendations that are empirically informed and judiciously designed, and with full recognition of the controversy that stems from the downward extension of psychopathy to adolescents.

ADOLESCENT PSYCHOPATHY AND RISK ASSESSMENT

Like psychopathy's use in court proceedings with adolescents, there has been a noticeable increase in the number of instruments designed to assess the construct of psychopathy with adolescents. Vaughn and Howard (2005) conducted a review and found 30 articles related to the assessment of psychopathy in youth, with nine separate measures of adolescent psychopathy. It is clear that forensic clinicians have an armamentarium of measures at their disposal to evaluate psychopathic traits in youth. However, forensic clinicians should not blindly reach and pick a measure believing they can get a reliable and accurate assessment of psychopathic traits; instead, the reliability and validity of a measure must be preeminent when clinicians decide what instrument they are going to use. A basic question must be, "How will this instrument hold under intense judicial scrutiny?" Or, conversely, "Does the use of this instrument assist the trier of fact and appropriately answer the referral question?" The evaluator, when making such decisions, should heavily rely on published research, theses and dissertations, and technical manuals. In having a firm grounding in empirical data, the evaluator can make the most appropriate decision on which instrument of psychopathy is most suited toward the particular referral question.

As discussed in the Viljoen et al. (2010) review, the most frequent referral question in which psychopathy has a good deal of utility is within the field of violence risk assessment.

We recognize that some scholars have advocated for the abolition of the use of psychopathy with adolescent populations. However, this view fails to acknowledge that forensic clinicians are increasingly being called upon to make predictions of adolescent violent behavior (Borum & Verhaagen, 2006) or general dangerousness and that psychopathy, at least in the short term, has demonstrated predictive validity for the task of making violence predictions. Forth and Book (2010) described how psychopathic traits have been moderately strong predictors of various forms of antisocial behavior in youth (see also Edens, Skeem, Cruise, & Cauffman, 2001). Forth and Book (2010) stated, "Research clearly demonstrates that psychopathic traits in children predict later antisocial behavior and aggression. Nevertheless, several aspects of this relationship require clarification" (p. 274). These scholars identified a need for further research on females and ethnic minorities, as well as the need for a greater number of prospective studies. The concerns echoed by Forth and Book are valid and demonstrate that much more research needs to occur before the field can express confidence in psychopathy's relationship to adolescent violence. Yet, overall research has indicated support for the relation between the presence of psychopathic traits and violence.

In research using the Psychopathy Checklist: Youth Version (Forth, Kosson, & Hare, 2003), multiple studies have found positive associations between psychopathic traits and violence. Corrado, Vincent, Hart, and Cohen (2004) followed 182 male adolescent offenders for an average of 14.5 months post release. Their results indicated the PCL:YV significantly predicted general and violent recidivism, with items of behavioral dysfunction accounting for more of the predictive power. Penney and Moretti (2007) found that PCL:YV scores predicted aggression in delinquent boys and girls. Gretton, Hare, and Catchpole (2004) employed the PCL:YV in a study of 157 males. Their results indicated the PCL:YV was associated with violent recidivism. Notably, the PCL:YV demonstrated incremental validity over other indices (e.g., early behavioral problems). Salekin et al. (2004) reported findings consistent with those of Penney and Moretti (2007) and Gretton et al. (2004) by reporting that psychopathy scores possessed incremental validity over symptoms of conduct disorder in predicting violent behavior. Salekin (2008), in a later study, reported the PCL:YV and a self-report instrument of psychopathy demonstrated incremental validity over 14 variables with empirical links to antisocial behavior.

Psychopathy as a predictor of antisocial behavior in adolescents is becoming increasingly established, but other factors warrant consideration in clinical and forensic work. One key consideration is how callous traits, commonly referred to as callous/unemotional (CU) traits, a key facet in psychopathy, are associated with violence and risk. Frick and White (2008) emphasized the importance of CU traits in understanding youth violence and risk. Adolescents with CU traits are more violent, began offending earlier, and have a greater number of police contacts (see also White & Frick, 2010). It has been hypothesized that CU traits have a distinct etiology from other facets of psychopathy (Viding, Jones, Frick, Moffitt, & Plomin, 2008). Given the strong evidence related to CU traits and violence, clinicians should pay special attention to the presence or absence of CU traits in their risk assessments, especially when these traits are combined with the impulsive and antisocial facets of psychopathy.

Echoing the concerns of Forth and Book (2010) regarding the need for more research with diverse populations (i.e., ethnic minorities and females), Edens and Cahill (2007) attributed their non-significant findings regarding the relationship between psychopathy and violence to the fact that their sample was ethnically diverse. This underexplored area warrants additional research, but lack of findings limits the generalizability of psychopathy assessments in ethnically diverse samples. The use of any measure of psychopathy with adolescent females has not been sufficiently established. In fact, extant research questions the utility that psychopathy has high clinical utility with adolescent females (Pajer, 1998). Practitioners should consider alternative ideas of violence when evaluating female adolescents. For instance, it may be more beneficial to evaluate relational or interpersonal aggression in females. Marsee, Silverthorn, and Frick (2005) found psychopathy, as measured by the Antisocial Process Screening Device (APSD; Frick & Hare, 2001), was predictive of relational aggression in adolescent females. In a more traditional study of violence, Odgers, Reppucci, and Moretti (2005) found limited predictive power when using the PCL:YV to predict violence and recidivism in a sample of 125 females (see also Schmidt, McKinnon, Chattha, & Brownlee (2006). Schrum and Salekin (2006) conducted an item response theory (IRT) analysis on the PCL:YV with 123 female adolescents. IRT methodology allows for individual item analysis for items on a particular measure. Although the authors expressed some confidence in their overall findings regarding the properties of the PCL:YV with female adolescents, their results also indicated personality-based items provided greater information than traditional Factor 2 items (lifestyle impulsivity and antisocial behavior).

Related to violence risk assessment are evaluations designed to assess the appropriateness of transferring a juvenile to adult court. These evaluations are becoming increasingly common and have remained highly controversial. Such evaluations typically focus on three concepts including (1) risk for dangerousness, (2) sophistication-maturity, and (3) amenability to treatment (see Salekin & Grimes, 2008). The evaluation of psychopathy is frequently used in transfer evaluations; however, psychopathy is also frequently misused in transfer evaluations (Marczyk, Heilbrun, Lander, & DeMatteo, 2005; Salekin & Debus, 2008). In reviewing the three criteria, it is clear that psychopathy is related only to risk of dangerousness, and as discussed later in this chapter, has less to do with the other two constructs associated with juvenile transfer. Other psychological instruments, such as the Risk-Sophistication-Treatment Inventory (RST-I, 2004) and its self-report version (RSTI-SR), are more suited to measure the three prongs of juvenile transfer decisions as well as to provide general information regarding juvenile disposition (Salekin, Salekin, Clements, & Leistico, 2005). Although the RST-I considers psychopathic features as part of its risk scale, it is only one component of risk and is somewhat independent of the maturity and amenability scales. This is necessary as judges appear to view treatment amenability and maturity as more salient to waiver decisions (Brannen et al., 2006). As such, it appears that psychopathy should play only a limited role in transfer evaluations.

Recommendations

In striking a balance between acknowledging both appropriate uses and limitations of employing psychopathy in violence risk assessments with adolescents, we provide a series of recommendations based on the empirical literature.

(1) At this time, violence risk for legal proceedings should be done with the PCL:YV. There are several methods available for evaluating psychopathic traits in adolescents; however, none of them have appropriate empirical support to be used in court. However, other types of self-report and informant-report measures can be used to buttress the findings of the PCL:YV and to ensure adequate coverage of relevant domains.

(2) Statements of risk must be considered in light of the limitations of psychopathy assessments. For example, even high scores on the PCY:YV do not mean that the adolescent will be violent. Instead, a high score on the PCL:YV places the adolescent at a higher risk for violence over a relatively short time period. Likewise, lower scores generally indicate lower risk and should be reported as such.

(3) Clinicians, when conducting risk assessments, should pay attention to evaluating for the presence or absence of CU traits. The presence of CU traits confers a higher risk, especially when these traits are combined with the impulsive and antisocial facets of psychopathy. As discussed in the first recommendation, evaluators should review multiple sources of data across multiple contexts.

(4) Using the PCL:YV to assess violence in females has not been empirically supported.

Studies of psychopathy with adolescent females have found some support for predicting relational aggression but limited support for traditional violence risk assessment.

(5) We are concerned about the use of psychopathy measurements outside of the context of risk assessment, especially for other psycholegal questions like competency to stand trial or criminal responsibility. As Maslow stated, "when your only tool is a hammer, every problem looks like a nail." Forensic clinicians are advised to understand that most psycholegal questions outside of risk assessment should not involve the evaluation of psychopathic traits.

(6) Clinicians using psychopathy to evaluate adolescents must maintain up to date scientific knowledge through constant review of the literature. In addition, clinicians must be aware of developmental issues and their potential impact on psychopathy and risk in adolescents. Specialized training is required to conduct risk assessments with adolescents.

USING THE LABEL "PSYCHOPATHY" IN THE COURTROOM: A CALL FOR CAUTION

The Viljoen et al. (2010) article identified the use of the diagnostic term *psychopathy* with adolescents as a primary area of concern. There are manifold reasons for concern over the use of diagnostic labeling. First, psychopathy has come to connote a static, life-long condition that is not amenable to treatment or intervention. Second, these negative connotations are prevalent to such a degree that any benefit to the trier of fact is likely to be overshadowed by the prejudicial effect of reporting the adolescent as a psychopath. A prime example is found in popular literature. Kellerman (1999), a fiction writer, wrote a self-described nonfiction book entitled, "Savage Spawn: Reflections of Violent Children" in which he evoked the term psychopathy to describe some violent youth. He wrote "Sometimes a psychopathic child's cruelty tops off at the level of school yard bullying. But often it doesn't, because domination, like any narcotic, breeds satiation and habituation. When first shoving, then hurting, and then raping cease to provide a sufficiently potent thrill, the game can swell, peaking at the ultimate control scheme" (p. 23). This type of rhetoric has the potential to do a great disservice to adolescents who are in need of treatment. There were concerns that "superpredators" would develop and the juvenile justice system would be ill equipped to handle them. Vitacco and Vincent (2006) discussed how negative connotations unduly skew perceptions of adolescents with behavioral problems potentially leading to unnecessary barriers to treatment and potentially placing some adolescents in untenable situations.

Studies (e.g., Cox, DeMatteo, & Foster, 2010; Lloyd, Clark, & Forth, 2010) indicated that introducing psychopathy into adult proceedings increases the likelihood of more severe sentences, including and up to the death penalty. Forensic practitioners should be aware of the potential problems associated with the labeling of an adolescents as "psychopathic." Several empirical studies have been completed that inform the field on issues related to concerns about labeling adolescent as psychopaths. Prior to the abolition of the death penalty with adolescents, Edens, Guy, and Fernandez (2003) found individuals reviewing case material were more likely to recommend the death penalty to a juvenile diagnosed as a "psychopath." However, the label of psychopath is not the only diagnosis for which there is potential for prejudice. When reviewing data collected from a sample of 260 probation officers, Murrie, Cornell, and Mccoy (2005) discovered diagnosing an adolescent with antisocial personality disorder had a more negative impact than a diagnosis of psychopathy. In looking at the responses of 891 potential venirepersons, Boccaccini, Murrie, Clark, and Cornell (2008) reported that descriptions of antisocial behavior and descriptions of the traits were more impactful than formal diagnoses on opinions of risk. Not surprisingly, judges are also influenced by the diagnostic labels found in reports. Murrie and colleagues (Murrie, Boccaccini, McCoy, & Cornell, 2007) found judges were more influenced by the descriptions of antisocial behavior than by a formal diagnosis of psychopathy. In a sample of 100 judges, Jones and Cauffman (2008) found psychopathy predicted judicial findings of dangerousness and less treatment amenability. Likewise, clinicians were more likely to judge an adolescent as high risk for violence if the adolescent had a history of antisocial behavior or was labeled as a psychopath (Rockett, Murrie, & Boccaccini, 2007).

The discussion of diagnostic labels is a critical one for the forensic practitioner who conducts risk assessment evaluations with youth. This debate places the forensic clinician in a clear quandary with a potential for conflict. On one hand, the practitioner is charged with providing the court with full information, often related to the functioning of the adolescent, including his or her antisocial behavior and future risk. On the other hand, the clinician has an ethical responsibility to ensure his or her reports are used appropriately. Attempting to balance these roles, we have compiled the following recommendations.

Recommendations

(1) Forensic evaluators should refrain from referring to or "diagnosing" a youth as a "psychopath" or "psychopathic."
(2) Evaluators should discuss the limitations of their measurements and specific developmental limitations (e.g., questions regarding temporal stability) when applying psychopathy to youth. By acknowledging limitations regarding the construct of youth psychopathy, practitioners can submit balanced reports to the court.
(3) Forensic evaluators must be prepared to educate the trier of fact and other decision makers about the nature of psychopathy in youth. By extension, this implies the clinician must have a well-developed understanding of psychopathy and its application to youth.

TREATMENT OF PSYCHOPATHIC TRAITS WITH YOUTHFUL OFFENDERS

There has been a previous trend to dismiss the presence of psychopathic traits in youth as immutable, meaning that their presence in the adolescent signaled a lifelong pathway of antisocial behavior or even violence. Yet, there has been tremendous movement away from this idea, most of it demonstrating the treatment of youth psychopathy occurring over the previous 10 years. A meta-analysis on treatment of psychopathy that included analyses on antisocial behavior in children and adolescents by Salekin (2002) provided optimism that treatment can have a positive effect on antisocial youth with psychopathic traits. In a later analysis, Salekin, Worley, and Grimes (2010) evaluated five studies of treatment relating to adolescent psychopathy, finding reported treatment gains and treatment benefit in three of the five studies. It is no longer appropriate to dismiss the treatability of even the most severe antisocial youth, even those with psychopathic traits.

The treatability of psychopathy in youth has been exemplified by the findings of Caldwell and his colleagues (Caldwell, McCormick, Umstead, & Van Rybroek, 2007; Caldwell, Skeem, Salekin, & Van Rybroek, 2006; Caldwell, Vitacco, & Van Rybroek, 2006) at the Mendota Juvenile Treatment Center (MJTC), a maximum-security hybrid (correctional and treatment facility) located in Madison, Wisconsin. MJTC has been involved in several ecologically valid treatment studies of adolescent psychopathy over the last several years. We refer to these studies as ecologically valid because they occur in a correctional-treatment center with adolescents whose average score on the PCL:YV is greater than 30 (i.e., the traditional cut-score for psychopathy with PCL-based measures). Beyond psychopathy, these adolescents were incarcerated on the basis of their violent behavior, with the average adolescent having been incarcerated for multiple violent offenses (Vitacco, Caldwell, Van Rybroek, & Gabel, 2007).

One issue often overlooked in the literature is the idea that psychopathic traits, even in adolescents, do not respond to traditional offender programming (O'Neill, Lidz, & Heilbrun, 2003; Spain, Douglas, Poythress, & Epstein, 2004). This is essential because correctional facilities often serve as the gateway for the antisocial adolescent to begin to receive intensive treatment. However, when specialized treatments are employed, the results are generally positive in regard to the treatment of youth psychopathy. For example, Caldwell, Skeem, et al. (2006) compared specialized and standard treatments to comparable offender groups at a secure detention facility. Adolescent offenders receiving the treatment specifically designed to reduce antisocial behavior evidenced fewer violent acts after release and recidivated at a slower rate. Caldwell et al. (2007) found that PCL:YV scores were unrelated to treatment success or recidivism upon discharge; however, behaviors within the institution were predictive of later success or failure. In other words, those adolescents who appeared to be engaged in treatment and demonstrated improved behavior within the institution performed best after release. A key consideration when implementing long-term treatment is the considerable cost often associated with such treatments. However, as shown by Caldwell, Vitacco, and VanRybroek (2006), intensive treatment that decreases violence and recidivism is cost effective compared to the long-term costs associated with incarceration. Cost savings did not include the additional money associated with victims of future violence, which only add to the overall costs associated with violence.

A final thought related to treatment, but with clear implications to risk assessment, deals with protective factors. Salekin, Lee, Schrum, and Kubak (2010) found that motivation to change is a key factor when treating psychopathic traits in youth. As such, a desire to desist from lifestyle problems (e.g., incarceration) associated with psychopathy may serve to benefit adolescents when attempting to make change. More recently, Salekin, Lester, and Sellers (2012) have shown that mental set can have an impact on the cognitive ability of conduct problem youth with callous unemotional traits. Another more recent study has shown that a mental models intervention is effective at reducing psychopathic traits over a 12 week intervention (Salekin, Tippey, & Allen, 2012). These three studies, taken together, suggest that some work could be done on a cognitive level with psychopathic individuals. Other research has focused on how positive peer relationships (Munoz, Kerr, & Besic, 2008), along with social competence (Barry, Barry, Deming, & Lochman, 2008), can confer some protection against antisocial behavior, even in the presence of psychopathy. Hawes and Dadds (2007) indicated that providing parenting is beneficial to working with antisocial youth. Forensic evaluators and clinicians must be cognizant for current research on treatment successes for youth with psychopathic traits and include them in their recommendations to the court.

Recommendations

(1) Research has shown that adolescents with high levels of psychopathy and antisocial behavior can improve and it is likely that the chances of improvement will increase even further with specialized treatment. It is not clear what works with respect to treatment of adolescent offenders and more research is needed in this area. General treatment often associated with correctional settings may not adequately treat the complexity of psychopathy, thus limiting its effect on reducing future violence and recidivism.

(2) Even within the context of risk assessment, evaluators should recommend specialized treatment and not be quick to dismiss the real possibility of change. In that context, when making placement recommendations, consideration of structured therapeutic programs has merit.

(3) When making treatment recommendations for high-risk youth, evaluators should recommend high-intensity treatment. Often referred to as the Risk-Need-Responsivity model, these types of treatment provide the highest level of intensity to individuals most at risk. We suggest that adolescents with high levels of psychopathic traits are at high risk for violent behavior.

NEW DIRECTIONS IN ADOLESCENT PSYCHOPATHY RESEARCH AND THEIR APPLICATION TO THE LEGAL SYSTEM

In looking at recent trends in developmental research on psychopathy, two research ideas have generated a high level of interest. The first is evaluating neural activity through the use of magnetic resonance imaging (MRI) and functional MRI (fMRI) technologies. These technologies have provided great insights into the developing brain and how there may be manifest differences in the brains of adolescents with psychopathic traits. The second involves evaluating how the hypothalamic–pituitary–adrenal (HPA) axis manifests differently in adolescents with high levels of psychopathic traits evidenced by differing levels of cortisol. These two approaches have applied integrated multidisciplinary techniques to psychopathy in a way that improves understanding of brain development and the etiology of psychopathy. Several neuroimaging studies have implicated structures in the brain associated with psychopathy including abnormal hippocampus (Boccardi et al., 2010), frontal and temporal lobes (Kiehl et al., 2004; Wahlund & Kristiansson, 2009) and corpus callosum (Raine et al., 2003), which could lead to deviant emotional responding, a key feature of psychopathy (Hoff, Beneventi, Galta, & Wik, 2009).

Cortisol is a stress hormone that has received significant attention, especially in relation to the development of callous traits (Shirtcliff et al., 2009). In suggesting the HPA axis is linked to CU traits, Shirtcliff et al (2009) put forth a pathway hypothesis combining environmental factors with biology that might be of some use to future researchers. Hawes, Brennan, and Dadds (2009) indicated that in a certain group of antisocial adolescents with lower levels of CU, environment adversity produces HPA dysregulation. However, in high-CU individuals, the antisocial behavior develops relatively independently from HPA dysregulation. O'Leary, Loney, and Eckel (2007) found blunted emotional responses to stress in males but not females related to the presence of psychopathic traits. Holi et al. (2006) found cortisol related to psychopathic traits in a small sample of young offenders. This promising area of research is growing, and new findings will begin to emerge from the field regarding the relationship between psychopathic traits and cortisol.

It is clear that this emerging area of research has potential to truly shape the manner in which the development of psychopathic traits is conceptualized. It has yet to be determined if this area of developing research will have an impact on how courts view psychopathy. Early results studying the impact of neuroimaging and neuroscience on psycholegal issues are mixed. For example, Gurley and Marcus (2008) reported that juries are more likely to find an individual not responsible with presentation of neuroimaging data, but Schweitzer et al. (2011) found virtually no influence for the presentation of neuroimaging data regarding the insanity defense. It is unknown how juries viewing adolescent brains in individuals with high levels of psychopathic traits would respond. On one hand, juries and judges could disregard evidence of treatability and view the adolescent as having a lifelong condition highlighted by a "damaged brain." Conversely, triers of fact could view the brain and see a reason for the psychopathic behavior and be more likely to give leniency. Still, another possibility is that researchers could discuss the plasticity of the brain and the potential for change even where there are structural and functional differences.

Subsequent research is needed to fully understand how triers of fact will respond to this area of research and if this research assists the trier of fact in making relevant legal decisions to questions of risk and dangerousness. In a comprehensive law review article, Maroney (2010) attempted to temper expectations regarding the power that neuroscience will have in the courtroom regarding adolescents. In discussing limitations of neuroscientific approaches, Maroney (2010) acknowledged that outside of a few high-profile cases that have appeared to raise the expectations for the promise of neuroscience, there have been infrequent applications to everyday, garden-variety legal cases. At present, we believe that the research field may still be too young to be utilized in the legal system. Many of the points for concern in this regard were articulated by Aronson (2007), who has stated that there are no definitive links between brain structure or function and deviant behavior or culpability. Moreover, there has been some concern regarding the task-to-inference connection, replication of certain brain functioning findings (e.g., Johnstone et al., 2005), and the stability of the findings across development.

Until more work is done in this area, we would recommend that psychologists be very cautious with respect to their statements regarding potential differences in brain structure and/or function. Despite our hesitancy regarding the readiness of the field to use this information for legal purposes, we believe that this new generation of research (and researchers) offers potential to reinvigorate the scientific study of adolescent psychopathy, bringing further knowledge to the etiology and course of the disorder and eventually may have potential to inform legal scholars and the law.

SUMMARY

This chapter described the positive and negative consequences regarding the application of psychopathy to court proceedings. There continues to be great promise for the use of psychopathy in the area of risk assessment (Forth & Book, 2010; Leistico et al., 2008; Vincent, 2006), and some utility for assisting the court in juvenile disposition decisions. However, this chapter also pointed out some of the problematic concerns with using psychopathy, including applications for which there are no empirical support. We hope that by reading this chapter and the discussions of limitations related to assessing psychopathy in youth, clinicians and attorneys will become acutely aware that evaluating psychopathy has no utility in most psycholegal evaluations of youth. When used in competency to stand trial or criminal responsibility, evaluations, the evaluator must be forcefully questioned regarding his or her choice to use psychopathy assessment (Grisso, 1998). Even in risk assessment cases, evaluators must strike a moderate tone and be highly cautious to not commit errors of misrepresentation by overstating the relationship between the presence of psychopathic traits and long-term prognosis. Evaluators must maintain strict boundaries of competence, and consistent with our recommendations, should not label a youth as a "psychopath." Instead, we suggest that clinicians recognize that evaluating psychopathy has significant benefits when used in appropriate and circumscribed fashions.

REFERENCES

Aronson, J. D. (2007). Brain imaging, culpability and the juvenile death penalty. *Psychology, Public Policy, and Law, 13*, 115–142.

Barry, T. D., Barry, C. T., Deming, A. M., & Lochman, J. E. (2008). Stability of psychopathic characteristics in childhood: The influence of social relationships. *Criminal Justice and Behavior, 35*, 244–262.

Boccaccini, M. T., Murrie, D. C., Clark, J. W., & Cornell, D. G. (2008). Research report: Describing, diagnosing, and naming psychopathy: How do youth psychopathy labels influence jurors?. *Behavioral Sciences & the Law, 26*(4), 487–510.

Boccardi, M., Ganzola, R., Rossi, R., Sabattoli, F., Laakso, M. P., Repo-Tiihonen, E., et al. (2010). Abnormal hippocampal shape in offenders with psychopathy. *Human Brain Mapping, 31*, 438–447.

Borum, R., & Verhaagen, D. (2006). *Assessing and managing violence risk in juveniles*. New York: Guilford.

Brannen, D. N., Salekin, R. T., Zapf, P. A., Salekin, K. L., Kubak, F. A., & DeCoster, J. (2006). Transfer to adult court: A national study of how juvenile court judges weigh pertinent Kent criteria. *Psychology, Public Policy, and Law, 12*(3), 332–355.

Caldwell, M., McCormick, D., Umstead, D., & Van Rybroek, G. (2007). Evidence of treatment progress and therapeutic outcomes among adolescents with psychopathic features. *Criminal Justice and Behavior, 34*, 573–587.

Caldwell, M., Skeem, J., Salekin, R., & Van Rybroek, G. (2006). Treatment response of adolescent offenders with psychopathy features: A 2-year follow-up. *Criminal Justice and Behavior, 33*, 571–596.

Caldwell, M., Vitacco, M. J., & Van Rybroek, G. (2006). Are violent delinquents worth treating? A cost-benefit analysis. *Journal of Research in Crime and Delinquency, 43*, 148–168.

Corrado, R., Vincent, G., Hart, S., & Cohen, I. (2004). Predictive validity of the Psychopathy Checklist: Youth Version for general and violent recidivism. *Behavioral Sciences & the Law, 22*, 5–22.

Cox, J., DeMatteo, D. S., & Foster, E. E. (2010). The effect of the Psychopathy Checklist—Revised in capital cases: Mock jurors' responses to the label of psychopathy. *Behavioral Sciences & the Law, 28*, 878–891.

DeMatteo, D., & Edens, J. F. (2006). The role and relevance of the Psychopathy Checklist-Revised in court: A case law survey of U.S. courts (1991–2004). *Psychology, Public Policy, and Law, 12*, 214–241.

Edens, J., & Cahill, M. (2007). Psychopathy in adolescence and criminal recidivism in young adulthood: Longitudinal results from a multiethnic sample of youthful offenders. *Assessment, 14*, 57–64.

Edens, J., Guy, L., & Fernandez, K. (2003). Psychopathic traits predict attitudes toward a juvenile capital murderer. *Behavioral Sciences & the Law, 21*, 807–828.

Edens, J., Skeem, J., Cruise, K., & Cauffman, E. (2001). Assessment of "juvenile psychopathy" and its association with violence: A critical review. *Behavioral Sciences & the Law, 19*, 53–80.

Forth, A. E., & Book, A. S. (2010). Psychopathic traits in children and adolescents: The relationship with antisocial behaviors and aggression. In R. T. Salekin, D. R. Lynam, R. T. Salekin, D. R. Lynam (Eds.), *Handbook of child adolescent psychopathy* (pp. 251–283). New York: Guilford.

Forth, A., Kosson, D. S., & Hare, R. D. (2003). *Manual for the Psychopathy Checklist-Youth Version.* Toronto, ON: Multi-health Systems.

Frick, P., & Hare, R. D. (2001). *Technical manual for the Antisocial Process Screening Device.* North Tonawanda, NY: Multi-Health Systems.

Frick, P. J., & White, S. F. (2008). Research review: The importance of callous-unemotional traits for developmental models of aggressive and antisocial behavior. *Journal of Child Psychology and Psychiatry, 49*, 359–375.

Gretton, H., Hare, R., & Catchpole, R. (2004). Psychopathy and offending from adolescence to adulthood: A 10-year follow-up. *Journal of Consulting and Clinical Psychology, 72*, 636–645.

Grisso, T. (1998). *Forensic evaluation of juveniles.* Sarasota, FL: Professional Resource Press/Professional Resource Exchange.

Gurley, J. R., & Marcus, D. K. (2008). The effects of neuroimaging and brain injury on insanity defenses. *Behavioral Sciences & the Law, 26*, 85–97.

Hawes, D. J., Brennan, J., & Dadds, M. R. (2009). Cortisol, callous-unemotional traits, and pathways to antisocial behavior. *Current Opinion in Psychiatry, 22*, 357–362.

Holi, M., Auvinen-Lintunen, L., Lindberg, N., Tani, P., & Virkkunen, M. (2006). Inverse correlation between severity of psychopathic traits and serum cortisol levels in young adult violent male offenders. *Psychopathology, 39*, 102–104.

Johnstone, T., Somerville, L. H., Alexander, A. L., Oakes, T. R., Davidson, R. J., Kalin, N. H., et al. (2005). Stability of amygdale BOLD response to fearful faces over multiple scan sessions. *NeuroImage, 25*, 1112–1123.

Jones, S., & Cauffman, E. (2008). Juvenile psychopathy and judicial decision making: An empirical analysis of an ethical dilemma. *Behavioral Sciences & the Law, 26*, 151–165.

Hare, R.D. (2003). *Technical manual for the Revised Psychopathy Checklist* (2nd ed.). North Tonawanda, NY: Multi-Health Systems.

Hare, R.D. (1991). *Technical manual for the Psychopathy Checklist.* North Tonawanda, NY: Multi-Health Systems.

Hart, S. D., Watt, K. A., & Vincent, G. M. (2002). Commentary on Seagrave and Grisso: Impressions of the state of the art. *Law and Human Behavior, 26*(2), 241–245. doi:10.1023/A:1014648227688.

Hawes, D. J., & Dadds, M. R. (2007). Stability and malleability of callous-unemotional traits during treatment for childhood conduct problems. *Journal of Clinical Child and Adolescent Psychology, 36*, 347–355.

Hoff, H., Beneventi, H., Galta, K., & Wik, G. (2009). Evidence of deviant emotional processing in psychopathy: A fMRI case study. *International Journal of Neuroscience, 119*, 857–878.

Kellerman, J. (1999). *Savage spawn: Reflection of violent children.* New York: Ballatine.

Kiehl, K. A., Smith, A. M., Mendrek, A., Forster, B. B., Hare, R. D., & Liddle, P. F. (2004). Temporal lobe abnormalities in semantic processing by criminal psychopaths as revealed by functional magnetic resonance imaging. *Psychiatry Research: Neuroimaging, 130*, 27–42.

Leistico, A., Salekin, R., DeCoster, J., & Rogers, R. (2008). A large-scale meta-analysis relating the Hare measures of psychopathy to antisocial conduct. *Law and Human Behavior, 32*, 28–45.

Lloyd, C. D., Clark, H. J., & Forth, A. E. (2010). Psychopathy, expert testimony, and indeterminate sentences: Exploring the relationship between Psychopathy Checklist-Revised testimony and trial outcome in Canada. *Legal and Criminological Psychology, 15*, 323–339.

Maroney, T. A. (2010). The false promise of adolescent brain science in juvenile justice. *Notre Dame Law Review, 85*, 89–120.

Marsee, M., Silverthorn, P., & Frick, P. (2005). The association of psychopathic traits with aggression and delinquency in non-referred boys and girls. *Behavioral Sciences & the Law, 23,* 803–817.

Murrie, D., Boccaccini, M., McCoy, W., & Cornell, D. (2007). Diagnostic labeling in juvenile court: How do descriptions of psychopathy and conduct disorder influence judges? *Journal of Clinical Child and Adolescent Psychology, 36,* 228–241.

Murrie, D., Cornell, D., & McCoy, W. (2005). Psychopathy, conduct disorder, and stigma: Does diagnostic labeling influence juvenile probation officer recommendations? *Law and Human Behavior, 29,* 323–342.

Muñoz, L. C., Kerr, M., & Bešić, N. (2008). The peer relationships of youths with psychopathic personality traits: A matter of perspective. *Criminal Justice and Behavior, 35,* 212–227.

Odgers, C., Moretti, M., & Reppucci, N. (2005). Examining the science and practice of violence risk assessment with female adolescents. *Law and Human Behavior, 29,* 7–27.

O'Leary, M. M., Loney, B. R., & Eckel, L. A. (2007). Gender differences in the association between psychopathic personality traits and cortisol response to induced stress. *Psychoneuroendocrinology, 32,* 183–191.

O'Neill, M., Lidz, V., & Heilbrun, K. (2003). Adolescents with psychopathic characteristics in a substance abusing cohort: Treatment process and outcomes. *Law and Human Behavior, 27,* 299–313.

Pajer, K. (1998). What happens to "bad" girls? A review of the adult outcomes of antisocial adolescent girls. *American Journal of Psychiatry, 155,* 862–870.

Penney, S., & Moretti, M. (2007). The relation of psychopathy to concurrent aggression and antisocial behavior in high-risk adolescent girls and boys. *Behavioral Sciences & the Law, 25,* 21–41.

Raine, A., Lencz, T., Taylor, K., Hellige, J. B., Bihrle, S., Lacasse, L., et al. (2003). Corpus callosum abnormalities in psychopathic antisocial individuals. *Archives of General Psychiatry, 60,* 1134–1142.

Rockett, J., Murrie, D., & Boccaccini, M. (2007). Diagnostic labeling in juvenile justice settings: Do psychopathy and conduct disorder findings influence clinicians? *Psychological Services, 4,* 107–122.

Salekin, R. T. (2002). Psychopathy and therapeutic pessimism: Clinical lore or clinical reality? *Clinical Psychology Review, 22,* 79–112.

Salekin, R. T. (2008). Psychopathy and recidivism from mid-adolescence to young adulthood: Cumulating legal problems and limiting life opportunities. *Journal of Abnormal Psychology, 117,* 386–395.

Salekin, R. T., & Grimes, R. D. (2008). Clinical forensic evaluations for juvenile transfer to adult criminal court. *Learning forensic assessment* (pp. 313–346). New York: Routledge/Taylor & Francis Group.

Salekin, R. T., Lee, Z., Schrum Dillard, C. L., & Kubak, F. A. (2010). Child psychopathy and protective factors: IQ and motivation to change. *Psychology, Public Policy, and Law, 16,* 158–176.

Salekin, R. T., Leistico, A. R., Neumann, C. S., DiCicco, T., & Duros, R. (2004). Psychopathy and comorbidity in a young offender sample: Taking a closer look at psychopathy's potential importance over disruptive behavior disorders. *Journal of Abnormal Psychology, 113,* 416–427.

Salekin, R. T., Lester, W. S., & Sellers, M. K. (2012). Psychopathy in youth and mental sets: Incremental and entity theories of intelligence. *Law and Human Behavior, 36,* 283–292.

Salekin, R. T., Rogers, R., & Ustad, K. L. (2001). Juvenile waiver to adult criminal courts: Prototypes for dangerousness, sophistication-maturity, and amenability to treatment. *Psychology, Public Policy, and Law, 7,* 381–408.

Salekin, R. T., Salekin, K. L., Clements, C., & Leistico, A. R. (2005). Risk-Sophistication-Treatment Inventory. In T. Grisso, G. M. Vincent, & D. Seagrave (Eds.): *Mental health screening and assessment in juvenile justice* (pp. 341–356). New York: Guilford.

Salekin, R. T., Tippey, J. G., & Allen, A. D. (2012). Treatment of conduct problem youth with interpersonal callous traits using mental models: Measurement of risk and change. *Behavioral Sciences and the Law, 30,* 470–486.

Salekin, R. T., Worley, C. B., & Grimes, R. D. (2010). Treatment of psychopathy: A review and brief introduction to the mental models approach. *Behavioral Sciences and the Law, 28,* 235–266.

Schmidt, F., McKinnon, L., Chattha, H., & Brownlee, K. (2006). Concurrent and predictive validity of the Psychopathy Checklist: Youth Version across gender and ethnicity. *Psychological Assessment, 18,* 393–401.

Schrum, C. L., & Salekin, R. T. (2006). Psychopathy in adolescent female offenders: An item response theory analysis of the Psychopathy Checklist:

Youth Version. *Behavioral Sciences & the Law, 24*, 39–63.

Schweitzer, N. J., Saks, M. J., Murphy, E. R., Roskies, A. L., Sinnott-Armstrong, W., & Gaudet, L. M. (2011). Neuroimages as evidence in a mens rea defense: No impact. *Psychology, Public Policy, and Law, 17*, 357–393.

Seagrave, D., & Grisso, T. (2002). Adolescent development and the measurement of juvenile psychopathy. *Law and Human Behavior, 26*, 219–239.

Shirtcliff, E. A., Vitacco, M. J., Graf, A. R., Gostisha, A. J., Merz, J. L., & Zahn-Waxler, C. (2009). Neurobiology of empathy and callousness: Implications for the development of antisocial behavior. *Behavioral Sciences & the Law, 27*, 137–171.

Spain, S., Douglas, K. S., Poythress, N. G., & Epstein, M. (2004). The relationship between psychopathic features, violence and treatment outcome: The comparison of three youth measures of psychopathic features. *Behavioral Sciences & the Law*.

Vaughn, M., & Howard, M. (2005). Self-report measures of juvenile psychopathic personality traits: A comparative review. *Journal of Emotional and Behavioral Disorders, 13*, 152–162.

Viding, E., Jones, A. P., Frick, P. J., Moffitt, T. E., & Plomin, R. (2008). Heritability of antisocial behaviour at 9: Do callous-unemotional traits matter?. *Developmental Science, 11*(1), 17–22.

Viljoen, J. L., MacDougall, E. M., Gagnon, N. C., & Douglas, K. S. (2010). Psychopathy evidence in legal proceedings involving adolescent offenders. *Psychology, Public Policy, and Law, 16*, 254–283.

Viljoen, J. L., McLachlan, K., & Vincent, G. M. (2010). Assessing violence risk and psychopathy in juvenile and adult offenders: A survey of clinical practices. *Assessment, 17*, 377–395.

Vincent, G. M. (2006). Psychopathy and violence risk assessment in youth. *Child and Adolescent Psychiatric Clinics of North America, 15*, 407–428.

Vincent, G. M., & Grisso, T. (2005). A developmental perspective on adolescent personality, psychopathology, and delinquency. In: *Mental health screening and assessment in juvenile justice* (pp. 22–43). New York: Guilford.

Vitacco, M. J., Caldwell, M., Van Rybroek, G., & Gabel, J. (2007). Psychopathy and behavioral correlates of victim injury in serious juvenile offenders. *Aggressive Behavior, 33*(6), 537–544.

Vitacco, M. J., Lishner, D., & Neumann, C. S. (2012). Assessment. In H. Hakkanen-Nyholm & J. Nyholm (Eds.), *Psychopathy and the law: A practitioner's guide* (pp. 19–38). West Sussex, UK: John Wiley and Sons.

Vitacco, M. J., & Vincent, G. M. (2006). Applying adult concepts to youthful offenders: Psychopathy and its implications for risk assessment and juvenile justice. *International Journal of Forensic Mental Health Services, 5*, 29–38.

Wahlund, K., & Kristiansson, M. (2009). Aggression, psychopathy and brain imaging—Review and future recommendations. *International Journal of Law and Psychiatry, 32*, 266–271.

Walsh, T., & Walsh, Z. (2006). The evidentiary introduction of Psychopathy Checklist-Revised assessed psychopathy in U.S. courts: Extent and appropriateness. *Law and Human Behavior, 30*, 493–507.

White, S. F., & Frick, P. J. (2010). Callous-unemotional traits and their importance to causal models of severe antisocial behavior in youth. In R. T. Salekin, D. R. Lynam (Eds.), *Handbook of child and adolescent psychopathy* (pp. 135–155). New York: Guilford Press.

PART THREE

DECISION MAKING AND PSYCHOPATHY

CHAPTER 6

The Decision-Making Impairment in Psychopathy: Psychological and Neurobiological Mechanisms

Michael Koenigs and Joseph P. Newman

Psychopathy is essentially a disorder of decision making. For decades researchers have aimed to identify the psychobiological mechanisms that underlie the psychopath's profound decision-making impairment. Still, a comprehensive answer to a simple question (Why do psychopaths do the things they do?) remains elusive. The implications of a veritable answer to this question are far-reaching, including how criminals are evaluated, sentenced, and potentially rehabilitated. In this chapter, we (1) discuss clinical and research evidence that justifies our conceptualization of psychopathy as a disorder of decision making; (2) characterize the attentional abnormalities associated with psychopathy to highlight a distinction between active and passive decision making and the implications of this distinction for culpability; (3) selectively review evidence from neuroscientific studies for the purpose of evaluating the relevance of different neurobiological models of psychopathy; and (4) conclude with a brief discussion of the potential implications of our review for legal considerations in psychopathy.

PSYCHOPATHY AS A DECISION-MAKING DISORDER

One of the earliest and most compelling depictions of psychopathic behavior was conveyed through the accumulation of narrative case descriptions in Hervey Cleckley's book *The Mask of Sanity* (Cleckley, 1941, 1976). Over the multiple editions of this text, Cleckley assimilated his clinical experiences with scores of psychopathic cases to distill the essential features of the disorder. As a testament to the acumen of his observations, many of the core elements of psychopathy he described decades ago are incorporated into the current standard for evaluating psychopathy. With respect to the decision-making capability of psychopaths, Cleckley emphasized a blatant and enduring impairment:

> To say the least, the pattern of [the psychopath's] actions over any fairly long range of time indicates little that the observer can understand as what a human being would consciously choose. (pp. 261–262)
>
> Despite his excellent rational powers, the psychopath continues to show the most execrable judgment about attaining what one might presume to be his ends...This exercise of execrable judgment is not particularly modified by experience, however chastening his experiences may be...It is my opinion that no punishment is likely to make the psychopath change his ways. (pp. 345–346)

A number of the laboratory paradigms that have been used to distinguish the performance of psychopathic and nonpsychopathic individuals highlight this decision-making deficit. Building on the clinical descriptions of psychopathy, researchers modeled the apparent insensitivity to punishment and impaired decision making (i.e., behavioral choices) in a laboratory paradigm known as "passive-avoidance learning."

In one version of this task, the subject attempts to complete a mental maze through a series of choice points. At each choice point the subject has four possible choices. Only one choice is correct (leading to advancement in the maze), while one of the three error choices is associated with an additional punishment (e.g., painful electric shock). Multiple studies demonstrate that compared to nonpsychopaths, who show preferential avoidance of the punished (shocked) choice, psychopaths are more likely to commit the punished choice, a so-called error in passive avoidance (Lykken, 1957; Schmauk, 1970).

These results lend themselves to at least two possible interpretations. One interpretation holds that the psychopaths' passive-avoidance errors arise from a fundamental deficit in the generation of emotion. If the punishment does not elicit a subjectively aversive feeling, or if the aversion tied to past punishment experiences is not evoked during the contemplation of a subsequent action in a similar context, then the psychopath may not exhibit learning on the basis of punishment. A second interpretation is that the psychopath may simply be oblivious to the association between certain choices and their associated consequences. In other words, psychopaths may rigidly attend to learning the correct choices in the maze (as per task instructions) and fail to process (and/or link with their responses) the incidental punishments that are not directly germane to their instructed goal. Indeed, these differing interpretations for the passive-avoidance data can be extrapolated to the psychopaths' decision making more generally. At this level, the question can be posed succinctly: Is the psychopath's decision-making impairment due primarily to a deficit in the generation of emotion or primarily to a deficit in the allocation of attention? Cleckley seemed to acknowledge both possibilities as explanations for his clinical observations. At one point he proposed that "…despite [the psychopath's] otherwise perfect functioning, the major emotional accompaniments are absent or so attenuated as to count for little" (p. 371). At another point he proposed that "…[the psychopath's] difference from the whole or normal or integrated personality consists of an unawareness and a persistent lack of ability to become aware of what the most important experiences of life mean to others" (p. 371). The former quote seems to suggest a primary deficit of emotional processing, whereas the latter seems to implicate processes specifically related to attention and awareness.

These two competing explanations have spawned a wealth of theoretical and empirical work. In clinical and real-world settings, it can be difficult, if not impossible, to discern the relative validity of these different interpretations. However, it is often possible to parse these causal factors in laboratory experiments. In the following sections, we consider two lines of research on the psychobiological mechanisms of psychopathic decision making: one highlighting the role of emotion and the other highlighting the role of attention. We conclude this section with a brief integration of the findings.

Psychopathic Decision Making as a Deficit of Emotion

One of the earliest and most influential theoretical accounts of the primacy of an emotional deficit in psychopathy was formalized by David Lykken in his "low-fear hypothesis" (Lykken, 1957, 1995), which proposes simply that a defect in the generation of fear may underlie psychopathic behavior. This idea has clear appeal in its parsimony and experimental tractability. Initial laboratory support for this proposal included the passive-avoidance data described previously, as well as a host of studies employing galvanic skin response (GSR) as an index of autonomic physiological arousal. In these studies, GSR is taken as an objective proxy for the experience of fear or anxiety (though some have argued that it is more accurately regarded as an index of arousal) (e.g., Patrick, Bradley, & Lang, 1993). In normal (nonpsychopathic) subjects, the repeated pairing of an emotionally neutral stimulus (e.g., an auditory tone) with an emotionally aversive stimulus (e.g., electric shock) leads to a state of "conditioned fear," in which the presentation of the previously neutral stimulus alone is sufficient to elicit a fear response (as measured with GSR). Psychopaths, however, exhibit abnormally low GSR to the conditioned stimulus (Hare & Quinn, 1971; Lykken, 1957). In subsequent years, psychophysiological recording techniques have been used to document attenuated autonomic reactivity in psychopaths for a variety of experimental conditions, including passive-avoidance learning (Schmauk, 1970), hearing loud noises (Hare, 1978), anticipation of an aversive stimulus such as loud noise or shock (Hare, Frazelle, & Cox, 1978; Ogloff & Wong, 1990; Tharp, Maltzman, Syndulko, & Ziskind, 1980), viewing scenes of distress (Blair, Jones, Clark, & Smith, 1997), and imagining fearful situations (Patrick, Cuthbert, & Lang, 1994).

Although prolific from an experimental standpoint, the low-fear hypothesis has clear limitations in explaining the full spectrum of psychopathic traits. In the domain of emotions alone, psychopaths exhibit conspicuously diminished guilt, shame, embarrassment, empathy, and love, among others (Cleckley, 1976; Lilienfeld & Arkowitz, 2007). It seems implausible that these various manifestations of restricted affect are reducible to a root deficiency in fear. Furthermore, the lack of a measurable physiological response in psychopaths could conceivably be due to a failure to appropriately engage or attend the threatening or aversive stimuli, a possibility that is discussed at greater length in the text that follows.

A more comprehensive account of the role of emotion in decision making has been proposed by Antonio Damasio in his "somatic marker hypothesis" (Damasio, 1994). The central idea of this hypothesis is that physiological processes, including those that constitute emotion, may act as signals to influence behavior. More specifically, through experience humans develop associations between various situations and the corresponding somatic states (i.e., emotions). The recurrence of a particular situation triggers the reactivation of neural patterns depicting the associated emotion, which marks potential outcomes as good or bad. The poor judgment and decision making of psychopaths could be seen as a consequence of weak somatic markers due to the underlying defect in emotional reactivity. This hypothesis thus posits a causal relationship between emotion and decision making, without necessarily stipulating the preeminence of any particular emotion, such as fear.

Psychopathic Decision Making as a Deficit of Attention

As previously mentioned, it is possible that the poor decision making and even the conspicuously blunted affect associated with psychopathy could arise from abnormalities in the allocation of attention. This hypothesis is consistent with early clinical descriptions of psychopaths, which note a lack of "the active, searching attention and organizing process that normally puts [relevant] information to use" (Shapiro, 1965, p. 149), as well as with psychopaths' own statements: "I always know damn well I shouldn't do these things...it's just that when the time comes I don't think of anything else. I don't think of anything but what I want now." (Grant, 1965). These descriptions suggest an inability to flexibly reallocate attention away from a dominant goal.

Experimental evidence for this type of attentional defect in psychopathy can be found in the early passive-avoidance learning studies. In the initial such study (Lykken, 1957), the instructed goal of the task was to complete the maze successfully; thus learning the one correct choice at each choice point was the "manifest" task, whereas learning to avoid the shocked incorrect choice was the "latent" task (in that it was not directly related to the goal of the task). Interestingly, psychopathic subjects performed normally on the "manifest" task despite their insensitivity to the "latent" task. In fact, unlike nonpsychopathic subjects, the majority of psychopathic subjects did not report any awareness that the shocks were contingent upon any particular incorrect choice (Schmauk, 1970). In a modified version of the passive-avoidance task, subjects were endowed with a sum of money at the start of the task, and one of the errors was punished with a loss of money rather than shock. Importantly, subjects were allowed to keep the money at the end of the test, and therefore learning the punished error was arguably an explicit goal of the test (i.e., a "manifest" task). In this condition, psychopaths demonstrated both good awareness of the shock contingency and normal passive-avoidance learning (Schmauk, 1970).

The task dependence of the psychopaths' defect in learning from punishment was further demonstrated in a go/no-go discrimination learning task (Newman & Kosson, 1986). In one condition of this task, correct responses were rewarded with monetary gain and incorrect responses were punished with monetary loss. A second condition featured only punishment to incorrect responses. Psychopaths made more commission errors (responding to the "no-go" stimuli) than nonpsychopaths in the reward + punishment condition, but exhibited normal performance in the punishment-only condition. In a related study of decision making, subjects selected cards from a deck, with each card indicating a gain or loss of money (Newman, Patterson, & Kosson, 1987). The subject was free to stop playing the cards at any time and collect whatever money he had accumulated. The key experimental manipulation was that initially the cards were almost always rewarding (and hence established a dominant response set), but as the task progressed the cards became increasingly more punishing. When no restrictions were placed on the subjects' responding, the psychopaths played more cards and won less money than nonpsychopaths. However, when subjects were able to view their cumulative net earnings and also required to pause briefly before choices, psychopaths performed normally. Taken together, these data suggest that the psychopath's inability to inhibit a punished behavior depends critically on the context of the situation, which argues against a pervasive insensitivity to punishment. More generally, the psychopath's decision making may reflect a deficit in the natural human tendency to reflect on the various consequences of previous actions. In a subsequent study, Newman and colleagues addressed this issue directly (Newman, Patterson, Howland, & Nichols, 1990). Once again using a go/no-go test with reward and punishment, they found that following a monetary loss (punishment), psychopaths did not pause as long as nonpsychopaths before initiating their next choice. Importantly, the duration of the post-punishment pause was shown to be predictive of learning to avoid the punished response.

Integrating the aforementioned clinical and experimental observations, Newman and colleagues developed a theoretical framework that highlights the role of attention in psychopathic decision making: the "response modulation hypothesis" (Newman & Lorenz, 2003; Patterson & Newman, 1993). The essence of response modulation is the "temporary suspension of a dominant response set and a brief concurrent shift of attention from the organization and implementation of goal-directed responding to its evaluation" (Pattersons & Newman, 1993, p. 717). In the context of decision making, this involves attending to the remote or secondary consequences of one's actions as well as to immediate or primary considerations to guide decision making. In recent years, Newman and colleagues have tested the role of attention in decision making among psychopaths in various ways. Here we summarize a number of the key studies.

One study investigated psychopaths' capacity for attentional focus using Stroop tests. In the classic color-word Stroop test, the subject sees a color word (e.g., "red") with the letters printed in a different color (e.g., green). Confronted with a written word, the natural or "pre-potent" impulse is to read the word. However, in the Stroop test the subject's task is to name the color of the letters. Longer reaction times are interpreted as greater interference of the distracter task (reading the word) on the instructed task (naming the color). Psychopaths exhibit normal interference effects in the color-word Stroop test (Hiatt, Schmitt, & Newman, 2004; Smith, Arnett, & Newman, 1992). In a modified version of the test, the picture-word Stroop, the subject sees an object word (e.g., "chair") superimposed on a picture outline of another object (e.g., table). The subject's task is to name the picture while ignoring the word, and again, longer reaction times are interpreted as greater interference. Unlike nonpsychopaths, psychopaths were insensitive to interference in this variant of the task (Hiatt et al., 2004; Newman, Schmitt, & Voss, 1997). In another modification of the test, the spatially segregated color-word Stroop test, the subject sees a color word (e.g., "red") printed in white letters and surrounded by a rectangular frame of a different color (e.g., green). The subject's task is to name the color of the frame while ignoring the word. Again, the psychopaths showed reduced interference effects (Hiatt et al.,

2004). These results demonstrate that psychopaths' attention is relatively impervious to distraction when the task-irrelevant information is spatially distinct from the deliberately attended goal-relevant stimuli. However, when the irrelevant information is spatially integrated within the attended stimuli the psychopaths perform normally. These results point to a dysfunction in attending/processing certain types of contextual information. Importantly, the behavioral differences documented in these studies occur for emotionally neutral stimuli, indicating that psychopaths' defects are not restricted to the affective domain.

A second study examined whether this apparent defect in the integration of contextual information is also evident in psychopaths' memory function (Glass & Newman, 2009). This study consisted of three memory tasks. In each memory task, subjects saw a series of words (some neutral, some emotionally arousing) presented one at a time and were instructed to remember each word as it appeared. Each of the three tests involved an additional piece of contextual information: word location, color of a rectangular frame around the word, or color of the font. Subjects were later tested on their ability to recall the words as well as the associated contextual information. As predicted by the response modulation hypothesis, psychopaths and nonpsychopaths exhibited similar performance on the word recall task (both groups demonstrated memory bias for the emotional words), but unlike nonpsychopaths, who also demonstrated memory bias for the contextual information associated with the emotional words, psychopaths exhibited no such bias for the contextual information. These data indicate a specific impairment of integrating contextual information in memory, rather than a global impairment in memory for emotionally salient information.

A third study investigated whether psychopaths' physiological responses of fear are also dependent on attentional focus (Newman, Curtin, Bertsch, & Baskin-Sommers, 2010). In this study, subjects underwent a fear conditioning paradigm in which red letters were sometimes associated with electric shocks, whereas green letters were never associated with shocks. Attentional focus was manipulated with one of three concomitant behavioral tasks: indicating the color of the letter, indicating whether the letter was lowercase or uppercase, or indicating whether or not the letter matched the letter that appeared two letters back. Thus the color identification task requires the subject to attend to the property of the stimulus that predicts the painful stimulus ("threat-focus"), whereas the other two tasks require the subject to attend to other aspects of the stimulus ("alternative-focus"). Startle responses to sudden bursts of loud noise were measured as electrical activity in the facial muscles mediating eyeblinks. As predicted by the response modulation hypothesis, psychopaths exhibited normal startle responses in the threat-focus condition, but significantly diminished startle responses in the alternative-focus condition. These data demonstrate that focus of attention is a critical factor in determining the psychopaths' physiological fear response (see also Arnett, Smith, & Newman, 1997).

Overall, the results presented in this section argue for the primacy of an attention-related defect in psychopathy. The evidence suggests that psychopaths are not globally impaired in generating fear responses or in fear-related learning, but rather that the emotional hyporesponsiveness and impaired decision making are consequences of the psychopaths' abnormal deployment of attention. This conclusion is further supported by the studies revealing attention-related defects for certain types of nonaffective information as well. In the following section we elaborate on the relationship between these cognitive deficits and the concept of intentionality.

SPECIFYING THE COGNITIVE DEFICIT IN PSYCHOPATHY: IMPLICATIONS FOR INTENTIONALITY

The decision-making perspective on psychopathy highlights the potential importance of multiple information-processing stages. Decisions vary in complexity and, as a result, vary in the range of information processing skills required. In general, however, decision making will be affected by the quality of a person's perception (i.e., attention to all relevant stimuli), the accessibility of relevant memories and prior learning, and the integration of all relevant considerations during response selection. Similarly, poor decision making may reflect problems in perception, memory, and/or response selection.

An important question highlighted by this information processing perspective on decision making concerns the extent to which psychopaths' poor decision making reflects deliberate and callous choices as opposed to a deficit in information processing that systematically prevents consideration of all relevant information. Writing about a similar consideration, Shapiro (1965) wrote "conscience and moral values are not elemental psychological faculties, but involve and depend on a number of cognitive and affective functions" (p. 163). According to this view, differences in moral conduct "are not primarily matters of moral scruple on the part of the normal person or the lack of them on the part of the psychopathic character; they are matters of interest and automatic cognitive tendency" (p. 166).

"In the normal person, the whim or the half-formed inclination to do something is the beginning of a complex process, although, if all is well, it is a smooth and automatic one" (p. 140). The process entails integrating current experience or whims with preexisting values and provides a perspective on behavior that goes beyond one's immediate concerns. In the absence of this perspective, it is difficult to develop long-term goals or resist impulses, so that a person's thoughts and goals tend to shift erratically. Moreover, in providing a context for the person's commitment to a course of action, the process of integrating current motivations with more stable goals enhances one's ability to tolerate frustration, endure boredom, and accept responsibility for one's behavioral choices. By contrast, psychopaths' difficulty integrating current whims with past experience interferes with their ability to appreciate the emotional and moral significance of events as well as their ability to objectify their own behavior and exercise critical judgment. In the following section we consider experimental evidence that informs the question of whether the psychopaths' decision-making deficit reflects automatic or deliberate processes.

Early versus Late Selective Attention Deficits in Psychopathy

An important distinction in attention research pertains to early- versus late-selective attention. Early selection typically reflects a perceptual bottleneck and results in an unintentional failure to process all relevant information. Conversely, late selection is typically intentional and involves the application of capacity limited resources to focus one's attention in a particular direction while ignoring "goal-irrelevant" information. The standard color-word Stroop task described previously is a classic example. Even though reading words is "pre-potent," late selection may be used to attend preferentially to the color of the stimulus and ignore the color word. Factors that impair cognitive capacity (e.g., alcohol consumption) will also impair the consistent application of late-section (Curtin & Fairchild, 2003), resulting in the disinhibited expression of the pre-potent but incorrect response. Early selection is generally thought to reflect limitations in perception that limit the processing of distracting peripheral information. In contrast to late selection, secondary information is not suppressed using limited capacity resources. Rather, such information simply receives less attention and, thus, has minimal impact on behavior regardless of limitations on cognitive processing resources (Arnett et al., 1997; Bishop, Jenkins, & Lawrence, 2007; Lavie, Hirst, de Fockert, & Viding, 2004).

The relevance of the early and late selection distinction for psychopathy concerns its implications for judging whether psychopaths deliberately ignore behaviorally relevant information post-perception (late selection) or fail to perceive behaviorally relevant information unintentionally (early selection). Although the consequences for self-regulation will generally be the same, the distinction may have important implications for their culpability. Two recent studies from the Newman laboratory are germane.

The first study employed a modified version of the Erikson Flanker task. In a flanker task, participants are instructed to focus on a character that appears in the "target location" and make one of two responses depending upon the category of the target (e.g., letter vs. number). The task typically employs three conditions: congruent trials in which the target and distracters belong to the same category; incongruent trials in which the target and distracters belong to the opposite categories; and control trials in which the targets are paired with a neutral stimulus (e.g., *). The principal dependent measure is interference, which is calculated by subtracting response times for control trials from response times for incongruent trials. This measure is useful for quantifying the extent to which conflicting peripheral information (distracters) modulates a person's response to the goal-relevant cues (targets).

Zeier, Maxwell, and Newman (2009) modified the standard flanker task to evaluate the effects of early versus late selection on response incongruent distracters (Zeier et al., 2009). In this task, targets were either a letter (H, G) or a number (5, 8) and were presented to the left or right of a central arrow (i.e., > or <) along with one other letter, number, or neutral stimulus (*). The direction of the arrow indicated the location of the target stimulus. For example, G < 5 would be a letter trial. To address the early versus late selection issue, each display was preceded by a cue display that either directed attention to the location of the target or directed attention to the location of both the target and distracter. In the exogenous cuing condition, for instance, an open square appeared at the location for 100 milliseconds before the target display and was placed so that the eventual target would fill the square. This cue elicits an involuntary attention response prior to presenting the target/distracter display. In the opposing condition, two squares appear for 100 milliseconds so that both the target and distracter locations are highlighted. The single cue condition facilitates early selection of the target location whereas the double cue procedure facilitates processing of the distracter as well as the target and thus increases the demand for late selection.

The results of Zeier et al. (2009) showed that psychopathic offenders displayed significantly less interference than nonpsychopathic offenders in the single-cue condition, whereas they displayed at least as much interference as nonpsychopathic controls in the double-cue condition. These findings show that psychopaths and controls are equally sensitive to information that conflicts with their goal-directed behavior once it has been perceived but that they are insensitive to the identical information if their attention is already focused on goal-relevant stimuli. By this account, their obliviousness to peripheral information appears to be involuntary (i.e., relatively automatic) rather than deliberate.

The second study evaluated sensitivity to fear-related distracters rather than response conflict but yields a similar conclusion. Following up on the fear conditioning study described previously (Newman et al., 2010), Baskin-Sommers and colleagues (2011) examined fear-potentiated startle either before or after the presentation of goal-relevant cues. The results of this study showed that deficits in fear-potentiated startle to threat cues were specific to the condition that presented threat cues after an alternative focus had been established. Thus, as in the Zeier et al. study, psychopaths appear to be normally responsive to secondary information unless their attention is already focused on goal-relevant stimuli. Such findings strongly suggest that their failure to inhibit inappropriate responses or weaker responses to peripheral emotion cues in laboratory studies is an unintentional consequence of an attentional abnormality that restricts processing of secondary information once their attention is engaged in goal-relevant processing. Analogously, it would seem to follow that psychopaths would be impaired in their ability to process all relevant information and exercise good decision making in real-world contexts once focused on achieving an immediate goal.

To this point, our brief review suggests that psychopathy may be usefully understood as a deficit in decision making and that this deficit is mediated (at least in part) by dysfunction in the allocation of attention that starts at an early stage of the information-processing stream and appears to be involuntary. More generally, the results offer empirical support for a circumscribed cognitive or psychological basis for psychopathic behavior. A separate but related question is whether there is evidence for a neurobiological basis for psychopathy.

IS THERE A NEUROLOGICAL BASIS FOR PSYCHOPATHY?

Of central importance for the present discussion is whether psychopathic behavior can be attributed directly to an organic biological defect. If a neurobiological substrate for psychopathy could indeed be identified and ultimately treated through clinical intervention (pharmacologic, neurosurgical, or otherwise), it would seem to have significant implications for how psychopaths are handled by our legal system. Here we consider neuroscientific evidence related to two brain areas that have been theorized as central components of the neural substrate for psychopathy: the amygdala and ventromedial prefrontal cortex.

As previously discussed, much attention in psychopathy research has been paid to the psychopaths' experience (or non-experience) of fear—recall the "low-fear hypothesis" (Lykken, 1957, 1995). If a lack of fear is indeed a contributing factor underlying psychopathy, there is ample reason to suspect that psychopathy would be associated with dysfunction within the amygdala, an almond-shaped subcortical structure within the anterior temporal lobe. A wealth of neuroscientific data implicates the amygdala in fear-related processing. The amygdala is necessary for the acquisition of conditioned fear responses (Bechara et al., 1995; LaBar, LeDoux, Spencer, & Phelps, 1995) as well as the recognition of facial expressions of fear (Adolphs, Tranel, Damasio, & Damasio, 1994) and the allocation of attention to fear-related information (Adolphs et al., 2005). In recent years, brain imaging techniques such as positron emission tomography (PET) and magnetic resonance imaging (MRI) have allowed researchers to determine whether psychopaths exhibit abnormalities in the amygdala in terms of structure or function. One structural brain imaging study reports lower amygdala volumes in psychopaths (Yang, Raine, Narr, Colletti, & Toga, 2009). Among the functional imaging studies that compare psychopaths with nonpsychopaths, three report abnormalities in amygdala activity, one finding abnormally low levels of activity during a word identification test (Kiehl et al., 2001), one finding abnormally low levels of activity during a fear conditioning task (Birbaumer et al., 2005), and the other finding abnormally high levels of activity in response to positive emotion pictures (Muller et al., 2003). A related pair of studies have demonstrated that greater levels of "psychopathic" traits among the normal population are associated with lower levels of amygdala activity during socioaffective tasks such as moral judgment (Glenn, Raine, & Schug, 2009) and social cooperation (Rilling et al., 2007). However, viewing fearful facial expressions, which reliably elicits amygdala activation in normal subjects (Adolphs, 2002), did not elicit an abnormally diminished amygdala response in psychopaths (Deeley et al., 2006). One caveat for interpreting these brain imaging results is that neuroimaging data are inherently correlational; brain imaging data alone cannot address whether any observed neural abnormality is a cause or consequence of the disorder. Taken together, these studies offer some intriguing preliminary data relating amygdala dysfunction to psychopathy, but as of yet the data do not seem to provide unequivocal support for a hyporesponsive amygdala as the primary neurobiological basis of psychopathy.

A second candidate brain region for the neuropathological basis of psychopathy is the ventromedial prefrontal cortex (vmPFC). The putative connection between vmPFC dysfunction and psychopathy has long been recognized in the field of behavioral neurology. In 1975, Blumer and Benson coined the term "pseudopsychopathy" to summarize the personality changes ("the lack of adult tact and restraints") observed in their vmPFC-damaged patients (Blumer & Benson, 1975). A decade later, Damasio began a series of clinical and laboratory evaluations of such patients that would offer novel insight into the neural mechanisms of emotion and decision making. Damasio and colleagues' case descriptions of vmPFC patients typically noted the following "psychopathic" traits: lack of empathy and guilt, generally blunted affect, poor long-term planning, irresponsibility, marked lack of insight or concern, and defective decision making despite seemingly intact intellect (Anderson, Barrash, Bechara, & Tranel, 2006; Barrash, Tranel, & Anderson, 2000; Eslinger &

Damasio, 1985). Expanding on these clinical similarities, a number of laboratory paradigms have demonstrated parallel deficits between psychopaths and vmPFC lesion patients. Examples include reversal learning (Budhani, Richell, & Blair, 2006; Hornak et al., 2004), gambling tasks (Bechara, Damasio, Tranel, & Damasio, 1997; Mitchell, Colledge, Leonard, & Blair, 2002) (but see also Losel & Schmucker, 2004; Schmitt, Brinkley, & Newman, 1999), smell identification (Jones-Gotman & Zatorre, 1988; Lapierre, Braun, & Hodgins, 1995), and autonomic physiological responses to emotional stimuli (Blair et al., 1997; Damasio, Tranel, & Damasio, 1990; Patrick et al., 1994). These intriguing similarities hint that vmPFC dysfunction may contribute to psychopathic behavior. However, one notable difference is that the clinical reports of vmPFC patients do not typically feature criminal or violent behavior to the same degree as in psychopaths. An important consideration in this regard is the age at lesion onset in the vmPFC lesion patients. The majority of vmPFC patients described in the literature suffered their brain damage in middle age or later adulthood from a medical condition such as stroke, tumor, or aneurysm. This means that adult-onset vmPFC lesion patients had "normal" psychosocial development through childhood and early adulthood that perhaps mitigates their antisocial behavior following the lesion. A valuable source of data to inform this issue is individuals who suffered vmPFC damage very early in life (i.e., before age 2). Only a handful of such cases have been reported (Anderson, Bechara, Damasio, Tranel, & Damasio, 1999; Anderson, Wisnowski, Barrash, Damasio, & Tranel, 2009), but the results clearly indicate that early vmPFC damage leads to a pattern of behavior through adolescence and young adulthood that is even more reminiscent of psychopathy. Unlike the adult-onset vmPFC lesion patients, the early-onset patients exhibit stereotypical "psychopathic" antisocial behaviors such as petty theft, physical assaults, sexual promiscuity, and chronic lying. Taken together, the data from lesion patients indirectly support developmental dysfunction within vmPFC as a putative neurobiological mechanism of psychopathy.

As a noteworthy aside, we point out that the "somatic marker hypothesis" (Damasio, 1994) (described earlier in the chapter), which figures prominently in psychopathy research (Losel & Schmucker, 2004; Schmitt et al., 1999; van Honk, Hermans, Putman, Montagne, & Schutter, 2002), was actually developed to account for the decision-making impairments observed in vmPFC patients.

While initial studies on the neurobiology of psychopathy focused largely on individual brain structures with well-established roles in social-affective processing (e.g., amygdala and vmPFC), more recent neuroimaging work has associated psychopathy with reduced connectivity between more widely distributed networks of brain areas (Ly et al., 2012; Philippi et al., in review). Interestingly, the comparatively weak functional connections identified in psychopaths in these studies correspond to cortical networks that have been implicated in shifting or maintaining attentional state (Dosenbach et al., 2007; Fox et al., 2005; Raichle et al., 2001). Hence, this emerging line of neuroimaging research, which highlights the coordinated activity within large-scale cortical networks as the neural basis for attention and information integration, suggests a putative neurobiological mechanism for the observed abnormalities in attentional processing in psychopathy.

In sum, the results of neuroscientific studies suggest intriguing neuropathophysiological models of psychopathy, but at present there are insufficient data to conclude with certainty that psychopathy arises as a direct consequence of dysfunction in a particular area (or areas) of the brain. However, the field of cognitive neuroscience has been making rapid progress in identifying brain-based mechanisms underlying information processing and decision-making competence. We are optimistic that continuous advancements in technological and theoretical precision will ultimately reveal a biological basis for psychopathy.

In particular, one area that we believe warrants more precise and rigorous investigation is the distinction between "primary" and "secondary" psychopaths. This distinction reflects the long-theorized possibility that the extreme affective and behavioral traits that characterize the disorder could arise through different causal mechanisms. In other words, psychopaths may consist of phenotypically similar, but etiologically distinct subtypes (Lykken, 1957, 1995). In the "primary" subtype, psychopathy is presumed to arise directly from some fundamental intrinsic deficit, likely involving innate dysfunction in basic affective and/or attentional mechanisms. By contrast, "secondary" psychopathy is thought to arise as an acquired disturbance of social and affective processing—an indirect consequence of environmental or psychosocial factors such as parental abuse, socioeconomic disadvantage, poor intellect, substance abuse, or neurotic anxiety (Blackburn, Logan, Donnelly, & Renwick, 2008; Cleckley, 1976; Karpman, 1946, 1948; Lykken, 1995; Skeem, Johansson, Andershed, Kerr, & Louden, 2007). Clearly this theoretical distinction could have profound implications for research on the psychobiological basis of the disorder.

If there are indeed multiple, distinct causal mechanisms for psychopathy, then one may expect the different etiological subtypes to exhibit distinct psychological and neurobiological profiles within the context of similarly flagrant antisocial behaviors. The question, then, is how to differentiate primary psychopaths from secondary psychopaths for the purposes of research. A number of previous studies have differentiated primary and secondary psychopaths based on levels of trait anxiety (Arnett et al., 1997; Blackburn, 1975; Brinkley, Newman, Widiger, & Lyman, 2004; Fagan & Lira, 1980; Hiatt et al., 2004). This practice is supported by ample theoretical and empirical work. In his seminal clinical descriptions, Cleckley stressed the importance of considering anxiety levels for the classification of psychopathy: "…primary] psychopaths are sharply characterized by the lack of anxiety…I do not believe that [primary] psychopaths should be identified with the psychoneurotic group" (Cleckley, 1976, p. 257). Following Cleckley's recommendation of distinguishing low-anxiety individuals from those with high (neurotic) levels of anxiety, a large and growing number of laboratory studies demonstrate abnormal behavioral results for low-anxious (primary) psychopaths but not necessarily for high-anxious (secondary) psychopaths (Arnett, Howland, Smith, & Newman, 1993; Arnett et al., 1997; Chesno & Kilmann, 1975; Fagan & Lira, 1980; Lykken, 1957; Newman, Kosson, & Patterson, 1992; Newman et al., 1990, 1997; O'Brien & Frick, 1996; Schmitt et al., 1999; Skeem et al., 2007; Smith et al., 1992; Zeier et al., 2009). Despite this substantial literature, none of the recent brain imaging studies on psychopathy has differentiated subjects on the basis of anxiety. Assuming that the low- and high-anxiety psychopathy subtypes do indeed correspond to etiologically distinct conditions with unique psychological and neurobiological profiles, then the combination of the two subtypes in a single group of psychopaths for the purposes of a research study may result in muddled and inconsistent results. Thus we believe that the primary/secondary distinction is an important, but too often overlooked, consideration for investigating the biological basis of psychopathy.

SUMMARY AND CONCLUSION

The aim of this chapter is to outline the current state of knowledge on the psychological and neurobiological mechanisms that underlie psychopathy so that legal experts may incorporate this information as they craft policy to best ensure public safety and welfare. We assert that it is reasonable and instructive to conceptualize psychopathy broadly as a disorder of decision making and to consider the specific deficits that may contribute to the overall decision-making impairment. To this end, we have reviewed theoretical and empirical work suggesting that diminished emotional reactivity and a defect in the flexible allocation of attention are likely critical factors. We propose that psychopathy may entail a primary deficit in attention, as an attentional defect could theoretically account for the observed abnormalities in processing both affective and nonaffective information. Importantly, the attentional deficit in psychopathy appears to operate at an early, relatively automatic stage of information processing, suggesting that psychopathic behavior may reflect a lack of decision-making competence rather than a deliberate intention to harm others. To make this conclusion more definitively, future research will need to be carefully designed to parse and specify the information- processing capability of psychopaths.

Regarding the brain mechanisms of psychopathy, there are multiple neural structures that have been proposed as likely candidates, notably the amygdala and vmPFC. Both brain areas have been linked to psychopathy, albeit largely through indirect evidence at this point. Although there is currently no clear consensus on the biological root of the disorder, technological advances in brain imaging have clearly made the pursuit of a neural basis for psychopathy a tractable field of inquiry. As the science progresses, we expect such data to more frequently arise in court cases, specifically with respect to questions of culpability, likelihood of future offense, and prospects for rehabilitation. To establish the appropriate framework for incorporating psychopathy research into the legal system, legal scholars will have to consult closely with scientists as they further elucidate the psychological and neurobiological mechanisms of psychopathy.

REFERENCES

Adolphs, R. (2002). Recognizing emotion from facial expressions: Psychological and neurological mechanisms. *Behavioral and Cognitive Neuroscience Reviews, 1*, 21–62.

Adolphs, R., Gosselin, F., Buchanan, T. W., Tranel, D., Schyns, P., & Damasio, A. R. (2005). A mechanism for impaired fear recognition after amygdala damage. *Nature, 433*, 68–72.

Adolphs, R., Tranel, D., Damasio, H., & Damasio, A. (1994). Impaired recognition of emotion in facial expressions following bilateral damage to the human amygdala. *Nature, 372*, 669–672.

Anderson, S. W., Barrash, J., Bechara, A., & Tranel, D. (2006). Impairments of emotion and real-world complex behavior following childhood- or adult-onset damage to ventromedial prefrontal cortex. *Journal of the International Neuropsychological Society, 12*, 224–235.

Anderson, S. W., Bechara, A., Damasio, H., Tranel, D., & Damasio, A. R. (1999). Impairment of social and moral behavior related to early damage in human prefrontal cortex. *Nature Neuroscience, 2*, 1032–1037.

Anderson, S. W., Wisnowski, J. L., Barrash, J., Damasio, H., & Tranel, D. (2009). Consistency of neuropsychological outcome following damage to prefrontal cortex in the first years of life. *Journal of Clinical and Experimental Neuropsychology, 31*, 170–179.

Arnett, P. A., Howland, E. W., Smith, S. S., & Newman, J. P. (1993). Autonomic responsivity during passive avoidance in incarcerated psychopaths. *Personality and Individual Differences, 14*, 173–185.

Arnett, P. A., Smith, S. S., & Newman, J. P. (1997). Approach and avoidance motivation in psychopathic criminal offenders during passive avoidance. *Journal of Personality and Social Psychology, 72*, 1413–1428.

Barrash, J., Tranel, D., & Anderson, S. W. (2000). Acquired personality disturbances associated with bilateral damage to the ventromedial prefrontal region. *Developmental Neuropsychology, 18*, 355–381.

Baskin-Sommers, A.R., Curtin, J.J & Newman, J.P. (2011). Specifying the attentional selection that moderates the fearlessness of psychopathic offenders. Psychological Science,22(2), 226–234. PMC3358698

Bechara, A., Damasio, H., Tranel, D., & Damasio, A. R. (1997). Deciding advantageously before knowing the advantageous strategy. *Science, 275*, 1293–1295.

Bechara, A., Tranel, D., Damasio, H., Adolphs, R., Rockland, C., & Damasio, A. R. (1995). Double dissociation of conditioning and declarative knowledge relative to the amygdala and hippocampus in humans. *Science, 269*, 1115–1158.

Birbaumer, N., Veit, R., Lotze, M., Erb, M., Hermann, C., Grodd, W., et al. (2005). Deficient fear conditioning in psychopathy: A functional magnetic resonance imaging study. *Archives of General Psychiatry, 62*, 799–805.

Bishop, S. J., Jenkins, R., & Lawrence, A. D. (2007). Neural processing of fearful faces: Effects of anxiety are gated by perceptual capacity limitations. *Cerebral Cortex, 17*, 1595–1603.

Blackburn, R. (1975). An empirical classification of psychopathic personality. *British Journal of Psychiatry, 127*, 456–460.

Blackburn, R., Logan, C., Donnelly, J. P., & Renwick, S. J. (2008). Identifying psychopathic subtypes: Combining an empirical personality classification of offenders with the psychopathy checklist-revised. *Journal of Personality Disorders, 22*, 604–622.

Blair, R. J., Jones, L., Clark, F., & Smith, M. (1997). The psychopathic individual: A lack of responsiveness to distress cues? *Psychophysiology, 34*, 192–198.

Blumer, D., & Benson, D. F. (1975). Personality changes with frontal and temporal lesions. In D. F. Benson & D. Blumer (Eds.), *Psychiatric aspects of neurological disease*(p. 151–170). New York: Grune & Stratton.

Brinkley, C. A., Newman, J. P., Widiger, T. A., & Lyman, D. R. (2004). Two approaches to parsing the heterogeneity of psychopathy. *Clinical Psychology: Science and Practice, 11*, 69–94.

Budhani, S., Richell, R. A., & Blair, R. J. (2006). Impaired reversal but intact acquisition: Probabilistic response reversal deficits in adult individuals with psychopathy. *Journal of Abnormal Psychology, 115*, 552–558.

Chesno, F. A., & Kilmann, P. R. (1975). Effects of stimulation intensity on sociopathic avoidance learning. *Journal of Abnormal Psychology, 84*, 144–150.

Cleckley, H. (1941). *The mask of sanity*. St. Louis, MO: Mosby.

Cleckley, H. (1976). *The mask of sanity* (5th ed.). St. Louis, MO: Mosby.

Curtin, J. J., & Fairchild, B. A. (2003). Alcohol and cognitive control: Implications for regulation of behavior during response conflict. *Journal of Abnormal Psychology, 112*, 424–436.

Damasio, A. R. (1994). *Descartes' error*. New York: Putnam.

Damasio, A. R., Tranel, D., & Damasio, H. (1990). Individuals with sociopathic behavior caused by frontal damage fail to respond autonomically to social stimuli. *Behavioural Brain Research, 41*, 81–94.

Deeley, Q., Daly, E., Surguladze, S., Tunstall, N., Mezey, G., Beer, D., et al. (2006). Facial emotion processing in criminal psychopathy: Preliminary functional magnetic resonance imaging study. *British Journal of Psychiatry, 189*, 533–539.

Dosenbach, N. U., Fair, D. A., Miezin, F. M., Cohen, A. L., Wenger, K. K., Dosenbach, R. A.,... Petersen, S. E. (2007). Distinct brain networks for adaptive and stable task control in humans. *Proc Natl Acad Sci U S A, 104*(26), 11073-11078.

Eslinger, P. J., & Damasio, A. R. (1985). Severe disturbance of higher cognition after bilateral frontal lobe ablation: Patient EVR. *Neurology, 35*, 1731–1741.

Fagan, T. J., & Lira, F. T. (1980). The primary and secondary sociopathic personality: Differences in frequency and severity of antisocial behaviors. *Journal of Abnormal Psychology, 89*, 493–496.

Fox, M. D., Snyder, A. Z., Vincent, J. L., Corbetta, M., Van Essen, D. C., & Raichle, M. E. (2005). The human brain is intrinsically organized into dynamic, anticorrelated functional networks. *Proc Natl Acad Sci U S A, 102*(27), 9673-9678.

Glass, S. J., & Newman, J. P. (2009). Emotion processing in the criminal psychopath: The role of attention in emotion-facilitated memory. *Journal of Abnormal Psychology, 118*, 229–234.

Glenn, A. L., Raine, A., & Schug, R. A. (2009). The neural correlates of moral decision-making in psychopathy. *Molecular Psychiatry, 14*, 5–6.

Grant, V. W. (1965). *The menacing stranger: A primer on the psychopath*. Oceanside, NY: Dabor Science.

Hare, R. D. (1978). Psychopathy and electrodermal responses to nonsignal stimulation. *Biological Psychology, 6*, 237–246.

Hare, R. D., & Quinn, M. J. (1971). Psychopathy and autonomic conditioning. *Journal of Abnormal Psychology, 77*, 223–235.

Hiatt, K. D., Schmitt, W. A., & Newman, J. P. (2004). Stroop tasks reveal abnormal selective attention among psychopathic offenders. *Neuropsychology, 18*, 50–59.

Hornak, J., O'Doherty, J., Bramham, J., Rolls, E. T., Morris, R. G., Bullock, P. R., et al. (2004). Reward-related reversal learning after surgical excisions in orbito-frontal or dorsolateral prefrontal cortex in humans. *Journal of Cognitive Neuroscience, 16*, 463–478.

Jones-Gotman, M., & Zatorre, R. J. (1988). Olfactory identification deficits in patients with focal cerebral excision. *Neuropsychologia, 26*, 387–400.

Karpman, B. (1946). Psychopathy in the scheme of human typology. *Journal of Nervous and Mental Disease, 103*, 276–288.

Karpman, B. (1948). The myth of the psychopathic personality. *American Journal of Psychiatry, 104*, 523–534.

Kiehl, K. A., Smith, A. M., Hare, R. D., Mendrek, A., Forster, B. B., Brink, J., et al. (2001). Limbic abnormalities in affective processing by criminal psychopaths as revealed by functional magnetic resonance imaging. *Biological Psychiatry, 50*, 677–684.

LaBar, K. S., LeDoux, J. E., Spencer, D. D., & Phelps, E. A. (1995). Impaired fear conditioning following unilateral temporal lobectomy in humans. *Journal of Neuroscience, 15*, 6846–9855.

Lapierre, D., Braun, C. M., & Hodgins, S. (1995). Ventral frontal deficits in psychopathy: Neuropsychological test findings. *Neuropsychologia, 33*, 139–151.

Lavie, N., Hirst, A., de Fockert, J. W., & Viding, E. (2004). Load theory of selective attention and cognitive control. *Journal of Experimental Psychology: General, 133*, 339–354.

Lilienfeld, S. O., & Arkowitz, H. (2007). What "psychopath" means. *Scientific American Mind*, April/May, 90–91.

Losel, F., & Schmucker, M. (2004). Psychopathy, risk taking, and attention: A differentiated test of the somatic marker hypothesis. *Journal of Abnormal Psychology, 113*, 522–529.

Lykken, D. T. (1957). A study of anxiety in the sociopathic personality. *Journal of Abnormal Psychology, 55*, 6–10.

Lykken, D. T. (1995). *The antisocial personalities.* Mahwah, NJ: Erlbaum.

Ly, M., Motzkin, J. C., Philippi, C. L., Kirk, G. R., Newman, J. P., Kiehl, K. A., & Koenigs, M. (2012). Cortical thinning in psychopathy. *Am J Psychiatry, 169*(7), 743-749.

Mitchell, D. G., Colledge, E., Leonard, A., & Blair, R. J. (2002). Risky decisions and response reversal: Is there evidence of orbitofrontal cortex dysfunction in psychopathic individuals? *Neuropsychologia, 40*, 2013–2022.

Muller, J. L., Sommer, M., Wagner, V., Lange, K., Taschler, H., Roder, C. H., et al. (2003). Abnormalities in emotion processing within cortical and subcortical regions in criminal psychopaths: Evidence from a functional magnetic resonance imaging study using pictures with emotional content. *Biological Psychiatry, 54*, 152–162.

Newman, J. P., Curtin, J. J., Bertsch, J. D., & Baskin-Sommers, A. R. (2010). Attention moderates the fearlessness of psychopathic offenders. *Biological Psychiatry, 67*, 66–70.

Newman, J. P., & Kosson, D. S. (1986). Passive avoidance learning in psychopathic and nonpsychopathic offenders. *Journal of Abnormal Psychology, 95*, 252–256.

Newman, J. P., Kosson, D. S., & Patterson, C. M. (1992). Delay of gratification in psychopathic and nonpsychopathic offenders. *Journal of Abnormal Psychology, 101*, 630–636.

Newman, J. P., & Lorenz, A. R. (2003). Response modulation and emotion processing: Implications for psychopathy and other dysregulatory psychopathology. In: R. J. Davidson, K. Scherer, & H. H. Goldsmith (Eds.), *Handbook of affective sciences* (pp. 904–929). Oxford: Oxford University Press.

Newman, J. P., Patterson, C. M., Howland, E. W., & Nichols, S. L. (1990). Passive avoidance in psychopaths: The effects of reward. *Personality and Individual Differences, 11*, 1101–1114.

Newman, J. P., Patterson, C. M., & Kosson, D. S. (1987). Response perseveration in psychopaths. *Journal of Abnormal Psychology, 96*, 145–148.

Newman, J. P., Schmitt, W. A., Voss, W. D. (1997). The impact of motivationally neutral cues on psychopathic individuals: Assessing the generality of the response modulation hypothesis. *Journal of Abnormal Psychology, 106*, 563–575.

O'Brien, B. S., & Frick, P. J. (1996). Reward dominance: Associations with anxiety, conduct problems, and psychopathy in children. *Journal of Abnormal Child Psychology, 24*, 223–240.

Ogloff, J. R., & Wong, S. (1990). Electrodermal and cardiovascular evidence of a coping response in psychopaths. *Criminal Justice and Behavior, 17*, 231–245.

Patrick, C. J., Bradley, M. M., & Lang, P. J. (1993). Emotion in the criminal psychopath: Startle reflex modulation. *Journal of Abnormal Psychology, 102*, 82–92.

Patrick, C. J., Cuthbert, B. N., & Lang, P. J. (1994). Emotion in the criminal psychopath: Fear image processing. *Journal of Abnormal Psychology, 103*, 523–534.

Patterson, C. M., & Newman, J. P. (1993). Reflectivity and learning from aversive events: Toward a psychological mechanism for the syndromes of disinhibition. *Psychological Reviews, 100*, 716–736.

Philippi, C., Motzkin, J., Newman, J. P., Kiehl, K. A., & Koenigs, M. (in review). Altered resting-state functional connectivity within and between cortical networks in psychopathy.

Rilling, J. K., Glenn, A. L., Jairam, M. R., Pagnoni, G., Goldsmith, D. R., Elfenbein, H. A., et al. (2007). Neural correlates of social cooperation and non-cooperation as a function of psychopathy. *Biological Psychiatry, 61*, 1260–1271.

Schmauk, F. J. (1970). Punishment, arousal, and avoidance learning in sociopaths. *Journal of Abnormal Psychology, 76*, 325–335.

Schmitt, W. A., Brinkley, C. A., & Newman, J. P. (1999). Testing Damasio's somatic marker hypothesis with psychopathic individuals: Risk takers or risk averse? *Journal of Abnormal Psychology, 108*, 538–543.

Shapiro, D. (1965). *Neurotic styles.* New York: Basic Books.

Skeem, J., Johansson, P., Andershed, H., Kerr, M., & Louden, J. E. (2007). Two subtypes of psychopathic violent offenders that parallel primary and secondary variants. *Journal of Abnormal Psychology, 116*, 395–409.

Smith, S. S., Arnett, P. A., & Newman, J. P. (1992). Neuropsychological differentiation of psychopathic and nonpsychopathic criminal offenders. *Personality and Individual Differences, 13*, 1233–1245.

Tharp, V. K., Maltzman, I., Syndulko, K., & Ziskind, E. (1980). Autonomic activity during anticipation of an aversive tone in noninstitutionalized sociopaths. *Psychophysiology, 17*, 123–128.

van Honk, J., Hermans, E. J., Putman, P., Montagne, B., & Schutter, D. J. (2002). Defective somatic markers in sub-clinical psychopathy. *NeuroReport, 13*, 1025–1027.

Yang, Y., Raine, A., Narr, K. L., Colletti, P., & Toga, A. W. (2009). Localization of deformations within the amygdala in individuals with psychopathy. *Archives of General Psychiatry, 66*, 986–994.

Zeier, J. D., Maxwell, J. S., & Newman, J. P. (2009). Attention moderates the processing of inhibitory information in primary psychopathy. *Journal of Abnormal Psychology, 118*, 554–563.

CHAPTER 7

Do Psychopaths Make Moral Judgments?

Jana Schaich Borg and Walter P. Sinnott-Armstrong

Psychopaths are infamous for their immoral behavior. Indeed, immoral (or, in clinical terms, "antisocial") behavior is one of the critical symptoms used to diagnose psychopathy. However, we still do not understand *why* psychopaths behave so immorally. One popular hypothesis, which we will call the moral judgment hypothesis, is that psychopaths behave immorally because they lack the ability to make moral judgments. This chapter reviews current scientific evidence supporting or opposing the claim that psychopaths lack this ability. The primary competing explanation for psychopaths' immoral behavior is the moral motivation hypothesis: Psychopaths can make moral judgments, but they simply do not care about the verdicts of those judgments (Cima, Tonnaer, & Hauser, 2010). Of course, both hypotheses might be true to some extent, as each hypothesis might hold for some moral judgments but not others or for some psychopaths but not others. Further, psychopaths might only dimly understand immorality and therefore not care about it as much as normal people, but still care about it to some extent. Despite such potential complications, the "moral judgment" and "moral motivation" hypotheses are usually presented as competing options for explaining psychopaths' behavior, and many debates have grown out of efforts to determine which of those simple accounts is most likely to be accurate (Malatesti & McMillan, 2010). For scientific disciplines, these debates have drawn significant attention because their collective resolution can dramatically affect models of moral judgment and behavior as well as clinical treatment plans for antisocial patients (Glenn, 2010).

The issue of whether psychopaths behave immorally because they lack the ability to make moral judgments or because they lack moral motivation is also important to law for several reasons. First, the issue's resolution might affect whether psychopaths should be considered criminally (or morally) responsible for their crimes (Malatesti & McMillan, 2010). The most common versions of the insanity defense state that a defendant's responsibility depends in part on whether the defendant "did not know that what he was doing was wrong"[1] or "lacks substantial capacity...to appreciate the wrongfulness of his conduct."[2] These legal rules are open to various interpretations,[3] but some interpretations make the ability to form moral judgments necessary for criminal responsibility. If psychopaths cannot recognize that what they are doing is wrong, then they would seem to qualify for some versions of the insanity defense. Second, the ability of psychopaths to make normal moral judgments might also be relevant to predictions of future crime. Someone who cannot tell that an act is immoral will probably be more likely to commit that act again than someone who believes that act is immoral. Thus, psychopaths' moral judgment abilities might be relevant to how psychopaths should be sentenced (or whether they should be paroled or detained).[4] Third, whether psychopaths cannot tell right from wrong or, instead, can tell the difference but do not care could direct treatment of psychopaths by specifying which deficit—cognitive, emotional, or motivational—needs to be treated.[5] It might also affect whether psychopaths should be legally required to undergo treatment, given that some of the moral judgment deficits psychopaths potentially have may end up being treatable while others may not. In these ways, for law as well as for science, it is important to determine whether psychopaths can make or appreciate moral judgments.[6]

The goal of this chapter is to provide a comprehensive review of the scientific evidence for and against the hypothesis that psychopaths are impaired in their moral judgments. In our view, contrary to popular opinion, the current literature does not provide evidence suggesting psychopaths have severe moral cognition deficits. However, as will become clear, few firm conclusions can be reached about moral cognition in psychopaths without further research. So far there are very little data examining moral judgment or decision making in psychopaths, partially because historically psychopathy research has been practically and financially difficult to implement. Another reason is that psychopaths are often pathological liars, so it is hard to determine what they really believe. Additional obstacles arise because different researchers have used inconsistent criteria for diagnosing psychopathy and because few scientific tests of moral judgment or belief are established and/or standardized. To interpret this limited evidence, it is critical that both psychopathy and moral judgments be defined carefully. We therefore begin by discussing both of these definitions in turn.

DEFINING PSYCHOPATHY

Psychopathy is primarily diagnosed using the Psychopathy Checklist–Revised (PCL-R).[7] The PCL-R is based on a semistructured interview that assesses interviewees on 20 personality dimensions that can be divided into two separate factors: Factor 1, which reflects affective and interpersonal traits, and Factor 2, which reflects antisocial and unstable lifestyle habits. Factor 1 can further be broken up into two facets: Facet 1, representing interpersonal traits, and Facet 2, representing affective traits. Factor 2 can be broken up into Facet 3, representing an impulsive lifestyle, and Facet 4, representing antisocial behavior. The interview is supplemented by a full background check that can verify or falsify information provided by the interviewee.

Although there is still much debate over whether psychopathy is a categorical disorder or a spectrum disorder, a psychopath is clinically defined as anyone who scores 30 or above on the PCL-R. Unfortunately, most published studies of moral judgment in psychopaths have very few, if any, participants who score above 30, and many studies re-define "psychopath" to indicate a significantly lower PCL-R score. Because of the distribution of scores, a sample of subjects scoring 26 or above will often contain a high percentage of subjects with scores below 30. In addition, as clinicians who have interviewed psychopaths can attest, there is something qualitatively very different about psychopaths who score 34 or above compared to those who score around 30. Hare notes this impression and describes those who score 34 or above as "high psychopaths" (Hare, 1991, 2003). Almost no studies of moral judgment in psychopaths have participants who score above 34. Thus, moral judgment has been assessed very little in clinical psychopaths, and even less in this group of highest-scoring psychopaths whom clinicians differentiate from other psychopaths.

Another analytical difficulty to keep in mind is that older studies, before the PCL-R, often assessed psychopathy using the Minnesota Multiphasic Personality Inventory (MMPI). In contrast to the PCL-R, the MMPI is based only on measures of self-reports provided by the interviewee, which are problematic given that psychopaths are known to be pathological liars.[8] Perhaps as a result, the relevant scores on the MMPI do not correlate well with scores on the PCL-R, especially with PCL-R Factor 1 (O'Kane, Fawcett, & Blackburn, 1996). Thus, studies that use the MMPI to assess psychopathy might not be measuring the same population as studies that use the PCL-R, making it challenging to compare the results of studies that use these different diagnostic tests.

Acknowledging these two issues, it is good practice to ask the following questions when assessing the literature currently available on psychopathy and moral judgment: (1) was the PCL-R (or an accepted derivative) used to assess psychopathy? If not, then the population being described might be dramatically different from a clinical psychopathic population. (2) If the PCL-R was used, how many participants scored a 30 or above (or 34 and above)? If none, then the population being described does not in fact contain any clinical psychopaths.

DEFINING MORAL JUDGMENT

Defining a moral judgment is both more challenging and more controversial than defining a psychopath, partly because there are many different common usages of the term. In our discussion, we use "moral judgment" to refer to the mental state or event of judging that some act, institution, or person is morally wrong or right, good or bad. Even if moral judgments are understood as mental states or events, however, there are many controversial issues surrounding what counts as a moral judgment.

One issue is that it is often unclear which mental states count as moral judgments as opposed to judgments of some nonmoral kind. A primary reason for this is that cultures can differ in what they see as morally relevant. For example, a woman's decision to wear a short-sleeved shirt would not be considered a moral matter in most Christian cultures, but would be considered morally important in many Muslim cultures. Furthermore, even when cultures agree on what has moral content, the same act (such as killing a family member to maintain a family's honor) can be seen as morally obligatory in some cultures while morally forbidden in others. Even those who argue for some universal morality or "moral sense" rarely specify in detail how to determine which parts of morality are fundamental or universal as opposed to culturally labile. This vagueness makes it difficult for scientists or the law to develop standard methods of assessing moral aptitude that are valid across cultures and/or populations. Most relevant to this chapter, even if we assess moral judgments within a single culture, it is still unclear whether current methods of assessment are appropriate for all levels of education or socioeconomic status. This issue needs to be kept in mind when interpreting studies of morality in psychopaths because many psychopathy studies are conducted in incarcerated populations with lower than average education and socioeconomic levels.

Another underappreciated complication is that morality covers a diversity of areas, so people can have severe deficits in some types of moral judgments without any deficits at all in other kinds of moral judgments. For example, some moral judgments are based on issues of harm and elicit feelings of anger (such as assault and stealing), whereas other moral judgments are based on issues of impurity (such as incest or cannibalism) and elicit feelings of disgust (Haidt, Koller, & Dias 1993; Parkinson et al., 2011; Schaich Borg, Lieberman, & Kiehl, 2008). Thus, a person lacking the ability to feel disgust might be able to understand moral judgments condemning assault and stealing but not moral judgments condemning incest and cannibalism. In addition to eliciting different emotions, moral judgments about different kinds of acts can also require different cognitive abilities. For example, some moral judgments require one to understand another person's intentions, while other moral judgments require one to calculate and weigh consequences (Cushman, 2008; Young, Camprodon, Hauser, Pascual-Leone, & Saxe, 2010). Thus, a person lacking the ability for theory of mind, such as someone with autism, might be incapable of responding appropriately to the first type of judgment but be perfectly capable of making the second type of judgment. A stroke patient who has trouble understanding quantities or doing basic math, in contrast, might have the reverse problem. It is, therefore, important both in theory and in practice for the law to focus on moral assessments that are appropriately matched to the emotional and cognitive requirements of the crime(s) under consideration.

A third complication is that the ability to make moral judgments requires several subsidiary abilities. To make a successful moral judgment, for example, one must detect morally relevant variables, understand their moral relevance, integrate that relevance into a sound weighing of options, interpret that weighing of options accurately, and then plan actions that are consistent with the outcome of this moral deliberation (Schaich Borg, Sinnott-Armstrong, Calhoun, & Kiehl, 2011). All of these subsidiary abilities directly contribute to one's overall ability to make a moral judgment or behave morally, but the law or other disciplines might care only about a specific subset of these abilities. If so, because most scientific studies of moral judgment are not designed to distinguish the contribution of each of these subsidiary abilities, it is important to recognize that the data surveyed in this chapter may reflect competencies or deficits in a subset of subsidiary abilities different from those that matter to responsibility or other practical or legal issues.

Finally, a fourth complication is that some popular conceptions of moral judgment conflate judgment with affect. Moral judgment is distinguishable from moral emotions and motivations. In this chapter, we discuss psychopaths' moral judgments or beliefs and the application of these judgments or beliefs to particular real or hypothetical situations, not the emotions that accompany those moral judgments. This distinction is important because it is possible to have moral judgments without morally relevant emotions. This happens, for example, when people are convinced by arguments that certain acts are immoral, but do not yet have emotions that are consistent with those arguments or their moral beliefs.[9] One illustration of this dissociation is people who sincerely judge that it is morally wrong to eat meat but still do not actually feel any associated compunction or guilt when they eat a hamburger. On the other hand, it is also possible to have morally relevant emotions without relevant moral judgments. People who were raised in the Mormon faith, for example, might feel guilty while drinking coffee without truly believing that there is anything morally wrong with drinking coffee. These people have real guilt feelings, but they do not endorse those feelings as justified, so they do not make a moral judgment in the sense that is relevant to this chapter. In sum, even if some colloquial references to being a "good person" imply that morality requires both good moral judgment and appropriate corresponding moral emotions, motivations, and actions, and even if some legal definitions of "appreciating the moral wrongness" of an act require one to feel bad about what one judges to be wrong, the ability to judge something to be morally wrong is distinguishable from the emotions such judgments elicit. This chapter surveys only data relevant to the former ability.

A Word on Psychopaths' Lack of Empathy

The distinction between moral judgment and moral emotion/motivation is particularly important for this chapter because it is well known that psychopaths have reduced empathy. Lack of empathy is one of the diagnostic criteria for psychopathy (in Facet 2), so most high-scoring psychopaths will have empathic deficits by definition. Four studies have tried to quantify these deficits by measuring adult psychopaths' galvanic skin responses while they were observing people in physical distress. The galvanic skin response technique does not measure empathy directly, but it does measure how much arousal one feels when observing another in distress, and that is hypothesized to be an important part of an empathetic response. Two of the four studies found that psychopaths show little to no change in skin resistance in response to observing a confederate get shocked (Aniskiewicz, 1979; House & Milligan, 1976). A third study found that psychopaths actually show increased changes in skin resistance in response to observing a confederate get shocked (Sutker, 1970). Yet, these first three studies assessed psychopathy with the MMPI, rather than the PCL-R. The only study to employ the PCL-R while examining galvanic skin responses to other people in physical distress supported the results of the first two MMPI studies. This study used 18 psychopaths (scoring 30 or higher on the PCL-R) and 18 nonpsychopaths (scoring 20 or lower on the PCL-R) and found that psychopaths did demonstrate significant galvanic skin responses to pictures of distress cues (that is, a picture of a group of crying adults or a close-up of a crying face), but these responses were much reduced compared to those in nonpsychopaths (Blair, Jones, Clark, & Smith, 1997). These studies together suggest that psychopaths are much less aroused than nonpsychopaths when they observe others in pain or distress, but they still might be aroused to some degree.[10]

We want to emphasize that this iconic symptom of psychopathy does not indicate that psychopaths cannot make moral judgments. Lack of empathy might help to explain some of psychopaths' behavior, because empathy does correlate with prosocial behavior, especially when few conscious moral rules are in place (e.g., Batson, 1991, 2011). However, neither psychopaths' lack of empathy nor their immoral behavior shows that psychopaths do not make normal moral judgments. Instead, psychopaths might make normal moral judgments but fail to translate this cognitive ability into normal emotions or motivations to avoid immoral actions. The ability to empathize is not functionally, neurologically, or psychologically the same as the ability to judge that something is morally wrong, nor is it the same as the ability to guide one's action in accordance with a moral judgment. A person who does not respond emotionally to another person in pain still might be able to make appropriate judgments about whether it is morally wrong to cause pain in another person. Indeed, a recent study found no correlation between empathy (as opposed to theory of mind) and awareness that a situation has moral or ethical implications, willingness to use utilitarian- or nonutilitarian-based rules in moral judgments, or the likelihood of agreement with a given verdict in a moral scenario (Mencl & May, 2009). Hence, psychopaths might make moral judgments without any feelings of empathy. This possibility makes it important for science as well as law to distinguish moral judgment from empathy and focus on moral judgment separately. That is what we do in this chapter.

HOW PSYCHOPATHS PERFORM ON MORAL JUDGMENT TESTS

With these definitions and clarifications in hand, we are ready to review the literature assessing moral judgments by psychopaths. Empirical studies of moral judgments vary dramatically in their assumptions and measurements. The field of Law and Neuroscience will need to decide whether any of these tasks adequately index the ability to "know" or "appreciate" what is wrong as referenced by the M'Naghten Rule or the Model Penal Code (cited in notes 1 and 2). To facilitate such reflection, this chapter provides details about the specific tests that are used to assess the relationship between psychopathy and morality as well as the results from the administration of such tests to psychopaths. The results detailed in the text are summarized in Table 7.1.

Although many relevant studies have been implemented in adolescents, the construct of psychopathy is not as well defined in adolescents, and the legal issues are also complicated by their juvenile status. Therefore, this review is confined to studies testing adults.

KOHLBERGIAN TESTS OF MORAL REASONING

Moral Judgment Interview

One of the first established measures of moral judgment was Lawrence Kohlberg's "Moral Judgment Interview" (MJI; Kohlberg, 1958). In the MJI, participants are asked to resolve complex moral dilemmas in an interview format during which trained experimenters ask directed open-ended questions. The goal is to determine not which moral judgment (conclusion, resolution, or verdict) is reached after deliberation but rather which types of reasons (justification or reasoning) people give to support their moral conclusions. In other words, the critical variable for Kohlberg is *why* participants judge something to be morally right or wrong in these dilemmas, not *what* they judge to be morally right or wrong.

The reasons that participants give for their resolutions in the MJI are divided into three major successive levels, each with two sublevels. The first level, called "pre-conventional" reasoning, comprises reasons based on immediate consequences to oneself (via reward or punishment). The second level, called "conventional" reasoning, comprises reasons based on the expectations of social groups and society. The third level, called "post-conventional" reasoning, comprises reasons based on relatively abstract moral principles independent from existing social rules, law, or authority. Importantly, post-conventional reasoning can reflect either utilitarian or deontological principles, or both. Kohlberg argues that individuals and cultures advance through these stages in a fixed order, and also that later levels of moral reasoning reflect better moral reasoning than earlier levels (Kohlberg, 1973).

Table 7.1 Published studies testing psychopaths' moral judgments

Study	Test administered	Psychopathy assessment used	PCL-R cut-off used in published analyses	Number of psychopaths scoring > 30 on the PCL-R	Summary of result
Link et al. (1977)	Moral Judgment Interview	MMPI	–	–	Psychopaths had *improved* moral reasoning compared to both control groups
O'Kane et al. (1996)	Defining Issues Test	PCL-R	Not available	1	*No correlation* between total PCL-R scores and P-scores
Lose (1997)	Defining Issues Test	PCL-R	Not available	0	*No significant differences* in the P-scores of the inmates with the five highest PCL-R scores compared to the inmates with the five lowest PCL-R scores
Blair (1995)	Moral/Conventional Test	PCL-R	Not available	Not available (but 31.6 was the mean score of the high-scoring psychopathic group)	Psychopaths *failed the test*, rating moral and conventional transgressions to be equally impermissible and equally serious
Blair et al. (1995)	Moral/Conventional Test	PCL-R	30	20	Psychopaths *failed the test*, rating moral and conventional transgressions to be equally impermissible (but not equally serious)
Aharoni et al. (2012)	Modified Moral/Conventional Test	PCL-R	30	5	Psychopaths *passed the test*, and PCL-R scores had no correlation with performance or participants' harm ratings
Glenn et al. (2009a, 2009b)	Subset of Trolley Scenarios (by Greene et al.)	PCL-R	26	Not available	*No significant differences* in the responses given by PCL-R groups
Cima et al. (2010)	Subset of Trolley Scenarios (by Greene et al.)	PCL-R	26	Not available	*No significant differences* in the responses given by PCL-R groups across all moral scenarios (but some differences in individual scenarios)
Pujol et al. (2011)	Subset of Trolley Scenarios (by Greene et al.)	PCL-R	15.8	Not available	*No significant differences* in the responses given by PCL-R groups across all moral scenarios (but some differences in individual scenarios)
Koenigs et al. (2012)	Subset of Trolley Scenarios (by Greene et al.)	PCL-R	30	24 (split into high- and low- anxious)	Psychopaths reported they would perform actions in *impersonal* moral scenarios *more often* than non-psychopaths; low-anxious psychopaths reported they would perform actions in *personal* moral scenarios *more often* than high-anxious psychopaths and non-psychopaths
Aharoni et al. (2011)	Moral Foundations Questionnaire	PCL-R	30	40	PCL-R scores *correlated negatively* with ratings for the harm and fairness foundations
Harenski et al. (2010)	Rating Moral Violations in Pictures	PCL-R	30	16	*No significant differences* in the ratings provided by PCL-R groups

Only one study has tested how adult psychopaths score on the MJI. Link, Scherer, & Byrne (1977) administered 4 of Kohlberg's dilemmas to 16 psychopathic inmates, 16 nonpsychopathic inmates, and 16 noninmate employees from the same Canadian facility. In this study, psychopathy was assessed using the MMPI, not the PCL-R. Contrary to some expectations, the authors reported that psychopaths had *improved* moral reasoning compared to both control groups. Psychopaths offered 36% ($p < .01$) and 5% ($p < .05$) more stage 5 post-conventional justifications than incarcerated nonpsychopaths and hospital employees, respectively, despite no significant differences in age, IQ, or education. However, it is not clear how this inmate population would score on the PCL-R, so it is inappropriate to draw any conclusions about how psychopaths, as they are understood today, would score on the MJI.

Defining Issues Test

The most widely used derivative of the Kohlberg MJI is the Defining Issues Test (DIT) (Rest, Cooper, Coder, Masanz, & Anderson, 1974). The DIT presents participants with six dilemmas derived from Kohlberg's MJI, the most famous of which is this (from Rest, 1979):

> HEINZ AND THE DRUG
> In Europe a woman was near death from a special kind of cancer. There was one drug that doctors thought might save her. It was a form of radium that a druggist in the same town had recently discovered. The drug was expensive to make, but the druggist was charging ten times what the drug cost to make. He paid $200 for the radium and charged $2,000 for a small dose of the drug. The sick woman's husband, Heinz, went to everyone he knew to borrow the money, but he could get together only about $1,000, which is half of what it cost. He told the druggist that his wife was dying, and asked him to sell it cheaper or let him pay later. But the druggist said, "No, I discovered the drug and I'm going to make money on it." So Heinz got desperate and began to think about breaking into the man's store to steal the drug for his wife.
> Should Heinz steal the drug? __Should Steal __Can't Decide __Should not steal

Each scenario is followed by a list of 12 considerations. Participants are asked to rate each consideration according to its importance for making their moral judgment, and then told to select the four most important considerations and rank-order them from 1 to 4. Each consideration specifically represents one of Kohlberg's six moral stages. Here are some examples:

- Whether the law in the case is getting in the way of the most basic claim of any member of society
- Whether a community's laws are going to be upheld
- What values are going to be the basis for governing how people act toward each other
- Whether stealing in such a case would bring about more total good for the whole society

The most commonly used metric in scoring the DIT is the *P*-score, ranging from 0 to 95, which represents the proportion of the four selected considerations that appeal to Kohlberg's Moral Stages 5 and 6. It has been suggested (Rest, 1979) that people can be divided broadly into those who make principled moral judgments (*P*-score > 50) and those who do not (*P*-score < 49). A correlation of 0.68 has been reported between the scores of the DIT and Kohlberg's more complex MJI (O'Kane et al., 1996). Of note, the verdicts of the judgments themselves are not assessed or reported in any way in the *P*-score of the DIT test.

Two studies have examined the relationship between PCL-R and DIT scores. O'Kane et al. (1996) found no correlation between total PCL-R scores and *P*-scores in 40 incarcerated individuals in a British prison once IQ was accounted for. However, the mean PCL-R score was fairly low (15) and only one inmate scored above 30. Further, no between-group comparisons were reported between low versus high scorers (O'Kane et al., 1996). These results were consistent with the results in an American sample of inmates. In the second sample, no inmates scored above 30 on the PCL-R, and no regression was reported between PCL-R scores and *P*-scores, but no significant differences were found in the *P*-scores of the five highest compared to the five lowest PCL-R-scoring inmates (Lose, 1997). Thus, although no significant relationship has been found between psychopathy and performance on the DIT, the DIT has yet to be administered to a sizable sample of high-scoring psychopaths.

Moral Judgment Task

A second adaptation of the MJI is the Moral Judgment Task (MJT) (Lind, 1978). The MJT is based on Kohlberg's stages of moral development, but it is adapted to try to differentiate the ability to reason about moral issues objectively (which Lind calls the "cognitive" part of moral aptitude) from preference for a certain basis of moral judgment (which Lind calls the "affective" part of moral aptitude, although this metric may not reflect what is often described as the "affective" part of moral reasoning today). The MJT asks participants to assess two moral dilemmas, one taken directly from Kohlberg's MJI and the other developed from a real-life situation:

WORKER'S DILEMMA
Due to some seemingly unfounded dismissals, some factory workers suspect the managers of eavesdropping on their employees through an intercom and using this information against them. The managers officially and emphatically deny this accusation. The union declares that it will only take steps against the company when proof has been found that confirms these suspicions. Two workers then break into the administrative offices and take tape transcripts that prove the allegation of eavesdropping.

DOCTOR'S DILEMMA
A woman had cancer and she had no hope of being saved. She was in terrible pain and so weakened that a large dose of a painkiller such as morphine would have caused her death. During a temporary period of improvement, she begged the doctor to give her enough morphine to kill her. She said she could no longer endure the pain and would be dead in a few weeks anyway. The doctor complied with her wish.

For each dilemma, participants are first asked to judge whether the actor's solution was right or wrong on a scale from –3 to +3, and then asked to rate six moral arguments that are consistent with the participants' judgment and six moral arguments that are against the participants' judgment on a scale from –4 ("I strongly reject") to +4 ("I strongly agree"). Each of the six arguments represents one of Kohlberg's six stages of moral orientation. Participants typically prefer Stage 5 moral arguments for Worker's Dilemma and Stage 6 moral arguments for Doctor's Dilemma.

Two metrics can be calculated based on the participants' responses in the Moral Judgment Task. The first metric, called the Competence score or "C-score," is unique to the Moral Judgment Task and is the most widely used. The C-score reflects the ability of a participant to acknowledge and appropriately weigh moral arguments, regardless of whether those arguments agree with the participant's own opinion or interests. More technically, it tests the percentage of an individual's total response variation that can be uniquely attributed to the individual's concern for the quality of a moral argument. People generally do not score very high on the moral competence test. Scores of 1–9 out of 100 are considered "very low" but not uncommon, while scores of 50 out of 100 are considered "extraordinarily high" (Lind, 2000 [original 1984]). The second metric that can be assessed in the Moral Judgment Task is moral "preference" for or "attitude" toward arguments of different moral development stages. This is calculated by simply averaging the participants' ratings of the arguments provided at each stage. The moral attitude metric is similar to the metric used by other Kohlberg test derivatives. The verdicts or outcomes of the moral judgments provided by participants in the Moral Judgment Task are not usually assessed or reported except to calculate the C-score.

Although the Defining Issues Test and the Moral Judgment Task may sound similar, their metrics have been shown to measure different aspects of moral aptitude. In particular, it is possible to score high on the Defining Issues Test scale but very low on the Moral Judgment Task scale, particularly if one subscribes to absolute moral rules with little tolerance for other views (Ishida, 2006).

No published studies have examined the relationship between psychopathy and performance on the Moral Judgment Task, but our research group[11] has administered the Moral Judgment Task to a population of 74 inmates in New Mexico (21 with PCL-R scores of 30 or above). This study is currently being prepared for publication and should not be interpreted strongly without peer review, but our analyses so far do not indicate any significant correlations between C-scores on the Moral Judgment Task and PCL-R Total scores or Factor scores.

TURIEL'S MORAL/CONVENTIONAL TEST

A very different perspective on moral judgment is provided by the Moral/Conventional Test, which is based on social domain theories pioneered by Elliot Turiel. In Turiel's view, moral judgments are seen as (1) serious, (2) based on harm to others, and (3) independent of authority and geography in the sense that what is morally wrong is supposed to remain morally wrong even if authorities permit it and even if the act is done in some distant place or time where it is common. Important to social domain theories, Turiel argues that moral violations and conventional violations are differentiated by most people early in life, whereas Kohlberg asserts that moral and conventional thought diverge later in life and only for individuals who reach "post-conventional" levels of moral reasoning. Turiel's view is supported by a wealth of research showing that children as young as 3 years old draw the distinction between moral and conventional transgressions (for reviews, see Turiel, 1983; Turiel, Killen, & Helwig, 1987). By age 4, children tend to say, for example, that although it is wrong to talk in class without raising one's hand and being called on, it would *not* be wrong for one to talk in class without being called on if the teacher said that they were allowed to do so. In contrast, the same children say that it *would* still be wrong to hit other kids even if the teacher said that they were allowed to do so. Four-year-olds also tend to report that violations such as hitting other kids are more serious than violations such as talking without being called on, and what makes it more wrong is the fact that hitting those kids causes harm. Harm to an individual is not what makes it wrong to talk without being called on, however. These findings have been interpreted as suggesting that reasoning relevant to moral transgressions might develop independently from reasoning relevant to conventional transgressions, and the two types of reasoning might additionally be regulated by separate cognitive systems.

To determine whether an individual distinguishes moral and conventional wrongdoing, Moral/Conventional tests present participants with short scenarios, usually about kids, describing events such as one child pushing another off of a swing or a child wearing pajamas to school. After the scenario is described, participants are asked these questions about the act (Y) that the agent (X) did in the scenario:

1. *Permissibility*: "Was it OK for X to do Y?"
2. *Seriousness*: "Was it bad for X to do Y?" and "On a scale of 1 to 10, how bad was it for X to do Y?"
3. *Justification*: "Why was it bad for X to do Y?"
4. *Authority-dependence*: "Would it be OK for X to Y if the authority says that X may do Y?"

Responses to Questions 1 and 4 are scored categorically with "Yes" or "OK" responses assigned a score of 0 and "No" or "Not OK" responses assigned a score of 1. Cumulative scores are then compared between scenarios describing conventional and moral transgressions. Seriousness as assessed in Question 2 is scored according to the value (between 1 and 10) the subject gave that transgression. The justifications given in Question 3 are scored according to standardized categories (e.g., Smetana, 1985).

Blair et al. published two classic studies of the moral/conventional distinction in adult psychopaths. In the first study, Blair administered the moral/conventional test to 10 high-PCL-R scoring patients (mean PCL-R score: 31.6) and ten lower PCL-R scoring patients (mean PCL-R score: 16.1) from high-security British psychiatric hospitals (Blair, 1995). Blair reported that high scorers did not demonstrate an appreciation for the distinction between moral and conventional transgressions (Blair, 1995). Specifically, 6 of 10 high-scorers drew no distinction at all (and 2 drew only a mild distinction), whereas 8 of 10 low-scorers drew a clear distinction between moral and conventional violations. Further, failure to make the moral/conventional distinction correlated with the "lack of remorse or guilt," "callous/lack of empathy," and "criminal versatility" items on the PCL-R.

Regarding specific dimensions of the distinction, high-scorers did cite conventions or authorities to explain why both moral and conventional violations are wrong, whereas low-scorers cited harm and justice considerations to explain why moral violations are wrong (Question 3). However, high-scorers failed to make the distinction on all three other dimensions of permissibility (Question 1), seriousness (Question 2), and authority independence (Question 3). Surprisingly, high-scorers did not rate moral and conventional violations to be permissible, not serious, and authority dependent. Instead, they rated conventional violations to be very serious and impermissible even if society and authorities said that the act was acceptable. In short, they treated conventional transgressions as moral, whereas they had been expected to treat moral transgressions as conventional.

These results were mostly replicated in a second study of 20 psychopaths (mean PCL-R score: 32) and 20 nonpsychopaths (mean PCL-R score: 11.2) from a high-security British psychiatric hospital or prison (Blair et al., 1995). In this second study, again, psychopaths did not make the moral/conventional distinction on the permissibility and authority independence dimensions (but unlike the first study did make the moral/conventional distinction on the seriousness dimension). Likewise, failure to make the moral/conventional distinction correlated with the "lack of remorse or guilt" and "criminal versatility" items on the PCL-R (but unlike the first study failure did not correlate with the "callous/lack of empathy" PCL-R item). It should be noted, however, that the differences between groups were less dramatic in this second study than in the first.

The explanation for these highly cited results is still unclear. One possible explanation is that adult psychopaths lack the ability to distinguish between moral and conventional wrongs and, indeed, really think that both moral and conventional norms have the same status that most people selectively associate with only moral norms. Another explanation, proposed by Blair (1995), is that psychopaths lack the ability to distinguish between moral and conventional wrongs and, instead, really think that both moral and conventional norms have the same authority-dependent status that most people selectively associate with only conventional norms. According to this explanation, psychopaths report that all violations have the authority independence associated with moral norms because they try to impress investigators—a behavior called "impression management"—perhaps to improve their treatment in prison or chances of release. A third possible explanation is that adult psychopaths actually *can* cognitively make the moral/conventional distinction (perhaps on the basis of seriousness and justification by harm), but they nonetheless think that both moral and conventional norms have the same authority-dependent status that most people associate with only conventional norms, so they purposely inflate their ratings of the authority independence of both moral and conventional transgressions for "impression management" purposes. A final possible explanation is that adult psychopaths actually *can* make the moral/conventional distinction, *do* believe that moral norms are more authority independent than conventional norms, but still purposely inflate their ratings of only conventional transgressions for impression management. Blair's data cannot differentiate among these hypotheses.

To help decide between the alternative explanations for Blair's results, our group presented our own set of moral and conventional violations to 109 inmates (including five who scored above 30 on the PCL-R) and 30 controls (Aharoni et al., 2012). The difference between our test and previous tests is that we told participants that eight of the listed acts were prerated to be morally wrong (that is, wrong even if there were no rules, conventions, or laws against them), and eight were prerated as conventionally wrong (that is, not wrong if there were no rules, conventions, or laws against them). Knowing how many violations fit into each category, then, participants were asked to label each violation according to whether or not the act would be wrong if there were no rules, conventions, or laws against it. The forced-choice nature of our task is important because it purposely aligned impression management with success on the test such that participants would have to label and rate the transgressions correctly in order to maximize investigators' positive impressions. This removed the incentive to rate all acts as morally wrong independent of convention and authority because participants knew that rating all of them as moral would misclassify half of the scenarios. Aharoni et al. found that the general population of inmates did quite well on this task (82.6% correct), though significantly worse than non-incarcerated controls (92.5% correct). Within the inmate population, PCL-R scores had no relation to how well inmates performed. PCL-R scores also had no relation to participants' ratings of the presence or absence of harm in each scenario, even though these ratings explained a significant proportion (26%) of the variance in moral classification accuracy. Although more studies are needed, this result suggests that, contrary to previous belief, psychopaths have the ability to distinguish moral from conventional transgressions, even if their ability is not obvious in all situations. What is still left to determine is whether psychopaths care about this identifiable difference in transgressions.

TESTS THAT USE PHILOSOPHICAL SCENARIOS

A new class of moral cognition tests inspired by centuries of moral philosophy was developed in the mid-1990s. These philosophers' dilemmas were not originally validated by psychological studies or reviewed by psychology experts like the tests discussed in the preceding sections. Instead, these dilemmas were validated by philosophical principles and vetted by philosophical experts. The goal of these dilemmas was usually to test proposed moral principles, but sometimes such dilemmas were also intended to evoke new moral intuitions that did not fit easily under any moral principle that had been formulated in advance. In a clinical context, thus, administering philosophically vetted dilemmas allows one to test (1) whether psychopaths have intuitions that reflect specific moral principles and, in theory, (2) whether psychopaths will report these same moral intuitions even if the intuition is not based on explicit moral principles.

Some of the most famous of the philosophical dilemmas are the "Trolley Scenarios" (Foot, 1978; Thomson, 1976, 1985) including these versions from Greene et al. (2001):

SIDE TRACK
You are at the wheel of a runaway trolley quickly approaching a fork in the tracks. On the tracks extending to the left is a group of five railway workmen. On the tracks extending to the right is one railway workman. If you do nothing, the trolley will proceed to the left, causing the deaths of the five workmen. The only way to save these workmen is to hit a switch on your dashboard that will cause the trolley to proceed to the right, causing the death of the workman on the side track.
Is it appropriate for you to hit the switch in order to avoid the deaths of the five workmen?

FOOTBRIDGE
A runaway trolley is heading down the tracks toward five workmen who will be killed if the trolley proceeds on its present course. You are on a footbridge over the tracks between the approaching trolley and the five workmen. Next to you on this footbridge is a stranger who happens to be very large. The only way to save the lives of the five workmen is to push this stranger off the bridge and onto the tracks below, where his large body will stop the trolley. The stranger will die if you do this, but the five workmen will be saved.
Is it appropriate for you to push the stranger on to the tracks in order to save the five workmen?

Trolley scenarios vary in details (see Greene et al., 2009) and experimenters may use different words such as "Is it appropriate to…?", "Is it permissible to…?", "Is it wrong to…?", or "Would you…?" to phrase their questions (O'Hara, Sinnott-Armstrong, & Sinnott-Armstrong, 2010). One way or another, though, these scenarios typically pit conflicting moral intuitions or principles against each other in an effort to determine which moral intuitions prevail.

Greene on Personal versus Impersonal Dilemmas

Petrinovich and O'Neill (1996) was the first group to use trolley scenarios in a scientific experiment, but enthusiasm for the experimental use of philosophical dilemmas took off when Joshua Greene et al. published a battery of 120 short scenarios consisting of 40 non-moral scenarios,

40 "impersonal" moral scenarios, and 40 "personal" moral scenarios (Greene et al., 2001). A moral violation was defined as personal (or "up close and personal") if it was (1) likely to cause serious bodily harm, (2) directed to a particular victim, and (3) the harm did not result merely from the deflection of a preexisting threat onto a different party. If a moral violation failed to meet these criteria, it was defined as impersonal. Pushing the man in the Footbridge scenario was said to exemplify personal moral violations, whereas hitting the switch in the Side Track scenario was said to exemplify impersonal moral violations (because a preexisting threat was deflected).

Greene et al. (2001) hypothesized that personal moral violations would automatically cause a negative emotional reaction that would subsequently cause participants to judge the violations as morally wrong or inappropriate, despite any benefits in saving more lives. This hypothesis was supported by evidence that people were more likely to judge personal violations wrong than impersonal violations, and personal moral scenarios elicited hemodynamic activity in brain regions (including the ventromedial prefrontal cortex) claimed to be involved in "emotion," while impersonal moral scenarios did not elicit activity in such areas. These behavioral and brain imaging results have been replicated and supported in subsequent studies by Greene's group (e.g., Greene et al., 2004, 2008), and their experimental scenarios have proven to elicit consistent and robust patterns of moral intuitions.

Based on the assumption that psychopaths have reduced emotional reactions, especially with regard to harm to others, a popular prediction was that psychopaths would be more likely than nonpsychopaths to judge acts described in personal moral scenarios as permissible or *not* wrong because psychopaths would lack the emotional response that makes most people reject acts that cause "up close and personal" harm. This prediction was galvanized by reports that patients with damage in the ventromedial prefrontal cortex—a region believed to have reduced functionality in psychopathy (Damasio, 1994; Koenigs, Baskin-Sommers, Zeier, & Newman, 2011; Koenigs, Kruepke, & Newman, 2012)—are more likely to judge personal harm for the greater good as morally permissible than patients without brain damage or with damage in other brain regions (Koenigs et al., 2007).

This prediction, however, has not been fully supported in psychopaths so far. In two reports from the same group (Glenn, Raine, Schug, Young, & Hauser, 2009; Glenn, Raine, & Schug, 2009), both high and low scorers on the PCL-R were more likely to say that it is "not appropriate" to perform acts described in personal moral scenarios than in impersonal moral scenarios, exactly as had previously been reported in healthy populations. There were no significant differences in the responses given by PCL-R groups. It is worth noting, however, that both studies only used a small subset of the original scenarios from Greene et al. (2001), and psychopathy scores ranged only as high as 32 with the cutoff for "high scorers" being 26 rather than the more standard cutoff of 30. Thus, these studies do not rule out the possibility that psychopaths with very high PCL-R scores might represent a discrete group with enough neural impairment to result in differences in personal moral judgment.

Another study by Cima et al. (2010) found similar results. Using a PCL-R cutoff of 26 to define "psychopath," this group studied 14 psychopathic offenders and 23 nonpsychopathic offenders from a forensic psychiatric center in the Netherlands as well as 35 healthy controls (Cima et al., 2010). It was not reported how many participants scored above 30. Each participant received 7 personal and 14 impersonal moral dilemmas from Greene et al. (2001, 2004) and was instructed to respond "Yes" or "No" to the question "Would you X?" after reading each dilemma. In an attempt to control for lying, participants were also given a Socio-Moral Reflection questionnaire (SMR-SF) about straightforward and familiar transgressions such as "How important is it to keep a promise to your friend?" and "How important is it not to steal?" that they answered on a 5-point scale from very unimportant to very important. The results showed no significant difference between psychopaths and the other groups in the percentage of endorsed acts of impersonal or personal harm, or in any principled subset of acts of personal harm (such as those serving the interest of the agent vs. other people or high-conflict vs. low-conflict dilemmas). Further, there were no significant correlations between PCL-R total scores, Factor 1 scores, or Factor 2 scores and either moral judgments on personal dilemmas or scores on the SMR-SF. Cima et al. do report that there were "four cases where the psychopaths judged the case more permissible by 20–40%," but these differences did not extend to the majority of personal dilemmas, and it was not specified which cases these were.

A third study also found similar results, with psychopaths more often than nonpsychopaths reporting that they would perform the actions described in a small random subset of specific individual personal moral dilemmas, but not all personal moral dilemmas as a whole (Pujol et al., 2011). However, the group of "psychopaths" from this study included participants with PCL-R scores ranging as low as 15.8 (but as high as 34), and it was not reported how many participants had PCL-R scores of 30 or above. Thus, although these null results line up nicely with the studies by Glenn et al., the use of lower PCL-R cutoffs leaves open the possibility that very high scorers (>30 or >34) might show differences that do not appear in any of these population samples. In addition, the lack of significant differences in the Cima et al. study might be due to the low number of personal dilemmas (7) or the low overall IQ of the participants (mean = 81.6).

A fourth study by Koenigs et al. (2012) reported somewhat conflicting results to the previous three studies. As the only study out of the four using a PCL-R cutoff of 30 to define "psychopath," this group studied 24 psychopathic offenders and 24 nonpsychopathic offenders from a medium-security correctional institution in the United States (Koenigs, Kruepke, Zeier, & Newman, 2012). Each participant received 14 personal and 10 impersonal moral dilemmas from Greene (2001, 2004) and was instructed to respond "Yes" or "No" to the question "Would you X?" after reading each dilemma. No nonmoral dilemmas were administered. Contrary to predictions and previous studies, Koenigs et al. (2012) found that psychopaths said they would perform the actions described in impersonal moral dilemmas significantly more often than nonpsychopaths. More consistent with the authors' predictions, they also found that psychopaths with low anxiety were more likely to say they would perform the actions described in personal moral dilemmas than either high-anxiety psychopaths or nonpsychopaths. Thus unlike previous studies, this study found that all psychopaths perform differently on impersonal moral scenarios than nonpsychopaths, but only low-anxious (which correlates with low emotionality) psychopaths perform differently on personal moral scenarios compared to nonpsychopaths. This was the only study to report such moral judgment differences using Greene and colleagues' moral dilemmas.

The authors' favored explanation for why low-anxious psychopaths gave different responses to personal moral scenarios than high-anxious psychopaths or nonpsychopaths is that low-anxious psychopaths, but not high-anxious psychopaths, may lack the emotional response that makes most people reject acts that cause "up close and personal" harm. However, this explanation does not account for why both low-anxious and high-anxious psychopaths said they would perform the actions described in impersonal moral scenarios more often than nonpsychopaths. Thus, an alternative explanation might be that low-anxious psychopaths were simply less anxious than high-anxious psychopaths or nonpsychopaths about telling interviewers what they truly believed they would do. Even more important to appreciate, the participants in the Koenigs et al. study on average reported that they would perform all of the actions described in any of the scenarios more often than in previous studies, which might indicate that differences in the way the test was administered (on paper vs. on a computer, with a guard in the room vs. no guard in the room, randomized vs. nonrandomized, etc.) may have affected participants' responses in ways that interacted with psychopathy and led to different overall results. Given these differences and without control scenarios to normalize to, it is difficult to determine how to interpret the Koenigs et al. (2012) results in the context of the previously reported negative results.

An even more striking complication with interpreting the studies described in the preceding text is that they all asked participants slightly different questions. Koenigs et al. (2012) asked their subjects "Would you [do the act] in order to [achieve the benefit]?" The same question was used in both the Cima et al. study and also the Pujo et al. study. However, the question of what someone *would* do asks for a prediction of actual behavior rather than a moral judgment about what they *ought* to do. Subjects could easily respond that they would not do an act even if they thought that act was not morally wrong. They could also respond that they would do an act even though they thought it was morally wrong, especially if they did not care about whether the act was morally wrong. Glenn et al. avoided this problem by asking their participants "Is it appropriate to X?" but in that case it was not clear whether participants gave answers according to what they thought was morally wrong or what they thought was required by local nonmoral conventions. Moreover, it is also not clear that any of these studies succeeded in controlling for lying. Hence, even if the aforementioned studies asked participants explicitly about what they thought was morally wrong, it would still be possible that psychopaths' responses to these scenarios would not reflect their real moral beliefs. In sum, the data are inconclusive in regard to whether psychopaths respond differently to impersonal or personal moral dilemmas than nonpsychopaths. Recent efforts in this area are intriguing, but further research will be needed to resolve the presently conflicting reports.

ROBINSON AND KURZBAN'S DESERVED PUNISHMENT TEST

Another kind of moral judgment concerns not which acts are morally wrong but, instead, how others should react to wrongdoing and, in particular, how much punishment is deserved by differing types of wrongdoing. These moral judgments involve not only the categorical judgment of whether some punishment is deserved, but also how much punishment is deserved and whether certain crimes deserve more punishment than others.

To test these moral judgments about punishment, Paul Robinson and Robert Kurzban used legal principles to construct descriptions of 24 standard crimes (including theft, fraud, manslaughter, murder, and torture) that collectively represent 94.9% of the offenses committed in the United States (Robinson & Kurzban, 2007). Here are three examples of their stimuli chosen to illustrate scenarios that are closer in rank ("Short Change Cheat" and "Clock Radio from Car") and far away in rank ("Burning Mother for Inheritance" compared to either of the first two scenarios):

SHORT CHANGE CHEAT
John is a cab driver who picks up a high school student. Because the customer seems confused about the money transaction, John decides he can trick her and gives her $20 less change than he knows she is owed.

CLOCK RADIO FROM CAR
As he is walking to a party in a friend's neighborhood, John sees a clock radio on the back seat of a car parked on the street. Later that night, on his return from the party, he checks the car and finds it unlocked, so he takes the clock radio from the back seat.

BURNING MOTHER FOR INHERITANCE
John works out a plan to kill his 60-year-old invalid mother for the inheritance. He drags her to bed, puts her in, and lights her oxygen mask with a cigarette, hoping to make it look like an accident. The elderly woman screams as her clothes catch fire and she burns to death. John just watches her burn.

Robinson and Kurzban also included scenarios describing 12 nonstandard crimes, such as prostitution and drug possession. In their first study, the scenarios were written on cards, and participants were asked to order the cards according to how much punishment they thought each crime deserved (they were also allowed to set aside acts that deserve no punishment at all). They were given a second chance to consider each pair of scenarios to confirm that they were ordered as wished before committing to their final ordering. Importantly, then, the moral judgments made in the Robinson and Kurzban studies reflect judgments of relative comparisons of specific criminal acts, unlike the moral judgments collected in the assessments described earlier in this chapter. The card study was followed up with a larger study over the Internet using similar instructions.

Robinson and Kurzban found that people's moral intuitions of deserved punishment for standard crimes are surprisingly specific and widespread. In a sample of 64 participants given the card test, 92% of the time subjects agreed that no punishment was deserved for four scenarios, and 96% of the time subjects agreed about how to rank pairs of the other twenty scenarios (Kendall's $W = .95$, $p < .001$). In the Internet replication using a sample of 246 subjects, 71% to 87% of the time (depending on the scenario) subjects agreed that no punishment was deserved for four scenarios, and 91.8% of the time subjects agreed about how to rank pairs of the other 20 scenarios (Kendall's $W = .88$, $p < .001$). These data suggest that Robinson and Kurzban's test provides a robust way to probe moral intuitions toward relative punishment.

We administered Robinson and Kurzban's test to 104 adult male inmates. The PCL-R scores for 3 of these inmates were not available, but 25 had a PCL-R score of 30 or higher. PCLR scores ranged from 3.2 to 36.8 with a mean of 22.5. Similar to Robinson and Kurzban's findings in nonincarcerated populations, our incarcerated sample had high agreement in rankings of deserved punishment, with a Kendall's W of .85 overall ($p < .001$). These results are currently being prepared for publication and should not be strongly interpreted until they are peer-reviewed, but so far the data suggest that there is no significant correlation between PCL-R total scores and intuitions of relative justice. However, on further inspection, this lack of correlation was due to the fact that Factor 1 and Factor 2 scores correlated with task performance in opposite directions (Factor 1 correlated positively and Factor 2 correlated negatively) and canceled themselves out in the PCL-R total score. Similar mutually repulsive effects of Factor 1 and Factor 2 have been observed in the prediction of self-reports of emotional distress (Hicks & Patrick, 2006; Patrick, 1994), fearfulness (Hicks & Patrick, 2006; Patrick, 1994), anger and hostility (Hicks & Patrick, 2006; Reidy, Shelley-Tremblay, & Lilienfeld, 2011), positive affect (Patrick, 1994), achievement (Verona, Patrick, & Joiner, 2001), stress reactivity (Verona et al., 2001), interpersonal aggression (Kennealy, Hicks, & Patrick, 2007), and attempted suicide history (Verona, Hicks, & Patrick, 2005). Thus, the literature justifies caution when interpreting studies of moral judgments in psychopaths that only examine PCL-R Total scores. Perhaps one reason why it is so hard to find effects of psychopathy on moral judgment in studies with small numbers of psychopathic participants is that Factor 1 psychopathic traits and Factor 2 psychopathic traits influence moral judgments in opposite directions and ultimately cancel each other out when neither Factor dominates. Future studies will need to include enough participants with varying PCL-R scores to ensure that different Factors (and Facets) of psychopathy don't have different, or conflicting, effects.

BEYOND HARM-BASED MORALITY

The tests discussed thus far focus on moral prohibitions against causing harm and in favor of punishing people who cause harm. This focus is justifiable, given that harm judgments are central to morality and that society is most affected by psychopaths through their tendency to cause harm to others. Nonetheless, morality pertains to more than just prohibitions on causing harm. In this section, therefore, we review what is known about how psychopaths perform in non-harm–based areas of morality.

Anthropology and evolutionary psychology have inspired tests of moral judgment that are complementary to those developed through philosophy. The theoretical backbone of these tests is that moral judgments should be defined by their function rather than their content. In particular, the Moral Foundations Theory of Jonathan Haidt defines morality to cover any mechanism (including values, rules, or emotions) that regulates selfishness and enable successful social life, regardless of what the contents of those mechanisms are (Haidt & Kesebir, 2010). This definition implies that many prohibitions or principles that others classify as nonmoral conventions or biases are instead moral in nature, just as much as rules of justice, rights, harm, and welfare. In total, Haidt delineates five areas or "foundations" of moral regulation: (1) Harm/care, (2) Fairness/reciprocity, (3) Ingroup/loyalty, (4) Authority/respect, and (5) Purity/sanctity. Haidt argues that these areas are common (though not necessarily universal) across cultures and have some clear evolutionary basis.

Haidt's Moral Foundations Questionnaire (MFQ) tests judgments in these five areas of morality in two ways. In the first part of the MFQ, participants are asked to determine whether something is right or wrong and indicate how relevant various considerations are (from "not at all" to "extremely relevant"). Each consideration represents a different moral foundation, as illustrated by these examples:

- Whether or not someone showed a lack of respect for authority
- Whether or not someone violated standards of purity and decency
- Whether or not someone was good at math
- Whether or not someone was cruel
- Whether or not someone was denied his or her rights

In the second part of the MFQ, participants are asked whether they strongly, moderately, or slightly agree or disagree with statements reflecting various moral foundations, including these examples:

- I would call some acts wrong on the grounds that they are unnatural.
- It can never be right to kill a human being.
- It is more important to be a team player than to express oneself.
- If I were a soldier and disagreed with my commanding officer's orders, then I would obey anyway because that is my duty.
- People should not do things that are disgusting, even if no one is harmed.

The final scores are the average, ranging from 0 to 5, of participants' responses across each of the five moral foundations.

Glenn, Iyer, Graham, Koleva, and Haidt (2009) investigated the relationship between MFQ scores and nonclinical psychopathic personality traits in the general population. To assess psychopathic traits, this study used the Levenson Self-Report Psychopathy Scale (SRPS)—a self-report questionnaire that mildly, but significantly, correlates with the assessments made by the PCL-R (total PCL-R scores correlate with total SRPS scores by .35; when the PCL-R and the SRPS are used to divide participants into groups of high, low, and middle scorers, the kappa coefficient is only .11).[12] SRPS scores negatively correlated with endorsement of the moral foundations of Harm and Fairness in this study, and correlated slightly negatively with endorsement of the Purity moral foundation. They also found a positive correlation between SRPS scores and endorsement of the In-group moral foundation, and failed to find any correlation between SRPS scores and endorsement of the Authority foundations. However, because this study was run in the general population, it is not clear whether clinical psychopaths will show the same pattern.

To investigate this issue in clinical psychopaths, our laboratory administered the MFQ to 222 adult male inmates in New Mexico (Aharoni, Antonenko, & Kiehl, 2011). Of these, 37 inmates had a PCL-R score of 30 or higher, with the highest score being 37.9 and the mean being 21.54. As in Glenn et al. (2009), we found that total PCL-R scores correlated negatively with ratings for the Harm ($\beta = .20$, $t(211) = 2.98$, $p < .01$) and Fairness ($\beta = .17$, $t(211) = 2.48$, $p < .05$) foundations but did not correlate with the Authority foundation. However, unlike Glenn et al. (2009), we did not find any correlation between total PCL-R scores and ratings of the Purity or In-group foundations.

Although the correlations between psychopathy scores and ratings of some moral foundations were significant in our population, two important twists need to be considered. On average, our inmate population rated the Harm and Fairness foundations as highly as did nonincarcerated populations used to develop and validate the MFQ Graham, et al. (2011) (average scores of 3.44 and 3.43 from our population compared to the previously reported scores of 3.42 and 3.55, respectively, from nonincarcerated populations). Curiously, however, the In-group, Authority, and Purity foundations were rated as much more important by our incarcerated population than by the nonincarcerated population of Graham et al. The average rating of the In-group foundation was 3.43 in our population compared to 2.26 in the nonincarcerated population of Graham et al., 3.15 compared to 2.27 for the Authority foundation, and 3.02 compared to

1.54 for the Purity foundation. These differences suggest moral foundations may play different roles in incarcerated populations than nonincarcerated populations, and these roles need to be understood better before conclusions about psychopaths' aptitude for moral judgment can be confidently drawn from the results of the Moral Foundations Questionnaire.

Equally important, PCL-R scores are not the only predictor of moral foundation ratings. Haidt and colleagues have shown in multiple populations that political orientation correlates with moral foundation ratings just as much and sometimes even in the same direction as psychopathy (Graham, Haidt, & Nosek, 2009). In fact, the more conservative a political ideology one identifies with, the more likely one is to rate the moral foundations of Harm and Fairness like a high-scoring psychopath. Given that populations in previously published studies valued the foundations of Authority and Purity much less than our population, it is harder to compare the patterns for these foundations between our two studies. The point here is definitely not to say that conservatives are psychopaths. Rather, the point is that the same amount of variance accounted for by psychopathy can also be accounted for by many other, socially acceptable, traits. Therefore, without further research, the data from our MFQ study have few clear implications for the debate over whether psychopaths have clinically abnormal impairments in moral judgment, because even if psychopaths have tendencies toward specific types of moral intuitions, their intuitions might still be within the normal range of intuitions held by a healthy population.

MORAL PICTURES WITH BRAIN SCANS

All of the studies so far depend on verbal self-reports of moral judgments. However, two items on the PCL-R are pathological lying and being conning/manipulative, so there is good reason to believe that psychopaths' self-reports are not trustworthy. Psychopaths might be able to infer how nonpsychopaths would respond to questionnaires, and in order to appear normal, respond in the same ways without believing or "appreciating" what they say. By analogy, atheists can often respond in the same way as religious believers to questions about what is sacrilegious, even though atheists do not believe that anything really is sacrilegious. If psychopaths report what they know to be other people's moral beliefs but they do not share those moral beliefs, by some definitions it could be argued that they do not really make normal moral judgments or even any moral judgments at all.

Moreover, even if psychopaths really do believe the moral judgments that they report, they still might not make those moral judgments in the same way as nonpsychopaths. In his seminal book *Mask of Sanity*, Hervey Cleckley observed that psychopaths verbally express emotions, often at the appropriate times, but they don't seem to actually experience or value emotions (Cleckley, 1976). In other words, they "know the words but not the music" (Johns & Quay, 1962). This suspicion is supported by findings indicating that psychopaths have reduced autonomic responses in the body and hemodynamic responses in the brain in response to emotional stimuli, even when they report that they feel the appropriate emotions (Blair et al., 2006; Kiehl, 2006). These findings raise the possibility that, even if psychopaths report normal moral emotions and normal moral judgments, those reports of moral judgments might be arrived at through very different, and less emotional, processes than in nonpsychopathic individuals.

These hypotheses receive some preliminary support from three recent studies that had psychopaths report moral judgments in a functional magnetic resonance imaging (fMRI) scanner. The first study gave a subset of Greene's moral scenarios to 17 participants, 4 of whom scored a 30 or above on the PCL-R (Glenn et al., 2009bc). Psychopaths' explicit moral judgments of these scenarios did not differ significantly from those provided by nonpsychopaths, but higher psychopathy scores did correlate with reduced activity in the left amygdala (Glenn et al., 2009b) and increased activity in the dorsolateral prefrontal cortex (Glenn et al., 2009c) in response to personal moral scenarios compared to impersonal moral scenarios.

The second study gave a modified subset of Greene's moral scenarios to 22 participants with

PCL-R scores between 15.8 and 34.4, though it was not reported how many scored above 30 (Pujol et al., 2011). Overall, psychopaths' explicit moral judgments of these scenarios did not differ significantly from those provided by nonpsychopaths, but again psychopathy scores correlated with differences in brain activity while making moral judgments. This time PCL-R Total scores correlated with decreased activity in the posterior cingulate and right angular gyrus. These results lend some support to the hypothesis that psychopaths make moral judgments differently than nonpsychopaths, even if the verdicts of their judgments are rarely abnormal.

In a third study, Harenski et al. showed pictures of moral violations (such as a Ku Klux Klan rally), emotional scenes without moral violations (such as an automobile accident), and neutral scenes that were neither moral nor emotional (such as an art class) to 16 psychopaths (PCL-R: 30+) and 16 nonpsychopaths (PCL-R: 7–18) (Harenski, Harenski, Shane, & Kiehl, 2010). While in an MRI scanner, participants were asked whether each picture represented a moral violation, and if so, how severe the moral violation was on a 1–5 scale, with 5 representing the highest violation severity (participants were instructed to give a rating of 1 if they thought the picture did not represent a moral violation). Similar to past behavioral studies, the psychopaths rated the depicted moral violations just as severely as nonpsychopaths (about 4 on a 5-point scale, with 5 representing the highest violation severity). However, the psychopaths had abnormal brain activity while making this rating of moral violations. In particular, compared to nonpsychopaths, psychopaths had reduced activity in the ventromedial prefrontal cortex and anterior temporal cortex. Moreover, amygdala activity was parametrically related to moral severity ratings in nonpsychopaths ($t(30) = 4.31$, $p = .014$) but not in psychopaths. Perhaps most interestingly, activity in the right posterior temporal/parietal cortex correlated negatively with moral severity ratings in psychopaths ($t(30) = 3.77$, $p = .035$) but had no such correlation in nonpsychopaths. This brain area has been associated with ascriptions of beliefs to other people (Saxe, 2006), so this difference in neural activity might be explained by the process of psychopaths thinking about what other people believe instead of forming or expressing their own moral beliefs. However, this interpretation is complicated by the fact that the correlation is negative rather than positive. Further research is underway in our laboratory to discover, map, and understand the neural differences between psychopaths and nonpsychopaths while they consider and express moral judgments.

These few brain studies of psychopaths support at least two inferences. First, psychopaths do not seem to reach moral judgments in the same way as nonpsychopaths, even if they do reach the same moral judgments. Second, brain imaging is a powerful tool for finding such differences between psychopaths and nonpsychopaths that ultimately may provide critical insight into whether psychopaths behave immorally because they lack the ability to make moral judgments or because they lack appropriate moral motivation.

CONCLUSIONS

The studies reviewed in this chapter support a tentative and qualified conclusion: If psychopaths have any deficits or abnormalities in their moral judgments, their deficits seem subtle—much more subtle than might be expected from their blatantly abnormal behavior. Indeed, the current literature, summarized in Table 7.1, suggests that psychopaths might not have any specific deficits in moral cognition, despite their differences in moral action, emotion, and empathy.

That said, this conclusion must be tentative. Too few studies on moral judgment in psychopaths are available, these studies include too few clinical psychopaths, and the findings of various studies conflict too much to warrant confidence. Further, we are aware of several studies in progress that are approaching different conclusions than the ones we survey here, so these conflicts and uncertainties about psychopaths' moral judgments are likely to continue. One particular concern is that very few individuals with PCL-R scores above 34 have been studied, and anecdotal clinical evidence suggests that this group might be significantly different from the participants in most studies on record. We also do not yet know whether psychopaths believe the moral judgments they report, given their predisposition to lie and manipulate. To gain insights into psychopaths' true beliefs, more studies will need to be done on psychopath's implicit moral attitudes. Further, there are not yet standardized assessments to study all kinds of moral judgments. All of these issues point to mechanisms by which future studies could uncover moral judgment deficits that have yet to be identified.

Filling in the gaps in our knowledge about moral judgment in psychopaths is important for Neuroscience, Philosophy, and Law. For Neuroscience, such knowledge is important for understanding the neural underpinnings of morality, as well as how to treat psychopathy. For Philosophy, such knowledge will provide a concrete test case for abstract theories about the content and nature of morality. For Law, if psychopaths cannot know or appreciate the moral wrongfulness of their acts, then they should not be held morally or criminally responsible, at least according to some legal scholars and some versions of the insanity defense.[13] Better ways of determining which psychopaths do not or cannot make normal moral judgments—and which of their moral judgments are abnormal—might help authorities make better predictions of which prisoners will commit more crimes if released and which treatment programs will help which prisoners.[14] For these practical and theoretical reasons, there is much to benefit from further research on how psychopaths make moral judgments.

NOTES

1. *Regina v. M'Naghten*, 10 Cl. & Fin. 200, 9 Eng. Rep. 718 at 722 (1843).
2. American Law Institute, *Model Penal Code* (Philadelphia: The American Law Institute, Final Draft, 1962), § 4.01(1).
3. See Sinnott-Armstrong and Levy (2010) as well as Chapters 15 by Litton and 16 by Pillsbury in this volume.
4. See Chapters 13 by Rice and Harris and 14 by Edens, Magyar, and Cox on recidivism as well as the Chapters 17 by Morse and 18 by Corrado on preventive detention in this volume.
5. See Chapter 12 on treatment by Caldwell in this volume.
6. In philosophy, this issue also affects (1) whether psychopaths are counterexamples to internalism, one version of which claims that anyone who believes an act is immoral will be motivated not to do it, and (2) whether psychopaths reveal limits on epistemic justification for moral claims by showing that rational people can understand but still not accept those claims.
7. See Chapter 2 on assessment by Forth, Bo, and Kongerslev in this volume.
8. See Chapter 3 on self-report measures by Fowler and Lilienfeld in this volume.
9. In philosophy, emotivists and sentimentalists claim that emotion and sentiment are somehow essential to moral judgment, but they can and must allow some moral judgments without any present emotion.
10. There is also some (though not consistent) evidence suggesting psychopaths might have selective impairments in recognizing and processing fearful, sad, and/or disgusted faces (Blair et al., 2004; Glass & Newman, 2006; Iria & Barbosa, 2009; Kosson, Suchy, Mayer, & Libby, 2002; Wilson, Juodis, & Porter, 2011).
11. All research by "our research group" referred to in this chapter was completed in a collective effort by Jana Schaich Borg, Rachel Kahn, Carla Harenski, Eyal Aharoni, Elsa Ermer, and the research staff in Dr. Kent A. Kiehl's laboratory.
12. See Chapter 3 by Fowler and Lilienfeld in this volume.
13. See Chapters 15 by Litton and 16 by Pillsbury in this volume.
14. See Chapters 13 by Rice and Harris, 14 by Edens, Magyar, and Cox, and 11 by Caldwell in this volume.

REFERENCES

Aharoni, E., Antonenko, O., & Kiehl, K. (2011). Disparities in the moral intuitions of criminal offenders: The role of psychopathy. *Journal of Research in Personality*, *45*(3), 322–327.

Aharoni, E., Sinnott-Armstrong, W., & Kiehl, K. (2012). Can psychopathic offenders discern moral wrongs? A new look at the moral/conventional distinction. *Journal of Abnormal Psychology*, *121*(2), 484–497.

American Law Institute. (1962). *Model penal code* (final draft). Philadelphia; The American Law Institute.

Aniskiewicz, A. S. (1979). Autonomic components of vicarious conditioning and psychopathy. *Journal of Clinical Psychology, 35*(1), 60–67.

Batson, D. (1991). *The altruism question.* New York: Psychology Press.

Batson, D. (2011). *Altruism in humans.* New York: Oxford University Press.

Blair, J., Jones, L., Clark, F., & Smith, M. (1997). The psychopathic individual: A lack of responsiveness to distress cues? *Psychophysiology, 34*(2), 192–198.

Blair, K. S., Richell, R. A., Mitchell, D. G. V., Leonard, A., Morton, J., & Blair, R. J. R. (2006). They know the words, but not the music: Affective and semantic priming in individuals with psychopathy. *Biological Psychology, 73*(2), 114–123.

Blair, R. J. R. (1995). A cognitive developmental approach to morality: Investigating the psychopath. *Cognition, 57*(1), 1–29.

Blair, R. J. R., Jones, L., Clark, F. & Smith, M. (1995). Is the psychopath "morally insane?" *Personality and Individual Differences, 19*(5), 741–752.

Blair, R. J. R., Mitchell, D. G. V., Peschardt, K. S., Colledge, E., Leonard, R. A., Shine, J. H., et al. (2004). Reduced sensitivity to others' fearful expressions in psychopathic individuals. *Personality and Individual Differences, 37*(6), 1111–1122.

Cima, M., Tonnaer, F., & Hauser, M. D. (2010). Psychopaths know right from wrong but don't care. *Social Cognitive and Affective Neuroscience, 5*(1), 59–67.

Cleckley, H. (1976). *The mask of sanity.* St. Louis, MO: Mosby.

Cushman, F. (2008). Crime and punishment: Distinguishing the roles of causal and intentional analyses in moral judgment. *Cognition, 108*(2), 353–380.

Damasio, A. (1994). *Descartes' error.* New York: Grosset/Putnam.

Foot, P. (1978). The problem of abortion and the doctrine of the double effect. In P. Foot (Ed.), *Virtues and vices and other essays on moral philosophy* (pp. 19–33). Oxford: Basil Blackwell.

Glass, S. J., & Newman, J. P. (2006). Recognition of facial affect in psychopathic offenders. *Journal of Abnormal Psychology, 115*(4), 815–820.

Glenn, A. L. (2010). How can studying psychopaths help us understand the neural mechanisms of moral judgment. *Cell Science Reviews, 6*(4), 30–35.

Glenn, A. L., Iyer, R., Graham, J., Koleva, S., & Haidt, J. (2009a). Are all types of morality compromised in psychopathy? *Journal of Personality Disorders, 23,* 384–398.

Glenn, A. L., Raine, A., & Schug, R. A. (2009b). The neural correlates of moral decision-making in psychopathy. *Molecular Psychiatry, 14*(1), 5–6.

Glenn, A. L., Raine, A., Schug, R., Young, L., & Hauser, M. (2009c). Increased DLPFC activity during moral decision-making in psychopathy. *Molecular Psychiatry, 14,* 909–911.

Graham, J., Haidt, J., & Nosek, B. (2009). Liberals and conservatives use different sets of moral foundations. *Journal of Personality and Social Psychology, 96,* 1029–1046.

Graham, J., Nosek, B. A., Haidt, J., Iyer, R., Koleva, S., & Ditto, P. H. (2011). Mapping the moral domain. *Journal of Personality and Social Psychology, 101*(2), 366–385.

Greene, J. D., Cushman, F. A., Stewart, L. E., Lowenberg, K., Nystrom, L. E., & Cohen, J. D. (2009). Pushing moral buttons: The interaction between personal force and intention in moral judgment. *Cognition, 111*(3), 364–371.

Greene, J. D., Morelli, S. A., Lowenberg, K., Nystrom, L. E., & Cohen, J. D. (2008). Cognitive load selectively interferes with utilitarian moral judgment. *Cognition 107,* 1144–1154.

Greene, J. D., Nystrom, L. E., Engell, A.D., Darley, J. M., & Cohen, J. D. (2004). The neural bases of cognitive conflict and control in moral judgment. *Neuron, 44,* 389–400.

Greene, J. D., Sommerville, R., Nystrom, L. E., Darley, J. M., & Cohen, J. D. (2001). An fMRI investigation of emotional engagement in moral judgment. *Science, 293*(5537), 2105–2108.

Haidt, J., & Kesebir, S. (2010). Morality. In S. Fiske, D. Gilbert & G. Lindzey (Eds.), *Handbook of social psychology,* (5th ed., pp. 797–832). Hoboken, NJ: Wiley.

Haidt, J., Koller, S. H., & Dias, M. G. (1993). Affect, culture, and morality, or is it wrong to eat your dog? *Journal of Personality and Social Psychology, 65*(4), 613–628.

Hare, R. D. (1991). *The Hare Psychopathy Checklist—Revised (PCL-R).* Toronto, Ontario, Canada: Multi-Health Systems.

Hare, R. D. (2003). *Manual for the Revised Hare Psychopathy Checklist* (2nd ed.). Toronto, Ontario, Canada: Multi-Health Systems.

Harenski, C. L., Harenski, K. A., Shane, M. S., & Kiehl, K. A. (2010). Aberrant neural processing of moral violations in criminal psychopaths. *Journal of Abnormal Psychology, 119*(4), 863–874.

Hicks, B. M., & Patrick, C. J. (2006). Psychopathy and negative emotionality: Analyses of suppressor effects reveal distinct relations with emotional distress, fearfulness, and anger-hostility. *Journal of Abnormal Psychology*, 115(2), 276–287.

House, T. H., & Milligan, W. L. (1976). Autonomic responses to modeled distress in prison psychopaths. *Journal of Personality and Social Psychology*, 34(4), 556–560.

Iria, C. & Barbosa, F. (2009). Perception of facial expressions of fear: comparative research with criminal and non-criminal psychopaths. *Journal of Forensic Psychiatry & Psychology*, 20(1), 66–73.

Ishida, C. (2006). How do scores of DIT and MJT differ? A critical assessment of the use of alternative moral development scales in studies of business ethics. *Journal of Business Ethics*, 67, 63–74.

Johns, J. H., & Quay, H. C. (1962). The effect of social reward on verbal conditioning in psychopathic and neurotic military offenders. *Journal of Consulting Psychology*, 26(3), 217–220.

Kennealy, P. J., Hicks, B. M., & Patrick, C. J. (2007). Validity of factors of the Psychopathy Checklist—Revised in female prisoners: Discriminant relations with antisocial behavior, substance abuse, and personality. *Assessment*, 14(4), 323–340.

Kiehl, K. A. (2006). A cognitive neuroscience perspective on psychopathy: Evidence for paralimbic system dysfunction. *Psychiatry Research*, 142(2–3), 107–128.

Koenigs, M., Baskin-Sommers, A., Zeier, J., & Newman, J. P. (2011). Investigating the neural correlates of psychopathy: A critical review. *Molecular Psychiatry 16(8)*, 792–799.

Koenigs, M., Kruepke, M., & Newman, J. P. (2010). Economic decision-making in psychopathy: A comparison with ventromedial prefrontal lesion patients. *Neuropsychologia*, 48(7), 2198–2204.

Koenigs, M., Kruepke, M., Zeier, J., & Newman, J. P. (2012). Utilitarian moral judgment in psychopathy. *Social Cognitive and Affective Neuroscience*, 7, 708–714.

Koenigs, M., Young, L., Adolphs, R., Tranel, D., Cushman, F., Hauser, M. (2007). Damage to the prefrontal cortex increases utilitarian moral judgements. *Nature*, 446, 908–911.

Kohlberg, L. (1958). *The development of modes of moral thinking and choice in the years ten to sixteen* [dissertation]. University of Chicago.

Kohlberg, L. (1973). The claim to moral adequacy of a highest stage of moral judgment. *The Journal of Philosophy*, 70(18), 630–646.

Kosson, D. S., Suchy, Y., Mayer, A. R. & Libby, J. (2002) Facial affect recognition in criminal psychopaths. *Emotion*, 2(4), 398–411.

Lind, G. (1978). Wie misst man moralisches Urteil? Probleme und alternative Möglichkeiten der Messung eines komplexen Konstrukts. [How does one measure moral judgment? Problems and alternative ways of measuring a complex construct]. In G. Portele (Ed.), *Sozialisation und Moral* (pp. 171–201). Weinheim: Beltz.

Lind, G. (2000 [original 1984]). *Content and structure of moral judgment*. Second corrected edition. Unpublished doctoral dissertation, University of Konstanz.

Link, N. F., Scherer, S. E., & Byrne, P. N. (1977). Moral judgement and moral conduct in the psychopath. *Canadian Psychiatric Association Journal*, 22(7), 341–346.

Lose, C. A. (1997). Level of moral reasoning and psychopathy within a group of federal inmates. *Dissertation Abstracts International B: The Sciences and Engineering*, 57(7-B), 4716.

Malatesti, L., & McMillan, J. (Eds.). (2010). *Responsibility and psychopathy: Interfacing law, psychiatry, and philosophy*. New York: Oxford University Press.

Mencl, J., & May, D. (2009). The effects of proximity and empathy on ethical decision-making: An exploratory investigation. *Journal of Business Ethics*, 85(2), 201–226.

O'Hara, R., Sinnott-Armstrong, W., & Sinnott-Armstrong, N. (2010). Wording effects in moral judgments. *Judgment and Decision Making*, 5(7), 547–554.

O'Kane, A., Fawcett, D., & Blackburn, R. (1996). Psychopathy and moral reasoning: Comparison of two classifications. *Personality and Individual Differences*, 20(4), 505–514.

Parkinson, C., Sinnott-Armstrong, W., Koralus, P. E., Mendelovici, A., McGeer, V., Wheatley, T. (2011). Is morality unified? Evidence that distinct neural systems underlie moral judgments of harm, dishonesty, and disgust. *Journal of Cognitive Neuroscience*, 23(10), 3162–3180.

Patrick, C. J. (1994). Emotion and psychopathy: Startling new insights. *Psychophysiology*, 31(4), 319–330.

Petrinovich, L., & O'Neill, P. (1996). Influence of wording and framing effects on moral intuitions. *Ethology and Sociobiology*, 17, 145–171.

Pujol, J., Batalla, I., Contreras-Rodriguez, O., Harrison, B. J., Pera, V., Hernandez-Ribas, R., et al. (2011). Breakdown in the brain network subserving moral judgment in criminal psychopathy. *Social Cognitive and Affective Neuroscience, 7*(8), 917–923.

Regina v. M'Naghten, 10 Cl. & Fin. 200, 9 Eng. Rep. 718 at 722 (1843).

Reidy, D. E., Shelley-Tremblay, J. F., & Lilienfeld, S. O. (2011). Psychopathy, reactive aggression, and precarious proclamations: A review of behavioral, cognitive, and biological research. *Aggression and Violent Behavior, 16*(6), 512–524.

Rest, J., Cooper, D., Coder, R., Masanz, J., & Anderson, D. (1974). Judging the important issues in moral dilemmas—An objective measure of development. *Developmental Psychology, 10*(4), 491–501.

Rest, J. R. (1979). *Revised manual for the Defining Issues Test.* Minneapolis: Moral Research.

Robinson, P. H., & Kurzban, R. (2007). Concordance & conflict in intuitions of justice. *Minnesota Law Review, 91,* 1829–1907.

Saxe, R. (2006). Uniquely human social cognition. *Current Opinion in Neurobiology, 16,* 235–239.

Schaich Borg, J., Lieberman, D., & Kiehl, K. A. (2008). Infection, incest, and iniquity: Investigating the neural correlates of disgust and morality. *Journal of Cognitive Neuroscience, 20*(9), 1–19.

Schaich Borg, J., Sinnott-Armstrong, W., Calhoun, V. D., & Kiehl, K. A. (2011). Neural basis of moral verdict and moral deliberation. *Social Neuroscience, 6*(4), 398–413.

Sinnott-Armstrong, W., & Levy, K. (2011). Insanity defenses. In J. Deigh & D. Dolinko (Eds.), *The Oxford Handbook of Philosophy of Criminal Law* (pp. 299–334). New York: Oxford University Press.

Smetana, J. G. (1985). Preschool children's conceptions of transgressions: The effects of varying moral and conventional domain-related attributes. *Developmental Psychology, 21,* 18–29.

Sutker, P. B. (1970). Vicarious conditioning and sociopathy. *Journal of Abnormal Psychology, 76*(3), 380–386.

Thomson, J. J. (1976). Killing, letting die, and the trolley problem. *The Monist, 59,* 204–217.

Thomson, J. J. (1985). The trolley problem. *Yale Law Journal 94,* 1395–1415.

Turiel, E. (1983). *The development of social knowledge: Morality and convention.* Cambridge, UK: Cambridge University Press.

Turiel, E., Killen, M., & Helwig, C. (1987). Morality: Its structure, functions, and vagaries. In J. Kagan & S. Lamb (Eds.), *The emergence of morality in young children* (pp. 155–243). Chicago: The University of Chicago Press.

Verona, E., Hicks, B. M., & Patrick, C. J. (2005). Psychopathy and suicidality in female offenders: Mediating influences of personality and abuse. *Journal of Consulting and Clinical Psychology, 73*(6), 1065–1073.

Verona, E., Patrick, C. J., & Joiner, T. E. (2001). Psychopathy, antisocial personality, and suicide risk. *Journal of Abnormal Psychology, 110*(3), 462–470.

Wilson, K., Juodis, M., & Porter, S. (2011). Fear and loathing in psychopaths: A meta-analytic investigation of the facial affect recognition deficit. *Criminal Justice and Behavior, 38,* 659–668.

Young, L., Camprodon, J. A., Hauser, M., Pascual-Leone, A., & Saxe, R. (2010). Disruption of the right temporoparietal junction with transcranial magnetic stimulation reduces the role of beliefs in moral judgments. *Proceedings of the National Academy of Sciences of the USA, 107*(15), 6753–6758.

PART FOUR
NEUROSCIENCE AND PSYCHOPATHY

CHAPTER 8

Functional Neuroimaging and Psychopathy

Nathaniel E. Anderson and Kent A. Kiehl

Psychopathy is a construct characterized by symptoms of emotional detachment and a propensity for disinhibited, impulsive behavior combined with a general callousness and lack of insight for the impact such behavior has on others (Cleckley, 1941). Consequently, psychopaths are prone to high rates of criminal behavior and demonstrate increased aggressive behavior (Leistico, Salekin, DeCoster, & Rogers, 2008; Salekin, Rogers, & Sewell, 1996). Further, these behavioral problems are alarmingly persistent, as rates of recidivism and especially violent recidivism are much higher for psychopaths than for nonpsychopaths (Porter, Birt, & Boer, 2001; Porter, Brinke, & Wilson, 2009), and therapeutic intervention and rehabilitation strategies with adult psychopaths often have proven to be ineffective (Rice, Harris, & Cormier, 1992; Seto & Barbaree, 1999).

Understanding the nature of psychopathy has become increasingly important to a diverse set of researchers because of the social, legal, and philosophical implications stemming from research suggesting that psychopathy has a neural basis (Blair, 2006; Kiehl, 2006) with strong genetic heritability (Larsson, Andershed, & Lichtenstein, 2006; Viding, Blair, Moffitt, & Plomin, 2005). For instance, it has become a pressing concern in forensic settings to decide whether a designation of psychopathy is a mitigating factor in determining responsibility and sentencing, and neuroscientists are on the front lines of this debate (Glenn, Raine, & Laufer, 2011; Harenski, Hare, & Kiehl, 2010). A critical component of this research has been the investigation of how the brains of psychopaths process information differently from those of healthy individuals, and there has been a particularly strong emphasis on identifying anatomical units of the brain that appear dysfunctional in psychopaths. Two prominent neurobiological theories of psychopathy put forth by Blair (2006, 2007) and Kiehl (2006) have emphasized disturbances in a tightly integrated network of brain regions responsible for processing emotional information, such as our basic responses to threat and the anticipation of punishment, and further integrating these basic responses into our higher-order thought processes, such as monitoring our own behavior and planning ahead. This chapter reviews the most prominent support for these theories, highlighting data that suggest functional differences in the brains of psychopaths.

BASIC FUNCTIONAL ORGANIZATION OF THE BRAIN

As this information is reviewed, it will be important to recognize a basic hierarchical organizational structure for how information is processed in the brain. Most of the highly complex, integrative cognitive processes depend on activity in the outer layers of the cerebral cortex (neocortex), while more fundamental, elementary processing occurs deeper in "subcortical" parts of the brain. In terms of basic anatomy, if we were to strip away most of the surrounding tissues of the head such as the hair, skin, and skull, the neocortex would be the mostly-visible, outer layer of the brain. It is phylogenetically the most recent anatomical development in the human brain and is, compared to other primates, the most highly developed in our species. It governs higher-order cognitive processing, combining information such as visual and auditory stimuli with working memory for use in complex ways. Subcortical structures—literally parts of the brain that exist beneath the superficial layers of the neocortex—are phylogenetically older and are involved in more basic processes of survival such as governing the body's reflexes, maintaining homeostasis, encoding memory, and responding to threat and reward. These subcortical processes largely operate "under the hood," that is, automatically and involuntarily, without the need for our conscious awareness. Subcortical processing organizes and feeds information up to higher levels of processing, where we can interact with it on a conscious-evaluative level. The neocortex, in turn, is capable of exerting some "top-down" control over subcortical structures, biasing these basic responses with inhibitory control or preferential activation. In this chapter, it will become clear that functional activity in both cortical and subcortical structures is implicated in psychopathy, as well as in brain regions that bridge the gap between the two.

METHODS FOR ASSESSING BRAIN FUNCTION

Some familiarity with the tools used to assess functional brain abnormalities is essential for applying these findings to theories of psychopathy. Perhaps the oldest and most basic method of examining functional units of the brain is by carefully keeping track of what processes are disrupted when specific parts of the brain have been damaged. Focal brain lesions can result from tumors, degenerative disease, the need for surgical removal of tissue due to epilepsy, or from direct injury such as the unfortunate circumstances leading to one of neuroscience's most famous historical case studies: Phineas Gage. The account, originally published in 1868 (reproduced in Harlow, 1993), described the dramatic personality transformation of the railroad worker who survived an accident that drove an iron rod through his head, damaging some frontal portions of his brain. It is often recounted that—in stark contrast to his behavior before this injury—Gage became raucous, aggressive, sexually promiscuous, and offensive in public after the injury, and this severely impacted his long-standing personal relationships and his capacity to maintain employment. This account is virtually ubiquitous in popular references describing neural underpinnings of personality and behavior; however, it has been emphasized that this account is surrounded by as much legend as fact, and many of the facts have been greatly exaggerated (Macmillan, 1999). Nonetheless, instances of focal brain lesions, while often tragic, can be valuable sources of information about localizing certain cognitive processes to specific parts of the brain. When combined with other systematic techniques of examining brain function, they can be particularly helpful for interpreting experimental results.

A prominent tool in contemporary neuroscience is magnetic resonance imaging (MRI), which exploits variations in the magnetic properties of different tissues or fluids in the body to create high-quality three-dimensional images of internal structures with unsurpassed spatial resolution (less than a cubic millimeter). The resolution of these images is suitable for discriminating detailed and highly specific neuroanatomical features. A more recent methodological development has been the use of MRI to create maps of functional neural activity, that is, images that display portions of the brain that are active during carefully controlled tasks, and this is referred to as functional MRI (fMRI). This technique takes advantage of subtle changes in the magnetic properties of the blood that coincide with local increases in neural activity. As the brain does its work, it consumes oxygen. Blood with oxygen in it has a different magnetic signal than does blood without oxygen in it, and the contrast representing relative ratios of oxygenated to deoxygenated blood can be translated into a digital image. This technique is referred to as blood oxygenation–level dependent (BOLD) contrast. The images that are produced can be superimposed over structural images obtained using standard MRI techniques to localize sources of neural activity to specific regions of the brain. Although this technique allows for highly specific spatial localization of neural activity, its specificity in time is limited by the fact that it ultimately relies on relatively slow changes in blood flow; therefore, experimental conditions have to be carefully controlled and meticulously timed to make meaningful estimates of how brain activity corresponds to specific cognitive–behavioral events.

Another adaptation of MRI technology called diffusion tensor imaging (DTI) is designed to measure the integrity of communicating pathways between different brain regions. Neurons that carry information relatively long distances are covered with a fatty sheath of insulation to promote fast, efficient transmission of these signals. These insulated bundles of neurons, or white matter, can be likened to cables connecting functional units of the brain. DTI measures the integrity of the white matter, which is thought to represent how efficiently certain brain regions are capable of sharing information. For example, if two telephones are in perfect working order, but the lines that connect the phones are damaged in some way, communication between those phones will be interrupted. The brain shares a great deal of information across multiple structures, so that it can be utilized simultaneously by distinct, parallel processing sites. This important property of *functional connectivity* relies on robust, reliable communication between brain regions, and it has become particularly relevant in psychopathy research in recent years.

Other neuroimaging methods employ radioactive isotopes attached to biologically useful compounds (such as glucose) that are injected into the bloodstream, concentrating where metabolic activity is elevated. Images depicting clusters of radiation represent localized neural activity (i.e. consumption of glucose). Both positron emission tomography (PET) and single photon emission computerized tomography (SPECT) utilize this technique and can be used to create similar maps of brain function, with somewhat reduced spatial resolution compared to fMRI. Although these tools have often been implemented in other realms of experimental psychology, there has never been a single PET study examining rigorously defined psychopaths, and only a couple of SPECT studies (Intrator et al., 1997; Soderstrom et al., 2002); therefore, these techniques are not discussed any further here. The vast majority of neuroimaging investigations of psychopathy have relied on MRI and fMRI when high spatial resolution is needed. Alternatively, psychopathy researchers concerned with the exact timing of cognitive events have primarily relied on electroencephalography (EEG).

EEG is another technique for measuring brain activity, as a participant performs various tasks designed to tap into specific cognitive functions. EEG records tiny electrical potentials on the surface of the scalp that result from the massed, synchronized neural activity occurring just beneath it. It is among the few methods available for measuring neural activity directly; however, it cannot provide detailed spatial maps of this activity owing to certain technical limitations in retracing the precise source of changes in electrical potentials measured outside the brain. Yet these changes fluctuate very quickly and can be sampled thousands of times per second, allowing for superb accuracy in the timing of neural activity, on the order of milliseconds. EEGs are recorded as complex waves, but these can be decomposed into individual frequency bands, which are indicative of general cognitive states (sleeping, relaxed, actively processing information). They can also be averaged across time with reference to specific repeated events in an experimental task, yielding a waveform with a characteristic shape called an event-related potential (ERP). ERPs are carefully examined for differences in the size and latency of specific peaks and troughs, and these features correspond to predictable differences in cognitive variables such as processing speed, shifts in attention, memory encoding, self-monitoring, and semantic evaluation. Indeed, a surprising amount of information can be gleaned from these electrocortical techniques, but decades of empirical investigations using EEGs and ERPs have endowed these methods with robust theories regarding relationships between cognitive function and the detectable changes in electrical activity measured at the scalp.

Each of these techniques has particular costs and benefits, as well as methodological issues that impact their interpretation and generalizability.

For instance, it is important to recognize that most investigations into brain function introduce an experimental manipulation intended to produce a certain pattern of neural activity, which may or may not differ across subjects; therefore, interpretation of the outcomes of such investigations should be made carefully, with strong consideration of the tasks implemented while brain activity is recorded. Ultimately, the strongest evidence for any given interpretation of these data, including neurobiological models of psychopathy, is provided by consistency across several or all of these methods of evaluation. With this in mind, some clear patterns have emerged that indicate that psychopaths have abnormalities in brain regions that support the processing of emotional information and the integration of this information into higher-order cognitive processes, such as the modification of ongoing behavior and the assessment of future consequences. These data are reviewed here in sections organized by some basic functional–anatomical units of the brain.

PREFRONTAL CORTEX

The cerebral cortex is divided into large topographical regions, and the frontal cortex comprises approximately one-third of the visible surface of the brain. The prefrontal cortex is the forward-most portion of the frontal cortex and is important for a wide variety of functions such as behavioral inhibition, the estimation of consequences and rewards, and utilizing stored memory for decision making—cognitive tasks that broadly fall under the umbrella of *executive functions*. However, only certain executive functions seem to be disrupted in psychopathy, whereas others such as working memory, problem solving, and verbal reasoning remain wholly intact in psychopaths (Dolan, 2011). This likely occurs due to fairly specific dysfunction in restricted portions of the prefrontal cortex, which evidence suggests may be limited to the lower-middle (ventromedial) aspect, overlapping with the orbitofrontal cortex, located just above the eyes. This portion of the prefrontal cortex has long been suspected for its role in the development of psychopathic behavior due in part to cases such as that of Phineas Gage (Dolan, 1999), described previously, whose purported behavior afterd brain injury imprecisely resembles some aspects of psychopathy. This phenomenon has sometimes been called *acquired sociopathy* or *pseudo-psychopathy* (Blumer & Benson, 1975; Damasio, 1994), and there have been several similar cases in more recent medical literature (e.g., Cato, Delis, Abildskov, & Bigler, 2004; Meyers, Berman, Scheibel, & Hayman, 1992). One should recognize, however, that the consequent symptoms precipitating from frontal lobe injury are highly variable and almost never represent the full spectrum of impairments consistent with psychopathy.

The suggestion of impairments in the ventromedial prefrontal cortex in psychopathy is consistent with many accounts of deficits following brain lesions, which regularly show relationships between disinhibited, impulsive behavior and damage in this region. Bechara, Damasio, Tranel, and Damasio (1997) utilized an experimental task that requires keen perception of changing reward and punishment contingencies and demonstrated that healthy controls were able to adopt advantageous strategies in this task even before they could explicitly describe the rules governing those strategies—an example of how subtle our estimation of future consequences and rewards can be. In stark contrast to healthy individuals, patients with ventromedial prefrontal damage continued to make disadvantageous decisions, even after they were explicitly aware of the most profitable strategy. Even still, it seems that the most destructive consequences of this sort result when damage to the ventromedial prefrontal cortex has occurred very early in development, often instigating life-long patterns of poor moral judgment and impulsivity (Anderson, Damasio, Tranel, & Damasio, 2000). Still others have associated specific biases toward a form of utilitarian moral judgments (e.g., sacrificing one life to save two) to damage in the ventromedial prefrontal cortex (Koenigs et al., 2007). Finally, damage to this region has been associated with an incapacity to feel regret (Camille et al., 2004). All of these characteristics are elements of the larger assemblage of traits defining the prototypical psychopath.

The orbitofrontal/ventromedial prefrontal cortex remains the most common prefrontal region implicated in neuroimaging investigations of psychopathy; but again, it is important to consider the tasks being implemented when functional brain activity is being assessed. For example, a *Prisoners Dilemma* task is commonly used in psychological experiments and requires participants to make decisions about maximizing their gains either through social cooperation or by defecting against a partner. It has been demonstrated that psychopaths evince lower orbitofrontal activity when choosing to cooperate and lower dorsolateral prefrontal activity when choosing to defect than do nonpsychopaths (Rilling et al., 2007). Others have demonstrated that, contrary to control subjects, psychopaths fail to engage the orbitofrontal cortex during tasks that require aversive conditioning, that is, learning to associate a specific behavior with punishment (Birbaumer et al., 2005; Veit et al., 2002). Further, this may be a persistent pattern evident from an early age. Although psychopathy is a designation ordinarily reserved for adults, persistent and severe conduct problems paired with callous–unemotional personality traits is widely considered the prodromal form of psychopathy in youth (Frick & White, 2008). It has been demonstrated that adolescents with these traits also exhibit reduced orbitofrontal activity during reinforcement stages of conditioning (Finger et al., 2011).

Reduced activity in this region of the prefrontal cortex has been reported in association with psychopathic traits in a wide variety of other tasks as well, including viewing pictures of faces with strong expressions of emotion (Gordon, Baird, & End, 2004), viewing pictures depicting moral violations (Harenski, Kim, & Hamann, 2009), and during a complex task requiring the integration of emotional information into ongoing behavioral outcomes (Müller et al., 2008). However, not all tasks elicit reduced prefrontal activity for psychopaths. For instance, the medial prefrontal areas were reportedly inactive when individuals scoring high on psychopathy measures retaliated against an opponent during a competitive reinforcement task, but they demonstrated relatively increased activity in this region when observing an opponent being punished (Veit et al., 2010). It has also been reported that psychopaths engage orbital and medial prefrontal cortex when performing a task requiring the estimation of other's emotions—a task that alternatively engages the supramarginal gyrus and the superior frontal gyrus in healthy individuals (Sommer et al., 2010)—these portions of the brain sometimes referred to as the "mirror neuron" system, have been associated with empathic perception. Outcomes such as these suggest that structures commonly labeled as dysfunctional in psychopaths are not necessarily inoperative, but may simply be utilized differently, processing information in divergent ways. They also highlight the importance of attending to specific task requirements when interpreting functional imaging reports, as ostensible abnormalities may be applicable only in the context of the specific cognitive function being assessed by a particular task.

AMYGDALA

Like the prefrontal cortex, the amygdala has featured prominently in theories of psychopathy. The amygdalae are two almond-shaped structures located bilaterally (one in each of the brain's two hemispheres) near the innermost portions of the temporal lobes. Many anatomical features of the brain exist bilaterally, and by convention they are often referred to in the singular (e.g., amygdala) even though it is understood that there are two of them, as one might refer to the function of "the kidney." The amygdala is a subcortical structure, again operating mostly in the absence of conscious awareness, and has a long-established role in emotional processing. This level of emotional processing is not the complex form of emotion often referred to in common parlance, but rather represents the elementary recognition of stimuli that are relevant to our safety and survival, such as those indicating potential threat and reward. More specifically, the amygdala is important in forming stimulus–reinforcement associations (e.g., learning that a rattle may be coming from a dangerous snake), conditioned fear responses (as when we unconsciously hesitate to touch the door knob that consistently delivers a shock of static electricity), and the initiation of emotional states (Davis, 1997; Davis & Whalen, 2001).

Damage to the amygdala often results in subtler behavioral consequences than damage to the prefrontal cortex; however, just as with damage to frontal brain regions, the most severe consequences of amygdala damage may develop over time only as certain kinds of reinforcement learning are impaired or prevented altogether. Adolphs and colleagues have reported several effects of amygdala damage including impaired declarative memory for emotional information (Adolphs, Cahill, Schul, & Babinsky, 1997) and impaired recognition of negative facial emotions (Adolphs et al., 1999). Bechara and colleagues (1999) have demonstrated that amygdala damage prevents acquisition of conditioned autonomic responses, that is, emotional states associated with punishment/reinforcement. More recently, it has been demonstrated that certain forms of social learning are dependent on amygdala function. Shaw et al. (2004) reported that damage to the amygdala early in life interrupts development of *theory of mind* reasoning; that is, the ability to entertain another person's point of view or state of being. However, damage to the amygdala later in life does not appear to interrupt this capacity. Owing to its involvement in threat detection, recognition of emotional information, and reinforcement learning, there has been a long-standing notion that the amygdala is dysfunctional in psychopaths, and indeed a great deal of research supports this conclusion.

Initial evidence for this perhaps begins with Lykken (1957), who demonstrated that inmates with psychopathic traits were impaired in their ability to learn to avoid electric shocks in an experimental task—a deficiency in forming stimulus–punishment associations, which only later was discovered to depend on the amygdala (Phillips & LeDoux, 1992). More recently, Patrick, Bradley, and Lang (1993) demonstrated that psychopaths lacked a common feature of the startle reflex, believed to be modulated by the amygdala (Davis, 1989). Specifically, healthy individuals show stronger startle reflexes when they are already in an aversive or anxious state (Lang, Bradley, & Cuthbert, 1990), but psychopaths do not. It has been demonstrated that focal lesions of the amygdala produce this same effect, disrupting normal patterns of emotion-modulated startle (Angrilli et al., 1996; Kim & Davis, 1993). These, and many other similar experiments, were clues that the amygdala may be dysfunctional in psychopaths, and recent neuroimaging data have supported this conclusion.

Kiehl and colleagues (2001) were the first to demonstrate with fMRI that criminal psychopaths exhibit abnormally reduced amygdala function, and this occurred during a memory task comparing emotional and neutral words. As is commonly reported in such investigations, there were no significant behavioral differences between groups, that is, psychopaths and nonpsychopaths performed equally well in the explicit recall for emotional or neutral words. Still, several brain regions are differentially activated during the task, suggesting that psychopaths and nonpsychopaths use alternative means of processing these stimuli. Abnormally low amygdala function in psychopaths has been reported in several other tasks as well; in fact, many of the tasks noted earlier that revealed lower prefrontal activity also revealed lower amygdala activity, which may be expected owing to the extensive white matter connections between the amygdala and prefrontal cortex. Relative to controls, psychopaths have demonstrated reductions in amygdala activity during aversive conditioning (Birbaumer et al., 2005; Rilling et al. 2007; Veit et al., 2002), when viewing pictures depicting moral violations (Harenski, Harenski, Shane, & Kiehl, 2010), pictures of facial affect (Gordon et al., 2004), generally aversive photographic stimuli (Harenski, Kim, & Hamann, 2009), and when viewing fearful faces (Dolan & Fullam, 2009). Youth with callous–unemotional traits and conduct disorder also show lower amygdala activity when engaged in a passive-avoidance learning (Finger et al., 2011). Based on such evidence, Blair (2006) developed a neurobiological model of psychopathy that has focused primarily on dysfunction in the amygdala, suggesting that apparent deficits in other brain regions may be a direct consequence of the hierarchical organization of the brain and reduced propagation of activity ordinarily projected to those regions (see also Blair, Mitchell, & Blair, 2005; Blair, 2007).

The amygdala, therefore, remains a prominent fixture in the investigation of psychopathy's neural underpinnings.

PARALIMBIC AND ADDITIONAL STRUCTURES

The amygdala and areas of the prefrontal cortex have well-understood relationships with stimulus–reinforcement learning and the estimation of rewards and consequences based on this information, but several other brain regions with subtler supportive roles in these processes have also been shown to be abnormal in psychopaths (for an extensive review see Kiehl, 2006). Referring back to the earlier discussion of the brain's hierarchical organization, there is a complex and tightly networked system for integrating core emotional responses into higher-order cognitive function that can be generally referred to as the paralimbic system (Broadmann, 1994; Mesulam, 2000). In addition to the amygdala and prefrontal cortex, several other brain regions influence these processes including the anterior cingulate and posterior cingulate, the anterior superior temporal gyrus (a.k.a. temporal pole), the insula, and the parahippocampal gyrus. The functions of these structures demonstrate their supportive role in several overlapping, highly integrative processes that remain relevant to the expression of psychopathic traits.

The cingulate cortex is part of the "limbic cortex," literally defining the border (limbus) between subcortical and cortical structures, the anterior cingulate (forward-most portion) contributes to affect-regulation, pain perception, and response-conflict/error monitoring (Kiehl, 2000), and damage here has consequences similar to those of damage to the orbitofrontal cortex, including hostility and irresponsibility (Hornak, 2003), as well as difficulty with conflict monitoring and cognitive control (di Pellegrino, Ciaramelli, & Ladavas, 2007). The posterior portion of the cingulate functions in episodic memory and self-reflective thought (Johnson et al., 2002) and its activity has been frequently associated with evaluation of emotional significance (Maddock, 1999). The temporal pole is found in the forward-most aspect of the temporal cortex—near the temples on each side of one's head. Activity here has been associated with complex social and emotional processing, including theory of mind reasoning and facial recognition, and damage here again can mimic the effects of orbitofrontal damage (Olson, Plotzker, & Ezzyat, 2007), with unstable mood and irritability being common symptoms (Glosser, Zwil, Glosser, O'Connor, & Sperling, 2000). The insula is a portion of the temporal lobe of the cerebral cortex folded inward toward the orbitofrontal area. It also plays a role in the integration of cognitive and affective information (Jones, Ward, & Critchley, 2010), and damage here has familiar consequences such as impaired assessment of risk and future consequences (Weller, Levi, Shiv, & Bechara, 2009). The parahippocampal gyrus is another closely associated structure in the medial temporal lobe and supports a wide range of memory-related functions. The deterioration of this structure is strongly associated with Alzheimer's disease and age-related memory loss (Burgmans et al., 2009; Echavarri et al., 2010).

While the orbitofrontal cortex and amygdala feature prominently in psychopathy-related abnormalities, abnormal task-specific functional activation patterns in psychopaths are also commonly found in these paralimbic structures. Kiehl and colleagues (2001) reported widespread activation differences in psychopaths (compared to noncriminal, nonpsychopathic controls) during an affective memory task, which, in addition to functional deficits in the amygdala, included reduced relative activity in the parahippocampal gyrus, anterior cingulate, and posterior cingulate. Abnormally low activity in the anterior cingulate cortex in psychopaths has also been reported during defection in the *Prisoners Dilemma* task (Rilling et al., 2007) and during acquisition of aversive conditioning (Birbaumer et al., 2005). Likewise, Veit and colleagues (2002) reported reduced relative activity (compared to controls) in the insula and anterior cingulate while psychopaths engaged in aversive conditioning. Abnormally low activity in the right temporal pole of psychopaths has also been reported during an emotion-modulation task (Müller et al., 2008) and when comparing abstract and concrete words (Kiehl et al., 2004). Further, several investigations examining youth with psychopathic traits have supported abnormal function in the cingulate cortex (Jones et al., 2009; Marsh et al., 2008), as well as the insula and parahippocampal region (Finger et al., 2011).

Although some neurobiological models of psychopathy have limited the scope of emphasis to dysfunction in the amygdala and orbitofrontal cortex (e.g., Blair, 2006, 2007), these regions share close connections with other paralimbic structures throughout the brain, which have demonstrable abnormalities in psychopaths as supported by the research noted in the preceding text. Owing to the hierarchical organization responsible for integrating basic emotional information into higher cognitive functions, it has been suggested that differences in relative activity levels of these paralimbic structures may be dependent on reduced activity in core limbic structures (see Blair, 2010; Kiehl et al., 2001), due to the existence of dense projections between these and paralimbic structures. However, reduced activity in higher-order paralimbic structures does not always correspond to reduced primary limbic activity. For instance, Müller and colleagues (2003) utilized a simple task, viewing pictures with varied emotional content, and reported a wide range of differences between psychopaths and controls in activity throughout paralimbic structures, which included relatively increased activity in the amygdala and insula during negative picture-viewing, but relative decreased activation in parts of the anterior cingulate and parahippocampal gyrus.

Outcomes such as this help to emphasize that activity in this circuit is not unidirectional. The complex connectivity and reciprocal influences that paralimbic structures have on each other can sometimes make interpretations tedious, and idiosyncratic features of an experimental design can obscure generalizable outcomes. For example, although the reductions in amygdala activity commonly found in psychopaths are generally interpreted as a decreased response to aversive-threatening stimuli, incidental increases in amygdala activity could be the consequence of reduced prefrontal inhibitory control over circuits responsible for emotional activation. Further, it is possible that these alternative conditions both exist as distinct conditions and may give rise to equally dangerous pathologies. Ultimately what may be necessary to clarify the roles of these anatomical components in psychopathy is the development of detailed path models of functional connectivity and how these paths may be altered in psychopaths. These may help to explain better some divergent findings in the literature as well. What we can be certain of is that healthy brains are able to adapt reflexively to a variety of emotionally relevant environmental cues, and when even isolated elements of the complex system governing these reactions are disrupted, the consequences can be severe. A summary of fMRI studies of psychopathy is presented in table 8.1

FUNCTIONAL CONNECTIVITY AND PSYCHOPATHY

No brain region operates in isolation. The widespread and complex interactions between various brain regions are the target of another brain imaging paradigm that specifically examines functional connectivity. This realm of investigation focuses on communication between specific neuroanatomical units. Functional connectivity may be assessed either by examining correlated activity between two brain regions over time or by assessing the structural integrity of the bundles of white matter connecting discrete units of the brain. As an example, recent research suggests that the impulsivity typically seen in juveniles that gives way to more controlled behavior in adulthood is associated with normal developmental patterns of functional connectivity between motor planning regions and regions responsible for executive control; however, highly impulsive (Factor 2 of psychopathy) juvenile offenders show patterns of connectivity suggesting delayed or disrupted maturation of these circuits (Shannon et al., 2011).

Referring back to the general principle that psychopaths have deficits using emotional information to regulate their behavior, it has been suggested that there may be defective

Table 8.1 Summary of fMRI studies in Psychopathy

Authors	Psychopaths	Comparison Groups	Task Description	Results (psychopaths vs.)
Sommer et al. 2010	N = 14 forensic patients PCL-R > 28 M = 28.6, SD = 1.2	N = 14 forensic patients PCL-R <15 M = 9.6, SD = 2.2	Attribution of emotional state to animated characters in a story	↑ OFC and Medial PFC (controls engage mirror neuron system)
Veit et al. 2010	N = 10 forensic patients PCL.SV mean 16.1(3.6) LSRS mean 62.5 (9.98)	N = 14 community sample from separate study Lotze et al (2007)	Competitive reaction time task vs. confederate, winners deliver punishment	↓ medial prefrontal during retaliation, ↑ medial prefrontal when witnessing punishment
Harenski et al. 2010	N = 16 incarcerated PCL-R > 30 M = 31.8, SD = 2.54	N = 16 incarcerated PCL-R < 18 M = 13.3, SD = 3.1	Viewing and rating pictures for severity of moral violations depicted	↓VMPFC and temporal pole distinguishing pictures, ↓ amygdala and ↑ posterior temporal for severe pictures
Harenski et al. 2009	N/A	N=10 females scored on PPI for continuous analysis M=123, SD=13.2	Instructed emotional regulation while viewing pictures depicting moral violations	Correlations with PPI: ↓ Medial PFC and Amygdala during pictures with moral violations. ↑ VLPFC and Superior PFC while regulating emotional response
Dolan & Fullam 2009	N = 12 violent, schizophrenic patients PCL.SV > 12.5 M = 17.4, SD = 2.3	N = 12 violent, schizophrenic patients PCL.SV <12.5 M = 8.4, SD = 3.3	Viewing pictures of facial affect, implicit processing of emotional expressions	↓ amygdala during fearful faces, ↑ amygdala for disgusted faces, all Ss ↑ VLPFC for negative emotions
Müller et al. 2008	N = 10 forensic patients PCL-R > 28 M = 30.5, R = 28-35	N = 12 healthy volunteers PCL-R < 4 R = 0–4	Emotion induction with affective pictures while categorizing shapes (Emotional Simon task)	↓ Temporal Pole, ↓ PFC during emotion/cognition interaction
Rilling et al. 2007	N/A	N = 30 mixed gender university sample for continuous analysis on PPI and Levenson psychopathy scales, no raw scores	Making decisions about social cooperation or defection in a strategy game (Prisoners Dilemma task)	Correlations with psychopathy scores: ↓ amygdala during aversive conditions, ↓ OFC when cooperating, ↓ DLPFC and ↓ anterior cingulate when defecting
Deeley et al. 2006	N = 6 forensic patients PCL-R > 25 M = 29.3, SD = 9	N = 9 healthy controls (No PCL-R data)	Viewing pictures of facial affect, implicit processing of emotional expressions	Overall: ↓ fusiform gyrus and ↓ extrastriate cortex. Fearful faces: ↓ fusiform gyrus Happy faces: ↓ fusiform gyrus and ↓ extrastriate cortex (vs. controls).

(continued)

Table 8.1 Continued

Authors	Psychopaths	Comparison Groups	Task Description	Results (psychopaths vs.)
Birbaumer et al. 2005	N = 10 criminal psychopaths Not currently incarcerated PCL-R Factor 1 > 10.5 M = 11.6, SD = 3.6 Total M = 24.9, SD = 5.2	N = 10 healthy non-criminal community volunteers PCL-R Factor 1 < 2	Aversive conditioning: neutral faces paired with painful pressure stimulus	Psychopaths had no activity in the limbic-prefrontal circuit (amygdala, OFC, anterior cingulate, insula) during aversive conditioning
Gordon et al., 2004	N = 10 college students PPI Factor 1 median split (values not provided)	N = 10 college students PPI Factor 1 median split (values not provided)	Viewing pictures of facial affect, attention to either emotion or identity of the faces	↓ Amygdala and ↓ medial PFC, but ↑ DLPFC during emotion task. No group differences for identity task
Kiehl et al. 2004	N = 8 incarcerated PCL-R > 28 M = 32.8, SD = 2.9	N = 8 non-incarcerated healthy controls PCL-R data not available	Differentiating between abstract and concrete words	↓ Temporal pole and surrounding cortical activity when differentiating between concrete and abstract
Müller et al. 2003	N = 6 incarcerated PCL-R > 30 M = 36.8, SD = 2.6	N = 6 non-incarcerated healthy controls PCL-R < 10 Descriptives not provided	Viewing photographic stimuli with positive, negative, and neutral valences (IAPS photos)	For *negative content*, ↓ anterior cingulate, parahippocampal and temporal gyrus, ↑ R.amygdala and R. PFC. For *positive content*, ↓ medial temporal and medial frontal, ↑ L.OFC
Veit et al. 2002	N = 4 criminal psychopaths PCL-R total M = 25.3, SD = 7.0	N = 4 social phobics (DSM) N = 7 healthy controls PCL-R data not available	Aversive conditioning: neutral faces paired with painful pressure stimulus	During acquisition and extinction, ↓ OFC, ↓ anterior cingulate, ↓ insula, ↓ amygdala activity (with only brief amygdala activation).
Kiehl et al. 2001	N = 8 incarcerated PCL-R > 28 M = 32.8, SD = 2.9	N = 8 incarcerated PCL-R < 23 M = 16.6, SD = 6.0 N = 8 community sample PCL-SV <20	Affective memory task for affective words	Affective - Neutral *vs.* all non-psychopaths: ↓ Amygdala, ↓ anterior cingulate, ↓ posterior cingulate, ↓ ventral striatum, ↓ inferior frontal gyrus. Vs. noncriminal controls: ↓ temporal pole, → parahippocampal, ↓ Amygdala

Note: All results are described as levels of brain activity in psychopaths compared to the corresponding control or comparison groups indicated.

connections between subcortical emotional centers and cortical units involved in behavioral regulation (van Honk & Schutter, 2006). Prominent among the possible connections disrupted is the *uncinate fasciculus*, a white-matter tract connecting the anterior temporal region (including the amygdala) with the orbitofrontal cortex. Impairments here could effectively disrupt the online retrieval of affect-related neural activity when it is relevant to decision-making. Recent neuroimaging investigations have found evidence to support this hypothesis, as reports indicate reduced structural integrity in the uncinate fasciculus of psychopaths (Craig et al., 2009; Motzkin, Newman, Kiehl, & Koenigs (2011), as well as relatively reduced correlated activity between ventromedial prefrontal cortex and the amygdala (Motzkin et al.). Functional connectivity is a promising avenue of research, yet a great deal remains to be learned about the relationship between white-matter integrity and functional activity. For instance, a recent report on neuropsychological outcomes of patients in whom the uncinate fasciculus was surgically removed (along with brain tumors) indicates that the most common resulting impairment is retrieving the names of objects and people (Papagno et al., 2011), which is not related to psychopathic traits in any discernible way. Research examining functional connectivity and white matter integrity is still fairly new, and a great deal remains to be learned about how these techniques will contribute to functional neuroimaging literature as a whole. More specific hypotheses and interpretations will surely become available through wider replication and as use of these techniques expands.

ELECTROCORTICAL MEASURES

As noted earlier, electrocortical measures such as EEG and ERP record small electrical potentials at the surface of the scalp, and although they offer a superb account of the temporal progression of brain activity, any estimation of the neural sources of this activity is constrained by what is called *the inverse problem*. Essentially, the human head acts as one large conductor, allowing neural signals from within the brain to propagate over the entire surface of the scalp, making it problematic to reconstruct the original source of this activity with much specificity. Still, modern techniques for modeling the variation in electrical conduction through the head, combined with very large numbers of electrodes for recording, have made it possible to make some meaningful estimates of the source activity in three-dimensional space (Michel et al., 2004; Scherg & Picton, 1991). Often, with EEG and ERP, however, the anatomical source of recorded activity is only a secondary concern next to the very reliable cognitive correlates of activity patterns measured at the scalp. For instance, a robust and exhaustively studied ERP component referred to as the P300 (or simply P3) has a widespread array of suggested neural sources (Halgren, Marinkovic, & Chauvel, 1998; Soltani, & Knight, 2000), but has very specific cognitive correlates related to attention and working memory (Donchin & Coles, 1988; Polich, 2007).

The term *P3* refers to the positive polarity of its deflection and its latency, approximately 300 milliseconds after an adequate eliciting stimulus. For a P3 to arise, the stimulus must be relatively rare in a series of varied stimuli and must be attended to by the observer (for an extensive review, see Polich, 2007). The P3 has often been associated with the orienting response, a mostly involuntary, automatic focusing of attention on salient features of our environment, which is accompanied by specific physiological responses (Sokolov, 1963). This makes the P3 potentially relevant to psychopathy research given theories that have suggested that psychopaths show abnormalities in attention and orienting response (Hare, 1968; Patterson & Newman, 1993). Several reports have indicated that psychopathy is associated with abnormal P3s (Anderson, Stanford, Wan, & Young, 2011; Carlson, Thai, & McLarnon, 2009; Kiehl et al., 1999, 2000, 2006; Rain & Venables, 1987, 1988), but the specific features of these abnormalities have varied between studies, suggesting complex relationships that may depend on specific facets underlying the broader psychopathy construct, such as affective deficits and antisocial traits (Carlson et al., 2009; Carlson & Thai, 2010).

As with the P3, investigations of other ERP components have revealed irregularities in functional neural activity in psychopaths. These include abnormal occurrences of N400 (Kiehl et al., 1999; Williamson, Harpur, & Hare, 1991), a component associated with semantic processing, especially when reading words, as well as unusual late-negative potentials in the frontal regions of the brain in response to task-relevant target stimuli (Kiehl et al., 1999, 2006; Forth & Hare, 1989). Similarities have also been drawn between ERP abnormalities in psychopaths and individuals with damage to parts of the temporal lobe and paralimbic structures. Patients with temporal lobe damage tend to have enlarged N2, reduced P3, and abnormal late frontocentral negativities (Yamaguchi & Knight, 1993), consequences similar to those in whom paralimbic structures, such as the amygdala and temporal pole, have been removed due to severe epilepsy (Johnson, 1993). Overall, it should be noted that though spatial localization of electrocortical measures is severely limited compared to data from fMRI, results from this methodology remain compatible with notions that psychopaths present with deficits in paralimbic regions of the brain. Further, electrocortical measures are endowed with fine enough temporal resolution to reveal that these abnormalities in information processing are evident very early in the processing stream, prior to stages when effortful, top-down cognitive control have recognizable effects on ERPs. That is to say, psychopaths' deficits are very likely the consequence of abnormalities in the way information is assembled on its way up to conscious-level processing.

KEY CONSIDERATIONS AND METHODOLOGICAL ISSUES

Investigating the neural underpinnings of psychopathy is an urgent endeavor that has rightfully attracted increasing efforts from the scientific community. Like any field undergoing rapid growth, neuroimaging in psychopathy is vulnerable to methodological issues and subsequent interpretive variation. An examination of the current literature will reveal a dizzying array of concepts and perspectives not easily addressed in this brief review. Several of these issues are discussed in greater detail in accompanying chapters, but at the risk of redundancy, a concise discussion of the most important issues to consider when generalizing outcomes across studies follows.

The construct of psychopathy has most often been defined in experimental settings using Hare's Psychopathy Checklist (PCL-R; Hare, 2003); this is considered the "gold standard" for measuring psychopathy in forensic settings and is scored based on a lengthy interview and collateral file review with a trained administer. Other tools have been developed to assess psychopathy in nonforensic settings, many of which are self-report measures such as Hare's Self-Report Psychopathy Scale (SRP; Hare, 1985), Levenson's Self Report Psychopathy Scale (LSRP; Levenson, Kiehl, & Fitzpatrick, 1995), and the Psychopathic Personality Inventory-Revised (PPI-R, Lilienfeld & Widows, 2005). Though each of these measurement devices is based primarily on the standard set of traits first delineated by Cleckley (1941), many of these tools only modestly correlate with each other; therefore, an informed caution should be applied when generalizing across studies using various measurement tools. As a related concern, the recommended cutoff for designating psychopathy with the PCL-R is a score of 30 or higher (out of a maximum 40); however, many investigators have adopted more liberal cutoffs that allow for larger numbers of subjects in this experimental category.

Another critical issue stems from longstanding confusion between the terms *psychopathy* and *antisocial personality disorder*, which are intended to be synonymous in the most current *Diagnostic and Statistical Manual for Mental Disorders* (DSM-IV-TR) (the most current diagnostic manual for psychiatric disorders in the United States). Psychopathy as a unique designation was essentially consolidated in the third version of this manual with *antisocial personality disorder*, which primarily emphasizes criminal activity and behavioral outcomes over the core personality traits outlined by Cleckley. However, psychopathy has remained an important clinical construct. In this context, psychopaths have sometimes been considered a subset of those who might otherwise only meet diagnostic criteria for antisocial personality disorder, but it should remain clear that antisocial personality disorder and psychopathy (as assessed by the Hare PCL-R) are NOT the same (Cunningham & Reidy, 1998; Hare, Hart, & Harpur, 1991). Psychopathy is strongly associated with several disorders including antisocial, narcissistic, borderline, and histrionic personality disorders (Blackburn, 2007). Psychopathy's distinguishing elements are the callous, unemotional traits and a lack of empathy, which many consider to be its primary features (Skeem & Cooke, 2010). Antisocial personality disorder does not account for these emotional elements; therefore, investigations into the nature of antisocial behavior or criminality per se should not be generalized, without qualification, to psychopathy.

The distinction between antisocial behavior and affective deficits is a prominent one in the assessment of psychopathy; in fact, many of the measurement tools designed for operationalizing psychopathy include factor scores that account for these elements separately (e.g., Hare, 2003; Lilienfeld & Widows, 2005). Initial factor analyses of the Psychopathy Checklist, for instance, revealed a two-factor structure (Harpur, Hakstian, & Hare, 1988), and this structure was duplicated in the updated version, the PCL-R (Hare, 1991). Factor 1 elements of the test corresponded to items measuring affective and interpersonal traits, whereas Factor 2 elements corresponded to antisocial behavior and lifestyle patterns. Further, it was apparent that these two factors had distinct intercorrelations with other personality measures, and that Factor 1 was more representative of the classic clinical description of psychopathic personality (Harpur, Hare, & Hakstian, 1989). Although this structure is still the most widely referenced in the literature, more recent analyses have offered alternative descriptions.

Cooke and Michie (2001) provided evidence that PCL-R structure was actually served better by a three-factor solution. For this three-factor model, the superordinate construct of psychopathy is broken down into an interpersonal factor, an affective factor, and a behavioral factor, preserving the distinction between affective traits and behavioral consequences. Items accounting for behavioral components were also limited, at the exclusionary expense of antisocial traits, which Cooke and Michie argued did not fit the model neatly in their analysis. Hare (2003) subsequently developed a model supporting a four-factor solution, accounting for an interpersonal factor, an affective factor, a lifestyle factor, and a separate antisocial factor. Here again, the core affective components are accounted for separately, and the author has differentiated between the impulsive lifestyle and deviant, rule-breaking tendencies seen in psychopaths.

A trend in recent neuroimaging literature has been to examine the unique influences of these underlying facets of psychopathy, which may ultimately help to account for some extant variability in the field. Along these lines, there is a long history of examining various subtypes of psychopathy, perhaps beginning with Karpman (1941) who suggested that *primary* psychopathy is the consequence of an intrinsic, idiopathic deficit (which we might now presume to be genetic) and *secondary* psychopathy as the result of indirect factors (i.e., trauma exposure or an impoverished upbringing). The distinction has evolved somewhat and it is often suggested that primary psychopaths are characterized by lower anxiety and poverty of emotional expression and tend to commit crimes that are fundamentally instrumental in nature; conversely, secondary psychopaths are more anxious, showing more emotional volatility, and commit more impulsive, reactionary crimes (Skeem et al., 2007). Many others have found reliable physiological and behavioral distinctions between *high-anxious psychopaths* and *low-anxious psychopaths* (e.g., Lykken, 1957; Newman, Kosson, & Patterson, 1992), supporting potentially divergent etiologies for dysfunction in these groups. Others have utilized statistical methods of accounting for patterns of how dimensions of the PCL-R cluster in very large samples of psychopathic inmates, and such methods have defined as many as four distinct subtypes of psychopathy and consistently support the primary/secondary distinctions (e.g., Hervé, 2003; Swogger & Kosson, 2007). Another distinction that has been made is between *successful* and *unsuccessful* psychopaths—that is, those with criminal records and those who either refrain from traditional criminal activity or have avoided conviction (Gao & Raine, 2010). The successful/unsuccessful psychopathy distinction may help resolve some divergent findings in the field, but more research in this area is needed.

These various considerations of the factors, facets, and subtypes of psychopathy emphasize the complexity of the psychopathy construct and the care needed in defining its elements consistently and appropriately. If such distinctions in neuroimaging prove useful, it would be consistent with trends in existing literature that suggest distinct physiological and behavioral correlates of these facets in the realm of peripheral autonomic arousal (Lykken, 1957), startle modulation (Patrick et al., 1993), and ERP measures (Carlson & Thai, 2010). Consumers of the literature in general should pay close attention to these variables, noting whether psychopathy is being treated as a unitary construct and being compared to various nonpsychopathic controls, or if comparisons are being made between different subtypes or features of psychopathy.

Many additional methodological considerations are potentially relevant but are beyond the scope of this chapter. Without lingering on more technical methodological issues, it is enough to say that the same rules apply here as in any field of scientific inquiry. When attempting to draw grand conclusions, it is important not to give undue credence to any single reported outcome. The strongest evidence is accumulated as a consensus among separate reports, from independent research groups, and even across diverse methodological protocols. Reported outcomes that continue to withstand the scrutiny of independent replication are likely the most valuable. The present review is intended to highlight some of these key features in psychopathy research. Although diversity exists in particular elements of their experimental design, some notable consistency has emerged across the broader landscape of reports examining functional differences in the brains of psychopaths.

SUMMARY AND CONCLUSIONS

Converging evidence from a number of experimental and neuroimaging modalities, across a diverse range of populations, supports the hypothesis that psychopathic traits are associated with abnormalities in the amygdala; orbital frontal cortex; and extended paralimbic structures, prominently the temporal pole (superior temporal gyrus) and anterior and posterior cingulate cortex. This evidence is further supported by the functional consequences of focal brain lesions, which can resemble elements of psychopathy and electrocortical measures, suggesting that temporolimbic dysfunction and abnormalities in orienting responses are features of psychopathy. The anatomical units implicated in these studies are part of a large network in the brain that serves in the evaluation of emotional information, including basic processing of threat and reward contingencies; the utilization of this information in the modification ongoing behavior; and the extension of these contingencies into the realm of higher cognitive processing. Several existing neurobiological models of psychopathy agree on the core functional–anatomical units involved, but differ mainly in the scope of their focus. For instance, Blair (2005, 2007) currently favors a model proposing dysfunction of amygdala and orbital frontal cortex whereas Kiehl's (2006) model additionally includes extended but closely related paralimbic structures.

The functional neuroimaging literature does not stand on its own, and combined with parallel lines of research, it has contributed a great deal to our understanding of psychopathy as a neurocognitive disorder with a persistent developmental trajectory and reliable psychophysiological and behavioral impairments. This research has informed our understanding of how emotional processing contributes to rational thought and behavioral control in healthy individuals and it demonstrates the immediate and long-term consequences of dysfunction in key brain regions that support these functions. If this information is to be applied effectively to inform more practical issues such as concerns of legal responsibility, there needs to be an interdisciplinary dialogue between neuroscience and the law, which can incorporate recent empirical findings into matters of legal ethics. A key feature of this conversation should include what impact psychopathy has in determining the scope of one's liability for a crime committed. A neurologically based incapacity to engage in effective moral reasoning and a deficiency in the appreciation of the impact of one's behavior on others may be important mitigating factors in the same way that other severe developmental conditions or the presence of a known brain injury may impact the assessment of one's rational capacity. There is much to be said in this debate, and a reasonable dialogue begins with an informed perspective on the science underlying these arguments.

REFERENCES

Adolphs, R., Cahill, L., Schul, R., & Babinsky, R. (1997). Impaired declarative memory for emotional material following bilateral amygdala damage in humans. *Learning and Memory, 4*, 291–300.

Adolphs, R., Tranel, D., Hamann, S., Young, A. W., Calder, A. J., Phelps, E. A., et al. (1999). Recognition of facial emotion in nine individuals with bilateral amygdala damage. *Neuropsychologia, 37*, 1111–1117.

Anderson, N. E., Stanford, M. S., Wan, L., & Young, K. A. (2011). High psychopathic trait females exhibit reduced startle potentiation and increased P3 amplitude. *Behavioral Sciences & the Law, 29*, 649–666.

Anderson, S. W., Damasio, H., Tranel, D., & Damasio, A. R. (2000). Long-term sequelae of prefrontal cortex damage acquired in early childhood. *Developmental Neuropsychology, 18*, 281–296.

Angrilli, A., Mauri, A., Palomba, D., Flor, H., Birbaumer, N., Sartori, G., et al. (1996). Startle reflex and emotion modulation impairment after a right amygdala lesion. *Brain: A Journal of Neurology, 119*, 1991–2000.

Bechara, A., Damasio, H., Damasio, A. R., & Lee, G. P. (1999). Differential contributions of the human amygdala and ventromedial prefrontal cortex to decision-making. *The Journal of Neuroscience, 19*, 5473–5481.

Bechara, A., Damasio, H., Tranel, D., & Damasio, A. R. (1997). Deciding advantageously before knowing the advantageous strategy. *Science, 275*, 1293–1294.

Birbaumer, N., Veit, R., Lotze, M., Erb, M., Hermann, C., Grodd, W., et al. (2005). Deficient fear conditioning in psychopathy: A functional magnetic resonance imaging study. *Archives of General Psychiatry, 62*, 799–805.

Blackburn, R. (2007). Personality disorder and antisocial deviance: Comments on the debate on the structure of the psychopathy checklist-revised. *Journal of Personality Disorders, 21*, 142–159.

Blair, R. J. R. (2006). The emergence of psychopathy: Implications for the neuropsychological approach to developmental disorders. *Cognition, 101*, 414–442.

Blair, R. J. R. (2007). The amygdala and ventromedial prefrontal cortex in morality and psychopathy. *Trends in Cognitive Science, 11*, 387–392.

Blair, R. J. R. (2010). Neuroimaging of psychopathy and antisocial behavior: A targeted review. *Current Psychiatry Reports, 12*, 76–82.

Blair, R. J. R., Mitchell, D., & Blair, K. (2005). *The psychopath: Emotion and the brain.* Malden, MA: Blackwell.

Blumer, D., & Benson, D. F. (1975). Personality changes with frontal lobe lesions. In D. F. Benson & D. Blumer (Eds.), *Psychiatric aspects of neurological disease* (pp. 151–170). New York: Grune and Stratton.

Brodmann, K. (1994). *Localisation in the cerebral cortex.* London: Smith-Gordon.

Burgmans, S., van Boxtel, M. P. J., van den Berg, K. E. M., Gronenschild, E. H. B. M., Jacobs, H. I. L., Jolles, J., et al. (2009). The posterior parahippocampal gyrus is preferentially affected in age-related memory decline. *Neurobiology of Aging, 32*, 1572–1578.

Camille, N., Coricelli, G., Sallet, J., Pradat-Diehl, P., Duhamel, J., & Sirigu, A. (2004). The involvement of the orbitofrontal cortex in the experience of regret. *Science, 304*, 1167–1170.

Carlson, S. R., & Thai, S. (2010). ERPs on a continuous performance task and self-reported psychopathic traits: P3 and CNV augmentation are associated with fearless dominance. *Biological Psychology, 85*, 318–330.

Carlson, S. R., Thai, S., & McLarnon, M. E. (2009). Visual P3 amplitude and self-reported psychopathic personality traits: Frontal reduction is associated with self-centered impulsivity. *Psychophysiology, 46*, 100–113.

Cato, M. A., Delis, D. C., Abildskov, T. J., & Bigler, E. (2004). Assessing the elusive cognitive deficits associated with ventromedial prefrontal damage: A case of a modern-day Phineas Gage. *Journal of the International Neuropsychological Society, 10,* 453–465.

Cleckley, H. (1941). *The mask of sanity: An attempt to reinterpret the so-called psychopathic personality.* Oxford: Mosby.

Cooke, D. J., & Michie, C. (2001). Refining the construct of psychopathy: Towards a hierarchical model. *Psychological Assessment, 13,* 171–188.

Craig, M. C., Catani, M., Deeley, Q., Latham, R., Daly, E., Kanaan, R., et al. (2009). Altered connections on the road to psychopathy. *Molecular Psychiatry, 14,* 946–953.

Cunningham, M. D., & Reidy, T. J. (1998). Antisocial personality disorder and psychopathy: Diagnostic dilemmas in classifying patterns of antisocial behavior in sentencing evaluations. *Behavioral Sciences & the Law, 16,* 333–351.

Damasio, A. R. (1994). *Descartes' error: Emotion, reason, and the human brain.* New York: Avon.

Davis, M. (1989). Neural systems involved in fear-potentiated startle. *Annals of the NY Academy of Sciences, 563,* 165–183.

Davis, M. (1997). Neurobiology of fear responses: The role of the amygdala. *Journal of Neuropsychiatry & Clinical Neurosciences, 9,* 382–402.

Davis, M., & Whalen, P. J. (2001). The amygdala: Vigilance and emotion. *Molecular Psychiatry, 6,* 13–34.

di Pellegrino, G., Ciaramelli, E., & Ladavas, E. (2007). The regulation of cognitive control following rostral anterior cingulate cortex lesions in humans. *Journal of Cognitive Neuroscience, 19,* 275–286.

Dolan, M. (2011). The neuropsychology of prefrontal function in antisocial personality disordered offenders with varying degrees of psychopathy. *Psychological Medicine, 42,* 1715–1725. published online Dec. 6th, 2011: DOI: 10.1017/S0033291711002686.

Dolan, M. C., & Fullam, R. S. (2009). Psychopathy and functional magnetic resonance imaging blood oxygenation level-dependent responses to emotional faces in violent patients with schizophrenia. *Biological Psychiatry, 66,* 570–577.

Dolan, R. J. (1999). On the neurology of morals. *Nature Neuroscience, 2,* 927–929.

Donchin, E., & Coles, M. G. H. (1988). Is the P300 component a manifestation of context updating? *Behavioral and Brain Sciences, 11,* 357–374.

Echavarri, C., Aalten, P., Uylings, H. B. M., Jacobs, H. I. L., Visser, P. J., Gronenschild, E. H. B. M., et al. (2010). Atrophy in the parahippocampal gyrus as an early biomarker of Alzheimer's disease. *Brain Structure and Function, 215,* 265–271.

Finger, E. C., Marsh, A. A., Blair, K. S., Reid, M. E., Sims, C., Ng, P., et al. (2011). Disrupted reinforcement signaling in the orbitofrontal cortex and caudate in youths with conduct disorder or oppositional defiant disorder and a high level of psychopathic traits. *American Journal of Psychiatry, 168,* 152–162.

Forth, A. E., & Hare, R. D. (1989). The contingent negative variation in psychopaths. *Psychophysiology, 26,* 676–682.

Frick, P. J., & White, S. F. (2008). Research Review: The importance of callous-unemotional traits for developmental models of aggressive and antisocial behavior. *Journal of Child Psychology and Psychiatry, 49,* 359–375.

Gao, Y., & Raine, A. (2010). Successful and unsuccessful psychopaths: A neurobiological model. *Behavioral Sciences & the Law, 28,* 194–210.

Glenn, A. L., Raine, A., & Laufer, W. S. (2011). Is it wrong to criminalize and punish psychopaths? *Emotion Review, 3,* 302–304.

Glosser, G., Zwil, A. S., Glosser, D. S., O'Connor, M. M., & Sperling, M. R. (2000). Psychiatric aspects of temporal lobe epilepsy before and after anterior temporal lobectomy. *Journal of Neurology, Neurosurgery, and Psychiatry, 68,* 53–58.

Gordon, H. L., Baird, A. A., & End, A. (2004). Functional differences among those high and low on a trait measure of psychopathy. *Biological Psychiatry, 56,* 516–521.

Halgren, E., Marinkovic, K., & Chauvel, P. (1998). Generators of the late cognitive potentials in auditory and visual oddball tasks. *Electroencephalography and Clinical Neurophysiology, 106,* 156–164.

Hare, R. D. (1968). Psychopathy, autonomic functioning, and the orienting response. *Journal of Abnormal Psychology, 73,* 1–24.

Hare, R. D. (1985). Comparison of procedures for the assessment of psychopathy. *Journal of Consulting and Clinical Psychology, 53,* 7–16.

Hare, R. D. (1991). *The Hare Psychopathy Checklist-Revised.* Toronto, Ontario, Canada: Multi-Health Systems.

Hare, R. D. (2003). *The Hare Psychopathy Checklist-2nd Edition Revised.* Toronto, Ontario, Canada: Multi-Health Systems.

Hare, R. D., Hart, S. D., & Harpur, T. J. (1991). Psychopathy and the DSM-IV criteria for antisocial personality disorder. *Journal of Abnormal Psychology, 100,* 391–398.

Harenski, C. L., Hare, R. D., & Kiehl, K. A. (2010). Neuroimaging, genetics, and psychopathy: Implications for the legal system. In L. Malatesti & J. McMillan (Eds.), *Responsibility and psychopathy: Interfacing law, psychiatry & philosophy* (pp. 125–154). New York: Oxford University Press.

Harenski, C. L., Harenski, K. A., Shane, M. S., & Kiehl, K. A. (2010). Aberrant neural processing of moral violations in criminal psychopaths. *Journal of Abnormal Psychology, 119,* 863–874.

Harenski, C. L., Kim, S. H., & Hamann, S. (2009). Neuroticism and psychopathy predict brain activation during moral and nonmoral emotion regulation. *Cognitive, Affective, and Behavioral Neuroscience, 9,* 1–15.

Harlow, J. M. (1993). Classic text no. 14: Recovery from the passage of an iron bar through the head. *History of Psychiatry, 4,* 271–281.

Harpur, T. J., Hakstian, A. R., & Hare, R. D. (1988). Factor structure of the psychopathy checklist. *Journal of Consulting and Clinical Psychology, 56,* 741–747.

Harpur, T. J., Hare, R. D., & Hakstian, A. R. (1989). Two-factor conceptualization of psychopathy: Construct validity and assessment implications. *Psychological Assessment: A Journal of Consulting and Clinical Psychology, 1,* 6–17.

Hervé, H. (2003). *The masks of sanity and psychopathy: A cluster analytical investigation of subtypes of criminal psychopathy* (Doctoral dissertation, University of British Columbia, British Columbia, Canada).

Hornak, J., Bramham, J., Rolls, E. T., Morris, R. G., O'Doherty, Bullock, P. R., & Polkey, C. E. (2003). Changes in emotion after circumscribed surgical lesions of the orbitofrontal and cingulate cortices. *Brain, 126,* 1691–1712.

Intrator, J., Hare, R., Stritzke, P., Brichtswein, K., Dorfman, D., Harpur, T., . . . & Machac, J. (1997). A brain imaging (single photon emission computerized tomography) study of semantic and affective processing in psychopaths. *Biological Psychiatry, 42(2),* 96–103.

Johnson, R. J. (1993). On the neural generators of the P300 component of the event-related potential. *Psychophysiology, 30,* 90–97.

Johnson, S. C., Baxter, L. C., Wilder, L. S., Pipe, J. G., Heiserman, J. E., & Prigatano, G. P. (2002). Neural correlates of self-reflection. *Brain, 125,* 1808–1814.

Jones, A. P., Laurens, K. R., Herba, C. M., Barker, G. J., & Viding, E. (2009). Amygdala hypoactivity to fearful faces in boys with conduct problems and callous unemotional traits. *American Journal of Psychiatry, 166,* 95–102.

Jones, C. L., Ward, J., & Critchley, H. D. (2010). The neuropsychological impact of insular cortex lesions. *Journal of Neurology, Neurosurgery, & Psychiatry, 81,* 611–618.

Karpman, B. (1941). On the need for separating psychopathy into two distinct clinical subtypes: Symptomatic and idiopathic. *Journal of Criminology & Psychopathology, 3,* 112–137.

Kiehl, K. A. (2006). A cognitive neuroscience perspective on psychopathy: Evidence for paralimbic system dysfunction. *Psychiatry Research, 142,* 107–128.

Kiehl, K. A., Hare, R. D., Liddle, P. F., & McDonald, J. J. (1999). Reduced P300 responses in criminal psychopaths during a visual oddball task. *Biological Psychiatry, 45,* 1498–1507.

Kiehl, K. A., Hare, R. D., McDonald, J. J., & Brink, J. (1999). Semantic and affective processing in psychopaths: An event-related potential (ERP) study. *Psychophysiology, 36,* 765–774.

Kiehl, K. A., Liddle, P. F., & Hopfinger, J. B. (2000). Error processing and the rostral anterior cingulate: An event-related fMRI study. *Psychophysiology, 37,* 216–223.

Kiehl, K. A., Smith, A. M., Hare, R. D., Mendrek, A., Forster, B. B., Brink. J., et al. (2001). Limbic abnormalities in affective processing by criminal psychopaths as revealed by functional magnetic resonance imaging. *Biological Psychiatry, 50,* 677–684.

Kiehl, K. A., Smith, A. M., Mendrek, A., Forster, B. B., Hare, R. D., & Liddle, P. F. (2004). Temporal lobe abnormalities in semantic processing by criminal psychopaths as revealed by functional magnetic resonance imaging. *Psychiatry Research: Neuroimaging, 130,* 297–312.

Kim, M., & Davis, M. (1993). Electrolytic lesions of the amygdala block acquisition and expression of fear-potentiated startle even with extensive training but do not prevent reacquisition. *Behavioral Neuroscience, 107,* 580–595.

Koenigs, M., Young, L., Adolphs, R., Tranel, D., Cushman, F., Houser, M., et al. (2007). Damage to the prefrontal cortex increases utilitarian moral judgments. *Nature, 446,* 908–911.

Lang, P. J., Bradley, M. M., & Cuthbert, B. N. (1990). Emotion, attention, and the startle reflex. *Psychological Review, 97,* 377–395.

Larsson, H., Andershed, H., & Lichtenstein, P. (2006). A genetic factor explains most of the variation in the psychopathic personality. *Journal of Abnormal Psychology, 115,* 221–230.

Leistico, A. R., Salekin, R. T., DeCoster, J., & Rogers, R. (2008). A large-scale meta-analysis relating the hare measures of psychopathy to antisocial conduct. *Law and Human Behavior, 32,* 28–45.

Levenson, M., Kiehl, K., & Fitzpatrick, C. (1995). Assessing psychopathic attributes in a noninstitutionalized population. *Journal of Personality and Social Psychology, 68,* 151–158.

Lilienfeld, S. O., & Widows, M. (2005). Professional manual for the Psychopathic Personality Inventory-Revised (PPI-R). Lutz, FL: Psychological Assessment.

Lykken, D. T. (1957). A study of anxiety in sociopathic personality. *Journal of Abnormal and Social Psychology, 55,* 6–10.

Macmillan, M. (1999). An odd kind of fame: Stories of Phineas Gage. Cambridge, MA: MIT Press, Bradford Books.

Maddock, R. J. (1999). The retrosplenial cortex and emotion: New insights from functional neuroimaging of the human brain. *Trends in Neurosciences, 22,* 310–316.

Marsh, A. A., Finger, E. C., Mitchell, D. G., Reid, M. E., Sims, C., Kosson, D. S., et al. (2008). Reduced amygdala response to fearful expressions in children and adolescents with callous-unemotional traits and disruptive behavior disorders. *American Journal of Psychiatry, 165,* 712–720.

Mesulam, M. M. (2000). Paralimbic (mesocortical) areas. In M. M. Mesulam (Ed.), *Principles of behvioral and cognitive neurology* (pp. 49–54). New York: Oxford University Press.

Meyers, C. A., Berman, S. A., Scheibel, R. S., & Hayman, A. (1992). Case report: Acquired antisocial personality disorder associated with unilateral left orbital frontal lobe damage. *Journal of Psychiatry & Neuroscience, 17,* 121–125.

Michel C., Murray, M., Lantz, G., Gonzalez, S., Spinelli, L., & Grave de Peralta Menendez, R. (2004). EEG source imaging. *Clinical Neurophysiology, 115,* 2195–2222.

Motzkin, J., Newman, J. P., Kiehl, K. A., & Koenigs, M. (2011). Reduced prefrontal connectivity in psychopathy. *Journal of Neuroscience, 31*(48), 17348–17357.

Müller, J. L., Sommer, M., Dohnel, K., Weber, T., Schmidt-Wilcke, M. D., & Hajak, G. (2008). Disturbed prefrontal and temporal brain function during emotion and cognition interaction in criminal psychopathy. *Behavioral Sciences & the Law, 26,* 131–150.

Müller, J. L., Sommer, M., Wagner, V., Lange, K., Taschler, H., Roder, C. H., et al. (2003). Abnormalities in emotion processing within cortical and subcortical regions in criminal psychopaths: Evidence from a functional magnetic resonance imaging study using pictures with emotional content. *Biological Psychiatry, 54,* 152–162.

Newman, J. P., Kosson, D. S., & Patterson, M. C. (1992). Delay of gratification in psychopathic and nonpsychopathic offenders. *Journal of Abnormal Psychology, 101,* 630–636.

Olson, I. R., Plotzker, A., & Ezzyat, Y. (2007). The enigmatic temporal pole: A review of findings on social and emotional processing. *Brain, 130,* 1718–1731.

Papagno, C., Miracapillo, C. Casarotti, A., Romero Lauro, L. J., Castellano, A., Falini, A., et al. (2011). What is the role of the uncinate fasciculus? Surgical removal and proper name retrieval. *Brain, 134,* 405–414.

Patrick, C. J., Bradley, M. M., & Lang, P. J. (1993). Emotion in the criminal psychopath: Startle reflex modulation. *Journal of Abnormal Psychology, 102,* 82–92.

Patterson, C. M., & Newman, J. P. (1993). Reflectivity & learning from aversive events: Toward a psychological mechanism for the syndromes of disinhibition. *Psychological Review, 100,* 716–736.

Phillips, R. G., & LeDoux, J. E. (1992). Differential contribution of amygdala and hippocampus to cued and contextual fear conditioning. *Behavioral Neuroscience, 106,* 274–285.

Polich, J. (2007). Updating p300: An integrative theory of P3a and P3b. *Clinical Neurophysiology, 118,* 2128–2148.

Porter, S., Birt, A., & Boer, D. P. (2001). Investigation of the criminal and conditional release profiles of Canadian federal offenders as a function of psychopathy and age. *Law and Human Behavior, 25,* 647–661.

Porter, S., Brinke, L., & Wilson, K. (2009). Crime profiles and conditional release performance of psychopathic and non-psychopathic sexual offenders. *Legal and Criminological Psychology, 14,* 109–118.

Raine, A., & Venables, P. H. (1987). Contingent negative variation, P3 evoked potentials and antisocial behavior. *Psychophysiology, 24,* 191–199.

Raine, A., & Venables, P. H. (1988). Enhanced P3 evoked potentials and longer P3 recovery times in psychopaths. *Psychophysiology, 25,* 30–38.

Rice, M. E., Harris, G. T., & Cormier, C. A. (1992). An evaluation of a maximum security therapeutic community for psychopaths and other mentally disordered offenders. *Law and Human Behavior, 16*, 399–412.

Rilling, J. K., Glenn, A. L., Jairam, M. R., Pagnoni, G., Goldsmith, D. R., Elfenbein, H. A., et al. (2007). Neural correlates of social cooperation and non-cooperation as a function of psychopathy. *Biological Psychiatry, 61*, 1260–1271.

Salekin, R. T., Rogers, R., & Sewell, K. W. (1996). A review and meta-analysis of the psychopathy checklist and psychopathy checklist—revised: Predictive validity of dangerousness. *Clinical Psychology: Science and Practice, 3*, 203–215.

Scherg, M., & Picton, T. W. (1991). Separation and identification of event-related potential components by brain electric source analysis. *Electroencephalography and Clinical Neurophysiology, 42*, 24–37.

Seto, M. C., & Barbaree, H. E. (1999). Psychopathy, treatment behavior, and sex offender recidivism. *Journal of Interpersonal Violence, 14*, 1235–1248.

Shannon, B. J., Raichle, M. E., Snyder, A. Z., Fair, D. A., Mills, K. L., Zhang, D., et al. (2011). Premotor functional connectivity predicts impulsivity in juvenile offenders. *Proceedings of the National Academy of Sciences of the USA, 108*, 11241–11245.

Shaw, P., Lawrence, E. J., Radbourne, C., Bramham, J., Polkey, C. E., & David, A. S. (2004). The impact of early and late damage to the human amygdala on "theory of mind" reasoning. *Brain: A Journal of Neurology, 127*, 1535–1548.

Skeem, J., Johansson, P., Andershed, H., Kerr, M., & Louden, J. E. (2007). Two subtypes of psychopathic violent offenders that parallel primary and secondary variants. *Journal of abnormal psychology, 116*(2), 395.

Skeem, J. L., & Cooke, D. J. (2010). Is criminal behavior a central component of psychopathy? Conceptual directions for resolving the debate. *Psychological Assessment, 22*, 433–445.

Soderstrom, H., Hultin, L., Tullberg, M., Wikkelso, C., Ekholm, S., & Forsman, A. (2002). Reduced frontotemporal perfusion in psychopathic personality. *Psychiatry Research, 114*, 81–94.

Sokolov, E. N. (1963). Higher nervous functions: The orienting reflex. *Annual Review of Physiology, 25*, 545–580.

Soltani, M., & Knight, R. T. (2000). Neural origins of the P300. *Critical Reviews in Neurobiology, 14*, 199–224.

Sommer, M., Sodian, B., Dohnel, K., Schwerdtner, J., Meinhardt, J., & Hajak, G. (2010). In psychopathic patients emotion attribution modulates activity in outcome-related brain areas. *Psychiatry Research Neuroimaging, 182*, 88–95.

Swogger, M. T., & Kosson, D. S. (2007). Identifying subtypes of criminal psychopaths: A replication and extension. *Criminal Justice and Behavior, 34*, 953–970.

van Honk, J., & Schutter, D.J.L.G. (2006) Unmasking feigned sanity: A neurobiological model of emotion processing in primary psychopathy. *Cognitive Neuropsychiatry, 11*, 285–306.

Veit, R., Flor, H., Erb, M., Hermann, C., Lotze, M., Grodd, W., et al. (2002). Brain circuits involved in emotional learning in antisocial behavior and social phobia in humans. *Neuroscience Letters, 328*, 233–236.

Veit, R., Lotze, M., Sewing, S., Missenhardt, H., Gaber, T., & Birbaumer, N. (2010). Aberrant social and cerebral responding in a competitive reaction time paradigm in criminal psychopaths. *NeuroImage, 49*, 3365–3372.

Viding, E., Blair, J. R., Moffitt, T. E., & Plomin, R. (2005). Evidence for substantial genetic risk for psychopathy in 7-year-olds. *Journal of Child Psychology and Psychiatry, 46*, 592–597.

Weller, J. A., Levin, I. P., Shiv, B., & Bechara, A. (2009). The effects of insula damage on decision-making for risky gains and losses. *Social Neuroscience, 4*, 347–358.

Williamson, S., Harpur, T. J., & Hare, R. D. (1991). Abnormal processing of affective words by psychopaths. *Psychophysiology, 28*, 260–273.

Yamaguchi, S., & Knight, R.T. (1993). Association cortex contributions to the human P3. In W. Haschke, A. I. Roitbak, E. J. Speckmann, E.J. (Eds.), *Slow potential changes in the brain* (pp. 71–84). Boston: Birkhauser.

CHAPTER 9

Structural Brain Abnormalities and Psychopathy

Marina Boccardi

The curious case of Phineas Gage (Harlow, 1848) launched modern neurology in general and the specific linkage of behavior to brain structures in particular (Neylan, 1999). Gage was a diligent railroad supervisor, husband, and family man until a freak accident changed his personality. A large tamping iron was blasted up behind Gage's left eye and out the top of his head, destroying parts of his brain adjacent to and just above his eyes. Gage survived the accident; indeed, he apparently never lost consciousness, but, as his physician famously reported, "Gage was no longer Gage."

Whereas Gage's intellect, memory, and other faculties were untouched by the accident, his personality was radically changed. Gage was described as irritable, even hot-tempered. He left his job and wife and traveled extensively. He apparently had problems making plans or executing long-term goals. He lived the remainder of his life drinking, philandering, and with no general purpose. Gage and many subsequent patients like him have been well studied. The condition associated with damage to the region above the eyes, the orbital frontal cortex, is known as "acquired" sociopathy or psychopathy (Blair & Cipolotti, 2000; Meyers, Berman, Scheibel, & Hayman, 1992; Saver & Damasio, 1991; Stip, 1995). The syndrome of acquired psychopathy is reminiscent of people with developmental or congenital psychopathy, that is, lack of comprehension of the emotional inputs that guide behavior, inability to achieve long-term goals and use abstract knowledge of rules and cause–effect relationships, and lack of empathic feelings in everyday contexts. Although more refined investigations have shown that acquired and developmental psychopathy do differ in important respects, such as instrumental aggression (Mitchell, Avny, & Blair, 2006), the fact that specific focal lesions lead to symptoms of psychopathy clearly suggests that some regions of the brain are more important than others with respect to psychopathic symptomatology.

Damasio and coworkers (Damasio, 1994; Damasio, Grabowski, Frank, Galaburda, & Damasio, 1994) have argued that damage to the orbital frontal cortex causes disruption in a circuit connecting subcortical *limbic* structures, regions largely devoted to the processing of emotions, to cortical regions involved in decision making. Brain structures such as the orbital frontal cortex, anterior insula, cingulate gyrus, amygdala, and other regions are responsible for the generation of proper emotional input to help guide decision making. These systems are also involved in the learning of conditioned rewarding stimuli, the reversal of such learning when environmental conditions change, the interruption of courses of actions that do not achieve the foreseen results, the shift to differently planned courses of action, and the awareness of that experience. These are also among the regions that allow a person to carry out moral reasoning (Adolphs 2001, 2002; Damasio, 1994; Damasio et al., 1994; Devinsky, Morrell, & Vogt, 1995; Tsuchiya & Adolphs, 2007; Vogt, Finch, & Olson, 1992).

Very similar observations have come from recent cases of acquired psychopathy due to degenerative dementia, specifically frontotemporal dementia (Stip, 1995). In frontotemporal dementia, degeneration involves the entire rostral limbic system (Boccardi et al., 2005), occurring in a very unique tissue that is fundamental to the paralimbic system (Seeley et al., 2008). The Von Economo neurons that are lost in frontotemporal dementia are located in the fronto-insular regions. These neurons are unique to the paralimbic regions of primates (Allman, Watson, Tetreault, & Hakeem, 2005; Watson, Jones, & Allman, 2006). Their exceptional size and connectivity likely facilitate the processing of social information, which requires fast and simultaneous computation of complex sets of perceptual and emotional stimuli, rules, and abstract knowledge. Patients with frontotemporal dementia are not typically violent. But, as in the case of Phineas Gage, patients with frontotemporal dementia acquire an inability to use emotional information and abstract knowledge, especially as it relates to social interaction (Mendez, Chow, Ringman, Twitchell, & Hinkin, 2000; Wittenberg et al., 2008). These latter symptoms led clinicians to speculate that frontotemporal dementia may be a condition similar to acquired psychopathy first described in patients with orbital frontal and anterior cingulate lesions (Stip, 1995).

In summary, studies of behavioral changes following traumatic or degenerative brain damage show that lesions disrupting specific circuits may lead to psychopathy-like symptoms. The prominent regions implicated are part of the paralimbic system described by Brodmann (1909) and in more recent literature (Devinsky et al., 1995; Vogt et al., 1992). These studies inspired the research of cerebral correlates of developmental psychopathy through radiological imaging.

Investigating structural brain correlates through the analysis of radiological images has only recently been undertaken in psychopathy. Magnetic resonance imaging (MRI) is the most informative and least invasive method capable of acquiring information about brain size, shape, density, and volume. To date, all of the structural brain imaging studies in psychopathy have used information derived using various MRI protocols.

Over the last 10 years, research on the structural brain correlates of psychopathy can be roughly divided into two groups. In the first group, brain size and shape was investigated by computing the global volume of cerebral lobes or structures. These initial analyses, reviewed in detail in the text that follows, provided gross information about brain abnormalities in psychopathy. In the second group, begun more recently, "local" analyses examining the volume, shape, and density of specific brain structures have been conducted.

Studies in the first group were much less sensitive and found inconsistent results relating gross brain morphology to psychopathy. For example, in one study psychopathy scores were associated with volume reductions of between 11% and 22% in the prefrontal cortex (Raine, Lencz, Bihrle, LaCasse, & Colletti, 2000; Yang et al., 2005), but, in other samples, findings were not significant (Barkataki, Kumari, Das, Taylor, & Sharma, 2006; Dolan, 2002) or could be explained by moderating or confounding variables (Laakso et al., 2002). Others reported changes involved the corpus callosum white matter, which displayed an increase of total volume (Raine et al., 2003). The corpus callosum is the main fiber bridge that connects the two hemispheres in the brain. Other studies reported that the temporal lobe was of smaller volume in psychopathy (Barkataki et al., 2006). The medial aspects of the temporal lobe include the hippocampus. Studies have reported that the volume of the hippocampus is abnormal in psychopathy, but again with inconsistent results (Barkataki et al., 2006; Laakso et al., 2000, 2001; Raine et al., 2004). In summary, examination of large regions, like the entire volume of the temporal or frontal lobes, has not identified reliable abnormalities in psychopathy. Let us now turn to more modern assessments of focal brain abnormalities in psychopathy.

The neuroscientist's arsenal of algorithms for parsing focal brain anatomy is rapidly evolving. Each year new and improved algorithms are developed that provide additional insights into understanding the complexity of brain maturation; aging; degeneration; and the morphological correlates of psychopathology, drug use, and other related disorders. Several popular algorithms have been applied to the study of focal brain abnormalities in psychopathy. Cortical pattern matching (CTM) provides information regarding group differences in regional brain size and shape. Voxel-based morphometry (VBM), the most popular structural brain imaging technique, provides information on each voxel's (three-dimensional digital unit, or cube, in computer renderings of the brain) density or concentration—allowing for analyses of the "integrity" of brain regions.

Studies examining small, focal regions of the brain revealed negative correlations of posterior sections of the hippocampus with Hare Psychopathy Checklist-Revised (PCL-R; Hare, 2003) scores (Laakso et al., 2001), or increased hippocampal asymmetry in anterior sectors (Raine et al., 2004).

Three studies have been carried out so far with VBM analysis, a technique that examines local tissue density. Of these, one (Müller et al., 2008) found reduced gray matter in the psychopathy group relative to controls in the cingulate, superior, and middle frontal gyri (including the supplementary motor area) and superior temporal gyrus. Tiihonen et al. (2008) found psychopathy was associated with decreased gray matter density in the orbitofrontal cortex, insula, fusiform gyrus, and cuneus, and confirmed increases in the white matter densities consistent with those described by lobar volume studies in antisocial and psychopathic subjects. Another VBM study found a similar pattern of differences, confirming lesser gray matter tissue in the medial orbitofrontal, frontopolar, anterior temporal, and insular cortices, and finding lesser tissue also in the superior temporal sulcus, associated with psychopathy (de Oliveira-Souza et al., 2008). Atrophy of paralimbic structures was also detected in a very recent VBM study (Ermer, Cope, Nyalakanti, Calhoun, & Kiehl, 2012) that employed a very large sample size ($N = 298$). Orbitofrontal, anterior and posterior cingulate, temporal pole, and medial temporal structures showed reduced gray matter density associated with psychopathy. The latter regions form the paralimbic system and were interpreted as supporting the hypothesis (Kiehl, 2006) that this latter system has developed abnormally in psychopathy.

The first CPM study, the technique using the most advanced registration method (Thompson et al., 2004-b), was carried out on ASPD subjects whose diagnosis and severity of psychopathy were not quantified using a specific scale (Narayan et al., 2007). The authors found reduced gray matter tissue in the mesial prefrontal cortex. Some tissue reduction was also found in the sensory–motor cortex, similarly to the VBM studies (de Oliveira-Souza et al., 2008; Müller et al., 2008; Tiihonen et al., 2008). In the second work conducted using the CPM technique, this time involving participants scored on psychopathy, the cingulate, mainly posteriorly; the superior frontal gyrus; cuneus; precuneus; and dorsolateral frontal cortex, again including motor–sensory gyri, displayed reduced gray matter density associated with psychopathy (Yang, Raine, Colletti, Toga, & Narr, 2009-a). Finally, CPM data on a Finnish sample of psychopathic offenders, free of other major psychiatric or Cluster A personality disorders, showed severe and wide hypotrophy along the whole paralimbic and medial cortical structures, including the orbitofrontal cortex, cingulate gyrus, precuneus, and cuneus (Boccardi et al., 2011).

Three studies have examined subcortical structures using the radial distance mapping (RDM) technique (Thompson et al., 2004-a). The first one investigated amygdalar morphology in an American population sample with the Hare Psychopathy Checklist-Revised (PCL-R) (Hare, 2003). In this sample, Yang and colleagues found psychopathy was associated with reduced amygdalar volume that could be approximately ascribed to abnormally small basolateral, lateral, cortical, and central nuclei (Yang, Raine, Narr, Colletti, & Toga, 2009b).

The same technique was used to investigate the amygdala morphology in the aforementioned Finnish sample (Boccardi et al., 2011). This analysis showed involvement of the same nuclei, but both enlargement and reduction effects were observed. The larger central, basolatero–ventromedial and lateral nuclei are consistent with the literature on impulsive aggression, as these nuclei are closely connected with the periaqueductal gray, coding for the instinctive fight-or-flight reactions, and with the relevant neurotransmitter systems. The basolateral nucleus of the amygdala was hypotrophic, suggesting abnormal neural development in the circuitry controlling emotion regulation, which is consistent with the clinical picture of psychopathic individuals.

Hippocampal morphology (Boccardi et al., 2010) appears to be abnormal in psychopathic offenders who display an 8-like shape of coronal hippocampal sections, which differs significantly from that observed in normal control subjects.

Other findings of abnormal volumes in the basal ganglia nuclei (Glenn, Raine, Yaralian, & Yang, 2010) are consistent with an aberrant sensitivity of these structures to mesolimbic dopamine (Buckholtz et al., 2010) and further support the idea of an involvement of the paralimbic system in the genesis of psychopathic personality and behavior (Kiehl, 2006).

OTHER ADVANCED NEUROIMAGING STUDIES

MRI sequences also permit investigations of the morphology of white-matter fiber tracts. White-matter tracts are the wires that connect brain regions together. These MRI sequences can detect the direction along which water molecules are free to move in the brain: well-structured white-matter fibers force these molecules to move strictly along the orientation of the fiber itself, and this can be detected and recorded by the scanner. This technique, diffusion tensor imaging (DTI), enables the reconstruction of the morphology of fiber tracts connecting far cerebral structures. Craig and colleagues showed that the uncinate fasciculus, that is, the tract connecting the amygdala and the orbitofrontal cortex, has reduced fractional anisotropy in psychopathy (Craig et al., 2009). This means that water molecules within this tract are freer to move in different directions than in controls, and display lesser constraint to move along well-structured white-matter fiber bundles. A similar result is considered a consequence of reduced integrity of the white matter.

SOURCES FOR HETEROGENEOUS FINDINGS

Individuals with psychopathy often present with a number of comorbid issues that, if not controlled, cloud interpretation of morphometric analyses. Common comorbid issues that need to be addressed include history of psychotropic medications, substance and alcohol use, intelligence, head injuries, and neuropsychological functioning. Age is also a major determinant of brain morphometry (Good et al., 2001), and few studies have attempted to control for all of these latter issues in psychopathy research.

Similarly, another important source of variance between studies is substantial differences in study inclusion and exclusion criteria. Some studies have included patients with major psychiatric conditions, primarily schizophrenia and Cluster A personality disorders. Isolating the effect of psychopathy from these other clinical conditions is problematic. Sources of variability between studies also include the methods used to quantify structural brain changes in psychopathy. Image processing schemes often operate on different assumptions, quantify different aspects of brain anatomy, and have different error rates.

However, the most basic methodological difference across studies consists in the variability in the assessment procedures of psychopathy, control groups used, and in the cutoff scores for group classification. The Hare PCL-R is the most appropriate and common instrument to assess psychopathy in the majority of studies. However, nearly all of these studies failed to use the recommended diagnostic cutoff of 30 (out of 40 on the Hare PCL-R) for psychopathy (Barkataki et al., 2006; Dolan, Deakin, Roberts, & Anderson, 2002; Laakso et al., 2000). Finally, studies rarely employ comparison groups that mitigate the potential impact of incarceration (i.e., an incarcerated nonpsychopathic control group) and substance/alcohol abuse.

Finally, it should be noted that the sample sizes in morphometric studies of psychopathy have to date been small (exception is Ermer, Cope, Nyalakanti, Calhoun, & Kiehl, 2012). Attempting to draw firm conclusions from this literature thus presents with challenges and more research is needed before definitive conclusions can be drawn between morphometric abnormalities in psychopathy.

THE MEANING OF THE AVAILABLE STRUCTURAL IMAGING FINDINGS

Despite the aforementioned limitations, studies have tended to converge on a number of brain structures as being implicated in psychopathy. Across these diverse study populations, methods, and assessment procedures, the morphology of the amygdala, hippocampus, and basal ganglia, as well as the first-order cortical regions connected to the latter structures, appear to be abnormal.

Abnormalities in these latter structures are consistent with theories of psychopathy drawn from functional and neurophysiological data (Blair, 2007; Kiehl, 2006). The amygdala is a structure believed to be critical to processing attention and emotion. Individuals with psychopathy are known to be impaired in attention and emotional processing. Hippocampal formation is largely believed to store episodic memories, and some research suggests that psychopathy is associated with aberrant memory processing, especially for emotional stimuli (Christianson et al., 1996). Finally, basal ganglia play a prominent role in reward seeking and substance abuse. Abnormalities in basal ganglia structures may also be linked to the excessive substance abuse seen in individuals with psychopathy (Smith & Newman, 1990).

Studies have also reported various abnormalities in regions with dubious relation to psychopathy including posterior visual regions (cuneus, fusiform gyrus: Boccardi et al., 2011; Müller et al., 2008; Tiihonen et al., 2008; Yang et al., 2009-a) and dorsolateral prefrontal regions associated with processing sensory–motor information (de Oliveira-Souza et al., 2008; Müller et al., 2008; Narayan et al., 2007; Tiihonen et al., 2008; Yang et al., 2009-a).

On the whole, it seems that a pattern of morphological peculiarities accompanies psychopathy. This pattern includes structures that can lead to acquired sociopathy when lesioned in traumatic brain injury or via neurodegenerative brain disorders.

ARE THESE BRAIN ABNORMALITIES "CORRELATES" OR "CAUSES" OF PSYCHOPATHIC BEHAVIOR?

It is important to recognize that a common habit of laymen is to interpret the biological correlates observed in psychopathy studies as the cause of the condition. For instance, we may interpret the abnormal hippocampal and amygdalar morphology in psychopathy as the cause of the observed deficient fear conditioning and thus for the known insensitivity of psychopaths to punishment. For this reason, we should talk much more often about "correlates" rather than "causes." It is very difficult in scientific fields to demonstrate causality. To date, we cannot say that the abnormal medial temporal morphology of psychopathy explains (or causes) the reduced fear conditioning.

That said, the use of structural MRI for evaluation of a charged person is a valuable tool for exploring whether the person has suffered from major trauma or other neurological damage, which seems to be observed frequently in samples of antisocial individuals (Schiltz, Witzel, Bausch-Hölterhoff, Gubka, & Bogerts, et al., 2009) and that may indeed be relevant in the case of psychopathic-like behaviors, as described at the beginning of this chapter.

In short, when outlining guidelines on the use of neuroscientific data in the courts, it might be advisable to consider the value of a proper refereeing of how specific analyses are carried out, analogous to peer review for scientific papers by expert reviewers, as well as investigations of the actual added value and consistency with other sets of data.

CONCLUSIONS

The research described in this chapter illustrates that the young discipline of investigating the cerebral morphologic correlates of psychopathy has depicted a pattern of difference from controls involving limbic and paralimbic regions. These features are consistent with the specific pattern of functional differences, and with the observed behavior of these people in society. Nonetheless, the temptation to use scientific data in social settings and in the legal system may be more complex than the appearance of certainty of scientific data may suggest, and numerous methodological aspects must be taken into account for a proper *translation* of these findings—as well as to communicate correctly and effectively between disciplines characterized by a different level of complexity.

Neuroscientific evidence can be used to achieve a deeper comprehension of psychopathy and to treat individual cases in a more appropriate way, but many intermediate steps should be pursued first, as is the case when transferring medical data to clinical settings. Indeed, other disciplines, which, like medicine, have standardized a procedure leading from the research level to the outline of guidelines for using the achieved knowledge in clinical practice, may provide a template that might be useful in the dialogue between neuroscience and the law.

ACKNOWLEDGMENTS

This chapter would not have been written without my collaboration with Giovanni Frisoni, Jari Tiihonen, and Paul Thompson, with whom I carried out advanced morphometry studies on a sample of subjects with psychopathy.

I thank Amedeo Santosuosso and Barbara Bottalico for their useful comments on legal implications.

This work has been partially funded by AFaR—Associazione Fatebenefratelli per la Ricerca (Project: "Use of advanced neuroimaging algorithms for the study of the features and risk factors of neurodegenerative and psychiatric disorders," 2009).

REFERENCES

Adolphs, R. (2001). The neurobiology of social cognition. *Current Opinion in Neurobiology, 11,* 231–239.

Adolphs, R. (2002). Neural systems for recognizing emotion. *Current Opinion in Neurobiology, 12,* 169–177.

Allman, J. M., Watson, K. K., Tetreault, N. A., & Hakeem, A. Y. (2005). Intuition and autism: A possible role for Von Economo neurons. *Trends in Cognitive Sciences, 9*(8), 367–373.

Barkataki, I., Kumari, V., Das, M., Taylor, P., & Sharma, T. (2006). Volumetric structural brain abnormalities in men with schizophrenia or antisocial personality disorder. *Behavioural Brain Research, 169*(2), 239–247.

Blair, R. J. (2007). The amygdala and ventromedial prefrontal cortex in morality and psychopathy. *Trends in Cognitive Sciences, 11*(9), 387–392.

Blair, R. J., & Cipolotti, L. (2000). Impaired social response reversal: A case of "acquired sociopathy." *Brain, 123* (Pt 6), 1122–1141.

Boccardi, M., Frisoni, G. B., Hare, R. D., Cavedo, E., Najt, P., Pievani, M., et al. (2011). Cortex and amygdala morphology in psychopathy. *Psychiatry Research: Neuroimaging, 193*(2), 85–92.

Boccardi, M., Ganzola, R., Rossi, R., Sabattoli, F., Laakso, M. P., Repo-Tiihonen, E., et al. (2010). Abnormal hippocampal shape in offenders with psychopathy. *Human Brain Mapping, 31*(3), 438–447

Boccardi, M., Sabattoli, F., Laakso, M. P., Testa, C., Rossi, R., Beltramello, A., et al. (2005). Frontotemporal dementia as a neural system disease. *Neurobiology of Aging, 26*(1), 37–44.

Brodmann, K. (1909). *Vergleichende Lokalisationslehre der Grosshirnrinde in ihren Prinzipien dargestellt auf Grund des Zellenbaues.* Leipzig: Johann Ambrosius Barth Verlag.

Buckholtz, J. W., Treadway, M. T., Cowan, R. L., Woodward, N. D., Benning, S. D., Li R., et al. (2010). Mesolimbic dopamine reward system hypersensitivity in individuals with psychopathic traits. *Nature Neuroscience, 13*(4), 419–421.

Christianson, S.-A., Forth, A. E., Hare, R. D., Strachan, C., Lidberg, L., & Thorell, L.-H. (1996). Remembering details of emotional events: A comparison between psychopathic and nonpsychopathic offenders. *Personality & Individual Differences, 20*(4), 437–443.

Craig, M. C., Catani, M., Deeley, Q., Latham, R., Daly, E., Kanaan, R., et al. (2009). Altered connections on the road to psychopathy. *Molecular Psychiatry, 14*(10), 946–53, 907.

Damasio, A. (1994). *Descartes' error: Emotion, reason, and the human brain.* New York: Putnam.

Damasio, H., Grabowski, T., Frank, R., Galaburda, A. M., & Damasio, A. R. (1994). The return of Phineas Gage: Clues about the brain from the skull of a famous patient. *Science, 264*(5162), 1102–1105.

De Oliveira-Souza, R., Hare, R. D., Bramati, I. E., Garrido, G. J., Azevedo Ignácio, F., Tovar-Moll,

F., et al. (2008). Psychopathy as a disorder of the moral brain: Fronto-temporo-limbic grey matter reductions demonstrated by voxel-based morphometry. *NeuroImage, 40*(3), 1202–1213.

Devinsky, O., Morrell, M. J., & Vogt, B. A. (1995). Contributions of anterior cingulate cortex to behaviour. *Brain, 118*(Pt 1), 279–306.

Dolan, M. C., Deakin, J. F., Roberts, N., & Anderson, I. M. (2002). Quantitative frontal and temporal structural MRI studies in personality-disordered offenders and control subjects. *Psychiatry Research. 116*(3): 133–149.

Ermer, E., Cope, L. M., Nyalakanti, P. K., Calhoun, V. D., & Kiehl, K. A. (2012). Aberrant paralimbic gray matter in criminal psychopathy. *Journal of Abnormal Psychology, 121*(3), 649–658.

Glenn, A. L., Raine, A., Yaralian, P. S., & Yang, Y. (2010). Increased volume of the striatum in psychopathic individuals. *Biological Psychiatry, 67*(1), 52–58.

Good, C. D., Johnsrude, I. S., Ashburner, J., Henson, R. N., Friston, K. J., Frackowiak, R. S. (2001). A voxel-based morphometric study of ageing in 465 normal adult human brains. *NeuroImage, 14*, 21–36.

Hare, R. D. (2003). *The Hare Psychopathy Checklist*-Revised (2nd edition). Toronto, Ontario, Canada: Multi-Health Systems.

Harlow, J. M. (1848). Passage of an iron rod through the head. *Boston Medical and Surgical Journal, 39*, 389–393.

Kiehl, K. A. (2006). A cognitive neuroscience perspective on psychopathy: evidence for paralimbic system dysfunction. *Psychiatry Research, 142*(2–3), 107–128.

Laakso, M. P., Gunning-Dixon, F., Vaurio, O., Repo-Tiihonen, E., Soininen, H., & Tiihonen, J. (2002). Prefrontal volumes in habitually violent subjects with antisocial personality disorder and type 2 alcoholism. *Psychiatry Research, 114*(2), 95–102.

Laakso, M. P., Vaurio, O., Koivisto, E., Savolainen, L., Eronen, M., Aronen, H. J., et al. (2001). Psychopathy and the posterior hippocampus. *Behavioural Brain Research, 118*(2), 187–193.

Laakso, M. P., Vaurio, O., Savolainen, L., Repo, E., Soininen, H., Aronen, H. J., et al. (2000). A volumetric MRI study of the hippocampus in type 1 and 2 alcoholism. *Behavioural Brain Research, 109*(2), 177–186.

Mendez, M. F., Chow, T., Ringman, J, Twitchell, G, Hinkin, C. H. (2000). Pedophilia and temporal lobe disturbances. *Journal of Neuropsychiatry and Clinical Neuroscience, 12*(1), 71–76.

Meyers, C. A., Berman, S. A., Scheibel, R. S., & Hayman, A. (1992). Case report: Acquired antisocial personality disorder associated with unilateral left orbital frontal lobe damage. *Journal of Psychiatry and Neuroscience, 17*(3), 121–125.

Mitchell, D. G., Avny, S. B., & Blair, R. J. (2006). Divergent patterns of aggressive and neurocognitive characteristics in acquired versus developmental psychopathy. *Neurocase, 12*(3), 164–178.

Müller, J. L., Gänssbauer, S., Sommer, M., Döhnel, K., Weber, T., Schmidt-Wilcke, T., et al. (2008).. Gray matter changes in right superior temporal gyrus in criminal psychopaths: Evidence from voxel-based morphometry. *Psychiatry Research, 163*(3), 213–222.

Narayan, V. M., Narr, K. L., Kumari, V., Woods, R. P., Thompson, P. M., Toga, A. W., et al. (2007). Regional cortical thinning in subjects with violent antisocial personality disorder or schizophrenia. *American Journal of Psychiatry, 164*(9), 1418–1427.

Neylan, T. C. (1999). Frontal lobe function: Mr. Phineas Gage's famous injury. *Journal of Neuropsychiatry and Clinical Neurosciences, 11*(2), 280–283.

Raine, A., Ishikawa, S. S., Arce, E., Lencz, T., Knuth, K. H., Bihrle, S., et al. (2004). Hippocampal structural asymmetry in unsuccessful psychopaths. *Biological Psychiatry, 55*(2), 185–191.

Raine, A., Lencz, T., Bihrle, S., LaCasse, L., & Colletti, P. (2000). Reduced prefrontal gray matter volume and reduced autonomic activity in antisocial personality disorder. *Archives of General Psychiatry, 57*(2), 119–127.

Raine, A., Lencz, T., Taylor, K., Hellige, J. B., Bihrle, S., Lacasse, L., et al. (2003). Corpus callosum abnormalities in psychopathic antisocial individuals. *Archives of General Psychiatry,60*(11), 1134–1142.

Saver, J. L., & Damasio, A. R. (1991). Preserved access and processing of social knowledge in a patient with acquired sociopathy due to ventromedial frontal damage. *Neuropsychologia, 29*(12), 1241–1249.

Schiltz, K., Witzel, J., Bausch-Hölterhoff, J., Gubka, U., & Bogerts, B. (2009). Structural brain pathology and violent crime: Revisiting a possibly underestimated factor. In *9th World Congress of Biological Psychiatry*, P-18–015, WFSBP, Paris, June 29.

Seeley, W. W., Crawford, R., Rascovsky, K., Kramer, J. H., Weiner, M., Miller, B. L., et al. Frontal paralimbic network atrophy in very mild behavioral variant frontotemporal dementia. *Archives of Neurology, 65*(2), 249–255.

Smith, S. S., & Newman, J. P. (1990). Alcohol and drug abuse-dependence disorders in psychopathic and nonpsychopathic criminal offenders. *Journal of Abnormal Psychology, 99*(4), 430–439.

Stip, E. (1995). Compulsive disorder and acquired antisocial behavior in frontal lobe dementia. *Journal of Neuropsychiatry and Clinical Neurosciences, 7,* 116.

Thompson, P. M., Hayashi, K. M., De Zubicaray, G. I., Janke, A. L., Rose, S. E., Semple J, et al. (2004). Mapping hippocampal and ventricular change in Alzheimer disease. *NeuroImage, 22*(4), 1754–1766.

Thompson, P. M., Hayashi, K. M., Sowell, E. R., Gogtay, N., Giedd, J. N., Rapoport, J. L., et al. (2004). Mapping cortical change in Alzheimer's disease, brain development, and schizophrenia. *NeuroImage, 23*(Suppl 1), S2–18.

Tiihonen, J., Rossi, R., Laakso, M. P., Hodgins, S., Testa, C., Perez, J., et al. (2008). Brain anatomy of persistent violent offenders: More rather than less. *Psychiatry Research, 163*(3), 201–212.

Tsuchiya, N., & Adolphs, R. (2007). Emotion and consciousness. *Trends in Cognitive Science, 11*(4), 158–167.

Vogt, B. A., Finch, D. M., & Olson, C. R. (1992). Functional heterogeneity in cingulate cortex: The anterior executive and posterior evaluative regions. *Cerebral Cortex, 2*(6), 435–443.

Watson, K. K., Jones, T. K., & Allman, J. M. (2006). Dendritic architecture of the Von Economo neurons. *Neuroscience, 141*(3), 1107–1112.

Wittenberg, D., Possin, K. L., Rascovsky, K., Rankin, K. P., Miller, B. L., & Kramer, J. H. (2008). The early neuropsychological and behavioral characteristics of frontotemporal dementia. *Neuropsychology Review, 18*(1), 91–102.

Yang, Y, Raine, A., Colletti, P., Toga, A. W., & Narr, K. L. (2009a). Abnormal temporal and prefrontal cortical gray matter thinning in psychopaths. *Molecular Psychiatry, 14*(6), 561–562, 555.

Yang, Y., Raine, A., Lencz, T., Bihrle, S., LaCasse, L., & Colletti, P. (2005). Volume reduction in prefrontal gray matter in unsuccessful criminal psychopaths. *Biological Psychiatry, 57*(10), 1103–1108.

Yang, Y., Raine, A., Narr, K. L., Colletti, P., & Toga, A. W. (2009b). Localization of deformations within the amygdala in individuals with psychopathy. *Archives of General Psychiatry, 66*(9), 986–994.

PART FIVE
GENETICS OF PSYCHOPATHY

CHAPTER 10

Quantitative Genetic Studies of Psychopathic Traits in Minors: Review and Implications for the Law

Essi Viding, Nathalie M. G. Fontaine, and Henrik Larsson

INTRODUCTION

Individuals with criminal psychopathy effect extreme harm to those around them. Not only do they represent a significant tax burden, but they also leave a trail of emotional havoc in their wake. Since most individuals with criminal psychopathy have started their antisocial behavior early in life, several researchers have proposed extending the study of core psychopathic personality characteristics into childhood (Frick & Viding, 2009). Despite some reservations about labeling children with psychopathic traits, research into this area has been seen as an important endeavor from the point of view of prevention and intervention (see Lynam, Caspi, Moffitt, Loeber, & Stouthamer-Loeber, 2007 for a discussion of this issue).

Psychopathy syndrome[1] in both childhood and adulthood involves affective-interpersonal impairment (e.g., lack of empathy, lack of guilt, shallow emotions, superficial charm) as well as overt impulsive, irresponsible, and antisocial behavior (Blair, Peschardt, Budhani, Mitchell, & Pine, 2006; Frick & Marsee, 2006). Symptom profile of psychopathy is discussed extensively in other chapters in this book. In short, individuals with psychopathy represent a subset of those who would meet diagnostic criteria for Conduct Disorder (CD) in childhood or Antisocial Personality Disorder (APD) in adulthood (Blair et al., 2006). However, formal diagnostic criteria for child and adult antisocial behavior syndromes do not distinguish subgroups of antisocial individuals on the basis of their callous-unemotional traits (CU) profile (*Diagnostic and Statistical Manual of Mental Disorders*—4th Edition [*DSM–IV*], 1994; Hart & Hare, 1997).

Psychopathic personality traits may predispose children to life-course persistent antisocial behavior of a particularly serious nature (Frick & Hare, 2001; Frick & Marsee, 2006). Given the early emergence of antisocial behavior in individuals with criminal psychopathy and its long-term impact—even as compared with other antisocial individuals—there has been recent interest in conducting genetically informative research into psychopathy. Such research has scope to be informative with regard to not just genetic risk, but also the nature of the environmental risk associated with psychopathy. The research base at the time of writing this chapter consisted of 14 twin studies, but only 4 molecular genetic studies at the time of writing had specifically focused on psychopathic traits (rather than a broader construct of antisocial behavior). Only one study at the time of writing the chapter had examined potential environmental risk factors for psychopathy in the context of genetically informative studies. Finally, only a subset of the studies has focused on children or adolescents.

Taking genetic information into account adds an important level of analysis when we attempt to understand psychopathic personality traits as a potential risk factor for persistent antisocial behavior. Genetic information can help us to validate whether the large group of children and youth with conduct problems can be meaningfully subtyped into more homogeneous groupings. For example, our group has studied whether the heritability of antisocial behavior varies as a function of co-occurring CU (e.g., Viding, Blair, Moffitt, & Plomin, 2005; Viding, Jones, Frick, Moffitt, & Plomin, 2008). Twin studies focusing on different components of psychopathic personality have, predictably, documented significant heritable influences, but, more interestingly, also a lack of shared environmental influences (e.g., Blonigen, Carlson, Krueger, & Patrick, 2003; Larsson, Andershed, & Lichtenstein, 2006). This information has practical implications for understanding environmental risks related to development of adult psychopathy (and the prevention thereof) and is discussed later in this chapter. Large-scale twin studies have also enabled the investigation of common versus unique genetic and environmental influences on various aspects of psychopathic personality, the study of the etiology of stability and change in psychopathic personality, and the study of etiologic overlap between psychopathic personality and antisocial/externalizing behaviors in children and youth. The strength of the child and youth studies lies in the fact that they address important etiological questions before the accrual of negative consequences that are relevant for samples of criminal psychopaths (e.g., extensive substance abuse) may have muddied the etiological picture. Genetically informative studies of child and youth psychopathy thus have a great potential to provide information with relevance for prevention and treatment. These themes are discussed in the last section of this chapter.

In this chapter we review the current knowledge base of the genetics of child/adolescent psychopathy. We also touch on the study of environmental risk factors in the context of genetically informative studies. We concentrate on twin study findings, as these have dominated the genetic research into psychopathy. To assist the reading of this chapter we first provide a brief discussion of the basic premise of the twin method. We have also included a brief overview of quantitative genetic findings of antisocial behavior and adult psychopathy. The existing twin data make six important points with regard to child/adolescent psychopathy, and these are discussed in turn: (1) Psychopathic personality traits show heritable and nonshared environmental influences in childhood/adolescence. (2) Same genetic and environmental influences are important in accounting for individual differences in psychopathic personality traits for both males and females (although the magnitude of their influence may vary between sexes). (3) Genetic factors are important in explaining covariance among different aspects of psychopathic personality. (4) Genetic factors account for stability of psychopathic personality across development. (5) Genetic factors contribute to the relationship between psychopathic personality traits and antisocial behavior. (6) Negative parental practices constitute a nonshared environmental risk factor for antisocial behavior, but not for core psychopathic, CU. In addition to the twin data, we also briefly and selectively review molecular genetic and gene–environment interplay research on antisocial behavior and point out what we consider to be relevant new research directions for as well as potential future research directions and the implications of "genetics of child/adolescent psychopathy" research to the law.

TWIN METHOD

The twin method is a natural experiment that relies on the different levels of genetic relatedness between identical and fraternal twin pairs to estimate the contribution of genetic and environmental factors to individual differences, or extreme scores in a phenotype of interest. Phenotypes include any behavior or characteristic that is measured separately for each twin, such as twins' score on an antisocial behavior checklist. Statistical model fitting techniques and regression analyses methods incorporating genetic relatedness parameter are used to investigate the etiology of the phenotype of choice. These techniques are not covered in detail in this chapter, but there are textbooks on the topic available for interested readers (Plomin, DeFries, McClearn, & McGuffin, 2008). The basic premise of the twin method is this: If identical twins, who share 100% of their genetic material, appear more similar on a trait than fraternal twins, who share on average 50% of their genetic material (like any siblings)—then we infer that there are genetic influences on a trait. Identical twins' genetic similarity is twice that of fraternal twins.' If nothing apart from genes influences behavior, then we would expect the identical twins to be twice as similar with respect to the phenotypic measure as fraternal twins. Shared environmental influences—environmental influences that make twins similar to each other—are inferred if fraternal twin similarity is more than half of the identical twin similarity (as expected from sharing 50%, as opposed to 100%, of their genes). Finally, if identical twins are not 100% similar on a trait (as would be expected if only genes influenced a trait), nonshared environmental influences are inferred—in other words, environmental influences that make twins different from each other. The nonshared environmental estimate also includes measurement error.

QUANTITATIVE GENETIC FINDINGS OF ANTISOCIAL BEHAVIOR AND ADULT PSYCHOPATHY

Various conceptualizations of antisocial behavior have received more attention in behavioral genetic research than psychopathy. A meta-analysis, including 51 twin and adoption studies, estimated that on average 41% of the variance on antisocial behavior was due to genetic factors (additive and nonadditive genetic effects), about 16% to shared environmental factors, and about 43% to nonshared environmental factors (Rhee & Waldman, 2002). The finding of a heritability estimate of 41% for adolescent and adult antisocial behavior is substantially lower than the estimate of 82% reported in a study on childhood antisocial behavior pervasive across settings (Arseneault et al., 2003).

Further, stability in antisocial behavior from childhood to adolescence (Eley, Lichtenstein, & Moffitt, 2003) and childhood-onset antisocial behavior have also been found to have a strong genetic component (Taylor, Iacono, & McGue, 2000). It is not known what proportion of the samples in these twin studies included psychopathic individuals, although it is plausible that they have represented a subset of individuals sampled and have therefore contributed to the estimates of heritable and environmental influences. Based on the phenotypic data on individuals with psychopathy, it is possible to argue that studies of early-onset, pervasive, and stable antisocial behavior may bear particular relevance for psychopathy.

Studies on adult samples examining the heritability of psychopathic traits have reported moderately strong influences of genetic and nonshared environmental factors (Waldman & Rhee, 2006). For example, Blonigen and colleagues (2003) used a self-report measure, the Psychopathic Personality Inventory (PPI; Lilienfeld & Andrews, 1996), in a sample of adult male twins to examine the genetic and environmental influences on psychopathic personality traits. Individual differences in all eight dimensions measured by the PPI (i.e., Machiavellian Egocentricity, Social Potency, Fearlessness, Coldheartedness, Impulsive Nonconformity, Blame Externalizing, Carefree Nonplanfulness, Stress Immunity) were associated with genetic and nonshared environmental effects. Genetic effects explained 29% to 56% of the variation of the respective dimensions of the PPI. Shared environmental effects were not found for any of the PPI facets (Blonigen et al., 2003).

PSYCHOPATHIC PERSONALITY TRAITS SHOW HERITABLE AND NONSHARED ENVIRONMENTAL INFLUENCES IN CHILDHOOD/ ADOLESCENCE

At the time this chapter was written, there are at least 14 published twin studies examining the etiology of child/youth psychopathy and its various components (see Table 10.1). These studies come from the United States, Sweden, and the United Kingdom. The samples used in these studies vary in size from moderate (398 twin pairs) to large (3,687 twin pairs), represent different age ranges (7–24 years old), and have used a range of instruments to assess different aspects of psychopathic personality. All studies, except one (Taylor, Loney, Bobadilla, Iacono, & McGue,

Table 10.1 Twin Studies Examining the Etiology of Child/Youth Psychopathy and Its Various Components

Authors	N and % boys	Age	Questionnaires and psychopathic dimensions assessed	Main findings
MTFS sample				
Taylor et al. (2003)	398 twin pairs; boys only	16–18 years old	Minnesota Temperament Inventory; CU and impulsive-antisocial (self-reports)	• Genetic effects accounted for around 40% of the variance in both CU and impulsive-antisocial dimensions. • Nonshared environmental effects explained the remaining variance. • Half of the covariation between the CU and impulsivity-antisocial dimensions was due to genetic factors (53%), and nonshared environmental factors explained the remaining part of the covariation (47%). • Support for some independence in the etiology underlying the two dimensions was found.
Blonigen et al. (2005)	626 twin pairs; 46% boys	17 years old	Multidimensional Personality Questionnaire; CU/grandiose-manipulative and impulsive-antisocial (self-reports)	• Genetic effects accounted for 45% of the variance in the CU/grandiose-manipulative dimension and 49% of the variance in the impulsive-antisocial dimension. • Nonshared environmental effects explained the remaining variance for both dimensions. • Genetic overlap between self-reported psychopathic personality traits and externalizing psychopathology was found.
Blonigen et al. (2006)	626 twin pairs; 46% boys	17 and 24 years old	Multidimensional Personality Questionnaire; CU/grandiose-manipulative and impulsive-antisocial (self-reports)	• Genetic effects accounted for 42% of the variance in the CU/grandiose-manipulative dimension and 49% of the variance in the impulsive-antisocial dimension (24 years old). • Nonshared environmental effects explained the remaining variance for both dimensions. • There was little evidence of quantitative sex differences. • Relatively high stability in these traits was substantially influenced by genetic factors (58% and 62% of the stability in CU/grandiose-manipulative and impulsive-antisocial dimensions, respectively).

TEDS sample

Study	Sample	Age	Measures	Findings
Viding et al. (2005)	Sample driven from more than 3,000 twin pairs; about 50% boys	7 years old	Items from the Antisocial Process Screening Device and the Strengths and Difficulties Questionnaire; CU (teachers' reports)	• Antisocial behavior in children with CU was under strong genetic influence ($h_g^2 = .81$) and no influence of shared environment, but antisocial behavior in children without elevated levels of CU showed moderate genetic influence ($h_g^2 = .30$) and environmental influence ($c_g^2 = .34$).
Viding et al. (2007)	Sample driven from more than 3,000 twin pairs; about 50% boys	7 years old	Items from the Antisocial Process Screening Device and the Strengths and Difficulties Questionnaire; CU (teachers' reports)	• High levels of CU were highly heritable at age 7 (67%). • Higher heritability of CU was found in males (67% M vs. 48% F). • Substantial genetic overlap between CU and antisocial behavior in both boys and girls was found.
Larsson, Viding & Plomin (2008)	4,430 twins; 53% boys	7 years old	Items from the Antisocial Process Screening Device and the Strengths and Difficulties Questionnaire; CU (teachers' reports)	• The heritability estimates of elevated levels of CU were high regardless of the presence of antisocial behavior ($h_g^2 = .80$, for CU with antisocial behavior; $h_g^2 = .68$, for CU without antisocial behavior).
Viding et al. (2008)	1,865 twin pairs; about 50% boys	7 years old	Items from the Antisocial Process Screening Device and the Strengths and Difficulties Questionnaire; CU (teachers' reports)	• Antisocial behavior was more heritable with than without concomitant CU. The heritability difference was even more important in magnitude when hyperactive symptoms were taken into account.
Viding et al. (2009)	2,254 twin pairs; 46% boys	7 and 12 years old	Items from the Antisocial Process Screening Device and the Strengths and Difficulties Questionnaire; CU teachers' and parents' reports	• Negative parental discipline was a nonshared environmental risk factor for antisocial behavior, but not for CU.
Fontaine et al. (2010)	9,462 twins; 47% boys	Between 7 and 12 years old	Items from the Antisocial Process Screening Device and the Strengths and Difficulties Questionnaire; CU (teachers' reports)	• High heritability (78%) was observed for boys on a stable high CU trajectory (between 7 and 12 years old). • Stable and high levels of CU in girls, however, appeared to be almost entirely driven by shared environmental influences (75%).

(continued)

Table 10.1 (Continued)

Authors	N and % boys	Age	Questionnaires and psychopathic dimensions assessed	Main findings
TCHAD sample				
Larsson et al. (2006)	1,090 twin pairs; 48% boys	16–17 years old	Youth Psychopathic Traits Inventory; CU, grandiosity-manipulation, and impulsivity-irresponsibility (self-reports)	• Heritability was estimated at 43% for CU, 51% for grandiosity-manipulation, and 56% for impulsivity-irresponsibility. • Nonshared environmental effects explained almost all of the remaining variance for the three dimensions. • There was no evidence of significant sex differences in genetic and/or environmental effects behind any of the three dimensions. • The three dimensions were significantly linked to a highly heritable common latent factor. • The heritability of the latent psychopathic personality factor was estimated at 63%.
Larsson et al. (2007)	2,387 twins; about 50% boys	13–14 and 16–17 years old	Youth Psychopathic Traits Inventory; CU, grandiosity-manipulation, and impulsivity-irresponsibility (self-reports)	• A common genetic factor contributed substantially to the three psychopathic personality dimensions (CU, grandiosity-manipulation, and impulsivity-grandiosity) and to antisocial behavior.
Forsman et al. (2007)	More than 2,000 twins; about 50% boys	Externalizing behavior (at age 8–9 and 13–14 years) and psychopathic personality traits (at age 16–17 years)	Youth Psychopathic Traits Inventory; CU, grandiosity-manipulation, and impulsivity-irresponsibility (self-reports)	• Externalizing behavior in childhood was associated with higher levels of psychopathic personality traits in adolescence among boys but not in girls. • Genetic factors were responsible for the association between externalizing behavior and psychopathic personality traits.

Forsman et al. (2008)	More than 2,000 twins; about 50% boys	At ages 16 and 19 years old	Youth Psychopathic Traits Inventory; CU, grandiosity-manipulation, and impulsivity-irresponsibility (self-reports)	• Test–retest correlation of the higher-order psychopathic personality factor was high ($r = .60$), and as much as 90% of the test–retest correlation was explained by genetic factors. • Evidence for specific genetic stability in the CU and impulsive-irresponsible dimension, suggesting etiologic generality and specificity for the stability of psychopathic personality between mid- and late adolescence.
Forsman et al. (2010)	2,255 twins; 48% boys	Antisocial behavior (from age 8–9 to age 16–17) and psychopathic personality (age 16–17 years and age 19–20 years)	Youth Psychopathic Traits Inventory; CU, grandiosity-manipulation, and impulsivity-irresponsibility (self-reports)	• Psychopathic personality in adolescence predicted antisocial behavior in early adulthood. • Persistent antisocial behavior (from childhood to adolescence) had an impact on psychopathic personality in early adulthood via genetic effects.

2003), include information from both sexes with a similar male/female composition (46%–48% males). In this section we concentrate on reviewing the univariate data on psychopathic personality (variously measured) that are available from these studies. The univariate findings document the relative importance of heritable and environmental influences on psychopathic personality in children and youth.

Taylor and colleagues used items from a self-report questionnaire, the Minnesota Temperament Inventory (Loney, Taylor, Butler, & Iacono, 2002, unpublished data), to assess the CU (e.g., "I don't experience very deep emotions") and impulsive-antisocial (e.g., "I am unreliable") dimensions of adolescent psychopathy (Taylor et al., 2003). This self-report questionnaire was distributed to a sample of 398 adolescent male twin pairs (16–18 years old) from the Minnesota Twin Family Study (MTFS sample). In this study, genetic effects accounted for around 40% of the variation in both the CU and the impulsive-antisocial factor. Nonshared environmental effects explained all of the remaining variance, whereas the influences of shared environment seemed to be of no importance (Taylor el al., 2003).

Blonigen, Hicks, Krueger, Patrick, and Iacono (2005; 2006) used factor scores on the primary scales of the self-report Multidimensional Personality Questionnaire (MPQ) to index the two main dimensions of the PPI (Lilienfeld & Andrews, 1996): the fearless-dominance dimension (hereafter, the "CU/grandiose-manipulative" dimension for consistency across the chapters in this book) and the impulsive-antisocial dimension. In their first report on this sample of adolescents, Blonigen and colleagues examined data from 626 17-year-old male and female twin pairs from the MTFS sample. Genetic influences accounted for 45% of the variance in the CU/grandiose-manipulative dimension and 49% of the variance in the impulsive-antisocial dimension. The remainder of the variance in each dimension was accounted for by nonshared environmental influences. Very similar findings were reported for the same sample at age 24. Genetic influences accounted for 42% of the variance in the CU/grandiose-manipulative dimension and 49% of the variance in the impulsive-antisocial dimension. The remainder of the variance in each dimension was accounted for by nonshared environmental influences.

Viding and colleagues studied the heritability of teacher-rated CU in the Twins Early Development Study (TEDS) sample of more than 3,500 7-year-old twin pairs (both boys and girls) (Viding, Blair, Moffitt, & Plomin, 2005; Viding, Frick, & Plomin, 2007). The CU scale used in the TEDS analyses consisted of seven items: three Antisocial Process Screening Device items (Frick & Hare, 2001), as well as four items from the Strengths and Difficulties Questionnaire (SDQ; Goodman, 1997). Individual differences in CU were strongly heritable (67% of the phenotypic variance was due to genetic influences) (Viding et al., 2007). The group difference in CU between those scoring at the top 10% of the TEDS sample and the rest of the TEDS children was also strongly heritable ($h_g^2 = .67$). The group heritability estimates were very similar regardless of whether the CU occurred with ($h_g^2 = .80$) or without ($h_g^2 = .68$) elevated levels of conduct problems (Larsson, Viding, & Plomin, 2008). None of the shared environmental estimates for CU in these different analyses differed significantly from zero. However, nonshared environmental influences made a significant contribution to CU at age 7, both across the continuum as well as at the extremes.

Larsson and colleagues (Larsson et al., 2006) used a sample of 1,090 16-year-old male and female twin pairs from the Twin study of CHild and Adolescent Development (TCHAD; Lichtenstein, Tuvblad, Larsson, & Carlstrom, 2007) to examine the heritability of three psychopathic trait dimensions: CU, grandiosity-manipulation, and impulsivity-irresponsibility. A 50-item self-report measure, the Youth Psychopathic traits Inventory (YPI), was used to measure these traits (Andershed, Kerr, Stattin, & Levander, 2002). For the grandiose/manipulative dimension, the heritability was estimated at 51%. Nonshared environmental effects explained almost all of the remaining variance, whereas the influence of shared environmental effects seemed to be of minimal importance. Similar patterns, with considerable heritable and little shared environmental influences were evident for the other two dimensions as well; that is, genetic effects accounted for 43% and 56% of the variation in the CU and impulsive/irresponsible dimension, respectively.

Various aspects of psychopathic personality in children/adolescents show moderate to strong genetic influence and moderate nonshared environmental influence. Shared environmental influences on psychopathic personality traits were not detected in any of the studies, suggesting that the environmental factors making members of the twin pair similar to each other do not account for individual differences in psychopathic personality in children/adolescents. These results are in line with previous adult data on psychopathic personality traits (Blonigen et al., 2003) as well as data on other personality dimensions (Bouchard & Loehlin, 2001) and suggest that there is moderate to strong genetic vulnerability to psychopathic personality. Importantly, these results do not preclude that environmental factors (e.g., parental characteristics) exert an influence on psychopathic personality traits. However, such factors may act in a child-specific manner (represented by nonshared environmental variance) or via the process of gene–environment interplay.

In addition to reporting univariate heritabilities for various aspects of psychopathic personality, most of the studies reviewed in the preceding text have also addressed multivariate questions of etiology of covariance between different aspects of psychopathic personality, origins of stability of psychopathic personality across development, and the etiologic relationship between psychopathic personality traits and antisocial behavior. The subsequent sections review findings from these analyses.

SAME GENETIC AND ENVIRONMENTAL INFLUENCES ARE IMPORTANT IN ACCOUNTING FOR INDIVIDUAL DIFFERENCES IN PSYCHOPATHIC PERSONALITY TRAITS FOR BOTH MALES AND FEMALES

Two studies (Larsson et al., 2006; Viding et al., 2007) have incorporated dizygotic opposite-sex twin pairs in their analyses to explore the potential role of qualitative sex differences (i.e., different genes and environments influencing phenotypic variation for males and females) in the etiology of psychopathy. Neither of these studies found support for qualitative sex differences in the genetic and environmental influences on psychopathic traits. Few studies have also assessed the impact of quantitative sex differences (i.e., the same genetic and environmental influences affecting males and females to a different degree). Most studies found little evidence of quantitative sex differences (e.g., Blonigen et al., 2006; Larsson et al., 2006), but there is also some support for a higher heritability of CU for males (Fontaine, Rijsdijk, McCrory, & Viding, 2010; Viding et al., 2007). For instance, Fontaine and colleagues (2010) found that among 9,462 youths from the TEDS sample, high heritability was observed for boys on a stable high CU trajectory (between 7 and 12 years old). Stable and high levels of CU in girls, however, appeared to be almost entirely driven by shared environmental influences. Replications of these findings are needed given the small number of children (around 3% of the sample) who followed the stable and high-CU trajectory.

The findings so far indicate that although males and females show mean differences in psychopathic personality trait scores at the phenotypic level, the etiological influences on individual differences in these traits are similar for both genders in adolescence. Our data suggest that for younger children, genetic influences may be more prominent for CU traits in boys than in girls, but this finding needs replication.

GENETIC FACTORS ARE IMPORTANT IN EXPLAINING COVARIANCE AMONG DIFFERENT ASPECTS OF PSYCHOPATHIC PERSONALITY

At the time this chapter was written two twin studies had employed models to study the origins of covariance among different aspects of psychopathic personality. The first study to address this issue was conducted by Taylor et al. (2003), who reported that half of the covariation between CU and impulsivity-antisocial behavior at 16 to 18 years of age was due to genetic factors (53%), while nonshared environmental factors explained the remaining part of the covariation (47%). This study also found support for some independence in the etiology underlying the two trait dimensions. That is, approximately half of the total genetic variance in CU was unique to that personality dimension, which clearly suggests that the CU and impulsive-antisocial dimensions can be differentiated etiologically.

Larsson, Andershed, and Lichtenstein (2006) used the YPI in a sample of 16-year-old twin pairs from the aforementioned TCHAD study to examine the etiology underlying the covariation between the three psychopathic trait dimensions (i.e., CU, grandiosity-manipulation, and impulsivity-irresponsibility). The authors used multivariate model fitting techniques to study the covariance shared among the three psychopathic personality dimensions simultaneously. They found that a common pathway model provided the best fit of the data, which indicates that the three dimensions of psychopathic personality are significantly linked to a highly heritable common latent factor. The relatively high heritability (63%) for the latent psychopathic personality factor makes it a novel target for future research.

In sum, studies investigating the etiology of covariance among different aspects of psychopathic personality suggest that common genes are important.[2]

GENETIC FACTORS ACCOUNT FOR STABILITY OF PSYCHOPATHIC PERSONALITY ACROSS DEVELOPMENT

At the time this chapter was written a few twin studies had explored the genetic and environmental contributions to the stability of psychopathic personality traits in adolescence. Blonigen et al. (2006) used scales derived from the MPQ to examine the genetic and environmental contribution to stability of CU/grandiose-manipulative (what they called fearless-dominance) and impulsive-antisocial dimensions from adolescence to early adulthood. The authors applied model fitting analyses on MPQ information from two time-points 7 years apart, when the twins were 17 and 24 years old. Their results indicate that the heritability of the fearless-dominance and impulsive-antisocial dimension remained consistent across time. Further, the decomposition of the covariance across time for the fearless-dominance and impulsive-antisocial dimension revealed that 58% of the stability in fearless-dominance and 62% of the stability in the impulsive-antisocial dimension was due to genetic influences. This result indicates that the relatively high stability in these traits is substantially influence by genetic factors.

Furthermore, using the TEDS sample, Fontaine and colleagues (2010) found that stable high trajectory of CU in childhood (between 7 and 12 years) was highly heritable in boys ($h^2 = .78$), but not in girls, in whom high and stable levels of CU were explained by shared environmental influences ($c^2 = .75$).

Forsman, Lichtenstein, Andershed, and Larsson (2008) measured intented, and impulsive-irresponsible dimensions with the YPI and examined genetic and environmental contribution to the stability of psychopathic personality between ages 16 and 19 in the TCHAD sample. The authors focused on a hierarchical model of psychopathic personality in which a higher-order general factor substantially explained the variation in the three psychopathic personality dimensions, both in mid- and late adolescence. Results showed that the observed test–retest correlation of the higher-order psychopathic personality factor was high ($r = .60$). In addition, as much as 90% of the test–retest correlation was explained by genetic factors. However, they also found evidence for specific genetic stability in the CU and impulsive-irresponsible dimension. For example, 13% of the unique genetic effects in the CU dimension at age 19 were shared with the corresponding effects at age 16. Thus, their model provides evidence for etiologic generality and etiologic specificity for the stability of psychopathic personality between mid- and late adolescence.

In sum, longitudinal twin studies have found that the stability in psychopathic personality traits in childhood (in males in particular) and from adolescence to early adulthood is substantially influenced by genetic factors. Nothing is known about how genetic and environmental factors contribute to psychopathic traits during the transition from childhood to adolescence. It would therefore be interesting to explore how genetic and environmental factors influence different developmental trajectories for antisocial children with and without CU. Another important research goal for future longitudinal twin studies of psychopathic personality traits is to test the developmental hypothesis about different patterns of gene–environment correlations and gene–environment interaction.

GENETIC FACTORS CONTRIBUTE TO THE RELATIONSHIP BETWEEN PSYCHOPATHIC PERSONALITY TRAITS AND ANTISOCIAL BEHAVIOR

Blonigen and colleagues (2005) used a sample of adolescent twin pairs (626 pairs of 17-year-old male and female twins) from the MTFS to examine the etiologic connection between psychopathic personality traits and broad domains of externalizing behavior (antisocial behavior, substance abuse). They found a moderate positive genetic correlation ($r_g = .49$) between the impulsive-antisocial dimension and externalizing behavior and a modest positive genetic correlation ($r_g = .16$) between CU/grandiose-manipulative dimension and externalizing behavior (stronger for males, $r_g = .35$).

Larsson et al. (2007) conducted another twin study using the aforementioned TCHAD sample. Theory driven multivariate twin models were applied to self-report assessments of psychopathic personality (YPI) and adolescent antisocial behavior. They found that a common genetic factor contributed substantially to the three psychopathic personality dimensions (grandiose-manipulative, CU, and impulsive-irresponsible dimension) and antisocial behavior measured at age 13 to 14 and 16 to 17 years. This is in line with the study by Blonigen et al. (2005) that reported a genetic overlap between self-reported psychopathic personality traits and externalizing psychopathology.

A study by Forsman and colleagues using the TCHAD sample represents another example of how twin research designs can be used to study the nature of association between psychopathic personality traits and various forms of externalizing psychopathology (Forsman, Larsson, Andershed, & Lichtenstein, 2007). This study used a longitudinal design to compare identical and fraternal twins discordant for parent-rated externalizing behavior in childhood (at age 8–9 and 13–14 years) on their self-report scores on psychopathic personality traits in adolescence (at age 16–17 years). This study found that externalizing behavior in childhood was associated with higher levels of psychopathic personality traits (YPI) in adolescence among boys but not in girls. The comparison of identical and fraternal twins suggested that genetic factors were responsible for the association between externalizing behavior and psychopathic personality traits. Another study by Forsman and colleagues (Forsman, Lichtenstein, Andershed, & Larsson, 2009) using the TCHAD sample ($N = 2,255$ twins) showed that psychopathic personality (measured with the YPI) in adolescence predicted antisocial behavior in early adulthood, over and above both concurrent and preexisting levels of antisocial behavior. Their study also showed that the association between adolescent psychopathic personality and adult antisocial behavior was explained mainly by genetic effects, a result that can be interpreted as a genetically influenced personality-driven process, wherein individuals are predisposed to higher risk of involvement in antisocial behavior because of their antisocially prone personality.

Viding, Frick, and Plomin (2007) examined the extent to which genetic influences contribute to the overlap between CU and antisocial behavior at age seven. The analyses in this study were conducted on the same TEDS sample of twins as used in Viding et al. (2005). Teachers provided ratings of CU and antisocial behavior and analyses were performed across the continuum of scores and at the extreme end of the distribution. The findings demonstrated, most importantly, that there is a substantial genetic overlap between CU and antisocial behavior in both boys and girls. Common genetic influences operate to bring about both of these problems, assessed as a dimension and even more so at the high extremes.

Finally, twin data have also been used to inform the utility of using psychopathic personality traits to subtype individuals with antisocial behavior. Viding et al. (2005) used information from the TEDS sample to investigate whether the etiology of teacher rated antisocial behavior differs as a function of teacher-rated CU. Antisocial behavior was assessed using the SDQ five-item conduct problem scale. The authors separated children with elevated levels of antisocial behavior (in the top 10% for the TEDS sample) into two groups based on their CU score (in the top 10% or not). Antisocial behavior in children with CU was under strong genetic influence ($h_g^2 = .81$) and no influence of shared environment. In contrast, antisocial behavior in children without elevated levels of CU showed moderate genetic influence ($h_g^2 = .30$) and substantial environmental influence ($c_g^2 = .34$, $e_g^2 = .26$). Viding, Jones, Frick, Moffitt, and Plomin (2008) have recently replicated the finding of different heritability estimates for the CU and non-CU groups using the 9-year teacher data from the TEDS study. Their results indicate that the heritability differences hold even after hyperactivity scores of the children are controlled for, suggesting that the result is not driven by any differences in hyperactivity between the two groups. Taken together, these findings clearly suggest that while the CU subtype is genetically vulnerable to antisocial behavior, the non-CU subtype manifests a more strongly environmental etiology to their antisocial behavior.

In sum, twin studies that have examined the overlap between psychopathic personality traits and antisocial behavior have all shown that common genetic influences account for much of the covariation between psychopathic personality traits and antisocial behavior. In addition, data from young twins suggest that early-onset antisocial behavior is more heritable for the group of children with concomitant CU and antisocial behavior. These findings are consistent with the notion that there are a common set of genes that influence psychopathic personality traits and antisocial behavior (as well as other disorders on the externalizing spectrum) and are in line with the hypothesis that a shared set of genes affect various externalizing psychopathology (Krueger et al., 2002). However, there are other feasible competing explanations for the observed finding. For example, the findings are congruent with genetic influences that act indirectly on antisocial (externalizing) behavior, via psychopathic personality traits (Goldsmith & Gottesman, 1996). Multilevel methodological approaches (e.g., measured genes, brain functioning, neuropsychological measures) can be helpful when the shared versus unique etiologic processes between psychopathic personality traits and antisocial behavior are explored, especially because twin studies conceptualize these processes only as latent genetic and environmental factors. For example, molecular genetic studies could concentrate on testing the hypothesis that risk genes that eventually emerge for psychopathic personality traits would also increase the risk for antisocial behavior (and conversely test theoretically plausible risk genes for antisocial behavior for an association with psychopathic personality traits).

MOLECULAR GENETIC STUDIES

Molecular genetic research relevant for psychopathy has used association analysis to explore the impact of candidate genes (i.e., a gene whose function suggests that it might be associated with a trait). Association studies are usually conducted using a case-control design to compare allelic frequencies in unrelated affected and unaffected individuals in the population. An allele in a gene is said to be associated with a trait if it occurs at a significantly higher frequency in the affected individuals compared to the unaffected group.

Despite the substantial literature demonstration heritable component of psychopathic traits, we know of only four published studies that have looked at the association between candidate genes and psychopathy measures. Although three previous studies have focused on the narrower phenotype of adult psychopathy, these studies were limited because they involved small samples of substance abusing adults and focused on only a few candidate genes. The first of these studies found no association between psychopathy and either of the Taq1 single nucleotide polymorphisms (SNPs) located in the 3′-untranslated region of the DRD2 (Smith et al., 1993). Two recent studies focused on a small sample of adult alcoholic patients and found an association between psychopathy and specific allelic variants of cannabinoid receptor type 1, fatty acid amide hydrolase (Hoenicka et al., 2006), as well as psychopathy scores and DRD2 C957T and ANKK1 Taq1A acting epistatically (Ponce et al., 2008). An additional study on a relatively small sample of adolescents with ADHD recently reported associations between the *val* allele of the catechol-*O*-methyltransferase gene, the low-activity allele of monoamine oxidase-A gene (*MAOA-L*), the short allele of the 5HTT gene (*5HTTs*) and "emotional dysfunction" scores of psychopathy (Fowler et al., 2009). The latter two of these associations were unexpected based on imaging genetic data suggesting that *MAOA-L* and *5HTTs* confer a pattern of amygdala reactivity (heightened) opposite of that typically seen in individuals with psychopathy (dampened) (e.g., Meyer-Lindenberg et al., 2006; Munafò, Brown, & Hariri, 2008—see below for further discussion on *MAOA* and imaging genetics).

Research on molecular genetics of psychopathy is thus still very much in its infancy. Molecular genetic research on antisocial behavior, on the other hand, has received more attention. In the text that follows we review research on molecular genetics of antisocial behavior and relate it to potential future investigations in the field of psychopathy. Genes regulating serotonergic neurotransmission, in particular *MAOA*, have been highlighted in the search for a genetic predisposition to antisocial behavior (Lesch, 2003). The *MAOA* gene contains a well-characterized functional polymorphism consisting of a variable number of tandem repeats in the promoter region, with high- (*MAOA-H*) and low-activity (*MAOA-L*) allelic variants. The *MAOA-H* variant is associated with lower concentration of intracellular serotonin, whereas the *MAOA-L* variant is associated with higher concentration of intracellular serotonin. Recent research suggests that genetic vulnerability to antisocial behavior conferred by the *MAOA-L* may become evident only in the presence of an environmental trigger, such as maltreatment (e.g., Caspi et al., 2002;

Kim-Cohen et al., 2006). Despite the demonstration of genetic influences on individual differences in antisocial behavior, it is important to note that no genes *for* antisocial behavior exist. Instead genes code for proteins that influence characteristics such as neurocognitive vulnerabilities that may in turn increase risk for antisocial behavior. Thus, although genetic risk alone may be of little consequence for behavior in favorable conditions, the genetic vulnerability may still manifest at the level of brain/cognition. Imaging genetics studies attest to genotype differences being associated with variation in brain structure and function in nonclinical samples (Meyer-Lindenberg & Weinberger, 2006). We can think of this as the neural fingerprint, ready to translate into disordered behavior in the presence of unfortunate environmental triggers.

Meyer-Lindenberg and colleagues recently provided the first demonstration of the *MAOA-L* genotype being associated with a pattern of neural hypersensitivity to emotional stimuli (Meyer-Lindenberg et al., 2006). Specifically they reported increased amygdala activity coupled with lesser activity in the frontal regulatory regions in *MAOA-L* as compared with *MAOA-H* carriers. Similar findings have been reported in subsequent studies (reviewed in Buckholtz & Meyer-Lindenberg, 2008). Buckholtz & Meyer-Lindenberg speculate that the brain imaging findings of poor emotion regulation in *MAOA-L* carries relate to threat reactive and impulsive, rather than psychopathic antisocial behavior. They based this speculation on the findings of individuals with psychopathy displaying underreactivity of the brain's emotional circuit, particularly the amygdala (Birbaumer et al., 2005; Kiehl et al., 2001).

As highlighted in the preceding text, although the behavioral genetics data suggest heritability of psychopathy, a molecular level account of the disorder remains in its infancy. Suggestions have been made that psychopathy might be related to anomalies in noradrenergic neurotransmission (Blair, Mitchell, & Blair, 2005). It is also interesting to note that some studies have reported increased vulnerability to antisocial behavior in the presence of the *MAOA-H* allele (e.g., Manuck et al., 1999). These may reflect false-positive findings, but it is also possible to speculate that the amygdala hypo- as opposed to hyperreactivity seen in individuals with psychopathy could be influenced by *MAOA-H* rather than *MAOA-L* genotype. This suggestion remains highly speculative, and as for any behavior, the genetic influences will not be limited to a single candidate gene. However, it is entirely possible that different alleles of the same gene may predispose to different types of antisocial syndromes.

In addition to *MAOA*, other genes influencing the reactivity of the brain's affect circuitry, such as *5HTT*, could also play a role in psychopathy. Finally, new technologies, such as DNA pooling,[3] are enabling genome-wide association studies that search for novel single nucleotide polymorphisms (SNPs) that may be associated with psychopathy syndrome. We are currently conducting such a study, and preliminary data are suggestive of several SNPs associated with psychopathy syndrome (Viding et al., 2010). Once it is established whether these findings replicate, such SNPs could be assessed in relation to psychopathy/antisocial behavior in several existing studies that span beyond our own collaborations. They could also be incorporated into imaging genetic investigations of psychopathy.

GENETIC STUDIES INFORMING ON ENVIRONMENTAL RISK

Although there is a growing body of evidence pointing toward some genetic risk for psychopathy, this risk is likely to act in conjunction of environmental factors. One way to examine the causal role of environmental risk factors in the development of any trait or behavior is to use behavioral genetic research designs. These designs make use of "natural experiments" to disentangle the normally inseparable contribution of genes and environments, as well as to study gene–environment interplay (Moffitt, 2005). These studies encompass both twin and adoption methods, and more recently studies of measured genes and environments (see the section on molecular genetics highlighting gene–environment interaction). Only one study to date, from our own group, has focused on environmental risk factors for psychopathic (CU) traits within a genetically informative MZ-differences design (Viding, Fontaine, Oliver, & Plomin, 2009; see later).

Since nonshared environment appears to be important for the etiology of psychopathic traits, a twin design comparing identical twins discordant for a proposed environmental risk factor is an interesting analytic avenue to investigate candidate nonshared environmental risk factors for psychopathy. Caspi and colleagues recently used this design to good effect when they demonstrated that discrepancies in maternal negative emotion/hostility toward two identical twins were related to discrepancies in future, teacher-rated antisocial behavior (Caspi et al., 2004). This finding held even after previous levels of antisocial behavior were controlled for and suggests that when two genetically identical individuals receive discrepant treatment, the individual who is the focus of more negative emotionality is also the individual at most risk for developing antisocial behavior. This type of design strongly demonstrates an environmental effect because effects of the child genotype can be ruled out by effectively comparing "genetic clones." We have recently replicated the findings reported by Caspi and colleagues (2004) on antisocial behavior using a sample of children (2,254 twin pairs, between 7 and 12 years old) from the TEDS sample (Viding et al., 2009). We found that the twins who had received more negative parental discipline at age 7 were more likely to have higher levels of conduct problems at age 12 than their co-twins. However, this finding was not replicated for CU. Although there was a phenotypic relationship between negative parental discipline and CU, we did not find a longitudinal environmental effect of differential negative parental discipline on differences in subsequent CU between the twins. This finding may be explained by the fact that the relationship between parental practices and CU in children could reflect passive and/or evocative gene–environment correlation (i.e., genetic endowment within families), which would not be revealed in a design that focuses on nonshared environment (i.e., MZ twin differences design). In sum, we found that during the transition to early adolescence, negative parental discipline was a nonshared environmental risk factor for antisocial behavior but not for CU.

Existing behavioral genetic research also indicates that children's genetically influenced antisocial behavior evokes negative parenting (e.g., Larsson, Viding, Rijsdijk, & Plomin, 2008). One process that may explain this finding is likely to be evocative gene–environment correlation, in which genetically influenced child predispositions elicit responses such as negative parenting that in turn influence development of antisocial behavior (Moffitt, 2005). Unfortunately, the role of evocative gene–environmental correlation in the context of psychopathic personality traits has not been studied in a behavioral genetic framework. Future studies may therefore benefit from including careful assessments of proximal environmental components (e.g., maltreatment, parent–child conflict, parental negativity, and inconsistent-harsh parenting) in the twin study design.

Both twin designs and designs of measured genes and environments have been used to demonstrate that gene–environment interaction (G×E) seems to operate to bring about antisocial behavior (e.g., Jaffee et al., 2005; Kim-Cohen et al., 2006; see preceding text for the discussion of the MAOA G×E findings). No studies to date have investigated G×E in relation to psychopathy in either adult or child samples. It is important to note that treatment of antisocial behavior could be thought of as a "reverse" G×E, wherein an individual's genetic makeup may limit the success of a particular intervention. Given some preliminary data indicating that children with CU may respond differently to parenting and interventions than their antisocial peers (e.g., Hawes & Dadds, 2005; Hipwell et al., 2007; Wootton et al., 1997), it would be of interest to understand how the genetic susceptibility to antisocial behavior in this population could be best moderated with a suitable environmental buffering.

GENETIC RESEARCH TO PSYCHOPATHIC TRAITS: IMPLICATIONS FOR THE LAW AND POLICYMAKING

All legal and policymaking changes must proceed with extensive ethical consultation and should consider several issues, including discrimination, stigma, and labeling (see Singh & Rose, 2009). The research base into genetic influences on child/adolescent psychopathic traits is still sparse and requires considerable replication and extension before any legal or policy recommendations can be based on these findings.

For example, current studies have used variety of assessment methods and no behavioral genetic study to date has employed the gold-standard instrument, Psychopathy Checklist–Revised (PCL-R). It should, therefore, be kept in mind that the findings from the current studies relate to nondiagnostic, questionnaire-based measures of psychopathic traits and that the assessment instruments have varied across studies. With this caveat in mind, the current studies do suggest genetic vulnerability to psychopathic personality traits, as well as etiological differences between children with/without CU type antisocial behavior. These findings tie in with much of the neurocognitive, behavioral, and treatment research on psychopathic traits. The existing, nongenetic research base strongly suggests that children/adolescents with psychopathic traits form a specific, particularly high-risk group of youngsters (Frick & Viding, 2009). Prevention, therefore, remains an important policymaking goal. In this context it is important to note two things: (1) Genetic vulnerability does not denote immutability. It suggests, however, that it is important to intervene as early as possible. (2) Behavioral genetic designs suggest that environment drives change. Genetic risk for developing psychopathy could be buffered by prevention and treatment, which could be considered as positive gene–environment interaction. Future research is needed to better inform what environmental factors may exert an influence on the change of psychopathic traits in youths.

The law and policymaking communities have expressed worries about scientific findings on "risk genes" in two ways: (1) Some are concerned that such findings can be used to argue for more lenient sentences for the carriers of "risk genes" and (2) others worry that they may result in prosecutors demanding harsher and longer ("preventative") sentences for the carriers of "risk genes." Although the current data suggest that there is genetic vulnerability to psychopathic traits, we need to remember that there are no genes *for* psychopathic traits. Instead we know that genes act in a probabilistic manner and in concert with environmental factors to make some individuals more vulnerable for developing psychopathic traits (see, e.g., Moffitt, 2005; Viding, Jones, & Larsson, 2008). Policymakers may be interested in probabilities, but the court of law is ultimately not probabilistic in the same way that science is; science and law have different goals (Eastman & Campbell, 2006; Mobbs, Lau, Jones, & Frith, 2007). The legal system usually considers one individual at a time, which is in direct contrast to behavioral genetic research that focuses on heritable and environmental causes of individual differences in the population. In other words, behavioral genetic research is not concerned with determining the risk for a single individual. Even if we find genes that show reliable and consistent association with individual differences in psychopathic traits and that can be genotyped for single individuals, we argue that these findings should not influence sentencing decisions. Any gene alone will be neither necessary nor sufficient to predispose someone to high levels of psychopathic traits, and as such, the responsibility for choosing to offend still resides with an individual (unless, based on psychiatric assessment, a mental incapacity plea can be entered). Most "risk genes" are common in the population and yet do not cause the majority of the individuals carrying them to offend.

SUMMARY

Genetic research into psychopathy is off to a very promising start, with several research groups in different countries publishing in the area. Overall, psychopathic traits appear to be moderately to strongly heritable and show little shared environmental influence. Same genetic and environmental influences appear important in accounting for individual differences in psychopathic personality traits for both males and females. Genetic factors are important in explaining covariance among different aspects of psychopathic personality and stability of psychopathic personality across development. Furthermore, common genes contribute to the relationship between psychopathic personality traits and antisocial behavior. Much more work using genetically informative study designs is required. First, the quantitative genetic research findings need more replications. Second, the molecular genetic research base on psychopathy needs to start in earnest. Third, environmental risk can be studied in a particularly informative way using behavioral genetic designs, and this strand of research needs to be developed at the present—beyond our initial monozygotic twin differences study that demonstrated that negative parental discipline does not operate as a nonshared environmental risk factor for the development of core psychopathic traits. New avenues, such as imaging genetic studies, should also be engaged in. The current behavioral genetic research, alongside research conducted using other methodologies, underscores the importance of developing specific policy for prevention of psychopathic traits. The current behavioral genetic findings are not "ready for the courtroom," but neither are similar findings on other phenotypes. Given the probabilistic nature by which most genetic risk on behavior operates, it is questionable whether genotype information can be reasonably considered as a mitigating or extenuating factor on sentencing decisions.

ACKNOWLEDGMENTS

E. Viding was supported by research grants from the ESRC (RES-062-23-2202) and British Academy (BARDA-53229) during the writing of this chapter.

NOTES

1. The authors wish to note that they are aware of the controversies of calling children/adolescents psychopaths. The terms "psychopathy syndrome," "individuals (children/youth) with psychopathy," and children with antisocial behavior and callous-unemotional traits (AB/CU+) are all used interchangeably throughout this chapter. However, the authors do not have a strong preference for a particular label for this group of youngsters.
2. Note that Blonigen et al. (2005) did not report a significant phenotypic correlation between CU/grandiose-manipulative and irresponsible-antisocial dimensions and as such data from this study are not included in

the discussion of the etiology of covariance between different dimensions of psychopathic personality. The authors suggest that their data are compatible with the view that largely distinct genetic influences are responsible for variation in the CU/grandiose-manipulative and irresponsible-antisocial dimensions. However, the lack of phenotypic and genetic correlations between the two dimensions (as derived from the MPQ) is not in line with reports using other instruments to index psychopathic personality. This discrepancy clearly warrants further investigation.

3. DNA pooling refers to a genetic screening method that combines DNA from many individuals in a single molecular genetic analysis to generate a representation of allele frequencies. A DNA pool can thus be generated for all cases and all controls and allele frequencies can be compared between these pools.

REFERENCES

American Psychiatric Association (1994). *Diagnostic and statistical manual of mental disorders.* Washington, DC: American Psychiatric Association.

Andershed, H., Kerr, M., Stattin, H., & Levander, S. (2002). Psychopathic traits in non-referred youths: Initial test of a new assessment tool. In: E. Blaauw & L. Sheridan (Eds.), *Psychopaths: Current international perspectives* (pp. 131–158). The Hague: Elsevier.

Arseneault, L., Moffitt, T. E., Caspi, A., Taylor, A., Rijsdijk, F. V., Jaffee, S. R., et al. (2003). Strong genetic effects on cross-situational antisocial behavior among 5-year-old children according to mothers, teachers, examiner-observers, and twins' self-reports. *Journal of Child Psychology & Psychiatry & Allied Disciplines, 44*(6), 832–848.

Birbaumer, N., Veit, R., Lotze, M., Erb, M., Hermann, C., Grodd, W., et al. (2005). Deficient fear conditioning in psychopathy: A functional magnetic resonance imaging study. *Archives of General Psychiatry, 62*(7), 799–805.

Blair, R. J. R., Mitchell, D. G. V., & Blair, K. S. (2005). *The psychopath: Emotion and the brain.* Oxford: Blackwell.

Blair, R. J. R., Peschardt, K. S., Budhani, S., Mitchell, D. G. V., & Pine, D. S. (2006). The development of psychopathy. *Journal of Child Psychology & Psychiatry, 47*(3), 262–275.

Blonigen, D. M., Carlson, S. R., Krueger, R. F., & Patrick, C. J. (2003). A twin study of self-reported psychopathic personality traits. *Personality and Individual Differences, 35*, 179–197.

Blonigen, D. M., Hicks, B. M., Krueger, R. F., Patrick, C. J., & Iacono, W. G. (2005). Psychopathic personality traits: Heritability and genetic overlap with internalizing and externalizing psychopathology. *Psychological Medicine, 35*, 637–648.

Blonigen, D. M., Hicks, B. M., Krueger, R. F., Patrick, C. J., & Iacono, W. G. (2006). Continuity and change in psychopathic traits as measured via normal-range personality: A longitudinal-biometric study. *Journal of Abnormal Psychology, 115*, 85–95.

Bouchard, T. J., & Loehlin, J. C. (2001). Genes, evolution, and personality. *Behavior Genetics, 31*, 243–273.

Buckholtz, J. W., & Meyer-Lindenberg, A. (2008). MAOA and the neurogenetic architecture of human aggression. *Trends in Neurosciences, 31*, 120–129.

Caspi, A., McClay, J., Moffitt, T., Mill, J., Martin, J., Craig, I. W., et al. (2002). Role of genotype in the cycle of violence in maltreated children. *Science, 297*(5582), 851–854.

Caspi, A., Moffitt, T. E., Morgan, J., Rutter, M., Taylor, A., Arseneault, L., et al. (2004). Maternal expressed emotion predicts children's antisocial behavior problems: Using monozygotic-twin differences to identify environmental effects on behavioral development. *Developmental Psychology, 40*(2), 149–161.

Eastman, N., & Campbell, C. (2006). Neuroscience and legal determination of criminal responsibility. *Nature Reviews Neuroscience, 7*, 311–318.

Eley, T. C., Lichtenstein, P., & Moffitt, T. E. (2003). A longitudinal behavioral genetic analysis of the etiology of aggressive and nonaggressive antisocial behavior. *Development and Psychopathology, 15*, 383–402.

Fontaine, N. M. G., Rijsdijk, F. V., McCrory, E. J. P., & Viding, E. (2010). Etiology of different developmental trajectories of callous-unemotional traits. *Journal of the American Academy of Child and Adolescent Psychiatry, 49*(7), 656–664.

Forsman, M., Larsson, H., Andershed, H., & Lichtenstein, P. (2007). Persistent disruptive childhood behavior and psychopathic personality in adolescence: A twin study. *British Journal of Developmental Psychology, 25*, 383–398.

Forsman, M., Lichtenstein P., Andershed H., & Larsson H. (2008). Genetic effects explain the stability of psychopathic personality from mid- to late adolescence. *Journal of Abnormal Psychology, 117*, 606–617.

Forsman, M., Lichtenstein P., Andershed H., & Larsson H. (2010). A longitudinal twin study of the direction of effects between psychopathic personality and antisocial behaviour. *Journal of Child Psychology & Psychiatry, 51*(1), 39–47.

Fowler, T., Langley, K., Rice, F., van den Bree, M., Ross, K., Wilkinson, L. S., et al. (2009). Psychopathy trait scores in adolescents with childhood ADHD: The contribution of genotypes affecting *MAOA, 5HTT* and *COMT* activity. *Psychiatric Genetics, 19*(6), 312–319.

Frick, P. J., & Hare, R. D. (2001). *Antisocial process screening device.* Toronto, Ontario, Canada: Multi-Health Systems.

Frick, P. J., & Marsee, M. A. (2006). Psychopathy and developmental pathways to antisocial behavior in youth. In: C. J. Patrick (Ed.), *Handbook of psychopathy* (pp. 353–374). New York: Guilford.

Frick, P. J., & Viding, E. (2009). Antisocial behavior from a developmental psychopathology perspective. *Development and Psychopathology, 21,* 1111–1131.

Goldsmith, H., & Gottesman, I. I. (1996). Heritable variability and variable heritability in developmental psychopathology. In: M. F. Lenzenweger & J. J. Haugaard (Eds.), *Frontiers of developmental psychopathology* (pp. 5–43). New York: Oxford University Press.

Goodman, R. (1997). The Strengths and Difficulties Questionnaire: A research note. *Journal of Child Psychology & Psychiatry & Allied Disciplines, 38*(5), 581–586.

Hart, S. D., & Hare, R. D. (1997). Psychopathy: Assessment and association with criminal conduct. In: D. M. S. J. B. (Eds.), *Handbook of antisocial behavior.* New York: Wiley.

Hawes, D. J., & Dadds, M. R. (2005). The treatment of conduct problems in children with callous-unemotional traits. *Journal of Consulting & Clinical Psychology, 73*(4), 737–741.

Hipwell, A. E., Pardini, D. A., Loeber, R., Sembower, M., Keenan, K., & Stouthamer-Loeber, M. (2007). Callous-unemotional behaviors in young girls: shared and unique effects relative to conduct problems. *Journal of Clinical Child and Adolescent Psychology, 36,* 293–304.

Hoenicka, J., Ponce, G., Jimenez-Arriero, M. A., Ampuero, I., Rodriguez-Jimenez, R., Rubio, G., et al. (2006). Association in alcoholic patients between psychopathic traits and the additive effect of allelic forms of the *CNR1* and *FAAH* endocannabinoid genes, and the 3' region of the *DRD2* gene. *Neurotoxicity Research, 11,* 51–60.

Jaffee, S. R., Caspi A, Moffitt, T. E., Dodge, K. A., Rutter, M., Taylor, A., et al. (2005). Nature X nurture: Genetic vulnerabilities interact with physical maltreatment to promote conduct problems. *Development and Psychopathology, 17,* 67–84.

Kiehl, K. A., Smith, A. M., Hare, R. D., Mendrek, A., Forster, B. B., Brink, J., et al. (2001). Limbic abnormalities in affective processing by criminal psychopaths as revealed by functional magnetic resonance imaging. *Biological Psychiatry, 50*(9), 677–684.

Kim-Cohen, J., Caspi, A., Taylor, A., Williams, B., Newcombe, R., Craig, I. W., et al. (2006). MAOA, maltreatment, and gene-environment interaction predicting children's mental health: New evidence and a meta-analysis. *Molecular Psychiatry, 11*(10), 903–913.

Krueger, R. F., Hicks, B. M., Patrick, C. J., Carlson, S. R., Iacono, W. G., & McGue, M. (2002). Etiologic connections among substance dependence, antisocial behavior and personality: Modeling the externalizing spectrum. *Journal of Abnormal Psychology, 111*(3), 411–424.

Larsson, H., Andershed, H., & Lichtenstein, P. (2006). A genetic factor explains most of the variation in the psychopathic personality. *Journal of Abnormal Psychology, 115,* 221–230.

Larsson, H., Tuvblad, C., Rijsdijk, F., Andershed, H., Grann, M., & Lichtenstein, P. (2007). A common genetic factor explains the association between psychopathic personality and antisocial behavior. *Psychological Medicine, 37,* 15–26.

Larsson, H., Viding, E., & Plomin, R. (2008). Callous unemotional traits and antisocial behavior: Genetic, environmental, and early parenting characteristics. *Criminal Justice and Behavior, 35*(2), 197–211.

Larsson, H., Viding, E., Rijsdijk, F. V., & Plomin, R. (2008). Relationships between parental negativity and childhood antisocial behavior over time: A bidirectional effect model in a longitudinal genetically informative design. *Journal of Abnormal Child Psychology, 36*(5), 633–645.

Lesch, K. P. (2003). The serotonergic dimension of aggression and violence. In: M.P. Mattson (Ed.), *Neurobiology of aggression: Understanding and preventing violence* (pp. 33–63). Totowa, NJ: Humana Press.

Lichtenstein, P., Tuvblad, C., Larsson, H., & Carlstrom, E. (2007). The Swedish Twin study of CHild and Adolescent Development: the TCHAD-study. *Twin Research and Human Genetics, 10,* 67–73.

Lilienfeld, S. O., & Andrews, B. P. (1996). Development and preliminary validation of a self-report measure of psychopathic personality. *Journal of Personality Assessment, 66*, 488–524.

Lynam, D. R., Caspi, A., Moffitt, T. E., Loeber, R., & Stouthamer-Loeber, M. (2007). Longitudinal evidence that psychopathy scores in early adolescence predict adult psychopathy. *Journal of Abnormal Psychology, 116*, 155–165.

Manuck, S. B., Flory, J. D., Ferrell, R. E., Dent, K. M., Mann, J. J., & Muldoon, M. F. (1999). Aggression and anger-related traits associated with a polymorphism of the tryptophan hydroxylase gene. *Biological Psychiatry, 45*(5), 603–614.

Meyer-Lindenberg, A., Buckholtz, J. W., Kolachana, B., Hariri, A. R., Pezawas, L., Blasi, G., et al. (2006). Neural mechanisms of genetic risk for impulsivity and violence in humans. *Proceedings of the National Academy of Sciences of the USA, 103*(16), 6269–6274.

Meyer-Lindenberg, A., & Weinberger, D. (2006). Intermediate phenotypes and genetic mechanisms of psychiatric disorders. *Nature Review Neuroscience, 7*(10), 818–827.

Mobbs, D., Lau, H. C., Jones, O. D., & Frith, C. D. (2007). Law, responsibility, and the brain. *PLOS Biology, 5*(4), 693–700.

Moffitt, T. E. (2005). The new look of behavioral genetics in developmental psychopathology: Gene–environment interplay in antisocial behaviors. *Psychological Bulletin, 131*, 533–554.

Munafò, M. R., Brown, S. M., & Hariri, A. R. (2008). Serotonin transporter (*5-HTTLPR*) genotype and amygdala activation: A meta-analysis. *Biological Psychiatry, 63*, 852–857.

Plomin, R., DeFries, J., McClearn, G., & McGuffin, P. (2008). *Behavioral Genetics* (5th ed.). New York: Worth.

Ponce, G., Hoenicka, J., Jimenez-Arriero, M.A., Rodriguez-Jimenez, R., Aragues, M., Martin-Sune, N., Huertas, E., & Palomo, T., (2008). DRD2 and *ANKK1* genotype in alcohol-dependent patients with psychopathic traits: Association and interaction study. *British Journal of Psychiatry, 193*, 121–125.

Rhee, S. H., & Waldman, I. D. (2002). Genetic and environmental influences on antisocial behavior: A meta-analysis of twin and adoption studies. *Psychological Bulletin, 128*(3), 490–529.

Singh, I., & Rose, N. (2009). Biomarkers in psychiatry. *Nature, 460*, 202–207.

Smith, S. S., Newman, J. P., Evans, A., Pickens, R., Wydeven, J., Uhl, G. R., et al. (1993). Comorbid psychopathy is not associated with increased D-sub-2 dopamine receptor TaqI A or B gene marker frequencies in incarcerated substance abusers. *Biological Psychiatry, 33*, 845–848.

Taylor, J., Iacono, W. G., & McGue, M. (2000). Evidence for a genetic etiology of early-onset delinquency. *Journal of Abnormal Psychology, 109*(4), 634–643.

Taylor, J., Loney, B. R., Bobadilla, L., Iacono, W. G., & McGue, M. (2003). Genetic and environmental influences on psychopathy trait dimensions in a community sample of male twins. *Journal of Abnormal Child Psychology, 31*(6), 633–645.

Viding, E., Blair, R. J. R., Moffitt, T. E., & Plomin, R. (2005). Evidence for substantial genetic risk for psychopathy in 7-year-olds. *Journal of Child Psychology & Psychiatry, 46*, 592–597.

Viding, E., Fontaine, N. M.G., Oliver, B. R., & Plomin, R. (2009). Negative parental discipline, conduct problems and callous-unemotional traits: A monozygotic twin differences study. *The British Journal of Psychiatry, 195*, 414–419.

Viding, E., Frick, P. J., & Plomin, R. (2007). Aetiology of the relationship between callous-unemotional traits and conduct problems in childhood. *British Journal of Psychiatry, 49*, s33–s38.

Viding, E., Hanscombe, K., Curtis, C. J. C., Davis, O. S. P., Meaburn, E. L., & Plomin, R. (2010). In search of genes associated with risk for psychopathic tendencies in children: A two-stage genome-wide association study of pooled DNA. Manuscript submitted for publication.

Viding, E., Jones, A., Frick, P. J., Moffitt, T. E., & Plomin, R. (2008). Heritability of antisocial behaviour at nine-years: Do callous-unemotional traits matter? *Developmental Science, 11*(1), 17–22.

Viding, E., Hanscombe, K. B., Curtis, C. J. C., Davis, O. S. P., Meaburn, E. L., & Plomin, R. (2010). In search of genes associated with risk for psychopathic tendencies in children: A two-stage genome-wide association study of pooled DNA. *Journal of Child Psychology and Psychiatry, 51*(7), 780–788.

Waldman, I., & Rhee, S. H. (2006). Genetic and environmental influences on psychopathy and antisocial behavior. In: C. Patrick (Ed.), *Handbook of psychopathy* (pp. 205–228). New York: Guilford.

Wootton, J. M., Frick, P. J., Shelton, K. K., & Silverthorn, P. (1997). Ineffective parenting and childhood conduct problems: The moderating role of callous-unemotional traits. *Journal of Consulting & Clinical Psychology, 65*(2), 301–308.

CHAPTER 11

The Search for Genes and Environments that Underlie Psychopathy and Antisocial Behavior: Quantitative and Molecular Genetic Approaches

Irwin D. Waldman and Soo Hyun Rhee

Many research approaches have been employed to understand the etiology of psychopathy and antisocial behavior. Among these approaches, behavior genetic designs have the advantage of disentangling the effects of nature and nurture and characterizing the relative magnitudes of genetic and environmental influences. This is an important first step in explaining etiology, as it precedes the search for specific candidate genes and environmental risk factors. Although it is impossible to disentangle genetic from environmental influences in family studies because genetic and environmental influences are confounded in nuclear families, twin and adoption studies are uniquely poised to disentangle genetic and environmental influences and to estimate the magnitude of both simultaneously given the inclusion in these studies of individuals who vary in their genetic relatedness and environmental similarity. In this chapter, we summarize two recent meta-analyses that we conducted of twin and adoption studies on genetic and environmental influences on psychopathy and antisocial behavior (Rhee & Waldman, 2002; Waldman & Rhee, 2006). We then present the results of several more recent twin studies of psychopathy. We conclude with suggested future directions for research on the genetic and environmental influences underlying psychopathy and antisocial behavior, including the selection of relevant candidate genes and environmental risk factors as specific etiological mechanisms, issues in testing for gene–environment interactions, and the use of endophenotypes to help find genes for psychopathy and antisocial behavior and explain their underlying biopsychological mechanisms.

Despite more than 100 published twin and adoption studies of antisocial behavior, it is difficult to draw clear conclusions regarding the magnitude of genetic and environmental influences on antisocial behavior given the current literature. This is mainly due to the extraordinary heterogeneity of the results in this research domain, with published estimates of heritability (i.e., the magnitude of genetic influences) ranging from nil (e.g., .00; Plomin, Foch, & Rowe, 1981) to very high (e.g., .71; Slutske et al., 1997). Various sources of these heterogeneous results have been proposed, including differences in the age of the sample (e.g., Cloninger & Gottesman, 1987), the age at onset of antisocial behavior (e.g., Moffitt, 1993), and the measurement of antisocial behavior (e.g., Plomin, Nitz, & Rowe, 1990). Given these hypotheses, a major focus of our meta-analyses was to examine the impact of these sources of heterogeneity, in addition to characterizing the overall magnitude of genetic and environmental influences on antisocial behavior and psychopathy. Specifically, we tested for the moderating effects of how antisocial behavior was operationalized and assessed, and the age and sex of the participants, on the magnitude of genetic and environmental influences on antisocial behavior. Clarifying the effects of these moderating characteristics will increase our understanding of the etiology of antisocial behavior by shedding light on possible sex and developmental differences in the etiology of antisocial behavior, as well as how its etiology varies by what type of antisocial behavior is being studied and how it is being assessed. Applying a set of inclusion and exclusion criteria (described in detail in Rhee & Waldman, 2002 and Waldman & Rhee, 2006) led us to include 51 twin and adoption studies in a series of meta-analyses in order to characterize the magnitude of genetic and environmental influences on antisocial behavior and psychopathy and to test several moderators that may explain heterogeneity in these influences.

BIOMETRIC MODEL-FITTING ANALYSES IN QUANTITATIVE GENETIC STUDIES

In quantitative genetic studies, alternative models that include different sets of causal influences are analyzed and contrasted for their fit to the observed data (i.e., twin or familial correlations or covariances). These models posit that antisocial behavior is caused by different types of influences, which include additive genetic influences (A—genetic influences wherein alleles from different genetic loci are independent and "add up" to influence the liability for a trait), nonadditive genetic influences (D—genetic influences wherein alleles interact with each other to influence the liability for a trait, either at a single genetic locus or at difference loci), shared environmental influences (C—environmental influences that are experienced in common by family members that make them similar to one another), and nonshared environmental influences (E—environmental influences that are experienced uniquely by family members that make them different from one another). The magnitude of additive genetic influences, non-additive genetic influences, shared environmental influences, and nonshared environmental influences are symbolized by a^2, d^2, c^2, and e^2, respectively.

In the meta-analyses we included two types of adoption studies, specifically (1) parent–offspring studies contrasting the correlation between adoptees and their adoptive parents and the correlation between adoptees and their biological parents and (2) sibling adoption studies contrasting the correlation between adoptive siblings and the correlation between biological siblings, as well as two types of twin studies, specifically (1) twin pairs reared together and (2) twin pairs reared apart. The effect sizes from each study were included in separate groups in the model-fitting program Mx (Neale, 1995), in which the correlations between pairs of relatives are explained in terms of the components of variance that are shared between relatives (A, C, or D). Nonshared environmental influences, or E, do not explain any part of the correlation between relatives because (by definition) nonshared environmental influences are not shared between relatives. The correlation between different types of relatives is explained by different sets of influences and their appropriate weights which reflect the genetic or environmental similarity between pairs of relatives (for more detail see Appendix B of Rhee & Waldman, 2002).

We conducted the meta-analyses in a sequence of steps. First, the analyses were conducted using all appropriate data, and five alternative models (the ACDE model, the ACE model, the AE model, the CE model, and the ADE model) were contrasted. The fit of each model was assessed using the χ^2 statistic and the Akaike Information Criterion (AIC), a fit index that reflects both the fit of the model as well as its parsimony (Loehlin, 1992). Among competing models, that with the lowest AIC and the lowest χ^2 relative to its degrees of freedom is considered to be the best fitting model.

One important caveat to such biometric model-fitting analyses is that is not possible to estimate c^2 and d^2 simultaneously (i.e., to test an ACDE model) with data from twin pairs reared together only because the estimations of c^2 and d^2 both rely on the same information (i.e., the difference between the monozygotic [MZ] and dizygotic [DZ] twin correlations). If additional correlations that provide other types of information, such as the correlations between adoptees

Adoptee-Biological Parent		Adoptee-Adoptive Parent		Biological Siblings		Adoptive Siblings	
Stem	Leaf	Stem	Leaf	Stem	Leaf	Stem	Leaf
.5	3	.5		.5	2	.5	
.4	09	.4		.4	26	.4	
.3		.3		.3	1	.3	7
.2		.2		.2		.2	
.1	02467	.1		.1		.1	119
0	0	0	19	0		0	0
−.1		−.1	2	−.1		−.1	

Figure 11.1 Distribution of effect sizes—adoption studies.

and their adoptive and biological parents, also are included in the analyses, the simultaneous estimation of c^2 and d^2 and tests of the ACDE model are possible. It thus was possible to test the ACDE model only when all of the twin and adoption study data were included in the meta-analysis.

Second, we tested whether the genetic and environmental influences on antisocial behavior were moderated by its operationalization (i.e., diagnoses, criminality, aggression, and general antisocial behavior), assessment method (i.e., self-report, report by others, objective test, reaction to aggressive material, and records), sex (i.e., male and female), and age (i.e., children, adolescents, and adults) by contrasting the fit of a model in which the estimates of genetic and environmental influences are constrained to be equal across levels of each moderator to the fit of a model in which those estimates are free to vary across levels of the moderator. If the fit of the two models significantly differs, this indicates the importance of the moderator for explaining heterogeneity in the estimates of genetic and environmental influences on antisocial behavior. It was not possible to test moderators within the context of the ACDE model because both twin and adoption studies were not always available across different types of studies.

META-ANALYSES OF ALL DATA

The results of analyses of the data from all of the samples meeting the inclusion criteria ($N = 52$ samples, including 55,525 pairs of participants) are presented in Table 11.1 and Figures 11.1 and 11.2. As shown in the figures, correlations for antisocial behavior are higher among relative pairs that are more strongly genetically related than among those that are not. For example, correlations are stronger between adoptees and their biological parents than with their adoptive parents and between full siblings than between adoptive siblings. Analogously, correlations for antisocial behavior are stronger between identical twins than they are for fraternal twins. More formal statistical results are presented in Table 11.1 and show that the full ACDE model fit best as compared with the other, more reduced models. Excluding the studies that examined psychopathy (seven samples) did not alter the results, as estimates of genetic and environmental influences on antisocial behavior did not differ after we excluded these studies. We also present the meta-analytic results of psychopathy studies alone in the text that follows.

ASSESSMENT OF POTENTIAL MODERATORS

In Table 11.2 we present the results of tests of the moderators of the magnitude of genetic and environmental influences on antisocial behavior (i.e., operationalization, assessment method, sex, and age). The χ^2 difference between a model in which the estimates of genetic and environmental influences are constrained to be equal and a model in

	MZ Twin Pairs		DZ Twin Pairs	
Steam	Leaf	Steam	Leaf
.9	0	.9	
.8	011435	.8	0
.7	000244448889	.7	
.6	1223666778	.6	002224
.5	222457899	.5	0012236679
.4	23334568	.4	02224456677788899
.3	1235679	.3	001444678
.2	2499	.2	012235579
.1		.1	01222344566789
0	9	0	0
−.1		−.1	8
−.2		−.2	
−.3		−.3	4

Figure 11.2 Distribution of effect sizes—twin studies.

Table 11.1 Standardized Parameter Estimates and Fit Statistics—Inclusion of All Data

	Parameter estimates				Fit statistics			
	a^2	c^2	e^2	d^2	χ^2	df	p	AIC
ACDE model	.32	.16	.43	.09	1394.46	146	<.001	1,102.46
ACE model	.38	.18	.44	—	1420.38	147	<.001	1,126.38
ADE model	.41	—	.42	.17	1590.58	147	<.001	1,296.58
AE model	.55	—	.45	—	1707.89	148	<.001	1,411.89
CE model	—	.45	.55	—	2364.90	148	<.001	2,068.90

Data from Studies of Psychopathy Only

	Parameter estimates				Fit statistics			
	a^2	c^2	e^2	d^2	χ^2	df	p	AIC
ACE model	.48	.04	.48	.00	26.29	13	.02	0.29
ADE model	.50	.00	.48	.02	26.76	13	.01	0.76
AE model	.52	—	.48	—	26.78	14	.02	−1.22
CE model	—	.42	.58	—	81.36	14	<.01	53.36

which the estimates of genetic and environmental influences are free to vary across the different levels of the moderator is shown for each moderator.

Operationalization

The χ^2 difference test was significant for operationalization, $\Delta \chi^2 (9) = 339.87$, $p < .001$, indicating significant differences in the magnitude of genetic and environmental influences on antisocial behavior operationalized as diagnoses (14 samples; 11,681 pairs of participants), criminality (5 samples; 34,122 pairs of participants), aggression (14 samples; 4,408 pairs of participants), and general antisocial behavior

Table 11.2 Standardized Parameter Estimates and Fit Statistics for the Best Fitting Models—Test of Moderators

	Fit statistics			
	χ^2	df	P	AIC
Operationalization				
Parameters constrained to be equal	1406.50	139	<.001	1128.50
Parameters free to vary	1066.63	130	<.001	806.63
χ^2 difference test	339.87	9	<.001	321.87
Assessment method				
Parameters constrained to be equal	1361.73	139	<.001	1083.73
Parameters free to vary	530.47	128	<.001	274.47
χ^2 difference test	831.26	11	<.001	809.26
Age				
Parameters constrained to be equal	1351.30	133	<.001	1085.30
Parameters free to vary	1107.35	127	<.001	853.35
χ^2 difference test	243.95	6	<.001	231.95
Sex				
Parameters constrained to be equal	870.61	66	<.001	738.61
Parameters free to vary	869.07	63	<.001	743.07
χ^2 difference test	1.53	3	.68	−4.47

(15 samples; 4,365 pairs of participants). The ACE model was the best fitting model for diagnosis ($a^2 = .44$, $c^2 = .11$, $e^2 = .45$), aggression ($a^2 = .44$, $c^2 = .06$, $e^2 = .50$), and antisocial behavior ($a^2 = .47$, $c^2 = .22$, $e^2 = .31$), whereas the ADE model was the best fitting model for criminality ($a^2 = .33$, $d^2 = .42$, $e^2 = .25$). Within the operationalization of diagnosis, significant differences were found between studies examining ASPD (8 samples; 17 groups; 5,019 pairs of participants) and CD (5 samples; 22 groups; 6,560 pairs of participants). Although the magnitude of shared environmental influences (i.e., c^2) was similar, the heritability estimate (i.e., a^2) was higher in studies examining CD ($a^2 = .50$, $c^2 = .11$, $e^2 = .39$), whereas the magnitude of non-shared environmental influences (i.e., e^2) was higher in studies examining ASPD ($a^2 = .36$, $c^2 = .10$, $e^2 = .54$).

Assessment Method

The χ^2 difference test indicated that assessment method is a moderator of the magnitude of genetic and environmental influences on antisocial behavior, $\Delta\chi^2(11) = 831.26, p < .001$. We contrasted estimates of genetic and environmental influences on antisocial behavior for self-report (23 samples; 13,329 pairs of participants), report by others (14 samples; 6,851 pairs of participants), records (5 samples; 34,122 pairs of participants), reaction to stimuli (2 samples; 146 pairs of participants), and objective assessment (1 sample; 85 pairs of participants). The ACE model fit best for self-report ($a^2 = .39$, $c^2 = .06$, $e^2 = .55$) and report by others ($a^2 = .53$, $c^2 = .22$, $e^2 = .25$), whereas the AE model fit best for reaction to aggressive stimuli ($a^2 = .52$, $e^2 = .48$). All of the studies using the assessment method of records were also studies examining criminality, hence the ADE model was the best fitting model ($a^2 = .33$, $d^2 = .42$, $e^2 = .25$). Model-fitting could not be conducted for the assessment method of objective test because of lack of information (i.e., only one study).

Age

Age also was a significant moderator of the magnitude of genetic and environmental influences on antisocial behavior in children (15 samples; 7,807 pairs of participants), adolescents (11 samples; 2,868 pairs of participants), and adults

(17 samples; 27,671 pairs of participants), $\Delta \chi^2 (6) = 243.95, p < .001$. The ACE model fit best for children ($a^2 = .46, c^2 = .20, e^2 = .34$), adolescents ($a^2 = .43, c^2 = .16, e^2 = .41$), and adults ($a^2 = .41, c^2 = .09, e^2 = .50$), although the magnitude of familial influences (a^2 and c^2) decreased with age, whereas the magnitude of nonfamilial influences (e^2) increased with age.

Sex

Analyses of sex as a moderator of estimates of genetic and environmental influences on antisocial behavior were limited to studies that examined antisocial behavior in both males (17 samples; 5,610 pairs of participants) and females (17 samples; 7,225 pairs of participants). Studies examining antisocial behavior in *either* males or females were excluded, given the fact that studies examining antisocial behavior in only one sex varied a great deal in the operationalization of antisocial behavior (e.g., dishonorable discharge in males and aggression in females) and the assessment method (e.g., official records in males and parent report in females). The test of sex differences in the magnitude of genetic and environmental influences on antisocial behavior was not significant, $\Delta \chi^2 (3) = 1.53, p = .68$. The ACE model was the best fitting model for both males ($a^2 = .43, c^2 = .19, e^2 = .38$) and females ($a^2 = .41, c^2 = .20, e^2 = .39$), such that the magnitude of genetic and environmental influences on antisocial behavior was virtually identical across sex. This result is consistent with those of traditional literature reviews (e.g., Widom & Ames, 1988) that concluded that the magnitude of genetic and environmental influences on antisocial behavior in males and females is similar and inconsistent with those of Miles and Carey (1997), who found higher heritability estimates for aggression in males.

ASSESSMENT OF CONFOUNDING AMONG MODERATORS

When there is confounding among moderators being tested in a meta-analysis, one concern is whether a particular variable appears to be a significant moderator only because it is confounded with another moderating variable. The possibility of confounding was assessed between the following pairs of moderators: operationalization and assessment method, age and operationalization, and age and assessment method. All analyses showed that each moderator is significant even after the effects of the possible confounding moderator are controlled statistically. For example, the model estimating separate parameter estimates for each level of operationalization and each level of assessment method fit significantly better than the model estimating separate estimates for each level of operationalization only, $\Delta \chi^2 (13) = 633.67, p < .001$, and the model estimating separate estimates for each level of assessment method only, $\Delta \chi^2 (12) = 112.56, p < .001$. Similarly, assessment method was a significant moderator after controlling for age, $\Delta \chi^2 (7) = 676.28, p < .001$; operationalization was a significant moderator after controlling for age, $\Delta \chi^2 (18) = 410.52, p < .001$; and age was a significant moderator after controlling for operationalization, $\Delta \chi^2 (15) = 335.44, p < .001$; and after controlling for assessment method, $\Delta \chi^2 (7) = 102.73, p < .001$. Thus, each moderator was found to be significant even after the other potentially confounding moderator was controlled for statistically.

META-ANALYSES OF BEHAVIOR GENETIC STUDIES OF PSYCHOPATHY

Subsequent to the meta-analysis of twin and adoption studies of antisocial behavior and the tests of moderators of the genetic and environmental influences thereof, we conducted a meta-analysis of studies of psychopathy (Waldman & Rhee, 2006). Although deletion of the data from these seven samples did not alter the results of the main meta-analysis, meta-analytic results for only these seven samples may differ. The seven studies included five samples of reared-together twins, one sample of reared-apart twins, and one adoption sample that provided correlations between adoptees and their biological parents and all examined psychopathy via self-report. It was not possible to estimate the full ACDE model given the small number of samples and the presence of only one parent–offspring adoption sample. As shown at the bottom of Table 11.1, the estimates of c^2 and d^2 were negligible and the ACE and ADE models fit no better than the AE model, which fit best. Additive genetic influences accounted for 52% of the variance and nonshared environmental influences accounted for the remaining 48% of the variance in self-reports of psychopathy. Additive genetic influences are clearly more important than shared environmental influences in explaining the etiology and familiality of psychopathy, as the CE model (which omits genetic influences) did not fit the data well. Inclusion of a twin study (Taylor et al., 2000) that assessed psychopathy using self-reports on the California Psychological Inventory (CPI) Socialization scale yielded identical results. Thus, the etiology of psychopathy appears to be highly similar to that of antisocial behavior in general, with the exception that there is no evidence for shared environmental influences and the magnitude of genetic influences is slightly higher. Given the small number of studies included in the meta-analysis, more quantitative genetic studies of psychopathy are needed to achieve more reliable estimates of the magnitude of genetic and environmental influences on psychopathy. Fortunately, several additional quantitative genetic studies of psychopathy have appeared over the past six years.

RESULTS OF MORE RECENT BEHAVIOR GENETIC STUDIES OF PSYCHOPATHIC TRAITS

Since we conducted our meta-analyses, several twin studies of psychopathic traits have been published and have yielded similar findings. In the first, Taylor et al. (2003) estimated genetic and environmental influences on the Antisocial and Detachment subscales of the Minnesota Temperament Inventory in two 16- to 18-year-old twin cohorts from the Minnesota Twin Family Study (cohort 1: 142 MZ and 70 DZ twin pairs; cohort 2: 128 MZ and 58 DZ twin pairs). It was possible to equate the estimates of genetic and environmental influences across cohorts and to fit the models for both scales in both cohorts simultaneously. The AE model was the best fitting model for both scales, with additive genetic influences accounting for 39% and 42% of the variance, and nonshared environmental influences accounting for the remaining 61% and 58% of the variance, in the Antisocial and Detachment scales, respectively. There was no evidence for shared environmental influences on either of the scales. Additive genetic influences accounted for 53% and nonshared environmental influences accounted for the remaining 47% of the covariation between the two psychopathy-related traits. Expressed another way, approximately 55% of the genetic influences and 79% of the nonshared environmental influences on Detachment were shared in common with those that influenced the Antisocial scale. This suggests that the majority of the nonshared environmental influences on Detachment are the same as those on Antisociality, whereas just greater than half of the genetic influences on Detachment are the same as those on Antisociality, implying that each psychopathy trait shows substantial unique genetic influences.

A second recent study (Blonigen et al., 2003) used an adult male twin sample from Minnesota (165 MZ and 106 DZ twin pairs) to estimate genetic and environmental influences on eight subscales and the total score from the Psychopathic Personality Inventory (PPI; Lilienfeld & Andrews, 1996). These subscales include Machiavellian Egocentricity, Social Potency, Fearlessness, Coldheartedness, Impulsive Nonconformity, Blame Externalization, Carefree Nonplanfulness, and Stress Immunity. For each of the subscales, the best fitting etiological model included genetic and nonshared environmental influences, with no evidence for shared environmental influences. The genetic influences accounted for 29% to 56% of the variance and were nonadditive for all of the subscales except for Social Potency and Blame Externalization, for which the genetic influences were additive. In two follow-up publications, Blonigen et al. (2005, 2006) estimated Fearless Dominance (F-D) and Impulsive-Antisocial (I-A) dimension scores by factor analyzing the Multidimensional Personality Questionnaire (MPQ) scales and conducted biometric analyses to estimate the genetic and environmental influences underlying these two dimensions. The authors found that both dimensions were similarly heritable ($a^2 = .45–.51$) in males and females, as well as at both age 17 and age 24 (F-D: $a^2 = .48$ and .42, and I-A: $a^2 = .46$ and .49, respectively) and that the remaining variance was due to nonshared environmental influences with no evidence for shared environmental influences. Common genetic influences predominantly contributed to the stability of both dimensions, as 58% of the stability in F-D and 62% of the stability in I-A was due to genetic influences.

In a third recent study in a sample of 3,487 7-year-old U.K. twin pairs, Viding et al. (2005, 2007) estimated genetic and environmental influences on extreme levels of antisocial behavior and callous-unemotional traits, as well as the extent to which genetic and environmental influences on antisocial behavior vary by callous-unemotional trait levels. The etiology of antisocial behavior included additive genetic and nonshared environmental influences, with minimal and nonsignificant shared environmental influences, and was highly similar for boys and girls ($a^2 = .61$ and .57, and $e^2 = .34$ and .35, respectively). In contrast, the etiology of callous-unemotional traits differed substantially by sex, such that in boys the etiology included additive genetic and nonshared environmental influences, again with minimal shared environmental influences ($a^2 = .67$ and $e^2 = .29$), whereas heritability was lower and shared environmental influences were more substantial for girls ($a^2 = .48$, $c^2 = .20$, and $e^2 = .32$). In the analyses of moderation, the heritability of antisocial behavior varied by callous-unemotional group status such that antisocial behavior was highly heritable ($h_g^2 = .81$) with no evidence for shared environmental influences for the high callous-unemotional group, whereas antisocial behavior was only modestly heritable ($h_g^2 = .30$), with similar levels of shared environmental influences ($c_g^2 = .34$) for the low callous-unemotional group. Similar results were found in a 2-year follow-up when the twins were 9 years old (Viding et al., 2008). These results suggest that callous-unemotional traits are moderately to highly heritable and may identify a highly heritable form of childhood antisocial behavior.

A fourth recent study was conducted by Larsson et al. (2006) in 1,063 Swedish adolescent twin pairs assessed using the Youth Psychopathy Inventory (YPI; Andershed et al., 2002) in which the researchers took a multidimensional approach to psychopathic traits by characterizing levels on the Grandiose-Manipulative (G-M), Callous-Unemotional (C-U), and Impulsive-Irresponsible (I-I) subscales. The researchers examined the extent to which these three subscales reflected a general overarching trait of psychopathy and estimated the magnitude of genetic and environmental influences on the higher-order trait of psychopathy—reflecting what all three subscales share in common—as well as on each of the three subscales uniquely. Larsson et al. (2006) found that the general psychopathy trait was most strongly indicated by G-M (factor loading = .75), followed by I-I (.49) and C-U (.33), and that the etiology of the overarching psychopathy trait included additive genetic and nonshared environmental influences, with no evidence for shared environmental influences ($a^2 = .63$ and $e^2 = .37$). For G-M, almost all of its additive genetic influences underlie general psychopathy rather than acting on G-M directly ($a^2 = .48$ vs. $a^2 = .01$), whereas approximately 60% of its nonshared environmental influences underlie general psychopathy and approximately 40% acted directly ($e^2 = .27$ vs. $e^2 = .17$). Shared environmental influences acted on G-M directly but were minimal and nonsignificant ($c^2 = .06$). For C-U, half of its additive genetic influences underlie general psychopathy and half acted on C-U directly ($a^2 = .21$ vs. $a^2 = .22$), whereas approximately 20% of its nonshared environmental influences underlie general psychopathy and approximately 80% acted on C-U directly ($e^2 = .12$ vs. $e^2 = .45$). There was no evidence for shared environmental influences on C-U. Finally, for I-I, approximately 60% of its additive genetic influences underlie general psychopathy and approximately 40% acted directly ($a^2 = .31$ vs. $a^2 = .22$), whereas approximately 40% of its nonshared environmental influences underlie general psychopathy and approximately 60% acted on I-I directly ($e^2 = .18$ vs. $e^2 = .28$). Similar to C-U and general psychopathy, there was no evidence for shared environmental influences on

I-I. In a 3-year follow-up of this sample, Forsman et al. (2008) showed that almost all of the stability of the overarching psychopathy trait was due to common additive genetic influences at both ages (i.e., 16 and 19 years old). Analogously, almost all of the stability of G-M was due to the common additive genetic influences on general psychopathy at both ages, whereas the stability of C-U and I-I also were due to additive genetic influences that acted directly on each subscale.

FUTURE DIRECTIONS

Candidate Genes and Environments for Antisocial Behavior and Psychopathy

Given the moderate heritability of psychopathy and antisocial behavior, researchers have begun to search for the specific genes that contribute to their etiology. There are two general strategies for finding genes that contribute to the etiology of a disorder or trait. The first is a genome scan, in which association is tested between a disorder or trait and DNA markers densely distributed across the genome. Thus, genome scans are exploratory searches for genes that contribute to the etiology of a disorder or trait. The fact that genes have been found using genome scans for many medical diseases is testament to the usefulness of this method. Unfortunately, the statistical power of association tests in genome scans is fairly low in single samples of typical moderate size, making it difficult to detect genes that account for less than approximately 5% of the variance in a disorder. Given this, the promise for genome scans of complex psychiatric disorders and psychological traits will not be actualized until large-scale multisite studies that include tens of thousands of participants are conducted. Few genome scans of antisocial behavior have been conducted (Dick et al., 2004; Stallings et al., 2005), with a dearth of replicated findings thus far.

The second strategy for finding genes that underlie a disorder or trait is the candidate gene approach, which in many ways is the polar opposite of genome scans. In contrast to the exploratory nature of genome scans, well-conducted candidate gene studies represent a targeted test of the role of specific genes in the etiology of a disorder or trait as the location, function, and etiological relevance of candidate genes is most often known or strongly hypothesized a priori. An advantage of well-conducted candidate gene studies in comparison with genome scans is that positive findings are easily interpretable because one already knows the gene's location, function, and etiological relevance, even if the specific markers in the candidate gene chosen for study are not functional and the functional markers in the candidate gene have yet to be identified. The candidate gene approach also has disadvantages, as only previously identified genes can be examined. One therefore cannot find genes one has not looked for or that have yet to be discovered, and given that there are relatively few strong candidate genes for psychiatric disorders, the same genes are investigated for almost all psychiatric disorders or psychological traits, regardless of how disparate the disorders may be in terms of their symptoms or underlying pathophysiology. In well-designed candidate gene studies, however, knowledge regarding the biology of the disorder is used to select genes based on the known or hypothesized involvement of their gene product in the etiology of the trait or disorder (i.e., its pathophysiological function and etiological relevance).

Candidate genes for neurotransmitter systems may include: (1) *precursor genes* that affect the rate at which neurotransmitters are produced from precursor amino acids (e.g., tyrosine hydroxylase for dopamine, tryptophan hydroxylase for serotonin); (2) *receptor genes* that are involved in receiving neurotransmitter signals (e.g., genes corresponding to the five dopamine receptors, *DRD1*, *D2*, *D3*, *D4*, and *D5*, and to serotonin receptors such as *HTR1 β* and *HTR2A*); (3) *transporter genes* that are involved in the reuptake of neurotransmitters back into the presynaptic terminal (e.g., the dopamine and serotonin transporter genes, *DAT1* and *5HTT*); (4) *deactivation genes* that are involved in the metabolism or degradation of these neurotransmitters (e.g., the genes for catechol-O-methyl-transferase, *COMT*, and for monoamine oxidase A and B [i.e., *MAOA* and *MAOB*]); and (5) genes that are responsible for the *conversion* of one neurotransmitter into another (e.g., dopamine beta

-hydroxylase, or *DβH*, which converts dopamine into norepinephrine).

With regard to antisocial behavior and psychopathy, genes underlying various aspects of the dopaminergic, serotonergic, and noradrenergic neurotransmitter pathways and the oxytocin and vasopressin neuropeptide systems may be conjectured based on several lines of convergent evidence that suggests a role for these neurotransmitter and neuropeptide systems in the etiology and pathophysiology of these traits and disorders. For example, given the considerable overlap between antisocial behavior and childhood attention-deficit/hyperactivity disorder (ADHD) (e.g., Lilienfeld & Waldman, 1990), candidate genes for ADHD also are relevant candidates for antisocial behavior and psychopathy. Several genes within the dopamine system have been found to be risk factors for ADHD (see Gizer, Ficks, & Waldman, 2009; Waldman & Gizer, 2006, for recent reviews). Given that the stimulant medications that are the most frequent and effective treatments for ADHD act primarily by regulating dopamine levels in the brain (Seeman & Madras, 1998; Solanto, 1984) as well as by affecting noradrenergic and serotonergic function (Solanto, 1998), dopamine genes are plausible candidates for ADHD. "Knockout" gene studies in mice, in which the effects of deactivating specific genes are examined, have further demonstrated the relevance of genes within these neurotransmitter systems. Such studies have strengthened the status as candidate genes for ADHD and aggression of genes within the dopaminergic system, including catechol-O-methyltransferase (*COMT*; Gogos et al., 1998), the dopamine transporter gene (*DAT1*; Giros, Jaber, Jones, Wightman, & Caron, 1996), and the dopamine receptor D3 and D4 genes (*DRD3* and *DRD4*; Accili et al., 1996; Dulawa, Grandy, Low, Paulus, & Geyer, 1999; Rubinstein et al., 1997).

Evaluations of these candidate genes in humans have yielded results that are quite mixed. Two studies of a commonly studied variable number of tandem repeats (VNTR) in the dopamine transporter gene (*DAT1*) in community samples produced contradictory results, with one showing no association with aggression or ODD or CD symptoms (Jorm et al., 2001) and the other showing association of the 9-repeat allele with externalizing problems (Young et al., 2002). Very few association studies of aggression and antisocial behavior with dopaminergic or noradrenergic genes have been conducted in clinic-referred samples. In a recent study in a sample of children referred for ADHD, *DAT1* but not *DRD4* was associated with both oppositional defiant disorder (ODD) and conduct disorder (CD) symptoms (Waldman et al., in preparation). In addition, Holmes et al. (2002) found that *DRD4* was related to ADHD only when it was accompanied by CD. Finally, a commonly studied functional single nucleotide polymorphism (SNP) in *COMT* was associated with antisocial behavior in a sample of children referred for ADHD as well as two nonreferred community samples, and this association was particularly strong in children of low birth weight (Caspi et al., 2008; Thapar et al., 2005). We hypothesize that dopaminergic genes found to be associated with ADHD in humans or to underlie activity and impulsivity in knockout (KO) mice will be associated more strongly with the impulsive rather than the callous-unemotional psychopathic traits, and with reactive rather than with proactive aggression, whereas genes related to aggression in the absence of impulsivity in mice will be more associated with proactive aggression and psychopathic traits.

Serotonergic genes also are plausible candidates for antisocial behavior and psychopathy given the demonstrated relations between serotonergic function and aggression and violence (Berman, Kavoussi, & Coccaro, 1997; Kruesi et al., 1990). Several researchers have tested associations of aggression and antisocial behavior with candidate genes in the serotonergic system, including the serotonin transporter gene (*5HTT*), the serotonin 1β receptor gene (*HTR1β*), the tryptophan hydroxylase genes (*TPH1* and *TPH2*), and the monoamine oxidase A gene (*MAOA*). Despite the effects on aggression in KO mice (for *5HTT*, *HTR1β*, and *MAOA*; Cases et al., 1995; Holmes et al., 2003; Saudou et al., 1994) and the examination of functional polymorphisms in these genes (for *5HTT* and *MAOA*), findings of association between markers in these genes and aggression and antisocial behavior phenotypes in humans also are quite mixed and no clear results have emerged (for *5HTT*: Baca-Garcia et al., 2004; Cadoret et al., 2003; Davidge et al., 2004; Retz et al., 2004; for *HTR1β*: New et al., 2001; for *TPH1*: Hennig et al., 2005; Manuck et al., 1999; New et al., 1998; Staner et al., 2002; for *MAOA*: Caspi et al., 2002; Foley et al., 2004; Haberstick et al., 2005; Huang et al., 2004; Kim-Cohen et al., 2006; Manuck et al., 2000; Nilsson et al., 2005; Young et al., 2006). Similar to the aforementioned dopaminergic genes, we hypothesize that serotonergic genes that affect impulsivity in KO mice will be associated more strongly with the impulsive rather than the callous-unemotional psychopathic traits and with reactive rather than with proactive aggression. In contrast, genes related to aggression in the absence of impulsivity in mice should be more associated with proactive aggression and psychopathic traits.

Genes that underlie the neuropeptides oxytocin and vasopressin also are viable candidates for aggression and antisocial behavior given these genes' role in underlying aggression in knockout mice (for oxytocin and its receptor [*OXT*, DeVries et al., 1997; *OXTR*, Takayanagi et al., 2005], for the vasopressin receptor 1b gene [*AVPR1b*, Wersinger et al., 2002]). In addition to these molecular genetic studies, the role of vasopressin in aggression is suggested by a cross-fostering study in mice (Bester-Meredith & Marler, 2001) and by pharmacological challenge studies in hamsters (Ferris et al., 2006) and humans (Coccaro et al., 1998; Thompson et al., 2004). These studies also have suggested important interactions between these neuropeptides and serotonergic function (Coccaro et al., 1998), suggesting the possibility of interactions between genes in these systems. We hypothesize that oxytocin and vasopressin system genes will be associated more strongly with aggression in males, but have no predictions as to which type of aggression or which features of psychopathic traits or other aspects of antisocial behavior these genes will be associated with.

Many environmental variables have been proposed as risk factors for antisocial behavior and psychopathy. The domains of these environmental risk factors include pre- and perinatal influences, such as obstetrical complications and maternal smoking and drinking; aspects of parenting such as supervision and monitoring, warmth, control, and harsh discipline; facets of family background such as family poverty, size, disruption, and marital and family status (married vs. divorced, single vs. dual parent); sibling and peer influences such as aggression and antisocial behavior, substance use/abuse, academic achievement and aspirations; and neighborhood characteristics such as economic inequality, neighborhood cohesion, crime rates, and collective efficacy. Unfortunately, much of the literature on the relation between environmental variables and antisocial behavior and psychopathy is virtually uninterpretable because of confounding between the environmental and genetic influences that underlie such relations. For example, the relation between harsh parental discipline and children's antisocial behavior could be due to a direct environmental influence; to common background environmental influences such as socioeconomic status; or to shared genetic influences, in which the same genes that underlie parents' tendencies to discipline harshly also underlie their children's antisocial behavior. The aforementioned candidate environmental influences on antisocial behavior can best be considered putative given these confounds pending their examination in genetically informative designs to validate their effects as truly environmental.

Several authors have recently proposed genetically informative designs and analyses that are able to discriminate among these causal possibilities (e.g., D'Onofrio et al., 2003), and studies have begun to examine the relations of putative environmental influences with antisocial behavior using such designs. For example, the association between children's prenatal alcohol exposure and their later conduct problems appears to be direct and independent of confounded background genetic and environmental influences (D'Onofrio et al., 2007), whereas a related study suggested that the relation between children's prenatal smoking exposure and their later conduct problems is not direct but rather is due to confounded background environmental influences (D'Onofrio et al., 2008).

Researchers also have begun to test the hypothesis that candidate genes and candidate environments do not act in a simple additive fashion but rather interact as risk factors for antisocial behavior and psychopathy. The first contemporary study of gene–environment interaction showed that the risk for adolescent antisocial behavior based on early childhood abuse varied by genotypes of the *MAOA* promoter polymorphism (Caspi et al., 2002). It is important to realize that such gene–environment interactions are symmetrical, i.e., that gene–environment interaction refers both to the moderation of the effects of environmental risk factors as a function of individuals' genotypes for a particular gene, as well as to the moderation of individuals' genetic predispositions for a disorder as a function of aspects of the environment that they have experienced. Given the relative ease of testing whether the effects of candidate genes, such as those mentioned previously, vary as a function of environmental risk factors for antisocial behavior and psychopathy, we expect many more studies of gene–environment interaction for antisocial behavior and psychopathy to emerge in the near future.

Although the interaction between early childhood abuse and *MAOA* promoter genotypes predicting adolescent antisocial behavior has been replicated across studies (Kim-Cohen et al., 2006, 2007), other high-profile gene–environment interactions such as the moderation of the relation between stressful life events and later depression by genotypes of the serotonin transporter gene promoter polymorphism (*5HTTLPR*; Caspi et al., 2003) have failed to replicate (Munafò et al., 2009; Risch et al., 2009). Thus, despite the considerable interest in, and surfeit of research on, gene–environment interactions in psychopathology, several critical issues remain unsolved and are potentially problematic in such studies. First, any given environmental measure is correlated with any number of potentially relevant environmental influences. Is the putative, measured environmental variable the "true" environmental risk factor, or is it merely a proxy for or correlate of the "true" environmental influence on the trait or disorder? This raises further questions of whether different environmental variables are substitutable for one another, act in combination, or are distinguishable in their effects on the disorder or trait of interest. A second problem is that most environmental measures have been found to be heritable in quantitative genetic studies. Thus, each environmental measure has its own etiology, part of which may include genetic influences. If the environmental measure is heritable, it is possible that it is not the environmental component of this variable that the candidate gene is interacting with, but rather (1) the same gene in the parents or (2) some other gene or genes that are influencing the environmental measure. If the measured environmental variable is heritable, the very same candidate gene may influence the target disorder in the children and the environmental measure in their parents. Measured environmental variables (e.g., maternal smoking during pregnancy) may only be proxies for other related environmental variables (e.g., mother's life stress) that represent the true environmental risk factors. Finally, these putative environmental influences actually may reflect underlying genetic influences that have effects on both the environmental variable and the disorder or trait. For example, mothers with genetic predispositions for impulsivity may be more likely to smoke during pregnancy and more likely to transmit the genes predisposing to impulsivity to their children, leading them to be more likely to manifest antisocial behavior and psychopathy. Future studies of gene–environment interactions for antisocial behavior and psychopathy should employ more rigorous research designs to better ensure that relations with environmental measures are not merely reflections of other correlated environmental variables or background common genetic influences.

Endophenotypes for Antisocial Behavior and Psychopathy

There is a considerable gap between candidate genes and the symptoms of disorders or traits such as antisocial behavior and psychopathy that they influence. It makes sense conceptually and empirically to find valid and useful intervening constructs that bridge this gap. Endophenotypes are such constructs and were first applied to psychiatric disorders by Gottesman and Shields with respect to the genetics of schizophrenia (Gottesman & Shields, 1972) as "internal phenotypes discoverable by a biochemical test or microscopic examination" (p. 637, Gottesman & Gould, 2003; Gottesman & Shields, 1972). More generally, endophenotypes are constructs posited to underlie psychiatric disorders and related traits and to be influenced more directly by the risk-inducing genes than are the overt symptoms of disorder. Endophenotypes are thus hypothesized to more closely reflect the immediate gene products (i.e., the proteins they code for) and are thought to be more strongly influenced than the manifest symptoms by the genes that underlie the disorder. The underlying structure of genetic influences on endophenotypes is thought to be simpler than that of complex disorders and traits in the sense that fewer individual genes should contribute to their etiology (Gottesman & Gould, 2003).

Researchers have outlined criteria for evaluating the validity and utility of putative endophenotypes (e.g., Castellanos & Tannock, 2002; Doyle et al., 2005; Gottesman & Gould, 2003; Waldman, 2005). These include: (1) The endophenotype is related to the disorder in the general population. (2) The endophenotype is heritable. (3) The endophenotype is expressed regardless of whether the disorder is present. (4) The endophenotype and disorder are associated within families (i.e., they "co-segregate"). (5) Just as endophenotypes are hypothesized to occur more frequently in diagnosed than in unaffected family members, they also should occur more frequently in the unaffected relatives of diagnosed family members than in randomly selected individuals from the general population (given that the endophenotype reflects the inherited liability to a disorder).

Several criteria are relevant to assessing the validity and utility of endophenotypes in addition to the aforementioned criteria. First, the genetic influences that underlie the endophenotype also must underlie the disorder or related trait, and at least some of the genetic influences that underlie the disorder or related trait must underlie the endophenotype. This last criterion is asymmetrical, in that a greater proportion of the genetic influences on the endophenotype will be shared with those on the disorder or related trait, rather than vice versa. This follows from the aforementioned hypothesis that endophenotypes are thought to be genetically simpler than are complex traits such as disorders (i.e., fewer genes are posited to contribute a greater magnitude to their etiology) (Gottesman & Gould, 2003). Second, endophenotypic measures must show association with one (or more) of the candidate genes that underlie the disorder or related trait. Third, the endophenotype should *mediate* association between the candidate gene and the disorder or related trait, meaning that the effects of a particular gene on a disorder or trait are expressed—either in full or in part—through the endophenotype. The prerequisites for this causal scenario are that the candidate gene influences both the disorder or trait and the endophenotype and that the endophenotype in turn influences the disorder or related trait. Fourth, the endophenotype should show association with a candidate gene over and above the gene's relation with the disorder or related trait (i.e., the endophenotype should incrementally contribute to association with the candidate gene), and thus aid in the search for genes underlying the etiology of disorders or related traits.

Several biological, psychophysiological, and psychological mechanisms are plausible as candidate endophenotypes for antisocial behavior and psychopathy. Putative biological endophenotypes may include serotonin and dopamine levels, given their aforementioned relations to antisocial behavior and psychopathy (Berman, Kavoussi, & Coccaro, 1997) and related disorders such as ADHD. Putative psychophysiological endophenotypes may include avoidance conditioning (Lykken, 1957) and startle probe response (Patrick, 1994), given findings of deficits in such variables in psychopathic relative to nonpsychopathic individuals. Given their demonstrated relations to aggression, antisocial behavior, or psychopathy, putative psychological endophenotypes may include hostile perceptual and attributional biases (Waldman, 1996), executive function deficits (Morgan & Lilienfeld, 2000), and deficits in the perception of facial displays of fear and perhaps sadness (Blair, 2006).

With regard to the latter, individuals high in psychopathic traits have been shown to be less accurate in their perceptions of others' fear and—in some studies—sadness (Blair, 2006), perhaps due to spending much less time looking at the eyes of the sender when they are expressing these emotions (Dadds et al., 2008). Future studies examining the extent to which these variables meet the criteria outlined earlier are necessary to evaluate the validity and utility of these putative endophenotypes for antisocial behavior and psychopathy, as well as whether finding genes for these psychological mechanisms will aid considerably in the search for genes that underlie aggression, antisocial behavior, or psychopathy.

CONCLUSIONS

In a recent meta-analysis, we found that there were moderate additive genetic ($a^2 = .32$), non-additive genetic ($d^2 = .09$), shared environmental ($c^2 = .16$), and nonshared environmental influences ($e^2 = .43$) on antisocial behavior (Rhee & Waldman, 2002). Of the potential moderators examined (i.e., operationalization, assessment method, zygosity determination method, and age) all but sex were found to account for some of the heterogeneity in genetic and environmental influences on antisocial behavior. A meta-analysis of behavior genetic studies of psychopathy, as well as recent twin studies thereof, suggested that genetic influences on psychopathic traits are appreciable and are approximately the same magnitude as those on antisocial behavior in general. Moderate nonshared but no shared environmental influences on psychopathic traits were found across studies. These studies also raise the possibilities that much of the genetic influences on the development of antisocial behavior are common to or mediated via psychopathic personality traits, and that there are substantial common genetic influences on psychopathic traits, antisocial behavior, and "normal range" personality traits such as negative and positive emotionality and daring or constraint (Benning et al., 2005; Lahey & Waldman, 2003; Waldman et al., 2011).

Future directions for behavioral and molecular genetic studies of antisocial behavior, aggression, and psychopathy include multivariate analyses testing for common versus specific associations with relevant candidate genes and environmental influences, rigorous tests of gene–environment interactions, and studies examining the validity of putative endophenotypes and their utility for finding genes that underlie antisocial behavior, aggression, and psychopathy.

ACKNOWLEDGMENTS

This work was supported in part by NIDA DA-13956 and NIMH MH-01818. We thank the authors who made data from unpublished studies available through personal communication, and Deborah Finkel, Jenae Neiderhiser, Wendy Slutske, and Edwin van den Oord for making the data from their studies available before their publication. Earlier versions of this chapter were presented at the meeting of the American Society of Criminology in 1996 and the meeting of the Behavior Genetics Association in 1997. Some of the material herein is similar to that published in Rhee and Waldman (2002) and Waldman and Rhee (2006).

REFERENCES

Accili, D., Fishburn, C. S., Drago, J., Steiner, H., Lachowicz, J. E., Park, B. H., et al. (1996). A targeted mutation of the D3 dopamine receptor gene is associated with hyperactivity in mice. *Proceedings of the National Academy of Sciences of the USA, 93,* 1945–1949.

Achenbach, T. M., & Edelbrock, C. S. (1983). *Manual for the Child Behavior Checklist and revised behavior profile.* Burlington: University of Vermont.

Andershed, H., Kerr, M., Stattin, H., & Levander, S. (2002). Psychopathic traits in non-referred youths: Initial test of a new assessment tool. In: E. Blaauw, & L. Sheridan (Eds.), *Psychopaths: Current international perspectives* (pp. 131–158). The Hague: Elsevier.

Baca-Garcia, E., Vaquero, C., Diaz-Sastre, C., García-Resa,E.,Saiz-Ruiz,J.,Fernández-Piqueras, J., et al. (2004). Lack of association between the serotonin transporter promoter gene polymorphism and impulsivity or aggressive behavior among suicide attempters and healthy volunteers. *Psychiatry Research, 126,* 99–106.

Benning, S. D., Patrick, C. J., Blonigen, D. M., Hicks, B. M., & Iacono, W. G. (2005). Estimating facets of psychopathy from normal personality traits: A step toward community-epidemiological investigations. *Assessment, 12*, 3–18.

Berman, M. E., Kavoussi, R. J., & Coccaro, E. F. (1997). Neurotransmitter correlates of human aggression. In: D. M. Stoff, J. Breiling, & J. D. Masur (Eds.), *Handbook of antisocial behavior* (pp. 305–313). New York: Wiley.

Bester-Meredith, J. K., & Marler, C. A. (2001). Vaspressin and aggression in cross-fostered California mice (*Peromyscus californicus*) and white-footed mice (*Peromyscus leucopus*). *Hormones and Behavior, 40*, 51–64.

Blair, R. J. R. (2006). The emergence of psychopathy: Implications for the neuropsychological approach to developmental disorders. *Cognition, 101*, 414–442.

Blonigen, D. M., Carlson, S. R., Krueger, R. F., & Patrick, C. J. (2003). A twin study of self-reported psychopathic personality traits. *Personality and Individual Differences, 35*, 179–197.

Blonigen, D. M., Hicks, B. M., Krueger, R. F., Patrick, C. J., & Iacono, W. G. (2005). Psychopathic personality traits: Heritability and genetic overlap with internalizing and externalizing psychopathology. *Psychological Medicine, 35*, 637–648.

Blonigen, D. M., Hicks, B. M., Krueger, R. F., Patrick, C. J., & Iacono, W. G. (2006). Continuity and change in psychopathic traits as measured via normal-range personality: A longitudinal-biometric study. *Journal of Abnormal Psychology, 115*, 85–95.

Cadoret, R. J., Langbehn, D., Caspers, K., Troughton, E. P., Yucuis, R., Sandhu, H. K., et al. (2003). Associations of the serotonin transporter promoter polymorphism with aggressivity, attention deficit, and conduct disorder in an adoptee population. *Comprehensive Psychiatry, 44*, 88–101.

Cases, O., Seif, I., Grimsby, J., Gaspar, P., Chen, K., Pournin, S., et al. (1995). Aggressive behavior and altered amounts of brain serotonin and norepinephrine in mice lacking MAOA. *Science, 268*, 1763–1766.

Caspi, A., Langley, K., Milne, B., Moffitt, T. E., O'Donovan, M., Owen, M. J., et al. (2008). A replicated molecular-genetic basis for subtyping antisocial behavior in ADHD. *Archives of General Psychiatry, 65*, 203–210.

Caspi, A., McClay, J., Moffitt, T. E., Mill, J., Martin, J., Craig, I. W., et al. (2002). Role of genotype in the cycle of violence in maltreated children. *Science, 297*, 851–854.

Caspi, A., Sugden, K., Moffitt, T. E., Taylor, A., Craig, I. W., Harrington, H., et al. (2003). Influence of life stress on depression: Moderation by a polymorphism in the *5-HTT* gene. *Science, 301*, 386–389.

Castellanos, F. X., & Tannock, R. (2002). Neuroscience of attention-deficit/hyperactivity disorder: The search for endophenotypes. *Nature Reviews Neuroscience, 3*, 617–628.

Cloninger, C. R., & Gottesman, I. I. (1987). Genetic and environmental factors in antisocial behavior disorders. In: S. A. Mednick, T. E. Moffitt, & S. A. Stack (Eds.), *The causes of crime: New biological approaches* (pp. 92–109). New York: Cambridge University Press.

Coccaro, E. F., Kavoussi, R. J., Hauger, R. L., Cooper, T. B., & Ferris, C. F. (1998). Cerebrospinal fluid vasopressin levels: Correlates with aggression and serotonin function in personality disordered subjects. *Archives of General Psychiatry, 55*, 708–714.

Dadds, M. R., el Masry, Y., Wimalaweera, S., & Guastella, A. J. (2008). Reduced eye gaze explains "fear blindness" in childhood psychopathic traits. *Journal of the American Academy of Child and Adolescent Psychiatry, 47*, 455–463.

Davidge, K. M., Atkinson, L., Douglas, L., Lee, V., Shapiro, S., Kennedy, J. L., et al. (2004). Association of the serotonin transporter and 5HT1Dβ genes with extreme, persistent, and pervasive aggressive behavior in children. *Psychiatric Genetics, 14*, 143–146.

DeVries, A. C., Young, S. W., & Nelson, R. J. (1997). Reduced aggressive behavior in mice with targeted disruption of the oxytocin gene. *Journal of Neuroendocrinology, 9*, 363–368.

Dick, D. M., Li, T. K., Edenberg, H. J., Hesselbrock, V., Kramer, J., & Foroud, T. (2004). A genome-wide screen for genes influencing conduct disorder. *Molecular Psychiatry, 9*, 81–86.

D'Onofrio, B. M., Turkheimer, E., Eaves, L. J., Corey, L. A., Berg, K., Solaas, M. H., et al. (2003). The role of the Children of Twins design in elucidating causal relations between parent characteristics and child outcomes. *Journal of Child Psychology and Psychiatry, 44*, 1130–1144.

D'Onofrio, B. M., Van Hulle, C. A., Waldman, I. D., Rodgers, J. L., Harden, K. P., Rathouz, P. J., et al. (2008). Smoking during pregnancy and offspring externalizing problems: An exploration of genetic and environmental confounds. *Development and Psychopathology, 20*, 139–164.

D'Onofrio, B. M., Van Hulle, C. A., Waldman, I. D., Rodgers, J. L., Rathouz, P. J., & Lahey, B. B. (2007). Causal inferences regarding prenatal alcohol exposure and childhood externalizing problems. *Archives of General Psychiatry, 64*, 1296–1304.

Doyle, A. E., Faraone, S. V., Seidman, L. J., Willcutt, E., Nigg, J. T., Waldman, I. D., et al. (2005). Are endophenotypes based on measures of executive functions useful for molecular genetic studies of ADHD? *Journal of Child Psychology and Psychiatry, 46*, 774–803.

Dulawa, S. C., Grandy, D. K., Low, M. J., Paulus, M. P., & Geyer, M. A. (1999). Dopamine D4 receptor-knock-out mice exhibit reduced exploration of novel stimuli. *Journal of Neuroscience, 19*, 9550–9556.

Ferris, C.F., Lu, S-F., Messenger, T., Guillon, C.D., Heindel, N., Miller, M., et al. (2006). Orally active vasopressin V1a receptor agonist, SRX251, selectively blocks aggressive behavior. *Pharmacology, Biochemistry, and Behavior, 83*, 169–174.

Foley, D. L., Eaves, L. J., Wormley, B., Silberg, J. L., Maes, H. H., Kuhn, J., et al. (2004). Childhood adversity, monoamine oxidase A genotype, and risk for conduct disorder. *Archives of General Psychiatry, 61*, 738–744.

Forsman, M., Lichtenstein, P., Andershed, H., & Larsson, H. (2008). Genetic effects explain the stability of psychopathic personality from mid-to late adolescence. *Journal of Abnormal Psychology, 117*, 606–617.

Giros, B., Jaber, M., Jones, S. R., Wightman, R. M., & Caron, M. G. (1996). Hyperlocomotion and indifference to cocaine and amphetamine in mice lacking the dopamine transporter. *Nature, 379*, 606–612.

Gizer, I. R., Ficks, C. A., & Waldman, I. D. (2009). Candidate gene studies of ADHD: A meta-analytic review. *Human Genetics, 126*, 51–90.

Gogos, J. A., Morgan, M., Luine, V., Santha, M., Ogawa, S., Pfaff, D., et al. (1998). Catechol-*O*-m ethyltransferase-deficient mice exhibit sexually dimorphic changes in catecholamine levels and behavior. *Proceedings of the National Academy of Sciences of the USA, 95*, 9991–9996.

Gottesman, I. I., & Gould, T. D. (2003). The endophenotype concept in psychiatry: Etymology and strategic intentions. *American Journal of Psychiatry, 160*, 636–645.

Gottesman, I. I., & Shields, J. (1972). *Schizophrenia and genetics: A twin study vantage point.* New York: Academic Press.

Haberstick, B. C., Lessem, J. M., Hopfer, C. J., Smolen, A., Ehringer, M. A., Timberlake, D., et al. (2005). Monoamine oxidase A (MAOA) and antisocial behaviors in the presence of childhood and adolescent maltreatment. *American Journal of Medical Genetics, 135*, 59–64.

Hennig, J., Reuter, M., Netter, P., Burk, C., & Landt, O. (2005). Two types of aggression are differentially related to serotonergic activity and the A779C TPH polymorphism. *Behavioral Neuroscience, 119*, 16–25.

Holmes, A., Murphy, D. L., & Crawley, J. N. (2003). Abnormal behavioral phenotypes of serotonin transporter knockout mice: Parallels with human anxiety and depression. *Biological Psychiatry, 54*, 953–959.

Holmes, J., Payton, A., Barrett, J., Harrington, R., McGuffin, P., Owen, M., et al. (2002). Association of DRD4 in children with ADHD and comorbid conduct problems. *American Journal of Medical Genetics, 114*, 150–153.

Huang, Y., Cate, S. P., Battistuzi, C., Oquendo, M. A., Brent, D., & Mann, J. J. (2004). An association between a functional polymorphism in the monoamine oxidase A gene promoter, impulsive traits and early abuse experiences. *Neuropsychopharmacology, 29*, 1498–1505.

Jorm, A. F., Prior, M., Sanson, A., Smart, D., Zhang, Y., & Easteal, S. (2001). Association of a polymorphism of the dopamine transporter gene with externalizing behavior problems and associated temperament traits: A longitudinal study from infancy to the mid-teens. *American Journal of Medical Genetics, 105*, 346–350.

Kim-Cohen, J., Caspi, A., Taylor, A., Williams, B., Newcombe, R., Craig, I. W., et al. (2006). MAOA, maltreatment, and gene-environment interaction predicting children's mental health: New evidence and a meta-analysis. *Molecular Psychiatry, 11*, 903–913.

Kim-Cohen, J., & Taylor, A. (2007). Meta-analysis of gene-environment interactions in developmental psychopathology. *Development and Psychopathology, 19*, 1029–1037.

Kruesi, M. J., Rapoport, J. L., Hamburger, S., Hibbs, E., Potter, W., Z., Lenane, M., et al. (1990). Cerebrospinal fluid monoamine metabolites, aggression, and impulsivity in disruptive behavior disorders of children and adolescents. *Archives of General Psychiatry, 55*, 989–994.

Lahey, B. B., & Waldman, I. D. (2003). A developmental propensity model of the origins of conduct problems during childhood and adolescence. In: B. B. Lahey, T. E. Moffitt, & A. Caspi (Eds.), *Causes of conduct disorder and juvenile delinquency* (pp. 76–117). New York: Guilford.

Larsson, H., Andershed, H., & Lichtenstein, P. (2006). A genetic factor explains most of the variation in the psychopathic personality. *Journal of Abnormal Psychology, 115,* 221–230.

Lilienfeld, S. O., & Andrews, B. P. (1996). Development and preliminary validation of a self report measure of psychopathic personality traits in noncriminal populations. *Journal of Personality Assessment, 66,* 488–524.

Lilienfeld, S. O., & Waldman, I. D. (1990). The relation between childhood attention-deficit hyperactivity disorder and adult antisocial behavior reexamined: The problem of heterogeneity. *Clinical Psychology Review, 10,* 699–725.

Loehlin, J. C. (1992). *Latent variable models: An introduction to factor, path, and structural analysis* (2nd ed.). Hillsdale, NJ: Erlbaum.

Lykken, D. T. (1957). A study of anxiety in the sociopathic personality. *Journal of Abnormal and Social Psychology, 55,* 6–10.

Manuck, S. B., Flory, J. D., Ferrell, R. E., Dent, K. M., Mann, J. J., & Muldoon, M. F. (1999). Aggression and anger-related traits associated with a polymorphism of the tryptophan hydroxylase gene. *Biological Psychiatry, 45,* 603–614.

Manuck, S. B., Flory, J. D., Ferrell, R. E., Mann, J. J., & Muldoon, M. F. (2000). A regulatory polymorphism of the monoamine oxidase-A gene may be associated with variability in aggression, impulsivity, and central nervous system serotonergic responsivity. *Psychiatry Research, 95,* 9–23.

Miles, D. R., & Carey, G. (1997). Genetic and environmental architecture of human aggression. *Journal of Personality and Social Psychology, 72*(1), 207–217.

Moffitt, T. E. (1993). Adolescence-limited and life-course-persistent antisocial behavior: A developmental taxonomy. *Psychological Review, 100*(4), 674–701.

Morgan, A. B., & Lilienfeld, S. O. (2000). A meta-analytic review of the relation between antisocial behavior and neuropsychological measures of executive function. *Clinical Psychology Review, 20,* 113–156.

Munafò, M. R., Durrant, C., Lewis, G., & Flint, J. (2009). Gene X environment interactions at the serotonin transporter locus. *Biological Psychiatry, 65,* 211–219.

Neale, M. C. (1995). *Mx: Statistical modeling.* Richmond, VA: Department of Psychiatry, Medical College of Virginia, Virginia Commonwealth University.

New, A. S., Gelernter, J., Goodman, M., Mitropoulou, V., Koenigsberg, H., Silverman, J., et al. (2001). Suicide, impulsive aggression, and HTR1B genotype. *Biological Psychiatry, 50,* 62–65.

New, A. S., Gelertner, J., Yovell, Y., Trestman, R. L., Nielsen, D. A., Silvermann, J., et al. (1998). Tryptophan hydroxylase genotype is associated with impulsive-aggression measures: A preliminary report. *American Journal of Medical Genetics, 81,* 13–17.

Nilsson, K. W., Sjöberg, R. L., Damberg, M., Leppert, J., Öhrvik, J., Alm, P. O., et al. (2005). Role of monoamine oxidase A genotype and psychosocial factors in male adolescent criminal activity. *Biological Psychiatry, 59,* 121–127.

Patrick, C. J. (1994). Emotion and psychopathy: Startling new insights. *Psychophysiology, 31,* 319–330.

Plomin, R., Foch, T. T., & Rowe, D. C. (1981). Bobo clown aggression in childhood: Environment, not genes. *Journal of Research in Personality 15,* 331–342.

Plomin, R., Nitz, K., & Rowe, D. C. (1990). Behavioral genetics and aggressive behavior in childhood. In: M. Lewis & S. M. Miller (Eds.), *Handbook of developmental psychopathology* (pp. 119–133). New York: Plenum.

Retz, W., Retz-Junginger, P., Supprian, T., Thome, J., & Rösler, M. (2004). Association of serotonin transporter promoter gene polymorphism with violence: Relation with personality disorders, impulsivity, and childhood ADHD psychopathology. *Behavioral Sciences and the Law, 22,* 415–425.

Rhee, S. H., & Waldman, I. D. (2002). Genetic and environmental influences on antisocial behavior: A meta-analysis of twin and adoption studies. *Psychological Bulletin, 128,* 490–529.

Risch, N., Herrell, R., Lehner, T., Liang, K-Y., Eaves, L., Hoh, J., et al. (2009). Interaction between the serotonin transporter gene (*5-HTTLPR*), stressful life events, and risk of depression: A meta-analysis. *JAMA, 301,* 2462–2471.

Rubinstein, M., Phillips, T. J., Bunzow, J. R., Falzone, T. L., Dziewczapolski, G., Zhang, G., et al. (1997). Mice lacking dopamine D4 receptors are supersensitive to ethanol, cocaine, and methamphetamine. *Cell, 90,* 991–1001.

Saudou, F., Amara, D. A., Dierich, A., LeMeur, M., Ramboz, S., Segu, L., et al. (1994). Enhanced aggressive behavior in mice lacking 5-HT1β receptor. *Science*, 265, 1875–1878.

Seeman, P., & Madras, B. K. (1998). Anti-hyperactivity medication: Methylphenidate and amphetamine. *Molecular Psychiatry*, 3, 386–396.

Slutske, W. S., Heath, A. C., Dinwiddie, S. H., Madden, P. A. F., Bucholz, K. K., Dunne, M. P., et al. (1997). Modeling genetic and environmental influences in the etiology of conduct disorder: A study of 2,682 adult twin pairs. *Journal of Abnormal Psychology*, 106(2), 266–279.

Solanto, M. V. (1984). Neuropharmacological basis of stimulant drug action in attention deficit disorder with hyperactivity: A review and synthesis. *Psychological Bulletin*, 95, 387–409.

Solanto, M. V. (1998). Neuropsychopharmacological mechanisms of stimulant drug action in attention-deficit hyperactivity disorder: A review and integration. *Behavioural Brain Research*, 94, 127–152.

Stallings, M. C., Corley, R. C., Hewitt, J. K., Krauter, K. S., Lessem, J. L., Mikulich, S. K., et al. (2005). A genomewide search for quantitative trait loci influencing antisocial drug dependence in adolescence. *Archives of General Psychiatry*, 62, 1042–1051.

Staner, L., Uyanik, G., Correa, H., Tremeau, F., Monreal, J., Crocq, M.-A., et al. (2002). A dimensional impulsive-aggressive phenotype is associated with the A218C polymorphism of the tryptophan hydroxylase gene: A pilot study in well-characterized impulsive inpatients. *American Journal of Medical Genetics*, 114, 553–557.

Takayanagi, Y., Yoshida, M., Bielsky, I. F., Ross, H. E., Kawamata, M., Onaka, T., et al. (2005). Pervasive social deficits, but normal parturition, in oxytocin receptor deficient mice. *Proceedings of the National Academy of Sciences of the USA*, 102, 16095–16101.

Taylor, J., Loney, B. R., Bobadilla, L., Iacono, W. G., & McGue, M. (2003). Genetic and environmental influences on psychopathy trait dimensions in a community sample of male twins. *Journal of Abnormal Child Psychology*, 31, 633–645.

Taylor, J., McGue, M., Iacono, W. G., & Lykken, D. T. (2000). A behavioral genetic analysis of the relationship between the socialization scale and self-reported delinquency. *Journal of Personality*, 68, 29–50.

Thapar, A., Langley, K., Fowler, T., Rice, F., Turic, D., Whittinger, N., et al. (2005). Catechol-O-methyltransferase gene variant and birth weight predict early onset antisocial behavior in children with attention deficit hyperactivity disorder. *Archives of General Psychiatry*, 62, 1275–1278.

Thompson, R., Gupta, S., Miller, K., Mills, S., & Orr, S. (2004). The effects of vasopressin on human facial responses related to social communication. *Psychoneuroendocrinology*, 29, 35–48.

Viding, E., Blair, J. R., Moffit, T. E., & Plomin, R. (2005). Evidence for substantial genetic risk for psychopathy in 7-year-olds. *Journal of Child Psychology and Psychiatry*, 46, 592–597.

Viding, E., Frick, P. J., & Plomin, R. (2007). Aetiology of the relationship between callous-unemotional traits and conduct problems in childhood. *British Journal of Psychiatry*, 190(Suppl. 49), 33–38.

Viding, E., Jones, A., Frick, P. J., Moffitt, T. E., & Plomin, R. (2008). Heritability of antisocial behaviour at nine-years: Do callous-unemotional traits matter? *Developmental Science*, 11, 17–22.

Waldman, I. D. (1996). Aggressive children's hostile perceptual and response biases: The role of attention and impulsivity. *Child Development*, 67, 1015–1033.

Waldman, I. D. (2005). Statistical approaches to complex phenotypes: Evaluating neuropsychological endophenotypes for ADHD. *Biological Psychiatry*, 57, 1347–1356.

Waldman, I. D., & Gizer, I. (2006). The genetics of ADHD. *Clinical Psychology Review*, 26, 396–432.

Waldman, I. D., & Rhee, S. H. (2006). Genetic and environmental influences on psychopathy and antisocial behavior. In: C. Patrick (Ed.), *Handbook of psychopathy* (pp. 205–228). New York: Guilford.

Waldman, I. D., Tackett, J. L., Van Hulle, C. A., Applegate, B., Pardini, D., Frick, P. J., & Lahey B. B. (2011). Child and adolescent conduct disorder substantially shares genetic influences with three socioemotional dispositions. *Journal of Abnormal Psychology*, 120, 57–70.

Wersinger, S. R., Ginns, E. I., O-Carroll, A. M., Lolait, S. J., & Young, W. S. III (2002). Vasopressin V1b receptor knockout reduces aggressive behavior in male mice. *Molecular Psychiatry*, 7, 975–984.

Widom, C. S., & Ames, A. (1988). Biology and female crime. In: T. E. Moffitt & S. A. Mednick (Eds.), *Biological contributions to crime causation* (pp. 308–331). Dordrecht: Martinus Nijhoff.

Young, S. E., Smolen, A., Corley, R. P., Krauter, K. S., DeFries, J. C., Crowley, T. J., et al. (2002). Dopamine transporter polymorphism associated with externalizing behavior problems in children. *American Journal of Medical Genetics, 114*, 144–149.

Young, S. E., Smolen, A., Hewitt, J. K., Haberstick, B. C., Stallings, M. C., Corley, R. P., et al. (2006). Interaction between MAO-A genotype and maltreatment in the risk for conduct disorder: Failure to confirm in adolescent patients. *American Journal of Psychiatry, 163*, 1019–1025.

PART SIX
TREATMENT OF PSYCHOPATHY

CHAPTER 12

Treatment of Adolescents with Psychopathic Features

Michael F. Caldwell

INTRODUCTION

The construct of psychopathy has typically been characterized as a constellation of personality traits that are believed to be stable over time (Cleckley, 1982; Frick, Kimonis, Dandreaux, & Farell, 2003; Forth, Kosson, & Hare, 2003; Hare, 1991; Lynam, 2008; Lynam, Caspi, Moffitt, Loeber, & Stouthamer-Loeber, 2007; Lynam, Stouthamer-Loeber, & Loeber, 2008; McCord & McCord, 1964; Muñoz & Frick, 2007; Salekin & Lochman, 2008). The failure of psychopaths to respond to treatment has been described as nearly a defining feature of psychopathy (Cleckley, 1982; Karpman, 1946). In his seminal work, Cleckley described being impressed by the psychopath's "lack of response to psychiatric treatment of any kind" (p. 433). Similarly, Karpman considered psychopaths incurable and most appropriately dealt with through indefinite institutionalization.

More recently, the study of psychopathy has expanded with the development of more reliable methods of assessment. Specifically, the development of the Psychopathy Checklist—Revised (PCL-R; Hare, 1991), and the related measure for adolescents, the Psychopathy Checklist: Youth Version (PCL:YV; Forth, et al., 2003), have marked major advances in the reliable assessment of psychopathic personality features in adults and adolescents. Although there continues to be some disagreement about how best to conceptualize the facets of psychopathy (Benning, Patrick, Hicks, Blonigen, & Krueger, 2003; Cooke & Michie, 1997), there is general agreement that psychopathy can be understood as a constellation of related and persistent traits. Those traits encompass an egocentric, grandiose, dominant, dishonest, and guiltlessly exploitative interpersonal style; a callous lack of emotional concern for others and a general poverty of emotional depth; an irresponsible, impulsive, hedonistic, and directionless lifestyle; and an accompanying pattern of antisocial behaviors.

Theories of what may cause psychopathic personality traits to emerge are still evolving. Measures of psychopathic features in children have been spurred by the perception that psychopathy does not emerge fully formed at the age of majority and by the hope that early identification may lead to more effective treatment. Among the measures that have been developed to apply the PCL-R template to children and adolescents, the Psychopathy Checklist: Youth Version (PCL:YV; Forth et al., 2003) may be the most direct adaptation. The PCL:YV is a 20-item checklist made up of items from the PCL-R that have been altered to be more appropriate for adolescents and scored similarly to the PCL-R. The PCL:YV has proven to be a reliable measure that has demonstrated an association with the aggressive, antisocial behavior that is the hallmark of the behavioral manifestation of psychopathy.

Several authors have expressed considerable concern about the appropriateness and accuracy of assessments of psychopathic characteristics in adolescents (Edens, Skeem, Cruise, & Cauffman, 2001; Forth et al., 2003; Ogloff & Lyon, 1998; Seagrave & Grisso, 2002; Steinberg, 2002; Zinger & Forth, 1998). Seagrave and Grisso (2002) have noted that the label of psychopathy may be inappropriately applied to juveniles in court cases with severe and long-term consequences. They echoed concerns raised by Edens et al. (2001) that transient features of adolescence (impulsivity, sensation and thrill seeking, irresponsible behavior, etc.) may be difficult or impossible to distinguish from stable psychopathic personality traits. Forth et al. (2003) offered extensive caveats to clinical use of the PCL:YV. They noted the potential for the label of "psychopath" to be used to exclude these youth from treatment on the assumption that the optimal utilization of scarce treatment resources requires limiting treatment for youth who are designated as psychopathic. They expressed the view that PCL:YV ratings should not be used in court dispositions, or to recommend against treatment, and that youth should not be given the label of "psychopath." Considering the weight that a diagnosis of psychopathic personality can carry in dispositions in the criminal justice system, such cautions are clearly warranted. At the same time, the objective of understanding the development of psychopathic traits clearly justifies the assessment of these characteristics for research purposes. Moreover, the development of effective treatment for psychopathic personality is not possible without valid and reliable assessment of the relevant characteristics.

Although preliminary treatment models have been proposed (Doren, 1987; Lösel, 1998; Lykken, 1995; McCord & McCord, 1964; Thorne, 1959; Toch & Adams, 1994; Quay, 1964; Wallace & Newman, 2004; Wong & Hare, 2005), treatment to date has typically taken the form of ad hoc programs designed to treat offenders in general, or specific subgroups of offenders (such as sex offenders or chemically dependent offenders), with no specific programming targeting the features of psychopathy.

This chapter reviews preliminary studies of the treatment responses of adolescent males treated at the Mendota Juvenile Treatment Center (MJTC). The program was designed to treat aggressive, severely behaviorally disordered delinquent boys who were unmanageable in other secure corrections settings. Although the program was not explicitly designed to treat psychopathic characteristics, aggression and severe behavioral disorders may be the most important manifest characteristics of psychopathy and the foundation of interest in the disorder. As a result, the program incorporated components that target important features of psychopathy. The findings of several initial outcomes studies show that treatment appears to improve treatment responsiveness and reduce institutional misconduct and post-release violence in youth with pronounced psychopathic features. This chapter describes the history and structure of the MJTC program and the preliminary outcomes studies of the effects of treatment on psychopathic youth.

STUDIES OF THE TREATMENT RESPONSE OF ADULT PSYCHOPATHS

Despite the widespread interest in psychopathy, only a few studies have examined the treatment response of psychopaths. To date, the results have not been encouraging. In an early study, Rice, Harris, and Cormier (1992) evaluated outcomes of an unusual therapeutic community for mentally disordered and psychopathic offenders. The program was loosely structured and largely patient run, including decisions regarding which patients were admitted to the program. Psychopathy was assessed retrospectively from records. A number of treatment process variables were also coded to match treated participants with an untreated sample. Psychopaths showed significantly poorer adjustment in the treatment program than nonpsychopaths. When matched with untreated psychopaths, treated psychopaths had similar general failure rates (defined as a new charge or parole failure) but significantly higher rates of violent failure (77% for treated vs. 55% for untreated psychopaths). Treated and untreated nonpsychopaths had similar failure rates. These results suggest that the program was generally ineffective, but may have been particularly poorly suited for psychopaths, perhaps even increasing their risk for violent behavior.

Seto and Barbaree (1999) reported similar results in a study of a prison-based sex offender treatment program for adult men. Treatment program behavior was coded from a combination of therapist-rated progress and ratings completed by research assistants based on treatment notes. Reoffense was recorded in two ways. For participants who had been conditionally released, any parole suspension or revocation, or a violation of a condition of parole relevant to their sex offense history (as judged by a research assistant) that did not result in any official action were coded as reoffense failures. For all participants, failure was also coded when they were sentenced for a new offense.

Parolees were nearly twice as likely to fail compared to participants who were not paroled (27.7% vs. 14.7%, respectively). Psychopathy Checklist—Revised scores were negatively related to treatment behavior ($r(270) = -.16$, $p < .01$). Although both psychopathy and treatment behavior predicted serious reoffending, more positive treatment behavior scores predicted higher rates of serious reoffense. When participants were clustered into four groups using the median PCL-R score (total score = 15), to create high and low PCL-R groups, and the median treatment behavior score to create better and worse treatment behavior groups, the higher PCL-R and better treatment behavior group had significantly higher reoffense rates.

An important caveat to this finding is the subsequent failure of the authors to replicate the findings using the same sample. Specifically, in a subsequent study using a more complete database to record recidivism over a longer follow-up period, Barbaree, Seto, and Langton (2001) found significant differences in reoffense rates related to psychopathy but not related to treatment behavior or progress.

The original Seto and Barbaree (1999) findings have generated considerable discussion about the value of sex offender treatment in general and whether treatment of psychopaths may aggravate their condition (Barbaree, 2005). However, the structure of the original study raises other questions. Specifically, participants who did well in treatment may have been more apt to receive conditional release. Participants who were on conditional release were coded as reoffending if they violated certain rules of their parole, a relatively sensitive standard. By contrast, individuals whose treatment behavior did not warrant conditional release could not fail with a parole rule violation and were only coded as having reoffended when sentenced for a criminal offense, a relatively insensitive standard. The much higher reoffense rate among parolees suggests that the different reoffense coding standards may have skewed the results against more treatment responsive and conditionally released participants.

In a large study of offenders released from the English Prison Service, Hare and his colleagues (Hare, Clark, Grann, & Thornton, 2000) reported that reconviction for general or violent offenses was related to PCL-R scores. Total PCL-R scores did not interact with participation in rehabilitative services, but when Factor 1 was dichotomized (with a cut score of 9), individuals with high Factor1 who had participated in short-term anger management or social skills groups were reconvicted at a significantly higher rate. The same results were found for participation in educational and vocational training programs. Although it is unclear how participation in an educational service or brief anger management group could significantly exacerbate psychopathic criminality, it is clear that these services had no discernible benefit for participants with relatively high PCL-R scores.

A somewhat more encouraging finding was reported by Oliver and Wong (2009). They found that adult sex offenders treated in a prison treatment program who showed positive change in treatment were at significantly lower risk of violent recidivism after controlling for PCL-R scores and static sexual recidivism risk. However, participants with high (25 and above) PCL-R scores were more apt to drop out of treatment and had higher rates of sexual and violent recidivism over a 9.9-year ($SD = 2.8$ years) follow-up period.

In sum, studies of the treatment response of adult psychopaths have reported reasonably consistent results that psychopaths are more likely to drop out of treatment, less likely to show positive changes in treatment, and unlikely to show less violent recidivism related to changes in treatment. Although participation in some prison rehabilitation services is related to poorer reoffense outcomes for adults with pronounced psychopathic features, the characteristics of extant studies limit what conclusions can be drawn about whether treatment can be effective with adult psychopaths.

TREATMENT STUDIES OF ADOLESCENTS WITH PSYCHOPATHIC FEATURES

Although there has been little empirical evidence that psychopaths will respond to treatment, several researchers have noted that there also is little basis for the assumption that psychopaths are incapable of benefiting from any form of treatment (D'Silva, Duggan, McCarthy, 2004; Salekin, 2002). Many of the studies showing poor treatment results suffer from less than rigorous designs and often evaluate treatment programs that were not designed to treat psychopathic characteristics or closely associated features (Falkenbach, Poythress, & Heide, 2003; Hare et al., 2000; O'Neill, Lidz, & Heilbrun, 2003; Rice et al., 1992; Rogers, Jackson, & Sewell, 2004; Spain, Douglas, Poythress, & Epstein, 2004). Psychopathic features may be more amenable to treatment in younger individuals, owing to the more fluid nature of personality in children. However, consistent with the adult literature, studies of adolescents with psychopathic features have documented poor treatment compliance and behavior. Studies of the treatment response of adolescents with psychopathic features have also provided little cause for optimism.

Several studies have reported that adolescents with psychopathic features engage in significantly more rule violations and institutional violence than other youth (Brandt, Kennedy, Patrick, & Curtin, 1997; Dolan & Rennie, 2006; Forth, Hart, & Hare, 1990; Kaplan & Cornell, 2004; Murdock-Hicks, Rogers, & Cashel, 2000; O'Neill, et al., 2003 Rogers, Johansen, Chang, & Salekin, 1997; Stafford & Cornell, 2003). Published studies have consistently reported that psychopathy and institutional violence and misconduct are significantly related, although unpublished studies have tended to find a weaker relationship (Edens, Campbell, & Weir, 2007).

The handful of studies that have examined the treatment response of youth with psychopathic features have reported poor outcomes. Spain et al. (2004) reported on institutional misconduct and treatment progress for 85 adolescent male offenders treated in residential treatment. Using the Antisocial Processes Screening Device (APSD; Frick & Hare, 2001) and the Child Psychopathy Scale (CPS; Lynam, 1997), along with the PCL:YV, to assess psychopathic characteristics, they reported that psychopathy correlated with institutional misconduct and that the self-report measures, but not the PCL:YV, correlated to the number of days it took youths to advance to the second treatment program privilege level.

Similarly, Falkenbach et al. (2003) reported that psychopathic features correlated with treatment program failure and reoffense 1 year after treatment in a group of 69 youth treated in a court diversion program. Rogers et al. (2004) reported more disruptive management problems and lower staff ratings of improvement among youth with psychopathic features treated in a state hospital. In a study of 64 adjudicated boys referred to a partial hospitalization substance abuse program, O'Neill et al. (2003) reported that scores on the PCL:YV were related to poorer treatment participation and program completion and lower ratings of clinical improvement. One year after discharge from the program, PCL:YV scores were related to new arrests.

These studies have established that youth with psychopathic features are more difficult to treat when compared to youth without psychopathic features. However, these results did not address whether psychopathic youth made any progress in treatment or whether treatment progress moderated the effects of psychopathy on their future violent behavior. Considerable evidence has established that pronounced psychopathic features are a reliable indicator that a youth is more behaviorally disturbed. The question of whether related features of that condition can be meaningfully altered through treatment remains largely unexamined. However, the consistent finding that youth with psychopathic features are more disruptive and aggressive in treatment suggests that effective treatment of these youth will require treatment programming specifically designed to manage these behaviors with a minimum of interruption to the treatment process. In addition, the evidence that these youth are more apt to disrupt and drop out of treatment suggests that they are likely to be difficult to engage fully in the treatment process.

BACKGROUND OF THE MENDOTA JUVENILE TREATMENT CENTER

In 1995, the Wisconsin legislature established the Mendota Juvenile Treatment Center (MJTC) as part of a broad reform of juvenile justice legislation. The program was designed to provide mental health treatment to the most disturbed juveniles held in the state's secured correctional facilities. The program has a unique structure. Although operated under the administrative code of the Department of Corrections as a secured correctional facility, the program is housed on the grounds of a state psychiatric hospital. Members of the psychiatric hospital staff are employed by the Department of Health Services, which operates the facility. This organizational design allows for a clinical–correctional hybrid structure in which the physical security in the facility can be used to control the aggressiveness of the youth and the fear and accompanying emotional distance that staff may feel when dealing with dangerous persons.

The treatment program originally consisted of three units with 14 or 15 single bedrooms. The clinical staffing of MJTC includes one psychologist, one social worker, and a half psychiatry position for every 15 youth. Day-to-day administration of the MJTC program is the responsibility of a psychiatric nurse manager who directly supervises the front-line staff members who deal with daily care (meals, hygiene, movement, etc.) of the youth. In this way, the day-to-day operation of the unit is administered and operated by trained mental health professionals who are integrated into the security structure of the unit. Schooling is provided on the unit through an accredited public school that is housed on the grounds of the psychiatric hospital and staffed with teachers who are employees of the Department of Health Services. Housing the unit on the grounds of the state psychiatric hospital also has the advantage of making a variety of adjunctive services available to the unit from the hospital. Speech and Language, Occupational Therapy, Dietitian, Chaplain, Dental, Medical, and a wide range of other adjunctive services are available on an as-needed basis.

The boys who are transferred to MJTC are selected by the staff of the two larger secure state juvenile corrections institutions (JCIs). Youth are selected for transfer by the sending institution without prescreening by MJTC staff. In nearly every case, these boys have not benefitted from repeated community and institutional interventions. There are no exclusion criteria that would prevent a youth from being transferred to MJTC. In fact, issues that might otherwise exclude youth from a treatment service (such as poor treatment motivation or cooperation, low IQ, institutional violence, or neurological deficits) may serve as the basis for the decision to transfer the youth to MJTC. The guiding principle used by the sending JCI is that the youth has not benefited from their treatment and rehabilitation services. This is nearly always due to the youth having extended or repeated security stays to control his aggressive behavior.

The population treated on MJTC typically consists of 51% African American, 38% White, 9% Hispanic, and 2% Asian or Middle Eastern male juveniles. The average age when a youth is released is 17 years and 1 month ($SD = 13$ months). Typically, these youth come from economically disadvantaged, violent, or disrupted homes. Many began their criminal careers early: 50% are involved in crime before their 10th birthday. Previous studies have found that MJTC youth averaged 13.3 ($SD = 9.9$) formally filed charges. Of these youth, 60% ($N = 85$) had been charged with three or more crimes against persons, 51% ($N = 72$) were committed for a violent felony offense, and 49% ($N = 69$) had hospitalized or killed a victim (Caldwell & Van Rybroek, 2004).

CONCEPTUAL AND PHILOSOPHICAL BASIS OF MJTC

The group of people involved in planning MJTC had backgrounds in treating aggressive and treatment refractory adult forensic patients, and other difficult-to-treat patients. As a result, several guiding principles that are important in treating difficult-to-treat patients served as a foundation for the approach used on MJTC.

Principles of Treating Intractable Patients

The general approach to treating any mental condition is to first identify the internal factors that cause the condition, define the relationships among those factors, and then apply a treatment protocol that will ameliorate those causes. This is a nearly universal approach to the treatment of all forms of disease and is based on a positivistic view of disease. When treatment fails, it may be for a number of reasons, but it is often because the individual has problems the causes of which are not as well understood as those of more treatment-responsive individuals.

When the causes of a condition are not well defined, or whenever the conventional treatment of a condition fails, future treatment can be guided by principles that do not rely on a full understanding of the underlying causes of the disorder. Instead this approach focuses on how the individual functions in the world and how the disorder, whatever its cause, defines the individual in his or her social ecology, and more specifically in the social system of the treatment setting.

An initial step in this approach is to recognize that the diagnosis is a social construct. Often with treatment refractory individuals, the diagnosis is vague, complex, or immutable. A diagnosis may or may not be an accurate description of the internal causes of a disorder, but it is always the way that the treatment/rehabilitation system has come to understand the person. In this approach the definition of the problem, and the treatment intervention, is focused on the interpersonal space and interactions that define how the person is understood and how he or she functions in the social matrix of the treatment/rehabilitation system.

An initial step in treating treatment refractory individuals is to formulate a workable definition of the problem to be treated. The definition of the problem must be an essential feature that defines how the person is understood in the social ecology of the treatment system and that compels treatment for the individual in that system. For the youth transferred to MJTC, for example, the defining characteristic in the correctional treatment system is that his disruptive and aggressive behavior prevents him from engaging in treatment services.

It is not important that the definition of the problem be all-inclusive at this point, but it is essential that the problem be defined in a specific way that makes it possible to determine whether interventions have an impact on the problem. This approach emphasizes the functioning of the individual within a social ecology. Since the treaters are also part of the social ecology, progress cannot be reliably measured by changes in the treaters subjective clinical impressions, opinions, or judgments.

A second guideline in treating treatment refractory individuals is to understand and avoid the pitfalls of previous treatment failures. Even when the mechanism of failure is unclear, it is important to avoid repeating a treatment regimen that is based on a formulation that has failed in the past. Although this may seem self-evident, treatment-refractory individuals have often received the same basic treatment repeatedly in the past in various settings.

In treating treatment-refractory individuals it is essential to measure changes in the problem that is the focus of treatment. Ideally, the measurement of the problem should be simple and readily observable. At the same time, the treatment is likely to be complex and subject to setbacks and external pressures. Promising treatment approaches are sometimes abandoned to respond to new developments in the course of treatment. Without reasonably reliable data concerning the impact of treatment on the problem it can be difficult or impossible to determine what should be retained and what should be abandoned in response to developments in the course of treatment.

There are many other aspects to treating specific treatment-refractory individuals. However, the three described in the preceding text are those that most influenced the planning of the MJTC program.

Theoretical Foundation

The conceptual foundation of the program rests on the Social Learning Theory of Albert Bandura (1977), the Social Control theory of Travis

Hirschi (1969) and Sampson and Laub (1997), and Defiance Theory developed by Lawrence Sherman (1993), and a more specific application of Defiance Theory in the Decompression model described by Caldwell and Van Rybroek (2002; Monroe, Van Rybroek, & Maier, 1988). In this approach, the problematic behavior of the youth is regarded as an interactive process between the treatment system and the youth. The staff internal processes and responses to the youth's aggression are as important to the treatment as the youth's internal processes and responses to the treatment system. The problematic interactive process blocks the youth from forming very basic social bonds with the treatment system. Because the youth lacks these bonds, punitive or deterrent sanctions by the treatment system generate a defiant response of increased aggression from the youth. In turn, if the treatment staff respond to the youth's aggression with more controlling and punitive measures, the result is further disenfranchisement from the treatment system and increased defiant aggression.

Extending from this conceptual framework, the program includes structured components intended to manage the staff's responses to the youth's negative behavior and to engender greater therapeutic engagement in the youth. In the process, the program attempts to aid the youth in developing basic social bonds within the ecology of the treatment program. The primary program components designed to address these issues are the Behavioral Assessment System and the Today–Tomorrow Program.

MJTC Treatment Components

MJTC Behavioral Assessment System

The MJTC Behavioral Assessment System is a program that provides a sensitive measure of treatment involvement and compliance with behavioral expectations. The assessment program is anchored in five guiding principles:

1. *Continuity*: Behavior is assessed and individual data are recorded continuously throughout the day. This assists in tapping into behaviors that are low in frequency but high in treatment relevance.
2. *Clarity*: Clearly observable benchmarks that describe preselected behaviors are recorded in a data collection system designed to minimize interpretation of the behaviors motivation, while focusing on the consistency of recording behaviors by all program staff.
3. *Simplicity*: Scales are scored with numeric ratings that require minimal effort to accurately rate the relevant behaviors. Most scales ask if a behavior has never occurred, or has occurred once or more than once. The staff members are trained to merely record the absence or presence of the preselected behaviors using a simple numeric coding system.
4. *Relevance*: Scales assess behavior that is relevant to treatment goals and scores reflect the overall level of functioning of the youth in the relevant areas.
5. *Integrity*: Staff members are fully trained and ratings are completed on a consensus basis that includes the shift supervisor, with clinical oversight provided by a senior clinical psychologist.

At the end of each shift each treatment staff member reviews each youth's behavior during the shift and record their observations on several 10-point scales. Ratings are discussed in a group format by the frontline staff and supervisor working on that shift so that a consensus point value is reached for each behavior rating scale that reflects the youth's behavior over the entire shift. Each rating scale employs easily observable behavioral anchors that reflect the individual's compliance with behavioral expectations and positive interpersonal interactions in treatment. Scales rate the youth's peer interactions, staff interactions, rule compliance, performance in treatment groups, and school comportment and are scored at the end of each shift. At the conclusion of each day, staff ratings of each adolescent's behavior throughout the day are combined with ratings from each school class and treatment group to produce an individualized single score for the day.

Information Management

Information is entered into a database at the end of each shift and is available for review immediately. The percentages of possible points earned by a given youth can be retrieved over any time period aggregated into a daily, weekly, or monthly graph. Point ratings for the subscales (Peer Interactions, Adult Interactions, and Rule Compliance) can be generated in the same way.

Similarly, a summary report can be generated at any time describing youths' treatment process, privilege levels, security or other special interventions, medication changes, and other important aspects of the youth's treatment, over any time period from the youth's admission date up to the most recent complete shift. This individually recorded information is placed on computer-generated graphs showing behavior improvement, or lack thereof. The data are used as an aid in 1:1 counseling sessions, to monitor treatment progress and make data-driven treatment decisions, and to determine privilege levels in the contingency management program (the Today–Tomorrow program).

MJTC Today–Tomorrow Program

The MJTC Today–Tomorrow Program is a contingency management program designed to facilitate the development of basic prosocial bonding skills and engagement in the treatment process. Social Control Theory (Hirschi, 1969; Sampson & Laub, 1997) holds that aggressive and antisocial behavior occurs when prosocial bonds to school, prosocial peers, family, and other conventional activities are weakened or broken. Defiance theory (Sherman, 1993) holds that in the absence of prosocial bonds, an individual may respond to deterrent sanctions with a defiant lack of shame and an increase in aggressive behaviors. This is more apt to occur when the individual considers the sanction to be unfair or directed at him or her personally rather than related to his or her behavior. The Decompression Model (Caldwell & Van Rybroek, 2001) holds that deterrent sanctions and defiant responses can become a recursive cycle when a deterrent sanction produces a defiant response (increased aggression), and that response is sanctioned, resulting in more defiance and further repetition of the cycle. With each iteration of the cycle, the individual gives up a little more investment in convention, and the person's life becomes "compressed" as nondefiant behaviors are squeezed out, while the use of punitive and restrictive sanctions increases. The treatment model holds that a programming focus based on developing basic prosocial bonds can gradually "decompress" such confined individuals and reorient their existing skills toward minimal prosocial bonding. Hirschi (1969) described prosocial bonds as developing through any of four mechanisms: (1) involvement in conventional activities, (2) commitment that develops through consistent involvement in conventional activities, (3) attachment with conventional others, and (4) belief in the moral unacceptability of aggressive conduct. The primary focus of the Today–Tomorrow treatment program is on the first two of these. The program attempts to foster involvement by adhering to six guiding principles:

1. *Transparency*: The program is easily understandable for the youth.
2. *Predictability*: Behavior must result in reliable point assignments, and point totals must reliably result in privileges. A given behavior must result in a predictable outcome consistently in order for the youth to view the system as fair.
3. *Equity*: The program must work the same way for every youth; the points must be clearly based on the behavior and not on the personality, history, or other individual characteristics of the youth.
4. *Immediacy*: Rewards for involvement in the program must occur rapidly. For the youth, the payoff must follow close behind the behavior, and consistency of acceptable behavior must provide rapid benefits.
5. *Redeemability*: Related to the Immediacy principle, poor behavior and treatment cooperation must have very short penalties. A setback on any one day can be redeemed with acceptable behavior the next day, providing an important incentive to remain engaged in the program.
6. *Achievability*: The rewards in the program must appear easily achievable and be desirable for the youth. The program should be designed to generate success for the youth.

Programs that document consistent failure to achieve the treatment goals should be altered in line with these principles.

Training and Program Integrity

All staff members are provided extensive orientation and training to the program at the beginning of their employment, and the program and ratings are supervised by a licensed psychologist who developed the system. In addition, all youth are oriented to the program on admission to MJTC by the assigned psychologist. In most cases, a list of attractive incentives is developed so that the youth can choose from a menu of privileges available at each privilege level.

Program integrity is assured in several ways. The end of shift meetings where ratings are generated are monitored by a licensed psychologist to ensure that the rating scales are being used as intended. The supervising psychologist also monitors privilege levels to ensure that the point totals are being applied to privilege levels correctly.

Direct Treatment Services

Because the MJTC program has no exclusion criteria, the program serves a highly unusual population that requires highly individualized treatment. Although the clinical treatment staff have been trained in the Aggression Replacement Training model (Goldstein, Glick, & Gibbs, 1998) and rely on a cognitive–behavioral treatment approach, the majority of clinical treatment services are delivered in frequent, brief 1:1 or very small group sessions. The content of those sessions may be a specific social or problem-solving skill, but the method of delivery will be individualized to that specific youth.

This is not to say that the program does not have a specific treatment orientation. The program incorporates multifaceted treatment services from an individualized Risk/Needs perspective (Andrews & Bonta, 2007). The program provides services to address the range of treatment needs that have a documented relationship with persistent offending. Service areas include (1) educational classes and tutoring; (2) psychiatric services including empirical medication management; (3) family identification and brief strategic therapy; (4) individual mental health counseling; (5) sex education/intimacy skills treatment, and sex offender treatment; (6) anger and emotion management; (7) social skills building treatment; (8) cognitive distortions/delinquent attitudes treatment; (9) alcohol and drug abuse treatment; (10) occupational therapy; (11) speech and language therapy services; (12) recreational therapy; and (13) spirituality and religious counseling to specific youth when appropriate, (14) aftercare planning and reintegration, and a variety of other specific services for youth with less common needs.

TREATMENT OUTCOME STUDIES

The program uses data that are generated through the routine clinical assessments and treatment processes on the unit. These include multidisciplinary initial evaluations, the ongoing Behavioral Assessment System, and a series of repeated measures that assess characteristics of the youth that are relevant to treatment progress. As part of the original mandate that created MJTC, the program was required to conduct routine outcomes studies. In line with that mandate the MJTC program has included a research component that has conducted several studies of the effects of treatment on recidivism. Several of these studies used a group of youth who received the majority of their treatment services elsewhere but were sent to MJTC for a brief stay (generally for a diagnostic evaluation or stabilization services), as the comparison group. Youth who received the majority of their corrections-based treatment on MJTC were categorized as the treatment group. The sending Juvenile Corrections Institution (JCI) exercised full control over which youth were selected to send to MJTC, and youth were not prescreened by MJTC staff. Similarly, a youth could not "fail" the MJTC program. The operating structure gave the sending institution the responsibility to select youth to return to their institution, and youth were returned to the sending institution when staff at that facility believed that the youth had improved sufficiently that they could benefit from the usual treatment services. Thus, youth that were viewed by the sending institution staff as more disruptive and treatment refractory were more likely to receive the majority of their treatment from MJTC, and youth viewed as having stabilized more readily were apt to have a briefer stay on MJTC and to receive the majority of their treatment in the usual JCI settings.

The expected outcome of this process would be that more disruptive and difficult to treat youth would spend the majority of their treatment time on MJTC (and would be categorized as "treatment" youth in the outcome studies), and more cooperative and amenable youth would have shorter stays (and thus be categorized in the "comparison" group). However, the possibility that the bias in the treatment group assignment did not go in the expected direction cannot be ignored. In these studies the program took steps to control for the potential effects of nonrandom treatment group assignment on treatment outcomes. In most studies, a Propensity Score Analysis procedure (Rosenbaum & Rubin, 1983, 1984; Rubin, 1997) was used to attempt to capture and control potential bias in treatment group assignment. In brief, this procedure involves creating a statistical model for exposure to treatment. This is done by entering relevant variables into a forward conditional hierarchical regression to predict treatment group assignment. The regression value for each participant is saved and represents the propensity for that individual to be assigned to the treatment group, contingent on the variables that entered the equation. The propensity score model can then be assessed to determine how well it predicts treatment group assignment. In theory, a model that predicts group assignment very well should also capture a substantial portion of whatever bias may be embedded in that assignment process. Once a propensity score has been developed it can be used in a variety of ways to minimize the effects of nonrandom group assignment bias.

Shortly after the program opened we began including the experimental version of the Psychopathy Checklist: Youth Version (PCL:YV, Forth et al., 2003) to assess features of psychopathy. The PCL:YV was scored on the basis of a semistructured admission assessment interview conducted by a psychologist and a detailed review of the youth's correctional record. The correctional record includes police and juvenile court records, social worker reports to the court, social service records, and treatment and school records. Typically these records included several community psychosocial assessments that included information about the family, the youth's treatment and supervision, and educational performance. The admission interview was typically conducted with at least one observer. The psychologist and observer would then score the PCL:YV independently, then discuss any differences to arrive at a consensus final score.

The PCL:YV is a rater-based instrument consisting of 20 items, each of which is rated for its degree of match to the youth (0, 1, or 2). The PCL:YV possesses reasonably good interrater reliability (Forth & Burke, 1998) and moderate predictive utility for violence (Edens et al., 2001; Gretton, Hare, & Catchpole, 2004; Gretton, McBride, Hare, O'Shaughnessy, & Kumka, 2001) and institutional misbehavior (Kaplan & Cornell, 2004; Stafford & Cornell, 2003). Forth and Burke (1998) reported acceptable levels of internal consistency across several studies (mean $r = .83$). In the most recent study conducted on MJTC, independent ratings of a subgroup of 50 offenders manifested acceptable rates of interrater reliability (total score ICC = .93).

In each study recidivism information was collected from open records of all criminal charges filed in a Wisconsin Circuit Court. The type of offense (misdemeanor, felony, violent, or nonviolent) and the date of the first offense in each category were recorded. In this way, the days to first failure in each offense category could be calculated for survival analyses.

Study 1

Several early internal studies found a significant relationship between treatment on MJTC and a lower prevalence of violent recidivism on release. Although these studies included few control variables and were not peer reviewed, the results were promising. To determine better if treatment was effective with this population we conducted a systematic study of the effectiveness of treatment with youth that had pronounced psychopathic features.

This initial study (Caldwell, Skeem, Salekin, & Van Rybroek, 2006) included youth who had been released from MJTC in the first 18 months of its operation and who had obtained PCL:YV scores of 27 or greater. This generated a sample size of 141 youth. Eighty-five of these youth had been assessed on MJTC and returned to the JCI to receive the majority of their treatment elsewhere and were categorized as "comparison" youth. An additional 50 youth had received the majority of their treatment on MJTC and were categorized as "treatment" youth. An additional six youth had resided on MJTC for less than 50% of their incarceration, but were recommended for elective release by MJTC staff after stays exceeding 6 months and were considered "treatment" youth.

On admission to MJTC each youth was engaged in an extensive initial assessment that included a social, family, mental health, behavioral, and correctional histories and psychodiagnostic, intellectual, academic, and personality testing. The PCL:YV was scored on admission based on these assessments, a semistructured interview, and a detailed review of records. On release, the days of treatment in each setting and the type of placement on release (unsupervised, supervised nonsecured, or secured setting) were recorded. For most of the analyses, participants were followed for a uniform 2-year period. However, survival analysis using the Cox proportional hazard procedure used the full follow-up time (mean = 44.0 months, SD = 12.3 months).

Consistent with previous studies the initial data showed that MJTC treatment youth reoffended at significantly lower rates than their comparison counterparts. Fifty-seven percent of the treatment youth were charged with any offense in the 2-year follow-up compared to 78% of the comparison group (χ^2 (1, $N = 141$) = 7.58, $p < .01$). All six of the youth that had been electively released from MJTC as "successful" recidivated. Similar results were found for violent recidivism; 21% of the treatment group and 49% of the comparison group were charges with a violent offense in the follow-up period (χ^2 (1, $N = 141$) = 3.93, $p < .05$). Two of the six "successful" youth were charged with a violent offense.

To assess whether these results reflected a selection bias we conducted a propensity score analysis. To do this we entered 12 covariates that were plausibly related to treatment assignment into a forward conditional logistic regression equation to predict treatment group assignment (MJTC treatment vs. comparison group). These variables were; race, grade achievement level, full-scale IQ, days of JCI treatment before transfer to MJTC, PCL:YV Factor 1 and Factor 2 scores, number of conduct disorder systems, total score on the Young Offender Level of Service Inventory (YO-LSI; Shields & Simourd, 1991), level of worst victim injury, age at first crime, age at onset of behavioral symptoms, and the number of prior charges. Of these variables only race and the days of JCI treatment before transfer to MJTC entered the equation. Although the resulting model reliably predicted group assignment, χ^2 (3, $N = 141$) = 30.04, $p < .0005$, the propensity value was only moderately accurate in predicting treatment group assignment (accuracy = 69%).

Having developed propensity scores, we then analyzed the recidivism data using a stepwise logistical regression procedure. In this analysis the treatment propensity score, type of setting on release (secured, supervised community, or unsupervised), and PCL:YV total score were entered as a block on the first step, and treatment group assignment was entered on the second step of an equation to predict the prevalence of each type of offense (general or violent).

These analyses revealed no significant treatment effect for general offenses (χ^2 (5, $N = 141$) = 11.70, n.s.) after controlling for the covariates. For violent offending, however, adding treatment group assignment to the covariates-only model resulted in a significant improvement in the prediction equation (χ^2 change (1, $N = 141$) = 6.40, $p < .01$, R^2 change = .03. The coefficient for the treatment variable was in the expected direction ($\beta = 1.18$, $SE = .48$, $p < .05$, odds ratio = 3.3), indicating that youth in the treatment group were significantly less likely to be charged with a violent offense than were the comparison group.

An additional Cox proportional hazard survival analysis was conducted to assess recidivism for youth who were placed in the community over the full follow-up time period (mean = 44.0 months, $SD = 12.3$ months). After the covariates (propensity score, PCL:YV, and type of release placement) were entered on the first step, MJTC treatment significantly predicted survival time to the first violent offense, χ^2 *change* (1, $N = 126$) = 6.45, $p < .05$.

This initial study showed a significant reduction in violent offending in the treatment group. The participants in the study were drawn from the first months that MJTC was in operation. Not surprisingly, this was a time of considerable transition and growing pains in the program. Nearly every clinical staff position turned over three or more times during these initial years. As a result, it is unlikely that these results could be attributed to an unusually gifted or consistent clinical staff. A more likely explanation is that the program philosophy, staffing patterns and levels, and the daily unit structure implemented by front-line staff accounted for the treatment benefit that these youth received in the initial years of the program.

Study 2

In a larger subsequent study of the treatment program we followed 248 youth who had been admitted to the program over a 4½-year period (Caldwell & Van Rybroek, 2004). Twenty-one (8.5%) of these youth overlapped in the previous study, but the remainder were treated on MJTC after the program had been in operation for more than 1 year. Once again, youth were categorized in the "treatment" group if the majority of their treatment in juvenile corrections was delivered on MJTC and in the "comparison" group if they had been assessed or stabilized on MJTC but received the majority of their treatment elsewhere. Using this standard, 101 youth were categorized as "treatment" youth and 147 youth were categorized as "comparison" youth.

This study was conducted in two distinct phases. The first examined the prevalence of new charges over a 4½-year follow-up period (mean = 54.7 months, SD = 17.8 months) and the impact of treatment on the length of time to the first offense in several categories. All participants were followed for a minimum of 2 years after release from secured custody. The results (Table 12.1) showed significantly lower prevalence rates for new charges in each category over the 2-year follow-up and the full follow-up time frames.

After developing a propensity value for each participant, we conducted a survival analysis using the hierarchical Cox proportional hazard procedure. To do this we entered the propensity score value on the first step, followed by the treatment group on the second step, to predict the incidence of new charges in each category. The addition of treatment status resulted in a significant improvement in the prediction equation for misdemeanor offenses, χ^2 (1, $N = 248$) = 6.10, $p < .05$, for felony offenses, χ^2 (1, $N = 248$) = 9.12, $p < .005$, for violent offenses, χ^2 (1, $N = 248$) = 8.76, $p < .005$, and for felony violence, χ^2 (1, $N = 248$) = 9.86, $p < .005$. Thus, after controlling for nonrandom group assignment, the data indicated that treatment was associated with a lower prevalence of offending and longer periods of offense-free adjustment in the community.

Table 12.1 Prevalence of New Criminal Charges over a 2-Year Follow-up for Treatment ($N = 101$) and Comparison ($N = 147$) Youth

	2-year follow-up			Full follow-up		
	Any failure	Felony failure	Violent failure	Any failure	Felony failure	Violent failure
Comparison, % ($N = 101$)	72.1	58.5	43.5	98.0	74.8	60.5
Treatment, % ($N = 147$)	49.5	30.7	20.8	64.4	44.6	35.6
χ^2	13.11***	18.58***	13.75***	11.32*	23.41***	14.85***
η^2	.230	.274	.235	.214	.307	.245

*$p < .05$; ***$p < .001$.

Table 12.2 Incidence Rates of New Criminal Charges for Matched Treatment (N = 101) and Comparison (N = 101) Youth

	Any offense*	Felony offense*	Violent offense**
Comparison ($N = 101$)	2.49	0.89	0.85
Treatment ($N = 101$)	1.09	0.48	0.25

*$p < .05$; **$p < .001$.

Cost–Benefits of Treatment

We then elected to undertake a more in-depth and rigorous analysis of these data, including an analysis of the cost–benefits of treatment (Caldwell, Van Rybroek, & Vitacco, 2006). In this analysis, we first created a matched sample by yoking each treatment youth to a matched comparison youth. To do this we first completed a propensity score analysis by entering 21 variables in a forward conditional logistic regression equation (Rosenbaum & Rubin, 1983, 1984; Rubin, 1997). Seven variables entered the equation (race, the days of JCI treatment before transfer to MJTC, PCL:YV total score, age at onset of conduct disorder symptoms, age of first arrest, the number of prior charged crimes, and the number of prior charged crimes against persons). The resulting propensity score predicted treatment group membership with 86% accuracy. Receiver Operating Characteristics (ROC) analysis also indicated that the propensity score predicted group membership well (Area Under the Curve = .88). Each treatment participant was then matched to a comparison group member based on propensity score using a nearest neighbor matching strategy (Dehejia & Wahba 2002; Gu & Rosenbaum, 1993). Propensity scores were matched within 5% for each participant pair. The resulting final sample contained 101 treatment youth matched to 101 comparison youth.

The two groups were closely matched with regard to PCL:YV scores. The comparison group mean PCL:YV score was 32.6 ($SD = 4.7$, median = 34.4, range = 20–38), while the treatment group mean PCL:YV score was 32.8 ($SD = 5.1$, median = 35.0, range = 15–38). The time at risk was calculated by subtracting the time each youth was incarcerated from the total follow-up time available for that youth.

Table 12.2 shows the mean incidence of new criminal charges for the treatment and comparison group for each offense category. Whereas the prevalence of serious and violent offending by the comparison group in the unmatched sample was approximately twice that of the treatment group (Table 12.1), youth in the matched comparison group averaged more than twice the number of charged offenses in the follow-up period (2.49 vs. 1.09 for the treatment group) and more than three times the number of violent offenses (0.85 vs. 0.25 for the treatment group) when compared to the treatment group.

We then conducted an examination of the relationship between treatment and the incidence of offending in each offense category. To do this we entered the time at risk on the first step of a hierarchical regression equation, followed by treatment group assignment to predict the incidence of each type of offending. The results are shown in Table 12.3. As was observed in the previous study of the incidence of offending, treatment appeared to be most effective at reducing the rate of more serious and violent offending.

After establishing that treatment was associated with significantly lower recidivism rates, costs were assigned to the treatment and outcome variables. To do this we calculated the daily cost for each day of treatment on MJTC and other JCI settings. We calculated the number of days each youth resided in each setting and totaled these figures to obtain a juvenile institutional treatment cost for each youth. Costs for criminal justice processing including the cost of arrest, prosecution, and defense costs were calculated using updated estimates from the work of Cohen (Cohen, 1988, 1998, 2000; Cohen, Miller, & Rossman, 1994) and reflect estimates from a national sample. Costs were assigned for each

Table 12.3 Results of Linear Regression of Matched Treatment Group Status (N = 202) to Predict Incidence of Offending After Controlling for Time at Risk

	R^2	df, N	F change	Significance of F change ($p =$)
All offenses	.08	2,200	4.0	.047
Felony offenses	.21	2,200	6.0	.015
Violent offenses	.32	2,200	18.33	>.0005

charged offense in the follow-up period. Lastly, we calculated the costs of prison confinement by identifying youth that entered prison and the daily bed cost for each prison in which they were held.

The daily costs were totaled up to the date of the end of the study period. Although many youth remained in prison after the study ended (and thus accumulated more costs) we elected not to attempt to amortize these costs because the small sample size raised the possibility of significant error if a few youth did not continue in prison as expected. Similarly, we elected not to estimate and assign victim costs, medical costs, and other costs that could not be directly observed because the small sample size posed a risk of disproportionate errors. However, omitting these costs certainly resulted in an underestimate in the cost–benefits of treatment because the treatment group generated fewer serious and violent offenses (that would involve more victim and medical costs). In addition, 10% of the comparison group members were convicted of homicide offenses (accounting for 16 deaths) whereas none of the treatment group was charged with a homicide. As a result, the comparison group youth certainly generated more victim and prison costs that continued to mount after the end of the study period.

The results of the cost–benefit analysis are shown in Table 12.4. Although the daily bed cost of the MJTC program was more than double that of the usual JCI, youth treated on MJTC improved their institutional adjustment and made faster progress in treatment than the comparison youth. As a result, most of the added costs associated with MJTC were recovered by shortening the length of stay of the youth. The additional marginal cost of treatment on MJTC was less than 5% of the cost of treatment in the usual JCI.

The benefits in the form of avoided recidivism cost per youth can be expressed in comparison to the initial added cost of MJTC treatment in excess of the costs incurred by the comparison group youth. Specifically, the added investment of $7,014 (the mean marginal cost of MJTC treatment) per MJTC treated youth generated mean marginal benefits (or recidivism costs avoided) of $50,390 per youth ($8,176 in avoided criminal justice processing costs plus $42,214 in avoided prison costs) over the 4.5-year follow-up period. This represents a cost–benefit ratio of 1 to 7.18; that is, the program produced benefits of $7.18 for every dollar of cost. This translates into an average return on the marginal investment of more than 130% per year, or more than 618% over the life of the study.

Examining Psychopathy, Treatment, and Recidivism

To examine the relationship between PCL:YV score, treatment status, and recidivism in some detail, we divided the entire sample at the median PCL:YV score (median = 32), thus creating a high and low PCL:YV subgroup in each treatment condition. We then conducted an ANCOVA in which high or low PCL:YV category and propensity-matched treatment group assignment were entered as fixed variables, and the time at risk was entered as a covariate, to predict the number of general and violent offenses. The results (Table 12.5) showed a main effect for treatment group in each analysis. Neither the PCL:YV total score nor the Treatment × PCL:YV score was significant in these analyses.

Table 12.4 Differences in Mean Cost Category for Comparison (N = 101) and Treatment (N = 101)

	Juvenile institution	Criminal justice	Prison	Net
Comparison (N = 101)	$154,917.79	$14,103.24	$47,366.97	$216,388.00
Treatment (N = 101)	$161,932.23	$5,927.07	$5,152.90	$173,012.20
Difference (comparison − treatment)	($7,014.44)	$8,176.17**	$42,214.07**	$43,375.80*

Table 12.5 ANCOVA of Matched Treatment Groups and PCL:YV Total to Predict the Mean Number of All Offenses and Violent Offenses (N = 202)

Effect	F	df (Contrast/error)	η^2	Significance ($p =$)
All offenses				
Time at risk	13.95	1/196	.069	.000
Treatment group (Treatment = 1, Comparison = 0)	5.79	1/196	.029	.017
PCL:YV total	2.10	1/196	.011	.149
Treatment × PCL:YV total	0.12	1/196	.001	.630
Violent offenses				
Time at risk	13.90	1/196	.066	.000
Treatment group (Treatment = 1, Comparison = 0)	4.97	1/196	.025	.027
PCL:YV total	0.18	1/196	.001	.670
Treatment × PCL:YV total	0.10	1/196	.000	.757

In a large study of inmates from the English Prison Service, Hare et al. (2000) reported that inmates who were divided on the basis of the PCL-R total score into high and low psychopathy groups showed no significant interactions between treatment and reconviction rates. However, when groups were divided at the median into high and low Factor 1 scores, treatment appeared to be related to higher reconviction rates for the high Factor 1 group. That is, inmates who had higher Factor 1 scores and who participated in rehabilitation services had higher reconviction rates than those who did not. To further examine the MJTC data in light of these findings, we divided the Factor 1 and Factor 2 scores at the median and examined their relationship with treatment and recidivism using the same ANCOVA procedure (Factor 1 median = 10, Factor 2 median = 14). These results for general offenses are included in Table 12.6. This analysis also found significant main effects for the treatment variable that were not moderated by the PCL:YV Factor scores.

We repeated this analysis to predict the number of violent offenses. As with the previous analyses, the results for violent offending (Table 12.7) showed a main effect for treatment for both factors. In addition, the main effect for both factors, and the interaction terms, were not significant (although the Factor 1 × Treatment interaction approached significance, $p = .082$).

These results are illustrated in Figures 12.1 through 12.4, which show the treatment outcome as a function of high and low PCL:YV total and Factor 1 scores after controlling for time at risk. As expected, in the comparison group, higher PCL:YV total and Factor 1 scores were associated with more general and violent

Table 12.6 ANCOVA of Matched Treatment Groups and PCL:YV Factor Scores to Predict the Mean Number of All Offenses (N = 202)

Effect	F	df (Contrast/error)	η²	Significance (p =)
Factor 1				
Time at risk	13.52	1/196	.065	.000
Treatment group (Treatment = 1, Comparison = 0)	4.48	1/196	.022	.036
Factor 1	1.17	1/196	.006	.781
Treatment × Factor 1	1.68	1/196	.009	.196
Factor 2				
Time at risk	15.45	1/196	.073	.000
Treatment group (Treatment = 1, Comparison = 0)	5.91	1/196	.029	.016
Factor 2	0.13	1/196	.001	.724
Treatment × Factor 2	2.09	1/196	.011	.150

Table 12.7 ANCOVA of Matched Treatment Groups and PCL:YV Factor Scores to Predict the Mean Number of Violent Offenses (N = 202)

Effect	F	df (Contrast/error)	η²	Significance (p =)
Factor 1				
Time at risk	15.52	1/196	.073	.000
Treatment group (Treatment = 1, Comparison = 0)	7.10	1/196	.035	.008
Factor 1	2.29	1/196	.011	.132
Treatment × Factor 1	3.01	1/196	.015	.082
Factor 2				
Time at risk	13.30	1/196	.063	.000
Treatment group (Treatment = 1, Comparison = 0)	5.55	1/196	.027	.019
Factor 2	0.23	1/196	.001	.635
Treatment × Factor 2	0.23	1/196	.001	.634

recidivism. However, treatment was associated with a significantly lower rate of offending in each analysis. In the comparison group, youth with higher Factor 1 scores accounted for more general and violent offenses. Importantly, youth with higher Factor 1 scores who received treatment had nearly the same incidence of general and violent offending as youth with lower Factor 1 scores. The finding for violent offenses and the PCL:YV total score was similar.

These results indicate that treatment can reduce the propensity for general and violent recidivism among youth with pronounced features of psychopathy. Neither the total PCL:YV score nor the factor scores appeared to present a barrier to the effectiveness of treatment. More importantly, the link between serious violence

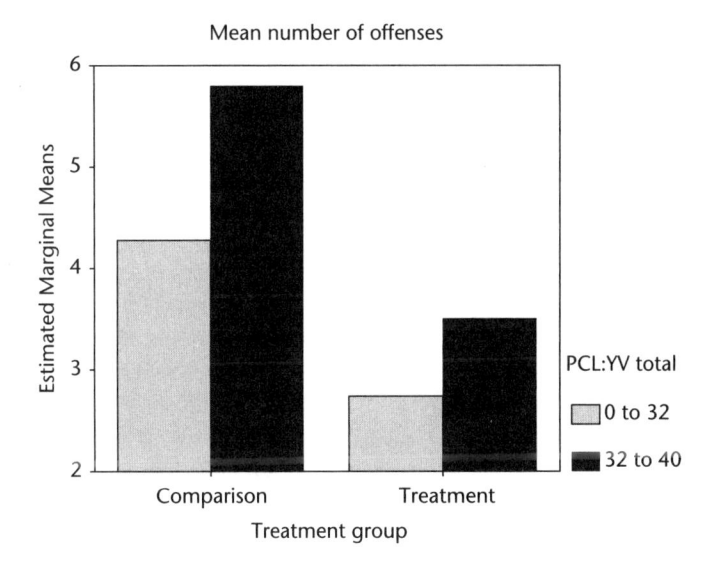

Figure 12.1 Mean number of criminal offenses for Treatment and Comparison groups sorted by high and low PCL:YV total score.

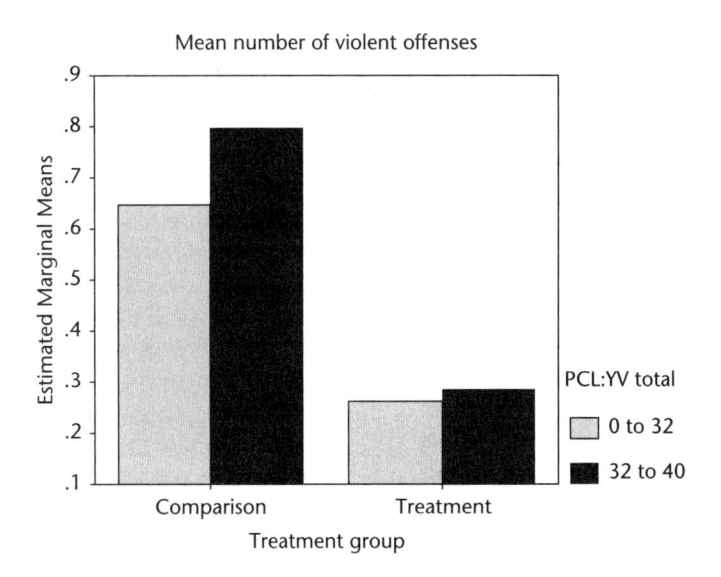

Figure 12.2 Mean number of violent offenses for Treatment and Comparison groups sorted by high and low PCL:YV total score.

and characteristics of psychopathy appears to have been disrupted by treatment for these youth. Although these results are encouraging, they tell us little about the process of change in treatment. To begin to address this issue, we conducted a study investigating the association between behavioral changes and treatment.

Study 3

Treatment Progress and Outcomes for Youth with Psychopathic Features

The findings in previous studies that the program appeared to reduce violent recidivism in youth with pronounced psychopathic features

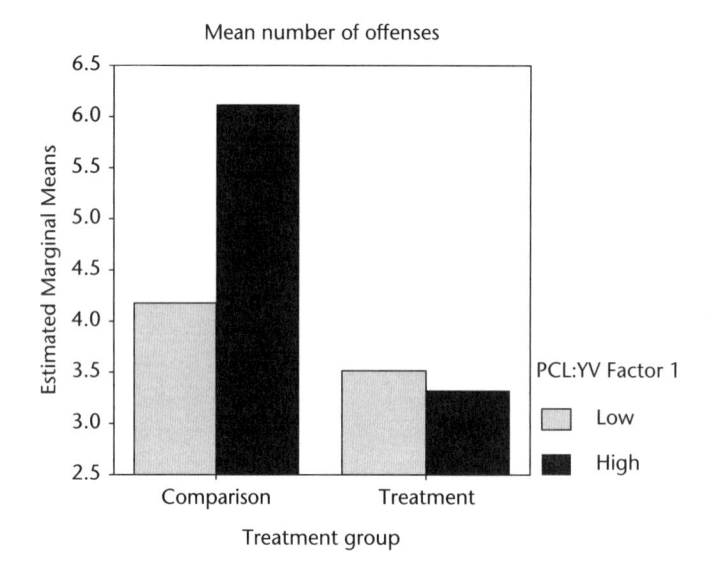

Figure 12.3 Mean number of criminal offenses for Treatment and Comparison groups sorted by high and low PCL:YV Factor 1 score.

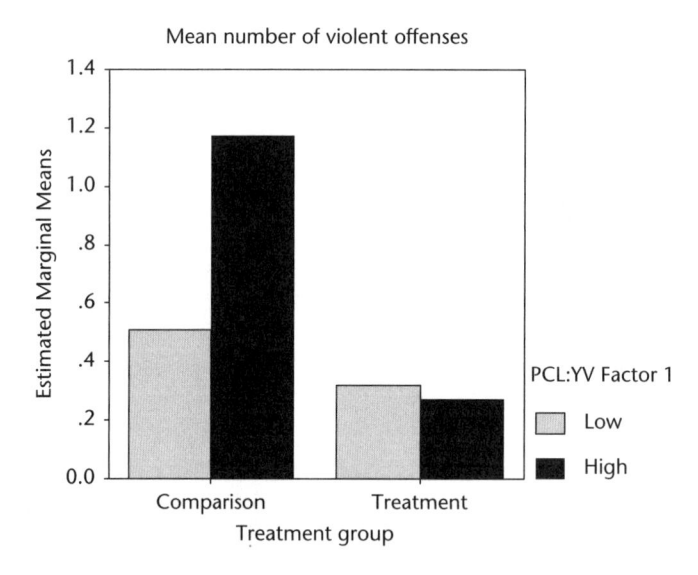

Figure 12.4 Mean number of violent offenses for Treatment and Comparison groups sorted by high and low PCL:YV Factor 1 score.

are important and warrant more detailed study. To determine if treatment is associated with changes in the other behavioral features of psychopathy we conducted an additional study focused on behavioral changes in the treatment process (Caldwell, McCormick, Umstead, & Van Rybroek, 2007).

In this study the treatment records of 86 youth who were consecutively admitted to MJTC were examined. The records of these youth had not been included in previous studies. All of the youth had been provided an extensive multidisciplinary assessment on admission to MJTC. Information garnered from that assessment

Table 12.8 Correlations Between Historical and Clinical Variables and PCL:YV Total Score

Variable	PCL:YV total
Number of CD symptoms	.63***
Number of violent CD symptoms	.55***
Worst victim injury code	.21*
Age of onset of behavioral problems	−.32**
Criminal versatility score	.39***
History of institutional violence	.29*
Baseline behavioral scores	−.40***
Baseline security scores	−.29**

*$p < .05$; **$p < .01$; ***$p < .001$.

and previous records was used to score the Psychopathy Checklist: Youth Version for each youth and to code pretreatment covariates.

As shown in Table 12.8, the PCL:YV was significantly associated with an array of indicators of the onset, severity, and persistence of behavioral problems up to the initial weeks of residence on MJTC.

The Behavioral Assessment System provides detailed information related to each youth's cooperation with treatment and behavioral control on the unit. The system generates a composite score that is compiled from up to ten individual scale scores that are rated throughout the day by the frontline treatment staff. For the purposes of this study, we aggregated daily data into weekly averages. These scales are sensitive to treatment involvement and cooperation and to incidents of minor misconduct.

In addition to the data from the Behavioral Assessment System we collected data on the number of days in each week that a youth was subject to any security restrictions. The Administrative Code of the Wisconsin Department of Corrections, under which MJTC operates, specifies behaviors for which a juvenile can be subject to security restrictions. Generally security restrictions involve restriction from certain activities or time isolated in the youth's room. Typically, misconduct that prompts a security restriction is more aggressive behavior or behavior that otherwise poses a significant security threat. We used the percentage of days each week that the youth was free from any form of security restriction as the dependent measure. In this way, higher scores in both dependent measures represented more positive behavior.

Scores from the Behavioral Assessment System covering the first 3 weeks that each youth was in full programming served as the Time 1, baseline behavioral score. Each youth's mean score from the final 3 weeks on MJTC served as the Time 2, final behavioral scores. Similarly, we calculated the percentage of the week each youth was free from any form of security restriction during the baseline and final time frame to serve as the security measures.

Psychopathy Traits and Treatment Progress

Previous studies have commonly found an association between measures of psychopathy and disruptive behavior in treatment and poor treatment progress (Brandt et al., 1997; Dolan & Rennie, 2005; Forth et al., 1990; Kaplan & Cornell, 2004; Murdock-Hicks et al., 2000; O'Neill et al., 2003; Rogers et al., 1997; Stafford & Cornell, 2003). We were specifically interested in whether high levels of psychopathic personality traits (measured by the PCL:YV), would block meaningful treatment progress. To examine this we divided the sample in half using the median PCL:YV score. This produced two groups of 43 youth with lower (mean= 26.3, $SD = 3.8$) and higher (mean = 34.3, $SD = 2.2$) PCL:YV scores. The lower PCL:YV group had PCL:YV scores between 15 and 31 while the high PCL:YV group had scores greater than 31. To examine the interaction between amount of treatment and PCL:YV scores we conducted a 2 × 2 repeated measures analysis of variance (ANOVA).

The results showed that youth with higher PCL:YV scores obtained lower Behavioral scores at both admission and release (Table 12.9). However, both groups showed similar improvement between admission and release. Specifically, the 2 × 2 repeated measures ANOVA showed a main effect for time (baseline and final behavioral scores) that was significant, $F(1, 84) = 196.10$, $p < .001$. Likewise, the main effect for the dichotomized PCL:YV variable was significant, $F(1, 84) = 8.35$, $p < .005$. However, the Time × Psychopathy interaction was not

Table 12.9 Repeated Measures ANOVA of High and Low PCL:YV Total Scores to Predict the Mean Baseline and Final Behavioral Scores

Effect	F	df (Contrast/error)	η^2	Significance ($p =$)
Treatment time	196.10	1/84	.700	.000
PCL:YV total group	8.35	1/84	.090	.005
Treatment time × PCL:YV	0.53	1/84	.006	.467

significant, $F(1, 84) = 0.53$, ns. These data indicate that both the high and low PCL:YV groups showed improved behavioral scores between the two time frames marking the beginning and end of treatment. Also, youth with higher PCL:YV scores showed lower behavioral scores in both time frames. However, youth with higher PCL:YV scores showed improvement that was similar to youth with lower PCL:YV scores between the beginning and end of treatment. Psychopathy as measured by the PCL:YV did not appear to block treatment progress in this group.

To determine if there were differences in response to treatment at the level of PCL:YV Factors we divided each of the PCL:YV Factors at the median score. We then conducted a similar 2 × 2 repeated measures ANOVA in which we examined the relationship between admission and final behavioral scores and high and low Factor 1 and 2 scores.

The results (Table 12.10) reveal that the time variable was significant in both analyses. This means that the youth in this study obtained significantly better behavioral scores in the final time period than at baseline. The analysis for both Factors also showed a significant main effect, indicating that these characteristics had a significant influence on behavioral scores. However, although there was no Time × Factor interaction for Factor 2, the interaction term for Factor 1 was significant. This indicates that the influence of Factor 1 varied with the time frame. The results for the security variable were similar in that there was a consistent main effect for time. However, the main effect for Factor 1 and both of the Time × Factor interactions were not significant, whereas the main effect for Factor 2 was significant. This suggests, again, that treatment was associated with improved behavior and that the link between more severe misconduct and the Factor 1 characteristics were not stable across the treatment time frames.

Reviewing the results for the behavioral data in Figures 12.5 and 12.6 shows that youth that had high Factor 1 scores improved more than those with low Factor 1 scores. By comparison, youth with high and low Factor 2 scores improved similarly. This finding suggests that the link between disruptive or uncooperative

Table 12.10 Repeated Measures ANOVA of High and Low PCL:YV Factor Scores to predict the Mean Baseline and Final Behavioral Scores ($N = 86$)

Effect	F	df (Effect/error)	η^2	Significance ($p =$)
Factor 1	5.40	1/84	.060	.023
Time	18.61	1/84	.181	.000
Time × Factor 1	6.31	1/84	.070	.014
Factor 2	7.61	1/84	.083	.007
Time	13.66	1/84	.140	.000
Time × Factor 2	0.27	1/84	.003	.602

TREATMENT OF ADOLESCENTS WITH PSYCHOPATHIC FEATURES

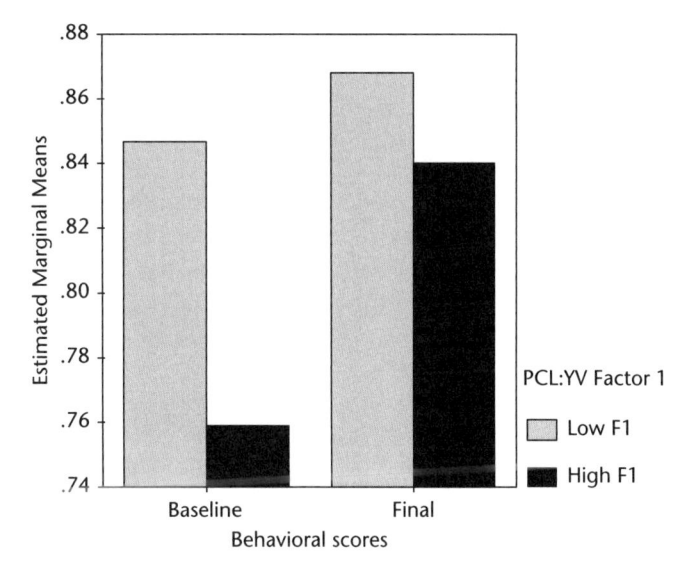

Figure 12.5 Repeated measures ANOVA of mean Baseline and Final behavioral scores by high and low PCL:YV Factor 1 scores.

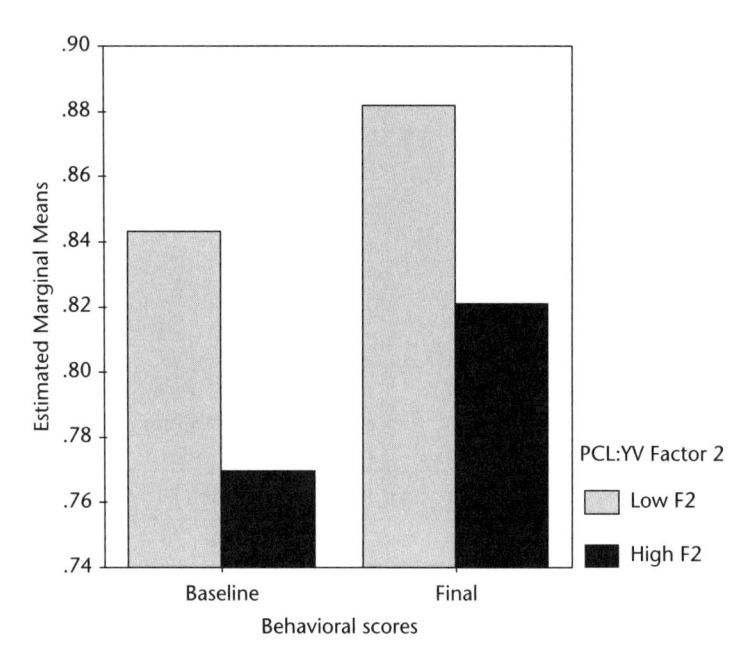

Figure 12.6 Repeated measures ANOVA of mean Baseline and Final behavioral scores by high and low PCL:YV Factor 2 scores.

treatment behavior and Factor 1 characteristics was less stable than with Factor 2. It is not surprising that Factor 2 characteristics (which draw from a number of historical, static variables), should have a consistent influence on behavior. However, the lack of stability in the influence of Factor 1 characteristics on behavior suggests that the influence on behavior of the interpersonal and emotional characteristics of psychopathy may be positively affected by treatment.

Although the MJTC program places an emphasis on the social ecology of the individual, there is no formal programming directed at changing the specific characteristics contained in Factor 1. Thus the mechanism of the change in the link between these characteristics and aggressive behavior is unknown and may result from any number of variables that are embedded in the treatment program.

Although the above analysis provides important information about the occurrence of change in treatment, the link between the amount of treatment and behavioral change was not examined. For this reason it was important to determine if the improvement in behavioral and security scores were related to the amount of treatment a youth received. To further examine the effect that psychopathic features and treatment exposure had on behavioral and security scores, we completed a hierarchical regression equation to predict the final security and behavioral scores. To do this we first identified variables that may have affected the length of treatment exposure using a forward conditional hierarchical regression equation. This procedure revealed that two variables were significantly related to the length of treatment (criminal versatility and age at admission to MJTC). We then controlled for these variables in later analyses.

We first entered PCL:YV score, age, criminal versatility, and the baseline behavioral score as a block into a hierarchical regression equation, followed by a second block that added the weeks of treatment, to predict the change in behavioral scores during treatment. This analysis (Table 12.11) revealed two important findings. First, despite the fact that the PCL:YV score had been strongly associated with past behavioral problems, and institutional misconduct up to the first weeks of admission to MJTC (including baseline behavioral scores), psychopathic personality traits were unrelated to improvement in behavioral scores during MJTC treatment. Second, after controlling for variables that may have affected length of stay, baseline behavioral scores, and PCL:YV score, the amount of treatment (measured in weeks of treatment) significantly predicted the final behavioral scores. The same results were found when examining the final security scores (Table 12.12).

Psychopathy Traits, Treatment Progress, and Violent Reoffense

Although significant improvement in institutional adjustment and aggression with treatment is an important finding, it was possible that youth with psychopathic features had made only superficial adjustments to their behavior in a self-interested effort to speed their release. Youth with more pronounced psychopathy features (particularly Factor 1 characteristics involving manipulation and deception) may have been more skilled at

Table 12.11 Stepwise Multiple Regression to Predict Change in Behavioral Scores ($N = 86$)

Variable	R^2 change	Standardized β	F change
Block 1	.2.93		8.38***
PCL:YV total		–.110	
Age		–.121	
Criminal versatility		.155	
Baseline score***		–.559	
Block 2	.085		10.95***
PCL:YV total		–.076	
Age*		.191	
Criminal versatility*		.233	
Baseline score***		–.536	
Weeks of treatment**		.313	

* $p < .05$; ** $p < .01$; *** $p < .001$.

Table 12.12 Logistic Regression to Predict Change in Security Restrictions ($N = 86$)

Variable	R^2 change	Standardized β	F change
Block 1	.552		24.92***
PCL:YV total		−.088	
Age		.023	
Criminal versatility		.038	
Baseline score***		−.760	
Block 2	.058		11.91**
PCL:YV total		−.072	
Age		.086	
Criminal versatility		.102	
Baseline score***		−.786	
Weeks of treatment**		.259	

*$p < .05$; **$p < .01$; ***$p < .001$.

learning how to con the system to obtain higher behavioral scores, while garnering no change in their underlying psychopathic traits. If this were the case, evidence of treatment progress in the form of improved behavioral scores should not be related to violent recidivism. Instead the PCL:YV scores, which have been demonstrated to predict violent recidivism on release (Edens et al., 2001; Falkenbach et al., 2003; O'Neill et al., 2003), and reliably predicted the onset, diversity, and intensity of aggression and behavioral problems up to the beginning of treatment, should continue to predict aggression in the follow-up period. To determine if these changes were sustained after release from the program, we conducted a 4-year follow-up study in which we collected data on new violent criminal charges. To examine this we initially entered the PCL:YV total and factor scores, along with four variables that have been related to persistent offending (the number of CD symptoms, the age of first arrest, age of admission to corrections, and criminal versatility) and the final behavioral score in a forward conditional Cox regression equation to predict new charges for a violent offense. Only the final behavioral score entered the equation (odds ratio = .05, Wald = 4.79, $p < .05$). To more directly compare the influence of PCL:YV scores and final behavioral scores on violent recidivism we conducted an analysis in which we entered the PCL:YV total score on the first step, followed by the final behavioral ratings on the second step of a hierarchical Cox regression analysis to predict violent recidivism. After controlling for the PCL:YV total score, the final behavioral score significantly predicted violent rearrest, $\chi^2 \Delta(1, 84) = 4.25, p < .05$. The PCL:YV score did not significantly contribute to the final prediction model (odds ratio = 1.04, Wald = .93, n.s.). However, the final behavioral score significantly predicted violent recidivism in the full prediction model (odds ratio = .03, Wald = 4.94, $p < .05$). Thus, the measure of positive treatment outcome (the final behavioral score) predicted violent recidivism whereas the measure of psychopathic personality features (the PCL:YV) did not.

DISCUSSION

Implications for Clinical Practice

In these studies of adolescent boys, psychopathic personality features were significantly associated in the expected way with the youth's history of behavioral problems up to the first few weeks on MJTC. From that point on, the youth's progress and cooperation in treatment and risk of future violent offending were determined by the amount of treatment they received. Furthermore, their level of treatment progress as measured by the final Behavioral scores predicted new violent offending while the PCL:YV score did not. These data further indicate that MJTC treatment appears to be effective in breaking the link between psychopathic features and violent behavior. Importantly, this effect appears to be more pronounced for youth with more severe psychopathic features (Figures 12.1 through 12.6) and has a greater impact on more severe violent offending.

The findings of behavioral changes related to Factor 1 of the PCL:YV raises a number of interesting questions. First, it is possible that youth with more substantial proclivities toward interpersonal manipulation and deceit also have more developed interpersonal skills. It is unlikely that, once acquired, a youth would lose the ability to lie and manipulate. Rather it is more likely that treatment may alter the way the youth decides to employ their interpersonal skills. This change may be entirely based on a self-interested calculation that interpersonal deception and exploitation is not ultimately as rewarding as more direct interactions are in the treatment setting. It is also possible that the observed changes reflect a more fundamental shift in the degree to which the youth values other humans. The proclivity for routine deception, exploitation, and manipulation of others requires a profound lack of appreciation of others as human beings. Adolescence is a time of extensive development in the social relatedness of the individual that includes the development of a sense of identity in relation to other people. Therefore, it may be a particularly fertile time for treatment to alter the social functioning of a youth. However, the treatment benefits observed in these studies are not restricted to the Factor 1 characteristics. Treatment was similarly effective when associated with all components of psychopathy that are measured by the PCL:YV, suggesting a more general shift in the relationship between violence and psychopathic features in treated youth.

Unfortunately, these studies shed no light on whether the internal characteristics of psychopathy in these youth were changed. The results of treatment can only be said to alter the functional manifestation of psychopathic features by decreasing the propensity for aggression. Exactly how treatment produced that result must remain speculative, for the time being. As a result, it is not clear what elements of the program are most vital in producing treatment benefits. However, several characteristics of the MJTC program are unusual and may be important in producing these results.

In an important work, Toch and Adams (1994) recommended a hybrid program for the treatment of disturbed violent offenders in a secured correctional setting. They described a model in which the program is separated from the rest of the prison; with staff in which security and treatment roles are blended; the security of the prison is retained, as is the philosophy and multidisciplinary treatment model of the mental health system; treatment goals are observable and focused on criminogenic needs; and the program incorporates a research component. Although MJTC was developed independent of these recommendations, the program is a very close match to this model. The results reported here lend support to the validity and generalizability of the Toch and Adams model for disruptive violent offenders.

Development of a serious treatment program for violent juveniles with psychopathic characteristics involves a paradigm shift away from traditional models of correctional care. In the MJTC program, the relationship between security measures and clinical intervention has been altered, so that the roles of security and clinical staff are less distinct. Important clinical functions (e.g., the Behavioral Assessment System and Today–Tomorrow contingency management program), are implemented by front-line staff who would generally have responsibilities exclusive to security in a traditional correctional model. Similarly, clinical staff members have responsibilities for the clinical and administrative supervision of front-line staff, and are responsible for carrying out security procedures in their work.

Significant administrative support is necessary to develop and maintain this type of approach. In working with dangerous individuals there is a tendency for the inevitable crisis to define staff attitudes and cause a drift toward more restrictive programming (Caldwell, 1992; Maier, Stava, Morrow, Van Rybroek, & Bauman, 1987). Draconian security measures have sometimes been developed under the guise of behavioral modification programs (Fellner, 2000; Toch, 2008). These approaches directly conflict with the Decompression Model, but direct and active administrative support is required to resist this pull and keep the program on track with its mission.

The program also places an emphasis on observable and relevant changes in behavior as the measure of treatment progress. The development of clinical insight, although valued, is viewed as relevant only to the extent that it results in positive behavior. There is a priority placed on engaging the youth in the treatment process and minimizing security-based interruptions to treatment. In addition, the program includes a research component to evaluate outcomes and the treatment process.

Implications for Policy

These data indicate that significant changes in the functioning of adolescents with psychopathic features can be obtained with appropriate treatment. The results suggest that treatment may be more effective with adolescents that have these features than with adults that do. Although appropriate treatment requires a substantial investment of time and resources, these studies also document that the effort can be very cost effective. These results suggest that juvenile justice systems that place a priority on developing appropriate specialized treatment programs for the most disruptive and aggressive juveniles in their systems can realize substantial benefits. Although the fiscal benefits are substantial, the human benefits of fewer violent offenses in the community are more important and compelling. Similarly, these results run counter to some recent policy trends that sentence juveniles with psychopathic features to very long-term confinement on the assumption that these traits are immutable.

One characteristic of the MJTC treatment program is the avoidance of extended punitive sanctions or restrictions. This is in keeping with a principle of treating intractable patients that asserts that treatment should avoid interventions that have failed in the past. Individuals with psychopathic traits in general, and the boys treated on MJTC specifically, tend to be insensitive to the corrective influence of deterrent sanctions. Instead, they respond to repeated punishments with the same, or more severe, antisocial behaviors. Assessing youth for psychopathic traits for the purpose of imposing more stringent sanctions, or to withhold treatment resources, runs counter to the results reported here. Rather, the assessment of psychopathy in adolescents should be in the service of developing, providing, and refining more effective specialized treatment services for these youth.

REFERENCES

Andrews, D., & Bonta, J. (2007). *Psychology of criminal conduct.* Southington, CT: Anderson.

Bandura, A. (1977). *Social learning theory.* Englewood Cliffs, NJ: Prentice Hall.

Barbaree, H. E. (2005). Psychopathy, treatment behavior, and recidivism: An extended follow-up of Seto and Barbaree. *Journal of Interpersonal Violence, 20,* 1115–1131.

Barbaree, H. E., Seto, M. C., & Langton, C. M. (2001, November). *Psychopathy, treatment behavior and sex offender recidivism: Extended follow-up.* Paper presented at the 20th annual conference of the Association for the Treatment of Sexual Abusers (ATSA), San Antonio, TX.

Benning, S. D., Patrick, C. J., Hicks, B. M., Blonigen, D. M., & Krueger, R. F. (2003). Factor structure of the Psychopathic Personality Inventory: Validity and implications for clinical assessment. *Psychological Assessment, 15,* 340–350.

Brandt, J. R., Kennedy, W. A., Patrick, C. J., & Curtin, J. J. (1997). Assessment of psychopathy in a population of incarcerated adolescent offenders. *Psychological Assessment, 9,* 429–435.

Caldwell, M. (1992). Incidence of PTSD among staff victims of patient violence. *Hospital and Community Psychiatry, 43*(8), 838–839.

Caldwell, M., McCormick, D., Umstead, D., & Van Rybroek (2007). Evidence of treatment progress and therapeutic outcomes among adolescents with psychopathic features, *Criminal Justice and Behavior, 34*(5), 573–587.

Caldwell, M. F., Skeem, J., Salekin, R., & Van Rybroek, G. (2006). Treatment response of adolescent offenders with psychopathy-like features. *Criminal Justice and Behavior, 33*(5), 571–596.

Caldwell, M., & Van Rybroek, G. (2001). Efficacy of a decompression treatment model in the clinical management of violent juvenile offenders.

International Journal of Offender Therapy and Comparative Criminology, 45(4), 469–477

Caldwell, M., & Van Rybroek, G. (2004). Reducing violence in serious and violent juvenile offenders using an intensive treatment program. *International Journal of Law and Psychiatry, 28,* 622–636.

Caldwell, M. F., Van Rybroek, G., & Vitacco, M. (2006). Are violent delinquents worth treating? A cost–benefit analysis. *Journal of Research in Crime and hDelinquency, 43*(2), 148–168.

Cleckley, H. (1982). *The mask of sanity* (5th ed.). St Louis, MO: Mosby.

Cohen, M. A. (1988). Pain, suffering, and jury awards: A study of the cost of crime to victims. *Law and Society Review, 22*(3), 537–555.

Cohen, M. A. (1998). The monetary value of saving a high—risk youth. *Journal of Quantitative Criminology, 14,* 5–56.

Cohen, M. A. (2000). Measuring the costs and benefits of crime and justice. In D. Duffee (Ed.), *Measurement and analysis of crime and justice* (Vol. 4., pp. 265–315). Washington, DC: National Institute of Justice.

Cohen, M. A., Miller, T., & Rossman, S. (1994). The costs and consequences of violent behavior in the United States. In: A. J. Reiss, Jr. & J. A. Roth (Eds.), *Understanding and preventing violence,* Vol. 4: *Consequences and control* (pp. 67–166). Washington, DC: National Academies Press.

Cooke, D. J., & Michie, C. (1997). An item response theory evaluation of Hare's Psychopathy Checklist. *Psychological Assessment, 9,* 2–13.

Dehejia, R., & Wahba, S. (2002). Propensity score matching methods for non-experimental causal studies. *Review of Economics and Statistics, 84(1),* 151–161.

Dolan, M., & Rennie, C. (2006). Reliability and validity of the psychopathy checklist: Youth version in a UK sample of conduct disordered boys. *Personality and Individual Differences, 40,* 65–75.

Doren, D. (1987). *Understanding and treating the psychopath.* New York: Wiley.

D'Silva, K., Duggan, C., & McCarthy, L. (2004). Does treatment really make psychopaths worse? A review of the evidence. *Journal of Personality Disorders, 18,* 163–177.

Edens, J., Campbell, J., & Weir, J. (2007). Youth psychopathy and criminal recidivism: A meta-analysis of the Psychopathy Checklist measures. *Law and Human Behavior, 31,* 53–75.

Edens, J. F., Skeem, J. L., Cruise, K. R., & Cauffman, E. (2001). Assessment of 'juvenile psychopathy' and its association with violence: A critical review. *Behavioral Sciences and the Law, 19,* 53–80.

Falkenbach, D., Poythress, N., & Heide, K. M. (2003). Psychopathic features in a juvenile diversion population: Reliability and predictive validity of two self-report measures. *Behavioral Sciences and the Law, 21,* 787–805.

Fellner, J. (2000). Out of sight: Super-maximum security confinement in the United States, *Human Rights Watch, 12*(1). Retrieved from:(http://www.hrw.org/reports/2000/supermax/index.htm). Retrieved January 2, 2005.

Forth, A. E., & Burke, H. C. (1998). Psychopathy in adolescence: Assessment, violence, and developmental precursors. In: D. J. Cooke, A. D. Forth, & R. D. Hare (Eds.), *Psychopathy: Theory, research and implications for society* (pp. 205–230). Dordrecht, The Netherlands: Kluwer.

Forth, A. E., Hart, S. D., & Hare, R. D. (1990). Assessment of psychopathy in male young offenders. *Psychological Assessment: A Journal of Consulting and Clinical Psychology, 2,* 342–344.

Forth, A. E., Kosson, D., & Hare, R. D. (2003). *Psychopathy Checklist—Youth Version.* Toronto, Ontario, Canada: Multi-Health Systems.

Frick, P. J., & Hare, R. D. (2001). *The antisocial process screening device (APSD).* Toronto, Ontario, Canada: Multi-Health Systems.

Frick, P. J., Kimonis, E. R., Dandreaux, D. M., & Farell, J. M. (2003). The 4 years stability of psychopathic traits in non-referred youth. *Behavioral Sciences and the Law, 21,* 1–24.

Goldstein, A., Glick, B., & Gibbs, J. (1998). *Aggression replacement training: A comprehensive intervention for aggressive youth.* Champaign, IL: Research Press.

Gretton, H., Hare, R., & Catchpole, R. (2004). Psychopathy and offending from adolescent to adulthood: A 10-year follow-up. *Journal of Consulting and Clinical Psychology, 72,* 636–645.

Gretton, H. M., McBride, M., Hare, R. D., O'Shaughnessy, R., & Kumka, G. (2001). Psychopathy and recidivism in adolescent sex offenders. *Criminal Justice and Behavior, 28,* 427–449.

Gu, X., & Rosenbaum, P. (1993). Comparison of multivariate matching methods: Structures, distances, and algorithms. *Journal of Computational and Graphical Statistics, 2,* 405–420.

Hare, R. D. (1991). *The Hare Psychopathy Checklist—Revised.* Toronto, Ontario, Canada: Multi-Health Systems.

Hare, R., Clark, D., Grann, M., & Thornton, D. (2000). Psychopathy and the predictive validity of the PCL—R: An international perspective. *Behavioral Sciences and the Law, 18,* 623–645.

Hirschi, T. (1969). *Causes of delinquency.* Berkeley: University of California Press.

Kaplan, S., & Cornell, D. (2004). Psychopathy and ADHD in adolescent male offenders. *Youth, Violence, and Juvenile Justice, 2,* 148–160.

Karpman, B. (1946). Psychopathy in the scheme of human typology. *Journal of Nervous and Mental Disease, 103,* 276–288.

Lösel, F. (1998). Treatment and management of psychopaths. In: D. J. Cooke, A. E. Forth, & R. D. Hare (Eds.), *Psychopathy: Theory, research, and implications for society* (pp. 303–354). Dordrecht, The Netherlands: Kluwer.

Lykken, D. T. (1995). *The antisocial personalities.* Hillsdale, NJ: Erlbaum.

Lynam, D. R. (1997). Pursuing the psychopath: Capturing the fledgling psychopath in a nomological net. *Journal of Abnormal Psychology, 106*(3), 425–438.

Lynam, D. R. (2008). The stability of psychopathy from adolescence into adulthood: The search for moderators. *Criminal Justice and Behavior, 35*(2), 228.

Lynam, D. R., Caspi, A., Moffitt, T. E., Loeber, R., & Stouthamer-Loeber, M. (2007). Longitudinal evidence that psychopathy scores in early adolescence predict adult psychopathy. *Journal of Abnormal Psychology, 116*(1), 155–165.

Lynam, D. R., Stouthamer-Loeber, M., & Loeber, R. (2008). The stability of psychopathy from adolescence into adulthood: The search for moderators. *Criminal Justice and Behavior, 35*(2), 228–243.

Maier, G., Stava, L., Morrow, B., Van Rybroek, G., & Bauman, K. (1987). A model for understanding and managing cycles of aggression among psychiatric inpatients. *Hospital and Community Psychiatry, 38*(5), 520–524.

McCord, W., & McCord, J. (1964). *The psychopath: An essay on the criminal mind.* Princeton, NJ: Van Nostrand.

Monroe, C. M., Van Rybroek, G. J., & Maier, G. J. (1988). Decompressing aggressive inpatients: Breaking the aggression cycle to enhance positive outcome. *Behavioral Sciences and the Law, 6,* 543–557.

Muñoz, L. C., & Frick, P. J. (2007). The reliability, stability, and predictive utility of the self-report version of the Antisocial Process Screening Device. *Scandinavian Journal of Psychology, 48*(4), 299–312.

Murdock-Hicks, M., Rogers, R., & Cashel, M. (2000). Predictors of violent and total infractions among institutionalized male juvenile offenders. *Journal of the American Academy of Psychiatry and the Law, 28,* 183–190.

Ogloff, J. P. R., & Lyon, D. R. (1998). Legal issues associated with the concept of psychopathy. In: D. J. Cooke, A. D. Forth, & R. D. Hare (Eds.), *Psychopathy: Theory, research and implications for society* (pp. 399–420). Dordrecht, The Netherlands: Kluwer.

Oliver, M., & Wong, S. (2009). Therapeutic responses of psychopathic sexual offenders: Treatment attrition, therapeutic change, and long-term recidivism. *Journal of Consulting and Clinical Psychology, 77*(2), 328–336.

O'Neill, M. L., Lidz, V., & Heilbrun, K. (2003). Adolescents with psychopathic characteristics in a substance abusing cohort: Treatment and process outcomes. *Law and Human Behavior, 27,* 299–313.

Quay, H. C. (1964). Dimensions of personality in delinquent boys as inferred from factor analysis of case history data. *Child Development, 35,* 479–484.

Rice, M. E., Harris, G., & Cormier, C. (1992). An evaluation of a maximum-security therapeutic community for psychopaths and other mentally disordered offenders. *Law and Human Behavior, 16,* 399–412.

Rogers, R., Jackson, R., Sewell, K., & Johansen, J. (2003). Predictors of treatment outcome in dually-diagnosed antisocial youth: An initial study of forensic inpatients. *Behavioral Sciences and the Law, 22,* 215–222.

Rogers, R., Johansen, J., Chang, J., & Salekin, R. (1997). Predictors of adolescent psychopathy: Oppositional and conduct-disordered symptoms. *Journal of the American Academy of Psychiatry and the Law, 25,* 261–271.

Rosenbaum, P., & Rubin, D. (1983). The central role of the propensity score in observational studies for causal effects. *Biometrika, 70,* 41–55.

Rosenbaum, P., & Rubin, D. (1984). Reducing bias in observational studies using subclassification on the propensity score. *Journal of the American Statistical Association, 79,* 516–524.

Rubin, D. (1997). Estimating causal effects from large data sets using propensity scores. *Annals of Internal Medicine, 127*, 757–763.

Salekin, R. T. (2002). Psychopathy and therapeutic pessimism: Clinical lore or clinical reality? *Clinical Psychology Review, 22*, 79–112.

Salekin, R. T., & Lochman, J. E. (2008). Child and adolescent psychopathy: The search for protective factors. *Criminal Justice and Behavior, 35*, 159–172.

Sampson, R., & Laub, J. (1997). *Crime in the making*. Cambridge, MA: Harvard University Press.

Seagrave, D., & Grisso, T. (2002). Adolescent development and the measurement of juvenile psychopathy. *Law and Human Behavior, 26*, 219–240.

Seto, M. C., & Barbaree, H. (1999). Psychopathy, treatment behavior, and sex offenders recidivism. *Journal of Interpersonal Violence, 14*, 1235–1248.

Sherman, L. W. (1993). Defiance, deterrence, and irrelevance: A theory of the criminal sanction. *Journal of Research in Crime & Delinquency, 30*,(4), 445–474.

Shields, I. W., & Simourd, D. (1991). Predicting predatory behavior in a population of incarcerated young offenders. *Criminal Justice and Behavior, 18*, 180–194.

Spain, S. E., Douglas, K. S., Poythress, N. G., & Epstein, M. (2004). The relationship between psychopathic features, violence and treatment outcome: The comparison of three youth psychopathic measures. *Behavioral Sciences and the Law, 22*, 85–102.

Stafford, E., & Cornell, D. G. (2003). Psychopathy scores predict inpatient aggression. *Assessment, 10*, 102–112.

Steinberg, L. (2002). The juvenile psychopath: Fads, fictions, and facts. *National Institute of Justice Perspectives on Crime and Justice 2000–2001 Lecture Series*, Vol. 5, Retrieved from: http://www.schmalleger.com/pubs/LS2001-2_2.pdf. (Accessed November 1, 2009).

Thorne, F. C. (1959). The etiology of sociopathic reactions. *American Journal of Psychotherapy, 13*, 319–330.

Toch, H. (2008). Punitiveness as "behavior management." *Criminal Justice and Behavior, 35*(3), 388–397.

Toch, H., & Adams, K. (1994). *The disturbed violent offender*. Washington, DC: American Psychological Association Press.

Wallace, J., & Newman, J. (2004). A theory-based treatment model for psychopathy. *Cognitive and Behavioral Practice, 11*, 178–189.

Wong, S., & Hare, R. D. (2005). *Guidelines for a psychopathy treatment program*. North Tonawanda, NY: Multi-Health Systems.

Zinger, I., & Forth, A. E. (1998, July). Psychopathy and Canadian criminal proceedings: The potential for human rights abuses. *Canadian Journal of Criminology*, ,237–276.

PART SEVEN
RECIDIVISM AND PSYCHOPATHY

CHAPTER 13

Psychopathy and Violent Recidivism

Marnie E. Rice and Grant T. Harris

There is abundant evidence that psychopathy is related to violent recidivism. Most of this research has utilized the Hare Psychopathy Checklist—Revised (PCL-R; Hare, 1991, 2003) or another of the Psychopathy Checklist (PCL) family of instruments (including the Psychopathy Checklist [PCL; Hare, 1985, unpublished manuscript], the Psychopathy Checklist Screening Version [PCL:SV; Hart, Cox, & Hare, 1995], and the Youth Version of the PCL [PCL:YV; Forth, Kosson, & Hare, 2003]) as the operational measure of the psychopathy construct. There are also a few studies of the ability of self-report, paper-and-pencil measures of psychopathy to predict violent recidivism (e.g., Salekin, 2008). There have also been several recent meta-analyses examining the relationship between Psychopathy Checklist scores and violent recidivism, and we begin with a summary of these. We then present data from our recent work examining the long-term predictive accuracy of the PCL-R and a risk assessment instrument containing PCL-R score as an item. We follow with a discussion of why measures of psychopathy do such a good job predicting violent recidivism and conclude by examining the legal implications of this work.

Leistico, Salekin, DeCoster, and Rogers (2008) conducted a meta-analysis of studies of the relationship between score on the PCL-R and violent recidivism. They included studies of both adults and juveniles of both sexes. They found 68 effect sizes yielding a mean weighted Cohen's *d* of .47, which, by standards in the behavioral sciences (Cohen, 1988), is a moderate effect size that also corresponds to an area under the receiver-operating characteristic (AUC) equivalent of .63 (see Rice & Harris, 2005).[1] Walters (2003) also conducted a meta-analysis of studies using one of the PCL family of instruments with adult or adolescent offenders. They included seven effect sizes (that partially overlapped with those in the Leistico et al., 2008 study described previously) and found a mean (weighted) AUC of .67. Campbell, French, and Gendreau (2009) conducted a meta-analysis of the PCL-R and other measures used for the prediction of violent recidivism among adult offenders and found 24 effect sizes (which, again, partially overlapped with those in the other meta-analyses) for the PCL-R. They reported a mean *r* weighted by sample size of .27 (confidence interval [CI] = .24–.30), which corresponds to an AUC equivalent of .66, again a moderate effect size. They also found that the PCL-R outperformed a tool developed specifically to predict violent recidivism (the HCR-20). There have also been two meta-analyses of the PCL:YV, both of which included effect sizes that partially overlapped with one another and with those reported in the two meta-analyses above that included adolescents. Olver, Stockdale, and Wormith (2009) reported 20 effect sizes, with a mean *r* weighted by sample size of .25 (95% CI = .21–.29), which corresponds to an AUC of .64. Similarly, Edens, Campbell, and Weir (2007) reported 15 effect sizes with a mean *r* weighted by sample size of .25 (95% CI = .20–.31), again with a corresponding AUC of .64. We also note results from the one individual study (Wormith, Olver,

Stevenson, & Girard, 2007) published since collection of studies for all these meta-analysis were completed, in which an AUC of .66 was found for the prediction of violent recidivism using the PCL-R for 61 Canadian male adult offenders. It is also of interest to note that, although the participants in almost all of the studies included in the meta-analyses reported in the preceding text were males, the few studies that included females (primarily the studies of adolescents) reported results that were, on average, similar to those for males.

It is evident that psychopathic traits fall into at least two distinguishable domains. Traits associated with callousness, lack of empathy, insincerity, shallow affect, cold-heartedness, and emotional unresponsiveness are correlated with, but different from, aggression, hostility, negative emotionality, impulsivity, and interpersonal exploitation. Thus, many people exhibit both aspects, but among many others, one factor predominates (Benning, Patrick, Hicks, Blonigen, & Krueger, 2003; Hicks & Patrick, 2006; Loney, Taylor, Butler, & Iacono, 2007; Patrick, Edens, Poythress, Lilienfeld, & Benning, 2006; Skeem, Johansson, Andershed, Kerr, & Louden, 2007). Both aspects exhibit high levels of heritability (Burt, McGue, Carter, & Iacono, 2007; Hicks et al., 2007; Larsson et al., 2007; Viding, Frick, & Plomin, 2007). The PCL-R Factor 1 (Interpersonal/Affective) and Factor 2 (Social Deviance) reflect these two domains. There is also evidence that each of the factors can be further divided into two facets (Hare, 2003).

Some of the meta-analyses of PCL measures of psychopathy described in the preceding text also examined the factors and/or facets of the PCL for their independent relationships with violent recidivism. The evidence is again quite consistent in showing that Factor 2 (social deviance) and Facet 4 (antisocial) best predict violence. Leistico et al. (2008) reported an AUC of .66 for Factor 2 compared to .61 for Factor 1 (and .63 for the total score). Walters (2003) reported an AUC of at least .65 for Factor 2 compared to .60 for Factor 1. Walters, Knight, Grann, and Dahle (2008) examined six adult samples and reported an AUC equivalent of .71 (a large effect) for the relationship between Facet 4 and violent recidivism, significantly higher than for any of the other three facets, and in regression, adding incremental predictive value to the other three facets combined. This was similar to an AUC equivalent of .74 for Facet 4 and .73 for Factor 2 (compared to .66 for PCL-R total score) among adults reported by Wormith et al. (2007) and an AUC equivalent of .64 for Facet 4 (with the same or nearly the same, for the total score and Facets 2 and 3) among juveniles reported by Salekin (2008). Edens et al. (2007) reported an AUC equivalent of .65 for the PCL:YV Factor 2 for the prediction of violent recidivism, compared to .61 for Factor 1 (and .64 for the total score).

In summary, the PCL family of measures of psychopathy yield scores that consistently yield moderate to large effect sizes for the prediction of violent recidivism among juvenile and adult offenders. Also, the studies indicate considerable generalizability across offender age, sex, country, method of scoring, setting, length of follow-up, and race or ethnicity (Leistico et al., 2008; see also Hare, Clark, Grann, & Thornton, 2000). Factor 2 and Facet 4 are especially related to violence and often yield stronger relationships than total PCL scores.

PSYCHOPATHY AND VIOLENCE RISK ASSESSMENT TOOLS

Although we have seen in the preceding text that the PCL-R and other PCL measures of psychopathy yield moderate to large effect sizes for the prediction of violent recidivism, the measures were derived for the purpose of measuring the underlying construct of psychopathy rather than measuring risk for future violence. Several instruments have now been developed specifically for the purpose of predicting future violence among forensic populations. Subsequent research has led to the development of tools that use the PCL as a component and that are specifically for violence prediction among forensic populations. The most widely studied actuarial violence risk assessment tool is the Violence Risk Appraisal Guide (VRAG; Harris, Rice, & Quinsey, 1993; Quinsey, Harris, Rice, & Cormier, 2006). There is also a nonactuarial scheme, the Historical-Clinical-Risk Management 20 (HCR-20; Webster, Douglas, Eaves, & Hart, 1997), incorporating the PCL.

The VRAG was developed using a follow-up of 613 male offenders who had been admitted to a maximum security hospital. Approximately 50 separate variables suggested by previous authors or shown in prior studies to be related to violence were considered for inclusion. The variables considered comprised some from each of the following domains: demographic, childhood history, criminal history, and psychiatric history. The final instrument was designed by selecting the best predictors of violent recidivism from each domain and then subjecting those items to a multiple linear regression to choose the smallest set of nonredundant items that achieved the best predictive accuracy. Using this method, score on the PCL-R (a psychiatric history variable) achieved the highest relationship with violent recidivism (a criminal charge or equivalent for a violent offense within a mean 7 years of opportunity). The 12 VRAG items are weighted by their bivariate relationship with violent recidivism in construction—score on the PCL-R has the highest weight. The VRAG has explicit scoring criteria (Quinsey et al., 2006) with no recommended adjustment based on clinical judgment. We are aware of approximately 60 subsequent evaluations of the VRAG's ability to predict violent recidivism among several subject groups, in ten countries, over widely varying follow-up durations (Harris, Rice, & Quinsey, 2010; www.mhcp-research.com/ragreps.htm), all yielding a mean weighted AUC greater than .72.

Interestingly, the nonactuarial ("structured professional") HCR-20 was fashioned after the PCL-R inasmuch as it contains 20 items, each scored 0, 1, or 2. Its items fall into three rational groups: historical (including the PCL-R or PCL-SV), clinical, and risk management and were selected by considering each VRAG item (and including most in some form) and the literature at large. Items were not selected on the basis of an empirical relationship with violence in a particular sample, but rather on the basis of the empirical and nonempirical literature on the topic combined with the clinical experience of the developers, with special attention to characteristics whose alteration were hypothesized to produce changes in violence risk. After scoring each item according to a manual, the overall result/judgment is not to be a numerical score but instead a low/medium/high opinion arrived at by considering all available information including idiosyncratic features of the case. Almost all the follow-up research on the HCR-20, however, is based on unadjusted total scores; as such, a more correct term for the HCR-20's research basis might be "mechanical" rather than "structured professional."

The meta-analysis by Campbell et al. (2009) compared the predictive accuracy of the VRAG, HCR-20, and PCL-R (and two other instruments) for the risk of violent recidivism. The PCL-R and VRAG were statistically significantly more accurate than the HCR-20 score,[2] and the VRAG performed significantly better than all others. Interesting as well was that the Level of Service Inventory (LSI-R; Andrews & Bonta, 1995), a tool for general criminal recidivism that includes assessment of psychopathy, also outperformed the HCR-20, but not the PCL-R (Campbell et al., 2009). This set of findings was somewhat to be expected because the VRAG was the only actuarial tool specifically developed for the prediction of violent recidivism, and because several HCR-20 items did not have an established empirical association with violent recidivism when selected. Other meta-analytic research on the performance of risk assessment tools for sex offenders has reported that actuarial tools outperform unaided clinical judgment, whereas total scores on "structured professional judgment" schemes were intermediate (Hanson & Morton-Bourgon, 2009).

We have recently conducted a long-term follow-up of 1,261 Canadian male offenders released between 1960 and 1995 (Rice, Harris, & Lang, submitted). All of the men have been subjects in previous follow-up studies (Rice & Harris, 1995; Harris et al., 2003). All were followed up again between 2004 and 2007 if possible. All of these men had been scored on the VRAG (including score on the PCL-R) based on information obtained from their clinical records, usually at the time of their index offenses.

To examine how well the VRAG performed with and without the PCL-R, we calculated the

VRAG score exactly according to the scoring criteria (Quinsey et al., 2006), and then again by omitting the PCL-R. The average time at risk for violent recidivism was 20 years for those who had no subsequent violent recidivism, and the violent failure rate for the entire sample was 50%. We were able to determine, for some of the subjects, whether or not they had died during the follow-up and if they had, they were counted as successes with a follow-up time that spanned the entire time between their release date and the study end date so long as they were not known to have been charged or convicted of any violent offense. A survival curve showing the proportion surviving (i.e., with no violent failure) by number of months at risk falls quite steeply and then gradually levels off, reaching the floor (46% survival rate) after 27 years of opportunity (Rice, Harris, & Lang, 2012).

To further elucidate the role of psychopathy in predicting violent recidivism, we compared the ROC curves for the VRAG, the PCL-R, and the VRAG scored without the PCL-R (Figure 13.1). The AUC for the VRAG including the PCL-R item was .752 (95% CI = .725 to .779), compared to .691 for the PCL-R alone (95% CI = .661 to .721) and .747 (95% CI = .719 to .775) for the VRAG scored without the PCL-R. Thus, in this large sample over a long follow-up, the PCL-R alone achieved its expected moderately high accuracy for the prediction of violent recidivism. In fact, its accuracy was at least as high as has been found in the meta-analyses described earlier. Nevertheless, the VRAG scored with or without the PCL-R did significantly better. Again, this is to be expected as, unlike the PCL-R, the VRAG was developed for just one purpose—the prediction of violent recidivism among men similar to (and in some cases, the same as) those upon whom the instrument was constructed.

We were surprised, however, that the VRAG performed as well over an average 21-year follow-up as it had in the 7-year construction follow-up. We were not surprised that the VRAG performed very nearly as well without the PCL-R as with it because its construction incorporated considerable redundancy inasmuch as each item was related to violent recidivism, each made a unique contribution to the multivariate prediction, and weights reflected individual relationships with outcome. As we discuss further later, some VRAG items (e.g., behavior problems evident in elementary school, not having lived with

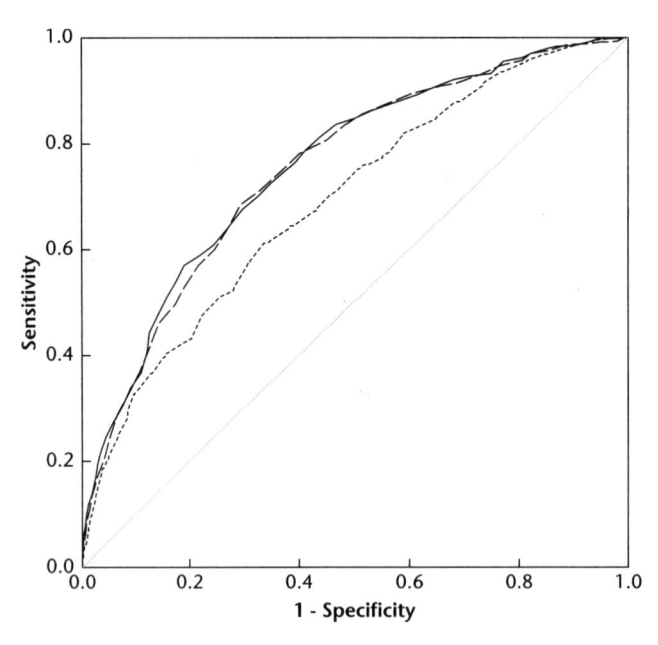

Figure 13.1 Receiver operating characteristic over 21-year mean follow-up for VRAG scored with PCL-R item (solid line), without PCL-R item (large-dashed line), and for PCL-R (small-dashed line).

both parents to age 16, alcohol problems, having a diagnosis of personality disorder, not having a diagnosis of schizophrenia, having a history of nonviolent offenses, and having failed on conditional release), as well as similar items on other risk assessment systems, all probably derive predictive power primarily from their being indicators of the phenomenon of psychopathy or life-course persistent antisociality.

We also examined how well the PCL-R and the VRAG predicted over specific follow-up times ranging from 6 months to 40 years. To conduct these analyses, we declared offenders who failed after various specific durations to be successes for that follow-up. To be counted as a success for a particular follow-up, an offender had to be known to have had at least that much time at risk and not have qualified as a violent recidivist before then. The results for the 6-month follow-up are shown in Figure 13.2. The base rate of violent recidivism for the 1,209 eligible offenders (i.e., violent recidivism within 6 months of opportunity or at least 6 months of opportunity without violent recidivism) was 7%. The AUC for the PCL-R was .740 (95% confidence interval [CI] =.687–.793) compared to .762 (95% CI = .710–.814) for the VRAG. By 10 years, the base rate was 47% for the 1,176 eligible offenders, and the AUC for the PCL-R was .685 (95% CI = .654–716) compared to .743 (95% CI = .716–.771) for the VRAG (Figure 13.3). At 20 years, the base rate was 63% for the 986 eligible offenders, and the AUCs for the PCL-R and VRAG were .703 (95% CI = .669–.737) and .783 (95% CI = .753–.812), respectively (Figure 13.4). At 30 years, the base rate was 78% for the 788 eligible offenders, and the AUCs for the PCL-R and the VRAG were .697 (95% CI = .654–.739) and .774 (95% CI = .735–.812), respectively. Finally, at 40 years, the violent recidivism base rate was 87% for the 712 eligible offenders, with AUCs for the PCL-R and VRAG of .699 (95% CI = .644–.754) and .804 (95% CI = .754–.853), respectively. Interestingly, the accuracy of both instruments at 6-months follow-up is nearly identical, but as the follow-up duration increases, the accuracy of the VRAG tended to increase, while accuracy for the PCL-R declined. It was also noteworthy that both instruments also predicted the number of violent reoffenses, the time until the first

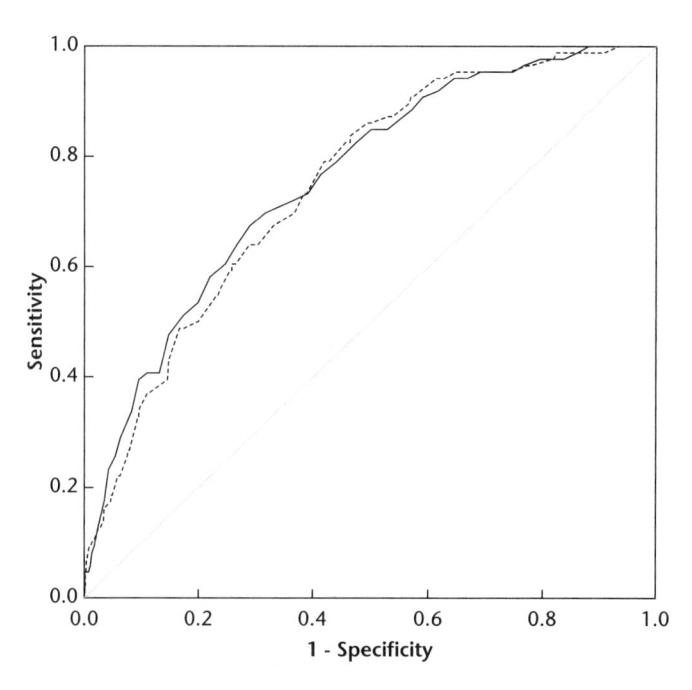

Figure 13.2 Receiver operating characteristic showing accuracy of VRAG (solid line) and for PCL-R (dashed line) over a 6-month follow-up.

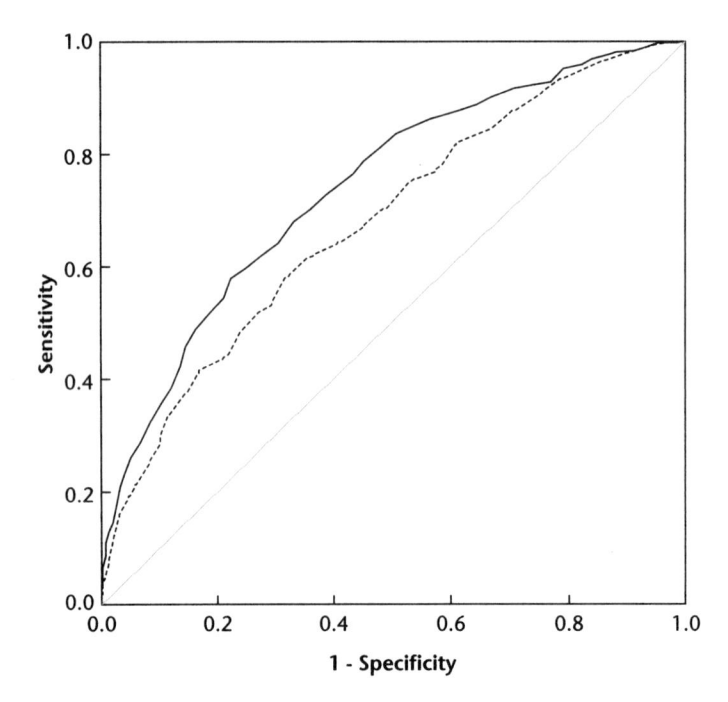

Figure 13.3 Receiver operating characteristic showing accuracy of VRAG (solid line) and for PCL-R (dashed line) over 10-year follow-up.

violent offense, the total months of opportunity to commit a new violent offense (i.e., a variable reflecting the time not incarcerated), and the severity of all violently recidivistic offenses (Rice, Harris, & Lang, 2012).

In some of our other research (Hilton, Harris, Rice, Houghton, & Eke, 2008) we examined the role of psychopathy in the prediction of repeated violence committed by men against a spouse. We had already developed an actuarial instrument (the Ontario Domestic Assault Risk Assessment [ODARA]; Hilton et al., 2004) for the prediction of wife assault recidivism. The instrument was developed for front-line police and criminal justice work using potential items easily obtained by officers attending domestic assault incidents. The ODARA contains 13 dichotomous items and has been shown to yield moderate to large effects in the prediction of recidivistic wife assault (Hilton & Harris, 2009; Hilton, Harris, & Rice, 2010; Hilton et al., 2004, 2008). Hilton et al. (2008) examined what variables including PCL-R scores, which could be obtained when more time and information were available, added to the predictive accuracy of the ODARA alone. The PCL-R performed just as well as the VRAG and ODARA (and better than three other nonactuarial instruments specifically designed for assessing domestic violence risk). In addition, the only item that consistently improved the prediction of wife assault recidivism (as well as its frequency and several indices of severity) after score on the ODARA was entered first was the PCL-R score. Thus, we developed a second actuarial instrument, the Domestic Violence Risk Appraisal Guide (DVRAG), for use in assessing wife assault risk among incarcerated domestic violence offenders or other cases where it is possible to score the PCL-R and it comprises the 13 ODARA items plus the PCL-R (Hilton et al., 2008).

WHY IS THE PCL-R SUCH AN EFFECTIVE RISK ASSESSMENT?

In several studies, PCL-R scores outperform and/or add to the predictive accuracy of optimal collections of items reflecting adult criminal history (e.g., Harris, Rice, & Cormier, 1991;

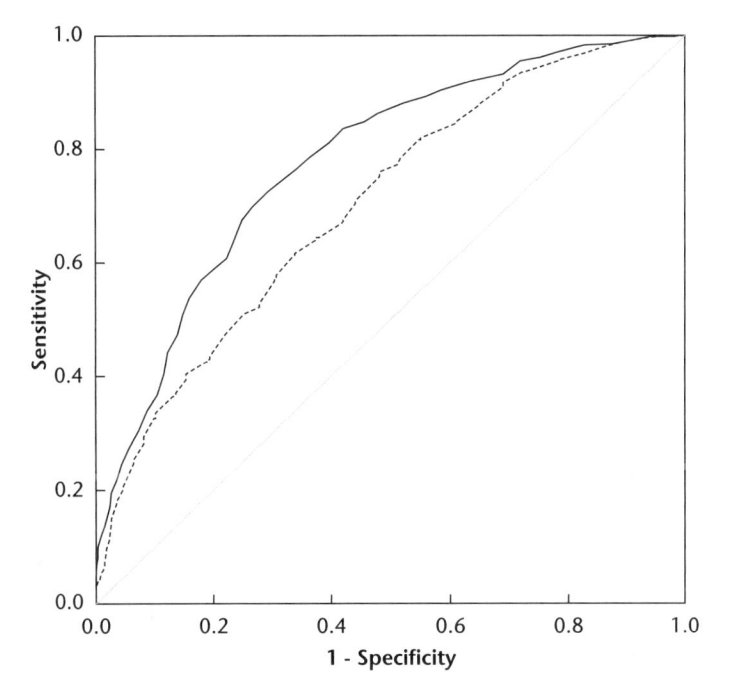

Figure 13.4 Receiver operating characteristic showing accuracy of VRAG (solid line) and for PCL-R (dashed line) over 20-year follow-up.

Harris et al., 1993; Hilton et al., 2008). We conclude, therefore, that the predictive performance of the PCL-R, Factor 2, and Facet 4 with respect to violent recidivism does not derive entirely from the PCL-R's inclusion of adult criminality (Harris & Rice, 2007a). Furthermore, the fact that the PCL-R is such an effective risk assessment (often outperforming tools specifically designed for the purpose and including measures of psychopathy) gives credence to the idea that aggression and antisociality are fundamental aspects of the phenomenon, and not merely sequelae (viz., Cooke, Michie, & Skeem, 2007; cf., Walters et al., 2008). That is, early aggression and antisociality (in the PCL-R, HCR-20, and VRAG), repeated criminal, violent, or aggressive behavior (PCL-R, ODARA, HCR-20, VRAG), substance abuse (VRAG, ODARA, HCR-20), failure on conditional release (PCL-R, ODARA, VRAG, HCR-20), and even separation from parents prior to the age of 16 (VRAG) are just as fundamental *indicia* of the underlying selfish and cheating psychopathic life strategy as are the observable behaviors that indicate impulsivity, irresponsibility, interpersonal exploitation, callousness, remorselessness, sensation-seeking, and so on.

More generally, psychopathy (clearly part of the more encompassing set of phenomena now often called externalizing traits; Dick, 2007; Farmer, Seeley, Kosty, & Lewinsohn, 2009; Forsman, Lichtenstein, Andershed, & Larsson, 2008; Hundt, Kimbrel, Mitchell, & Nelson-Gray, 2008; Pryor, Miller, Hoffman, & Harding, 2009) has been shown to be a life-course phenomenon with considerable temporal stability and predictable long-term outcomes. Thus, several recent studies of psychopathy, as an extreme form of externalizing traits, make the idea that psychopathy is something that is socially acquired simply untenable (see Baker, Jacobson, Raine, Lozano, & Bezdjian, 2007; Blair, Peschardt, Budhani, Mitchell, & Pine, 2006; Johansson et al., 2008; Larsson, Andershed, & Lichtenstein, 2006; Larsson, Viding, & Plomin, 2008). Such externalizing traits, together with some callous and unemotional traits, exhibit a distinct developmental trajectory detectable in individuals as young as age 3 (Glenn, Raine, Venables, & Mednick, 2007; Moffitt, 1993; Shaw, Bell, & Gilliom, 2000; Shaw, Gilliom, Ingoldsby, & Nagin, 2003; Vizard, Hickey, & McCrory, 2007; Vaughn & DeLisi, 2008). There is also clear evidence that this pattern of externalizing traits represents a stable, life-course phenomenon (Loney et al., 2007; Lynam, Caspi, Moffitt, Loeber, & Stouthamer-Loeber, 2007 Lynam, Charnigo, Moffitt, Raine, Loeber, & Stouthamer-Loeber, 2009; Lynam, Loeber, & Stouthamer-Loeber, 2008). This research does indicate that the expression of psychopathy is influenced by aspects of the environment (e.g., Knutson DeGarmo, & Reid, 2004), especially that which is "nonshared," but it does seem clear again that antisocial conduct and even aggression is just as fundamental an aspect of the phenomenon of psychopathy as the so-called affective aspects. Although there is much more to learn, we suggest a hypothesis that seems reasonably consistent with available scientific evidence.

PSYCHOPATHY'S OUTCOMES SUGGEST IT BE CONSIDERED A LIFE HISTORY STRATEGY

By definition, psychopaths exhibit (among other traits) early starting and life-course persistent antisociality and aggression. Considerable research shows that such life-course persistently antisocial individuals comprise a distinct path or trajectory starting (and identifiable) at a very early ages (see Lalumière, Harris, Quinsey, & Rice, 2005; Lalumière, Mishra, & Harris, 2008; Quinsey, Skilling, Lalumière, & Craig, 2004 for reviews of this evidence). Such individuals experience poorer outcomes in almost every aspect of life—poorer academic and vocational achievement, lower socioeconomic attainment, less successful intimate and social relationships, greater risk of addiction, worse lifespan health, and shorter longevity (Douglas, Herbozo, Poythress, Belfrage, & Edens, 2006; Farrington, Ttofi, & Coid, 2009; Huesmann, Dubow, & Boxer, 2009; Laub & Valliant, 2000; Nieuwbeerta & Piquero, 2008; Pulkkinen et al., 2009; Robins & O'Neal, 1958; Shepherd, Shepherd, Newcombe, & Farrington, 2009; Ullrich, Farrington, & Coid,

2008). About the only aspect of life that seems to be an exception here is the number of sexual partners and offspring, where life-course persistently antisocial individuals (or psychopaths) appear to be at least as "successful" as everyone else (Farrington et al., 2009; Harris, Rice, Hilton, Lalumière, & Quinsey, 2007; Pulkkinen, Lyyra, & Kokko, 2009), whereas that is not true of such serious disorders as schizophrenia, for example (MacCabe, Koupil, & Leon, 2009). This observation leads in turn to the question of the ultimate causes of life-course persistent antisociality, especially psychopathy.

The use of the term ultimate causes here is to distinguish between them, on the one hand, and proximal and developmental causes, on the other. It is understood by biologists that the valid articulation of both proximate and ultimate causes is required for complete scientific explanation. The proximate causes of psychopathy would, of course, refer to neurological and physiological mechanisms (and their interplay with external stimuli) that immediately produce psychopathic behavior. In addition, developmental causes refer to ontogeny—how the phenomenon arises (and is influenced by the developmental environment) within the maturing individual. Psychologists and social scientists are particularly familiar with proximate and developmental levels of explanation. Ultimate causation refers to adaptation and evolution—in the case of psychopathy, how the heritable basis of psychopathy arose and why it has persisted in the human population.

The nature of several psychopathic characteristics, the fact that it is a heritable condition, and the possibility that psychopathy has conferred no net reproductive costs to psychopaths of previous generations, leads directly to consideration of psychopathy as a life history strategy. The concept of life history strategy is established in biology where, in many species, individuals adopt different long-term approaches to the problems life presents (see Lalumière et al., 2005 for examples). These strategies can be conditional (or facultative) such that the individual adopts a life strategy contingent on environmental conditions that prevail, usually early in life (e.g., Belsky, Steinberg, & Draper, 1991; Buss

& Greiling, 1999; Giudice, Angeleri, & Manera, 2009). The ability to "select" one strategy over another is, of course, genetically based, but all the variance in outcome is attributable to aspects of the environment; therefore such facultative strategies exhibit no heritability as typically measured. Because psychopathy does exhibit considerable heritability, it cannot be a fully facultative life history strategy. Thus, if psychopathy is a life history strategy, it must be obligate, at least to a substantial degree. The best understood obligate life history strategy in humans is sex. Male and female humans adopt clearly different strategies in life, especially related to reproduction and sexual behavior (Daly & Wilson, 1983). Almost all aspects of each (of male vs. female) strategy are affected by features of the physical and social environment, but because mammals cannot (in any biological sense) change sex, observable differences persist throughout life and, among humans, male versus female sex comprises an obligate life-history strategy.

Elsewhere we and others have described in more detail the hypothesis that psychopathy has been a heritable reproductive strategy in which its component traits facilitate a short-term, risk-taking approach to reproductive fitness (Barr & Quinsey, 2004; Frank, 1988; Harpending & Sobus, 1987; Harris, Skilling, & Rice, 2001; Harris & Rice, 2006; Krupp, Sewall, Lalumière, Sheriff, & Harris, 2012; Mealey, 1995; Quinsey et al., 2004). The hypothesis further suggests that psychopathy has resulted from a particular type of selection, known as frequency dependent selection, that favors traits only when they occur at low prevalence (Buss & Greiling, 1999; Dall, Houston, & McNamara, 2004; Keller, 2008). In the case of psychopathy, simulation research suggests that cheating and defecting strategies thrive when cheaters are relatively unlikely to be encountered; otherwise vigilence and reprisal from noncheaters drives the prevalence back to low levels (Lalumière et al., 2008). Many psychopathic traits quite clearly consist of "skills" if the exploitation of others is the strategy (e.g., Wheeler, Book, & Costello, 2009). As well, many laboratory tasks show psychopaths performing worse than nonpsychopaths but in other tasks, psychopaths perform better (e.g., Schmauk,

1970). The hypothesis here is that psychopathy is a particular approach to life—an approach involving high activation, risk taking, selfishness, attraction to reward, insensitivity to punishment, quick decision-making, insensitivity to others' suffering, untrustworthiness, irresponsibility, aggression, manipulation, and deception of others (Barr & Quinsey, 2004; Frank, 1988; Harpending & Sobus, 1987; Harris et al., 2001; Mealey, 1995). It is important to note that, in theory, this psychopathic life history strategy is not superior to any other—indeed, it entails large costs. The point is, theoretically speaking, that it has persisted in the human population because it has been sufficiently reproductively successful to remain.

Elsewhere (Lalumière et al., 2005; Seto & Quinsey, 2006), we and others have addressed the possibility that psychopathy could be a rare genetic disorder similar to cystic fibrosis, for example. Briefly stated, psychopathy is too prevalent; it occurs with similar prevalence in too many cultures and ethnicities, and it is not associated with clear neurodevelopmental perturbations known to be associated with other serious disorders (Harris, Rice, & Lalumière, 2001; Lalumière, Harris, & Rice, 2001). The available evidence cannot reject the life-history hypothesis of psychopathy. As mentioned earlier, the heavy genetic component (and the concept of a life history strategy) does not mean, however, that the expression of psychopathy is expected to be insensitive to features of the environment.[3]

In our view, some of the most intriguing findings arise in the context of sexual behavior. As mentioned earlier, persistently antisocial people, and psychopaths in particular, exhibit earlier and more frequent sexual behavior (see Harris et al., 2007; Seto & Quinsey, 2006 for summaries), and high mating effort and early onset of sexual intercourse exhibit high levels of heritability. More recently, research has indicated that precocious, coercive sexuality is especially characteristic of the psychopathic life strategy (Harris et al., 2007). Even more recent findings suggest that sexual coercion and aggression or precocious sexuality (or both) are part-and-parcel to this early starting, life-course persistent antisociality—psychopathy, we hypothesize (Cale, Lussier,

Proulx, 2009; Johansson et al., 2008; McCrory, Hickey, Farmer, Vizard, 2008; Ramrakha et al., 2007; Schofield, Bierman, Heinrichs, & Nix, 2008). All this evidence suggests that psychopathy is a special kind of disorder—a disorder substantially and proximally caused by genes that, in biological terms, confer no fertility costs at low prevalence. As such, by some definitions at least (Wakefield, 1992), psychopathy does not meet the definition of a disorder. Its conceptualization as an alternative life history strategy is, we suggest, more consistent with the available evidence, and more consistent with its known proximate and ultimate causes. More important for present purposes however, we conclude that measures of psychopathy are such effective violence risk assessments because they measure the likelihood that an individual is executing the psychopathic life history strategy, a life strategy in which persistent antisociality and aggression (including, and perhaps especially, sexual aggression) are often inherent components to the strategy.

LEGAL IMPLICATIONS OF RESEARCH ON PSYCHOPATHY AND VIOLENT RECIDIVISM

In this final section, we illustrate (using Canadian law as an example) how we think the concept of psychopathy can contribute to public safety. Canada, like many or most other countries, allows for preventive detention and/or long-term community supervision of those at high risk of future violence. Canada has several different pieces of legislation that pertain to preventive detention including the insanity defense (not criminally responsible on account of mental disorder), civil commitment, dangerous offender, and long-term offender. In each of these laws, one of the requirements an individual must meet to be eligible to be subject to the legislation is that he or she must present a danger to others (except in the case of civil commitment where danger to self may also qualify). Because the PCL-R (and tools such as the VRAG that contain the PCL-R) allow the best currently available prediction of who is likely to be violent upon release, there is no doubt that public safety could be best served by the use of measures of psychopathy. For example, in our research, 86% of those 85 offenders with a PCL-R score higher than 30 were known to have committed a new violent offense within the average 20 years of opportunity, and more than 50% were known to have committed a new violent offense within 25 months of opportunity. Using the VRAG, more than 87% of those 119 offenders in one of the two highest risk categories (8 and 9) were known to have committed a new violent offense in the average 20 years at risk, and more than 50% were known to have committed a new violent offense within 25 months of opportunity. If applied judiciously (e.g., without increasing the rate of incarceration) to detain only those of highest risk, these tools could prevent many new violent offenses without detaining high numbers of offenders who would not commit new violent offenses.

It should be noted as well that we see no evidence of an inverse association between the duration of incarceration or the age at the time of assessment and the magnitude of the predictive relationship between formal assessments and recidivism (cf. Manchak, Skeem, & Douglas, 2008).

There are some practical problems with the use of the PCL-R, the VRAG, and other formal assessments in making decisions about preventive detention and/or long-term community supervision. The first pertains to reliability in forensic practice. Although it has been shown that both the PCL-R and the VRAG can be scored with high reliability (Hare, 1991, 2003; Quinsey et al., 2006), there have also been instances where interrater agreement on scoring has been poor (Edens, 2006; Murrie et al., 2009), especially in adversarial settings. A second potential problem is whether psychopathy is a mental disorder. Most Canadian laws (and those of most other jurisdictions) pertaining to preventive detention require that the individual being considered for such detention have a mental disorder that causes the individual (to be likely) to be violent. As discussed earlier, our research leads us to believe that psychopathy does not constitute a true disorder (by some definitions, e.g., Wakefield, 1992, at least), and thus, despite its accuracy in predicting future violence, would not meet current legal criteria to enable its use, if such definitions were adopted by the courts.

RESISTANCE TO THE FORENSIC APPLICATION OF PSYCHOPATHY

Based on the research summarized in this chapter, we conclude that the measurement of psychopathy in the context of formal violence risk assessment systems can afford real value in the operation of criminal justice. That is, some means must be used to determine the level of custody (e.g., community vs. incarceration), its conditions, and its duration. Somehow, candidates must be identified for long-term treatment and other interventions aimed at reducing the risk of violence in adult and juvenile offenders (Harris & Rice, 2003). Simply put, these tasks cannot be effectively accomplished without reference to a means to rank cases with respect to their actual risk of subsequent violent crime. Resources for custody, supervision, and intervention must be apportioned in accordance with that ranking. Any other policy must overall lead to unnecessary loss of offenders' liberty, avoidable violent crime, inefficient use of resources, or all three. Although existing assessments including the PCL-R are not perfectly accurate, there is persuasive evidence that, when implemented properly (e.g., Harris et al., 2010), they yield sufficiently large predictive effects to be useful. For example, releasing offenders indiscriminately with respect to the risk represented by PCL-R scores, as opposed to releasing the same proportion but detaining those with the highest scores, would, over the long run, inevitably result in avoidable harm (Harris & Rice, 2007b).

Because existing formal violence risk assessment systems (and the PCL-R) do not yield perfect associations with measures of violent recidivism, some potential users are, however, troubled about "error" or "imprecision" (e.g., Berlin, Galbreath, Geary, & McGlone, 2003; Cooke & Michie, 2010[4]; Edens, 2006, Hart, Michie, & Cooke, 2007). We believe these concerns are sometimes blended with personal disapproval of some criminal laws, such as those allowing preventive detention, for example (e.g., Wollert, 2006). Under circumstances of less than perfect certainty (and dealing with what feels to be unjust legislation), some forensic practitioners are uncomfortable, perhaps also feeling measures of psychopathy or the term itself to be prejudicial and stigmatizing.[5] A more straightforward tack is to simply declare the task of violence risk assessment to be wrong or hopeless and advise the enterprise be abandoned (Hart et al., 2007; Wollert, 2006). A similarly simple idea is to advise vagueness, perhaps only characterizing violence risk as low, medium, or high (Webster et al., 1997; Webster, Hucker, & Bloom, 2002).

A more subtle approach (compatible with the latter idea) is to recommend the incorporation of more clinical information in the form of dynamic risk factors or idiosyncratic considerations and to discuss interventions expected to lower violence risk (Douglas & Kropp, 2002; Edens, 2006; Webster et al., 2002). We suggest this last remedy affords comfort because it seems so reasonable—after all, the several other VRAG items add to the predictive value of the PCL-R, surely dynamic factors would do even better. It seems self-evident that remedying dynamic factors lowers risk, so that recommending such remedies avoids a feeling of permanently condemning the offender. If a violent reoffense did occur, there would be comfort in knowing that everyone appreciates how difficult it is to help offenders change and that no guarantees were given in the first place.

The problem with all these solutions to quite understandable forensic discomfort is that they are wasteful or cause worse decisions. Most obviously, abdicating the responsibility for risk-related decisions simply shifts it to someone else. Vagueness about the risk merely introduces avoidable confusion (Hilton, Carter, Harris, & Sharpe, 2008). Allowing practitioners to insert unspecified amounts of unaided clinical intuition or apply unspecified adjustments to the scores of formal assessments does not improve validity (Hanson & Morton-Bourgon, 2009). Because they inserted clinical adjustments, forensic practitioners, even when they had PCL-R and VRAG scores, provided opinions about violence risk unrelated (or weakly related) to the tools they had, so that their opinions (but not the tools) were unrelated to later recidivism (Hilton

& Simmons, 2001; McKee, Harris, & Rice, 2007). Decisions would have been much more accurate and valid had they relied solely on the formal assessments. Finally, there is no empirical basis yet to know which therapy-induced changes in which measures of which "dynamic" factors[6] especially psychopathic traits, produce known reductions in violence risk (Douglas & Kropp, 2002; Douglas & Skeem, 2005).

Under all realistically conceivable circumstances, criminal justice decisions about violence risk have to be made daily and cannot be postponed until predictive effects are larger or persuasive demonstrations of violence-reducing therapies published. We regard this state of affairs as temporary, but until more is known about assessing violence risk and its effective treatment, we believe the only scientifically and ethically defensible course is to use the PCL-R, and actuarial violence risk assessments that incorporate it, without modification or adjustment based on hypothesized additional factors.

SUMMARY AND CONCLUSIONS

Meta-analyses consistently find moderate (Cohen, 1988; Rice & Harris, 2005) concurrent and predictive associations between psychopathy (especially PCL scores) and violent crime. When scores on the PCL are evaluated with other variables, they are usually the strongest, or among the strongest, predictors of violence, and we are unaware of another variable or clinical construct for which there is stronger evidence. Predictive effects are sometimes modest, but, given that forensic decisions shall be made, such decisions must be based on the largest empirical effects available. Assessments including psychopathy (e.g., HCR-20, VRAG) sometimes, but not always, yield higher associations with violence; clearly, this buttresses the PCL as a predictor of violence.

The evidence is clear that those who score high on measures of psychopathy are at high risk for violent recidivism over periods as short as 6 months or as long as 40 years. Factor 2 and Facet 4 of the PCL-R are especially linked to future violence. There is evidence that psychopathy (especially Factor 2) might best be considered an alternative human life history strategy, as opposed to a mental disorder. There is also evidence that coercive precocious sexuality may be a central feature of psychopathy. Finally, although there are valid empirical, practical, emotional, and political concerns about the application of measures of psychopathy in the delivery of criminal justice, there is good evidence that the preventive detention (and other long-term interventions) of those with high psychopathy (and/or actuarial risk) scores can increase public safety beyond that achievable without the use of such scores.

NOTES

1. AUC, as used in this chapter, refers to area under the relative operating characteristic, a statistic that corresponds to the likelihood that a randomly selected recidivist will have a higher score than a randomly selected nonrecidivist (Swets, Dawes, & Monahan, 2000).

2. Contrary to the methods used by Campbell et al. (2009), we use the more common method for determining significant differences (i.e., the 95% confidence interval of one measure does not include the mean obtained for another).

3. The most straightforward expectation (Nettle, 2005) from frequency-dependent selection is that of discrete entities (or taxa; Meehl & Golden, 1982). Several studies have indicated that at least some aspects of psychopathy are underlain by a discontinuity (Ayers, 2000: Coid & Yang, 2008; Harris, Rice, & Quinsey, 1994; Skilling, Harris, Rice, & Quinsey, 2002; Skilling, Quinsey, & Craig, 2001; Vasey, Kotov, Frick, & Loney, 2005). On the other hand, several have failed in this regard (Edens, Marcus, Lilienfeld, & Poythress, 2006; Guay, Ruscio, Knight, & Hare, 2007; Walters et al., 2007). It is noteworthy then that obligate and facultative strategies can be mixed:

> [A] combination is also possible and may be more likely. Individuals whose strategies are condition-dependent, for example, may also show heritable variation in the thresholds or "switch points" for changing from one strategy to another,...Different individuals may attend to different cues in switching from one strategy to another, or might switch at different points along a single cue gradient. Thus, there can be continuous

heritable variation that is both adaptively patterned and condition-dependent. Just as two alternative strategies can be maintained by frequency-dependent selection, this heritable variation can also be maintained by frequency-dependent selection." (Buss & Greiling, 1999, p. 231)

This combined strategy seems more consistent with findings that aspects of juvenile environment, especially parental neglect or abuse, affect the expression of life-course persistent antisociality (e.g., Brumbach, Figueredo, & Ellis, 2009; Cadoret et al., 2003; Hicks, South, DiRago, Iacono, & McGue, 2009). As well, this combined form of frequency dependent selection could be responsible for findings suggesting there are several discrete types of psychopathy or antisociality (e.g., Broidy et al., 2003; Caspi et al., 2002; Hervé, 2007; Hicks, Markon, Newman, Patrick, & Krueger, 2004; Skeem, Poythress, Edens, Lilienfeld, & Cale, 2003; Wareham, Dembo, Poythress, Childs, & Schmeidler, 2009) when standard taxometric methods cannot always detect that psychopathy itself is a type.

4. There are no norms for point estimates of the likelihood of violent recidivism (or the probability of any other criminal outcome) using the PCL-R making most of a recent critique by Cooke and Michie (2010) substantively moot, in our view.

5. Are measures of psychopathy or the term psychopath prejudicial and stigmatizing? According to rules of evidence, preferred evidence has probative value if it has the "tendency to make the existence of any fact more probable or less probable than it would be without the evidence." PCL-R scores tend to be related (AUC at least .65) to the likelihood of violent recidivism. Prejudicial refers to that which would cause triers to ignore probative evidence (e.g., inflaming jurors' emotions by showing unnecessarily graphic crime scene photos). Laypeople's judgments are consistent with psychopathic traits' relevance to criminal justice (e.g., Edens, Guy, & Fernandez, 2003)—laypeople attend to the probative value of psychopathy. We are not aware of findings showing PCL-R scores or the term "psychopath" led to ignoring probative evidence. In our view, it would not be prejudicial to detain offenders because of high PCL-R scores because these are demonstrably related to the likelihood of subsequent violence. In addition, ignoring opinion about "dynamic" violence risk factors and amenability for therapy would not constitute prejudice because, at present, there is no evidence that such considerations are relevant, once PCL-R scores are known. Indeed, given a high PCL-R score, opinion about "dynamic" factors and anticipated benefits of therapy is, in our considered judgment, not yet probative. "Stigmatize" means to treat so as to invoke shame or disgrace and we assume that, in professional contexts, diagnostic labels may be employed without such implication or inference. Professionals avoid identifying people as their disabilities (as in "retardate" or "cripple"). We trust professional readers appreciate why we regard the term "psychopath" as not in that category, and that some "people first" language, as in "persons who have committed offenses and yield high scores on measures of psychopathic traits" can be needlessly cumbersome.

6. At present in the context of violence risk assessment, PCL-R scores are "static" in that all tests of violence prediction have been based on measurement at one point in time. There is nothing to prevent researchers from evaluating the association between changes in PCL-R items and subsequent violent recidivism. For example, Quinsey et al. (1997) reported that exacerbations in some antisocial traits preceded violent reoffending by forensic psychiatric patients. It is entirely possible that there is a dynamic relationship between psychopathic characteristics and violent recidivism, but this sort of research has yet to be conducted.

REFERENCES

Andrews, D. A., & Bonta, J. (1995). *Level of Service Inventory–Revised*. Toronto, Ontario, Canada: Multi-Health Systems.

Ayers, W. A. (2000). Taxometric analysis of borderline and antisocial personality disorders in a drug and alcohol dependent population. *Dissertation Abstracts International: Section B: The Sciences and Engineering, 61,* 1684.

Baker, L. A., Jacobson, K. C., Raine, A., Lozano, D. I., & Bezdjian, S. (2007). Genetic and environmental bases of childhood antisocial behaviour: A multi-informant twin study. *Journal of Abnormal Psychology, 116,* 219–235.

Barr, K. N., & Quinsey, V. L. (2004). Is psychopathy pathology or a life strategy?: Implications for social policy. In: C. Crawford & C. Salmon (Eds.), *Evolutionary psychology, public policy and personal decision* (pp. 293–317). Mahwah, NJ: Erlbaum.

Belsky, J., Steinberg, L., & Draper, P. (1991). Childhood experience, interpersonal development, and reproductive strategy: An evolutionary theory of socialization. *Child Development, 62*, 647–670.

Benning, S. D., Patrick, C. J., Hicks, B. M., Blonigen, D. M., & Krueger, R. F. (2003). Factor structure of the psychopathic personality inventory: Validity and implications for clinical assessment. *Psychological Assessment, 15*, 340–350.

Berlin, F. S., Galbreath, N. W., Geary, B., & McGlone, G. (2003). The use of actuarials at civil commitment hearings to predict the likelihood of future sexual violence. *Sexual Abuse: A Journal of Research and Treatment, 15*, 377–382.

Blair, R. J., Peschardt, K. S., Budhani, S., Mitchell, D. G., & Pine, D. S. (2006). The development of psychopathy. *Journal of Child Psychology and Psychiatry, 47*, 262–275.

Broidy, L. M., Nagin, D. S., Tremblay, R. E., Bates, J. E., Brame, B., Dodge, K. A., et al. (2003). Developmental trajectories of childhood disruptive behaviors and adolescent delinquency: A six-site, cross-national study. *Developmental Psychology, 39*, 222–245.

Brumbach, B. H., Figueredo, A. J., & Ellis, B. J. (2009). Effects of harsh unpredictable environments in adolescence on development of life history strategies. *Human Nature, 20*, 25–51.

Burt, S. A., McGue, M., Carter, L. A., & Iacono, W. G. (2007). The different origins of stability and change in antisocial personality disorder symptoms. *Psychological Medicine, 37*, 27–38.

Buss, D. M., & Greiling, H. (1999). Adaptive individual differences. *Journal of Personality, 67*, 209–243.

Cadoret, R. J., Langbehn, D., Caspers, K., Troughton, E. P., Yucuis, R., Sandhu Farrington, D. P., et al. (2003). Associations of the serotonin transporter pro-moter polymorphism with aggressivity, attention deficit and conduct problems in an adoptee population. *Comprehensive Psychiatry, 44*, 88–101.

Cale, J., Lussier, P., & Proulx, J. (2009). Heterogeneity in antisocial trajectories in youth of adult sexual aggressors of women. *Sexual Abuse: A Journal of Research and Treatment, 21*, 223–248.

Campbell, M. A., French, S., & Gendreau, P. (2009). A meta-analytic comparison of instruments and methods of assessment. *Criminal Justice and Behavior, 36*, 567–590.

Caspi, A., McClay, J., Moffitt, T. E., Mill, J., Martin, J., Craig, I. W., et al. (2002). Role of genotype in the cycle of violence in maltreated children. *Science, 297*, 851–854.

Cohen, J. (1988). *Statistical power analysis for the behavioral sciences* (2nd ed.). Hillsdale, NJ: Erlbaum.

Coid, J., & Yang, M. (2008). The distribution of psychopathy among a household population: Categorical or dimensional? *Social Psychiatry and Psychiatric Epidemiology, 43*, 773–781.

Cooke, D. J., & Michie, C. (2010). Limitations of diagnostic precision and predictive utility in the individual case: A challenge for forensic practice. *Law and Human Behavior, 34*, 259–274.

Cooke, D. J., Michie, C., & Skeem J. (2007). Understanding the structure of the Psychopathy Checklist-Revised. *The British Journal of Psychiatry, 190*, 39–50.

Dall, S. R., Houston, A. I., & McNamara, J. M. (2004). The behavioural ecology of personality: Consistent individual differences from an adaptive perspective. *Ecology Letters, 7*, 734–739.

Daly, M., & Wilson, M. (1983). *Sex, evolution and behaviour* (2nd ed.). Belmont, CA: Wadsworth.

Dick, D. M. (2007). Identification of genes influencing a spectrum of externalizing psychopathology. *Current Directions in Psychological Science, 16*, 331–335.

Douglas, K. S., Herbozo, S., Poythress, N. G., Belfrage, H., & Edens, J. F. (2006). Psychopathy and suicide: A multisample investigation. *Psychological Services, 3*, 97–116.

Douglas, K. S., & Kropp, P. R. (2002). A prevention-based paradigm for violence risk assessment: Clinical and research applications. *Criminal Justice and Behavior, 29*, 617–658.

Douglas, K. S., & Skeem, J. L. (2005). Violence risk assessment: Getting specific about being dynamic. *Psychology, Public Policy, and Law, 11*, 347–383.

Edens, J. F. (2006). Unresolved controversies concerning psychopathy: Implications for clinical and forensic decision making. *Professional Psychology: Research and Practice, 37*, 59–65.

Edens, J. F., Campbell, J. S., & Weir, J. M. (2007). Youth psychopathy and criminal recidivism: A meta-analysis of the Psychopathy Checklist Measures. *Law and Human Behavior, 31*, 53–75.

Edens, J. F., Guy, L. S., & Fernandez, K. (2003). Psychopathic traits predict attitudes toward a juvenile capital murderer. *Behavioral Sciences and the Law, 21*, 807–828.

Edens, J. F., Marcus, D. K., Lilienfeld, S. O., & Poythress, N. G. (2006). Psychopathic, not psychopath: Taxometric evidence for the dimensional structure of psychopathy. *Journal of Abnormal Psychology, 115*, 131–144.

Farmer, R. F., Seeley, J. R., Kosty, D. B., & Lewinsohn, P. M. (2009). Refinements in the hierarchical structure of externalizing psychiatric disorders: Patterns of lifetime liability from mid-adolescence through early adulthood. *Journal of Abnormal Psychology, 118*, 699–710.

Farrington, D. P., Ttofi, M. M., & Coid, J. W. (2009). Development of adolescence-limited, late-onset, and persistent offenders from age 8 to age 48. *Aggressive Behavior, 35*, 150–163.

Forsman, M., Lichtenstein, P., Andershed, H., & Larsson, H. (2008). Genetic effects explain the stability of psychopathic personality from mid- to late adolescence. *Journal of Abnormal Psychology, 117*, 616–617.

Forth, A. E., Kosson, D. S., & Hare, R. D. (2003). *Psychopathy Checklist: Youth Version (PCL:YV)*. Toronto, Ontario, Canada: Multi-Health Systems.

Frank, R. H. (1988). *Passions within reason: The strategic role of emotions*. New York: Norton.

Giudice, M. D., Angeleri, R., & Manera, V. (2009). The juvenile transition: A developmental switch point in human life history. *Developmental Review, 29*, 1–31.

Glenn, A. L., Raine, A., Venables, P. H., & Mednick, S. A. (2007). Early temperamental and psychophysiological precursors of adult psychopathic personality. *Journal of Abnormal Psychology, 116*, 508–518.

Guay, J. P., Ruscio, J., Knight, R. A., & Hare, R. D. (2007). A taxometric analysis of the latent structure of psychopathy: Evidence for dimensionality. *Journal of Abnormal Psychology, 116*, 701–716.

Hanson, R. K., & Morton-Bourgon, K. E. (2009). The accuracy of recidivism risk assessments for sexual offenders: A meta-analysis of 118 prediction studies. *Psychological Assessment, 21*, 1–21.

Hare, R. D. (1991). *Hare Psychopathy Checklist-Revised (PCL-R)*. Toronto, Ontario, Canada: Multi-Health Systems.

Hare, R. D. (2003). *Hare PCL-R* (2nd ed.). Toronto, Ontario, Canada: Multi-Health Systems.

Hare, R. D., Clark, D., Grann, M., & Thornton, D. (2000). Psychopathy and the predictive validity of the PCL-R: An international perspective. *Behavioral Sciences and the Law, 18*, 623–645.

Harpending, H. C., & Sobus, J. (1987). Sociopathy as an adaptation. *Ethology and Sociobiology, 8*, 63–72.

Harris, G. T., & Rice, M. E. (2003). Actuarial assessment of risk among sex offenders. *Annals of the New York Academy of Sciences, 989*, 198–210.

Harris, G. T., & Rice, M. E. (2006). Treatment of psychopathy: A review of empirical findings. In: C. Patrick (Ed.), *The handbook of psychopathy* (pp. 555–572). New York: Guilford.

Harris, G. T., & Rice, M. E. (2007-a). Psychopathy research at Oak Ridge: Skepticism overcome. In: H. Hervé & J. C. Yuille (Eds.), *The psychopath: Theory, research, and practice* (pp. 57–76). Mahwah, NJ: Erlbaum.

Harris, G. T., & Rice, M. E. (2007-b). Adjusting actuarial violence risk assessments based on aging and the passage of time. *Criminal Justice and Behavior, 34*, 297–313.

Harris, G. T., Rice, M. E., & Cormier, C. A. (1991). Psychopathy and violent recidivism. *Law and Human Behavior, 15*, 625–637.

Harris, G. T., Rice, M. E., Hilton, N. Z., Lalumière, M. L., & Quinsey, V. L. (2007). Coercive and precocious sexuality as a fundamental aspect of psychopathy. *Journal of Personality Disorders, 21*, 1–29.

Harris, G. T., Rice, M. E., & Lalumière, M. (2001). Criminal violence: The roles of psychopathy, neurodevelopmental insults and antisocial parenting. *Criminal Justice and Behavior, 28*, 402–426.

Harris, G. T., Rice, M. E., & Quinsey, V. L. (1993). Violent recidivism of mentally disordered offenders: The development of a statistical prediction instrument. *Criminal Justice and Behavior, 20*, 315–335.

Harris, G. T., Rice, M. E., & Quinsey, V. L. (1994). Psychopathy as a taxon: Evidence that psychopaths are a discrete class. *Journal of Consulting and Clinical Psychology, 62*, 387–397.

Harris, G. T., Rice, M. E., & Quinsey, V. L. (2010). Allegiance or fidelity? A clarifying reply. *Clinical Psychology: Science and Practice, 17*, 83–90.

Harris, G. T., Rice, M. E., Quinsey, V. L., Lalumière, M. L., Boer, D., & Lang, C. (2003). A multi-site comparison of actuarial risk instruments for sex offenders. *Psychological Assessment, 15*, 413–425.

Harris, G. T., Skilling, T. A., & Rice, M. E. (2001). The construct of psychopathy. In: M. Tonry

(Ed.), *Crime and justice: An annual review of research* (pp. 197–264). Chicago: University of Chicago Press.

Hart, S. D., Cox, D. N., & Hare, R. D. (1995). *Manual for the Hare Psychopathy Checklist: Screening Version (PCL:SV)*. Toronto, Ontario, Canada: Multi-Health Systems.

Hart, S. D., Michie, C., & Cooke, D. J. (2007). Precision of actuarial risk assessment instruments: Evaluating the "margins of error" of group v. individual predictions of violence. *The British Journal of Psychiatry, 190*, 60–65.

Hervé, H. (2007). Psychopathy across the ages: A history of the Hare psychopath. In: H. Hervé, & J. C. Yuille (Eds.), *The psychopath: Theory, research and practice*, (pp. 31–56). Mahwah, NJ: Erlbaum.

Hicks, B. M., Blonigen, D. M., Kramer, M. D., Krueger, R. F., Patrick, C. J., Iacono, W. G., et al. (2007). Gender differences and developmental change in externalizing disorders from late adolescence to early adulthood: A longitudinal twin study. *Journal of Abnormal Psychology, 116*, 433–447.

Hicks, B. M., Markon, K. E., Newman, J. P., Patrick, C. J., & Krueger, R. F. (2004). Identifying psychopathy subtypes on the basis of personality structure. *Psychological Assessment, 16*, 276–288.

Hicks, B. M., & Patrick, C. J. (2006). Psychopathy and negative emotionality: Analyses of suppressor effects reveal distinct relations with emotional distress, fearfulness, and anger-hostility. *Journal of Abnormal Psychology, 115*, 276–287.

Hicks, B. M., South, S. C., DiRago, A. C., Iacono, W. G., & McGue, M. (2009). Environmental adversity and increasing genetic risk for externalizing disorders. *Archives of General Psychiatry, 66*, 640–648.

Hilton, N. Z., Carter, A. M., Harris, G. T., & Sharpe, A. J. B. (2008). Does using non-numerical terms to describe risk aid violence risk communication? Clinician agreement and decision-making. *Journal of Interpersonal Violence, 23*, 171–188.

Hilton, N. Z., & Harris, G. T. (2009). How nonrecidivism affects predictive accuracy: Evidence from a cross-validation of the Ontario Domestic Assault Risk Assessment (ODARA). *Journal of Interpersonal Violence, 24*, 326–337.

Hilton, N. Z., Harris, G. T., & Rice, M. E. (2010). *Risk assessment for domestically violent men: Tools for criminal justice, offender intervention, and victim services*. Washington, DC: American Psychological Association.

Hilton, N. Z., Harris, G. T., Rice, M. E., Houghton, R. E., & Eke, A. W. (2008). An in depth actuarial assessment for wife assault recidivism: The Domestic Violence Risk Appraisal Guide. *Law and Human Behavior, 32*, 150–163.

Hilton, N. Z., Harris, G. T., Rice, M. E., Lang, C., Cormier, C. A., & Lines, K. J. (2004). A brief actuarial assessment for the prediction of wife assault recidivism: The Ontario Domestic Assault Risk Assessment. *Psychological Assessment, 16*, 267–275.

Hilton, N. Z., & Simmons, J. L. (2001). Actuarial and clinical risk assessment in decisions to release mentally disordered offenders from maximum security. *Law and Human Behavior, 25*, 393–408.

Huesmann, L. R., Dubow, E. F., & Boxer, P. (2009). Continuity of aggression from childhood to early adulthood as a predictor of life outcomes: Implications for the adolescent-limited and life-course-persistent models. *Aggressive Behavior, 35*, 136–149.

Hundt, N. E., Kimbrel, N. A., Mitchell, J. T., & Nelson-Gray, R. O. (2008). High BAS, but not low BIS predicts externalizing symptoms in adults. *Personality and Individual Differences, 44*, 563–573.

Johansson, A., Santtila, P., Harlaar, N., von der Pahlen, B., Witting, K., Algars, M., et al. (2008). Genetic effects on male sexual coercion. *Aggressive Behavior, 34*, 190–202.

Keller, M. C. (2008). The evolutionary persistence of genes that increase mental disorders risk. *Current Directions in Psychological Science, 17*, 395–399.

Knutson, J. F., DeGarmo, D. S., & Reid, J. B. (2004). Social disadvantage and neglectful parenting as precursors to the development of antisocial and aggressive child behaviour: Testing a theoretical model. *Aggressive Behavior, 30*, 187–205.

Krupp, D. B., Sewall, L. A., Lalumière, M. L., Sheriff, C., & Harris, G. T. (2012). Nepotistic patterns of violent psychopathy: evidence for adaptation? *Frontiers in Psychology*. Retrieved from www.frontiersin.org. doi:10.3389/fpsyg.2012.00305.

Lalumière, M. L., Harris, G. T., Quinsey, V. L., & Rice, M. E. (2005). *The causes of rape: Understanding individual differences in the male propensity for sexual aggression*. Washington, DC: American Psychological Association.

Lalumière, M. L., Harris, G. T., & Rice, M. E. (2001). Psychopathy and developmental instability. *Evolution and Human Behavior, 22*, 75–92.

Lalumière, M. L., Mishra, S., & Harris, G. T. (2008). In cold blood: The evolution of psychopathy. In: J. Duntley & T. K. Shackelford (Eds.), *Evolutionary forensic psychology* (pp. 176–197). New York: Oxford University Press.

Larsson, H., Andershed, H., & Lichtenstein, P. (2006). A genetic factor explains most of the variation in the psychopathic personality. *Journal of Abnormal Psychology, 115*, 221–230.

Larsson, H., Tuvblad, C., Rijsdijk, F., Andershed, H., Grann, M., & Lichtenstein, P. (2007). A common genetic factor explains the association between psychopathic personality and antisocial behavior. *Psychological Medicine, 37*, 15–26.

Larsson, H., Viding, E., & Plomin, R. (2008). Callous-unemotional traits and antisocial behaviour. *Criminal Justice and Behavior, 35*, 197–211.

Laub, J. H., & Vaillant, G. E. (2000). Delinquency and mortality: A 50-year follow-up study of 1,000 delinquent and nondelinquent boys. *The American Journal of Psychiatry, 157*, 96–102.

Leistico, A. M., Salekin, R. T., DeCoster, J., & Rogers, R. (2008). A large-scale meta-analysis relating the Hare measures of psychopathy to antisocial conduct. *Law and Human Behavior, 32*, 28–45.

Loney, B. R., Taylor, J., Butler, M. A., & Iacono, W. G. (2007). Adolescent psychopathy features: 6-year temporal stability and the prediction of externalizing symptoms during the transition to adulthood. *Aggressive Behavior, 33*, 242–252.

Lynam, D. R., Caspi, A., Moffitt, T. E., Loeber, R., & Stouthamer-Loeber, M. (2007). Longitudinal evidence that psychopathy scores in early adolescence predict adult psychopathy. *Journal of Abnormal Psychology, 116*, 155–165.

Lynam, D. R., Charnigo, R., Moffitt, T. E., Raine, A., Loeber, R., & Stouthamer-Loeber, M. (2009). The stability of psychopathy across adolescence. *Development and Psychopathology, 21*, 1133–1153.

Lynam, D. R., Loeber, R., & Stouthamer-Loeber, M. (2008). The stability of psychopathy from adolescence into adulthood. *Criminal Justice and Behavior, 35*, 228–243.

MacCabe, J. H., Koupil, I., & Leon, D. A. (2009). Lifetime reproductive output over two generations in patients with psychosis and their unaffected siblings: The Uppsala 1915–1929 birth cohort multigenerational study. *Psychological Medicine, 39*, 1667–1676.

Manchak, S. M., Skeem, J. L., & Douglas, K. S. (2008). Utility of the revised Level of Service Inventory (LSI-R) in predicting recidivism after long-term incarceration. *Law and Human Behavior, 32*, 477–488.

McCrory, E., Hickey, N., Farmer, E., & Vizard, E. (2008). Early-onset sexually harmful behaviour in childhood: A marker for life-course persistent antisocial behaviour? *Journal of Forensic Psychiatry and Psychology, 19*, 382–395.

McKee, S. A., Harris, G. T., & Rice, M. E. (2007). Improving forensic tribunal decisions: The role of the clinician. *Behavioral Sciences and the Law, 25*, 485–506.

Mealey, L. (1995). The socio-biology of sociopathy: An integrated evolutionary model. *Behavioral and Brain Sciences, 18*, 523–599.

Meehl, P. E., & Golden, R. R. (1982). Taxometric methods. In: P. Kendall & J. Butcher (Eds.), *Handbook of research methods in clinical psychology* (pp. 127–181). New York: Wiley.

Moffitt, T. E. (1993). Adolescence-limited and life-course persistent antisocial behaviour: A developmental taxonomy. *Psychological Review, 100*, 674–701.

Murrie, D. C., Boccaccini, M. T., Turner, D. B., Meeks, M., Woods, C., & Tussey, C. (2009). Rater (dis)agreement on risk assessment measures in sexually violent predator proceedings: Evidence of adversarial allegiance in forensic evaluation? *Psychology, Public Policy and Law, 15*, 19–53.

Nettle, D. (2005). An evolutionary approach to the extraversion continuum. *Evolution and Human Behavior, 26*, 363–373.

Nieuwbeerta, P., & Piquero, A. R. (2008). Mortality rates and causes of death of convicted Dutch criminals 25 years later. *Journal of Research in Crime and Delinquency, 45*, 256–286.

Olver, M. E., Stockdale, K. C., & Wormith, J. S. (2009). Risk assessment with young offenders: A meta-analysis of three assessment measures. *Criminal Justice and Behavior, 36*, 329–353.

Patrick, C. J., Edens, J. F., Poythress, N. G., Lilienfeld, S. O., & Benning, S. D. (2006). Construct validity of the psychopathic personality inventory two-factor model with offenders. *Psychological Assessment, 18*, 204–208.

Pryor, L. R., Miller, J. D., Hoffman, B. J., & Harding, H. G. (2009). Pathological personality traits and externalizing behaviour. *Personality and Mental Health, 3*, 26–40.

Pulkkinen, L., Lyyra, A. L., & Kokko, K. (2009). Life success of males on nonoffender, adolescence-limited, persistent, and adult-onset antisocial pathways: Follow-up from age 8 to 42. *Aggressive Behavior, 35*, 117–135.

Quinsey, V. L., Harris, G. T., Rice, M. E., & Cormier, C. A. (2006). *Violent offenders: Appraising and managing risk* (2nd ed.). Washington, DC: American Psychological Association.

Quinsey, V. L., Skilling, T. A., Lalumière, M. L., & Craig, W. M. (2004). *Juvenile delinquency: Understanding the origins of individual differences.* Washington, DC: American Psychological Association.

Quinsey, V. L., Coleman, G., Jones, B., & Altrows, I. F. (1997). Proximal antecedents of eloping and reoffending among supervised mentally disordered offenders. *Journal of Interpersonal Violence, 12*, 794–813.

Ramrakha, S., Bell, M. L., Paul, C., Dickson, N., Moffitt, T. E., & Caspi, A. (2007). Childhood behaviour problems linked to sexual risk taking in young adulthood: A birth cohort study. *Journal of the American Academy of Adolescent Psychiatry, 46*, 1272–1279.

Rice, M. E., & Harris, G. T. (1995). Violent recidivism: Assessing predictive validity. *Journal of Consulting and Clinical Psychology, 63*, 737–748.

Rice, M. E., & Harris, G. T. (2005). Comparing effect sizes in follow-up studies: ROC, Cohen's d and r. *Law and Human Behavior, 29*, 615–620.

Rice, M. E., Harris, G. T., & Lang, C. (2012).The Violence Risk and Sex Offender Risk Appraisal Guides: Validation and Revision. Manuscript submitted for publication.

Robins, L. N., & O'Neal, P. (1958). Mortality, mobility, and crime: Problem children thirty years later. *American Sociological Review, 23*, 162–171.

Salekin, R. T. (2008). Psychopathy and recidivism from mid-adolescence to young adulthood: Cumulating legal problems and limiting life opportunities. *Journal of Abnormal Psychology, 117*, 386–395.

Schmauk, F. J. (1970). Punishment, arousal, and avoidance learning in sociopaths. *Journal of Abnormal Psychology, 76*, 325–335.

Schofield, H. L., Bierman, K. L., Heinrichs, B., Nix, R. L., & Conduct Problems Prevention Research Group. (2008). Predicting early sexual activity with behaviour problems exhibited at school entry and in early adolescence. *Journal of Abnormal Child Psychology, 36*, 1175–1188.

Seto, M. C., & Quinsey, V. L. (2006). Toward the future: Translating basic research into prevention and treatment strategies. In: C. J. Patrick (Ed.), *Handbook of psychopathy* (pp. 589–601). New York: Guilford.

Shaw, D. S., Bell, R. Q., & Gilliom, M. (2000). A truly early starter model of antisocial behavior revisited. *Clinical Child and Family Psychology Review, 3*,155–172.

Shaw, D. S., Gilliom, M., Ingoldsby, E. M., & Nagin, D. S. (2003). Trajectories leading to school-age conduct problems. *Developmental Psychology, 39*, 189–200.

Shepherd, J. P., Shepherd, I., Newcombe, R. G., & Farrington, D. (2009). Impact of antisocial lifestyle on health: Chronic disability and death by middle age. *Journal of Public Health Advance Access, 31*, 506–511.

Skeem, J., Johansson, P., Andershed, H., Kerr, M., & Louden, J. E. (2007). Two subtypes of psychopathic violent offenders that parallel primary and secondary variants. *Journal of Abnormal Psychology, 116*, 395–409.

Skeem, J. L., Poythress, N., Edens, J. F., Lilienfeld, S. O., & Cale, E. M. (2003). Psychopathic personality or personalities? Exploring potential variants of psychopathy and their implications for risk assessment. *Aggression and Violent Behavior, 8*, 513–546.

Skilling, T. A., Harris, G. T., Rice, M. E., & Quinsey, V. L. (2002). Identifying persistently antisocial offenders using the Hare Psychopathy Checklist and DSM antisocial personality disorder criteria. *Psychological Assessment, 14*, 27–38.

Skilling, T. A., Quinsey, V. L., & Craig, W. A. (2001). Evidence of a taxon underlying serious antisocial behaviour in boys. *Criminal Justice and Behavior, 28*, 450–470.

Swets, J. A., Dawes, R. M., & Monahan, J. (2000). Psychological science can improve diagnostic decisions. *Psychological Science in the Public Interest, 1*, 1–26.

Ullrich, S., Farrington, D. P., & Coid, J. W. (2008). Psychopathic personality traits and life-success. *Personality and Individual Differences, 44*, 1162–1171.

Vasey, M. W., Kotov, R., Frick, P. J., & Loney, B. R. (2005). The latent structure of psychopathy in youth: A taxometric investigation. *Journal of Abnormal Child Psychology, 33*, 411–429.

Vaughn, M. G., & DeLisi, M. (2008). Were Wolfgang's chronic offenders psychopaths? On the convergent validity between psychopathy and career criminality. *Journal of Criminal Justice, 36*, 33–42.

Viding, E., Frick, P. J., & Plomin, R. (2007). Aetiology of the relationship between callous-unemotional traits and conduct problems in childhood. *The British Journal of Psychiatry, 190,* 33–38.

Vizard, E., Hickey, N., & McCrory, E. (2007). Developmental trajectories associated with juvenile sexually abusive behaviour and emerging severe personality disorder in childhood: 3-year study. *The British Journal of Psychiatry, 190,* S27–S32.

Wakefield, J. C. (1992). Disorder as harmful dysfunction: A conceptual critique of DSM-III-R's definition of mental disorder. *Psychological Review, 99,* 232–247.

Walters, G. D. (2003). Predicting criminal justice outcomes with the Psychopathy Checklist and Lifestyle Criminality Screening Form: A Meta-analytic comparison. *Behavioral Sciences and the Law, 21,* 89–102.

Walters, G. D., Gray, N. S., Jackson, R. L., Sewell, K. W., Rogers, R., Taylor, J., et al. (2007). A taxometric analysis of the Psychopathy Checklist: Screening Version (PCL:SV): Further evidence of dimensionality. *Psychological Assessment, 19,* 330–339.

Walters, G. D., Knight, R. A., Grann, M., & Dahle, K. P. (2008). Incremental validity of the Psychopathy Checklist Facet Scores: Predicting release outcome in six samples. *Journal of Abnormal Psychology, 117,* 396–405.

Wareham, J., Dembo, R., Poythress, N. G., Childs, K., & Schmeidler, J. (2009). A latent cross factor approach to identifying subtypes of juvenile diversion youths based on psychopathic features. *Behavioral Sciences and the Law, 27,* 71–95.

Webster, C. D., Douglas, K. S., Eaves, D., & Hart, S. D. (1997). HCR-20: *Assessing risk for violence* (Version 2). Vancouver, BC: Mental Health Law & Policy Institute, Simon Fraser University.

Webster, C. D., Hucker, S. J., & Bloom, H. (2002). Transcending the actuarial versus clinical polemic in assessment risk for violence. *Criminal Justice and Behavior, 29,* 659–665.

Wheeler, S., Book, A., & Costello, K. (2009). Psychopathic traits and perceptions of victim vulnerability. *Criminal Justice and Behavior, 36,* 635–648.

Wollert, R. (2006). Low base rates limit expert certainty when current actuarials are used to identify sexually violent predators. *Psychology, Public Policy, and Law, 12,* 56–85.

Wormith, S., Olver, M. E., Stevenson, H. E., & Girard, L. (2007). The long-term prediction of offender recidivism using diagnostic, personality, and risk/need approaches to offender assessment. *Psychological Services, 4,* 287–305.

CHAPTER 14

Taking Psychopathy Measures "Out of the Lab" and into the Legal System: Some Practical Concerns

John F. Edens, Melissa S. Magyar, and Jennifer Cox

Psychopathic personality disorder (psychopathy) is a psychopathological syndrome that has been the focus of considerable attention in the mental health field for many decades (Patrick, 2006). Since Cleckley's (1941) seminal description of psychopathic traits, a wealth of scientific research has accumulated examining the validity and utility of various methods used to operationalize this disorder (e.g., structured rating scales, self-report inventories). Much of this work has either direct or indirect implications for the applied utilization of psychopathy scales in "real-world" settings such as the courts and correctional systems, which seem increasingly interested in using this disorder to inform various types of legal decision making. In fact, references to psychopathy (and related terms such as antisocial personality disorder) appear to be growing in both judicial opinions and legislation; psychopathy also has been the focus of expert testimony in a variety of criminal and civil trials (DeMatteo & Edens, 2006; Edens & Cox, 2012; Ogloff & Lyon, 1998; Viljoen, MacDougall, Gagnon, & Douglas, 2010). Moreover, assessments of psychopathy often directly bear on some of the most heated topics in criminal and mental health law (Edens & Petrila, 2006). These debates address fundamental legal issues such as the culpability of criminal defendants, the restraint of liberty (e.g., through civil commitment or extended incarceration), and the imposition of capital punishment (Edens, Petrila, & Buffington-Vollum, 2001; Morse, 2008; Schopp & Slain, 2000; Zinger & Forth, 1998). These debates also raise important practical questions concerning the appropriate role of mental health expertise, psychological testing, and diagnostic labels in adversarial legal settings (Edens & Petrila, 2006).

Findings from several recent surveys of mental health and legal professionals and case law reviews suggest that psychopathy measures, particularly Hare's (1991/2003) Psychopathy Checklist-Revised (PCL-R),[1] are for the most part widely accepted and used in forensic and legal contexts (Archer et al., 2006; Edens & Cox, 2012; Jackson & Hess, 2007; Lally, 2003; Viljoen, MacDougall et al., 2010; Viljoen, McLachlan, & Vincent, 2010). For example, Lally (2003) surveyed American Board of Forensic Psychology diplomats and found the PCL-R the only instrument considered to be "recommended" for both violence risk evaluations and sexual violence evaluations. As well, Viljoen, McLachlan et al. (2010) reported that forensic clinicians assessing violence risk among adult offenders ($N = 130$) and juveniles ($N = 85$) included a PCL measure as a component of the risk assessment 65% and 26% of the time, respectively.

Given the growing interest in taking psychopathy, particularly the PCL measures, "outside the lab" to inform decision making, this chapter reviews various practical issues and concerns related to the assessment of this construct in applied settings,[2] particularly forensic contexts in which the ramifications of being labeled psychopathic are the most profound. These include concerns about (1) the field reliability of psychopathy scores in applied contexts, such as criminal and civil commitment cases; (2) the predictive validity of psychopathy scores in regard to "bad outcomes" (e.g., future violence) in the criminal justice system, particularly in relation to its use in controversial contexts (e.g., with juveniles, with ethnic minorities, in capital murder cases); and (3) the implications of ascribing psychopathic traits to justice-involved individuals, in terms of the potentially prejudicial impact such traits may have on perceptions of these persons. Space constraints preclude an exhaustive review of several of the issues highlighted in this chapter. Many of these topics have been addressed in greater detail in other recent publications (e.g., Campbell, French, & Gendreau, 2009; DeMatteo & Edens, 2006; DeMatteo, Edens, & Hart, 2010; Douglas, Edens, & Vincent, 2006; Edens, 2006; Edens & Petrila, 2006; Edens, Skeem, & Kennealy, 2009; Edens & Vincent, 2008; Leistico, Salekin, DeCosta, & Rogers, 2008; Olver, Stockdale, & Wormith, 2009); where appropriate, readers are referred to these more comprehensive reviews.

HOW RELIABLE ARE PSYCHOPATHY SCORES "OUTSIDE THE LAB"?

It is common in the scientific literature to see global references to measures of psychopathy, particularly the PCL scales, as *reliable and valid* (e.g., Hare & Neumann, 2008). Although much of this chapter focuses on validity, especially *predictive* validity, it is equally if not more important to consider the issue of reliability—particularly when moving from the lab (i.e., controlled research investigations) to the real world (i.e., evaluations performed to inform clinical and/or legal decision making). In relation to the PCL rating scales, general statements that they are "reliable" seem to be predicated on an accumulated body of research studies suggesting adequate levels of various forms of psychometric reliability (e.g., internal consistency of test items, stability of ratings across examiners; see, e.g., Campbell, Pulos, Hogan, & Murry, 2005; Hare, 2003) when assessing various offender populations. It is important to remember, however, that reliability (as well as validity) is not a static psychometric property inherent within any test (DeMatteo & Edens, 2006; more generally, see American Educational Research Association, American Psychological Association, & National Council on Measurement in Education 1999; Messick, 1995) and that adequate reliability in one context does not necessarily imply that similar results would be obtained in another quite different setting. In addition, unlike self-report inventories, the reliability of a rating scale is clearly a function of the individual conducting the ratings in addition to the individual being assessed.

Reliability in Applied Settings: Some Conceptual Issues

Using psychopathy measures for research purposes is quite different from using them to inform clinical or legal decision-making (Edens, 2006; Edens & Petrila, 2006; Zinger & Forth, 1998). For example, when conducting a research study using one of the PCL measures, providing evidence of interrater reliability is likely considered to be a prerequisite for publication in a peer-reviewed scientific journal. The absence of interrater reliability data, or evidence of poor reliability of the obtained ratings, generally would not bode well in terms of the likelihood of a positive editorial decision. As such, almost by definition peer-reviewed research studies involving the PCL measures (or any other rating scale) are likely to report at least minimally acceptable levels of reliability because that is one of the criteria on which the manuscript was initially evaluated in relation to its suitability for publication. When considering the relevance of published reliability statistics to real world practice, this raises two interesting and related questions: (1) How many PCL studies have failed to be published (or have been discontinued before completion) because adequate reliability could not be established across the raters scoring the instrument? (2) To what extent do the published reliability statistics from controlled research studies *generalize* to what would be expected in

PCL assessments conducted in more applied settings, such as when two or more examiners assess the same defendant in a criminal case?

In regard to (2), it is worth noting that research and forensic contexts differ from each other in a number of important ways that could impact the reliability of any psychological test or rating scale. For example, research and forensic contexts clearly differ in terms of the informed consent/disclosure process employed (e.g., the right to refuse to participate, the privacy and confidentiality of any information obtained during the assessment) and the "stakes" involved in the outcome of the evaluation for the examinee (e.g., essentially none in research settings; potential deprivations of liberty in civil commitment cases or even life in capital murder trials). In addition, research and forensic evaluations may also differ considerably in terms of (1) the degree of impression management, self-disclosure and candor of the examinee,[3] (2) the availability of and/or access to relevant file/collateral information needed to score the items,[4] and (3) external pressures on examiners (e.g., from retaining attorneys) to eschew the degree of impartiality expected from "objective" expert witnesses and align themselves with a particular "side" of a case—potentially resulting in adversarial allegiance (Shuman & Greenberg, 2003).

One final potential difference between research and applied settings that bears important implications in regard to the generalizability of reliability statistics from the former to the latter relates to the source of interview information—more specifically, whether separate raters observed the same interview or conducted independent interviews to score the PCL. Particularly in adversarial settings (e.g., criminal cases) where PCL scores are at issue, independent examiners almost certainly will conduct separate interviews.[5] A recent review of rater agreement data from more than 120 PCL studies (Rufino, Heinonen, Boccaccini, Murrie, & Edens, 2009) found that approximately half provided little or no information as to whether separate interviews were used when establishing interrater reliability; only a select few (see, e.g., Levenson, 2004; Rutherford, Cacciola, Alterman, McKay, & Cook, 1999; Tyrer et al., 2005) clearly indicated that they were based on cases in which both evaluators conducted an independent interview. These types of studies are much more similar to real-world scenarios, and the reliability statistics they report would seem to be more realistic indicators of the reliability of repeated PCL scores in applied settings. Notably, intra-class correlation (ICC) values have been lower in these studies than in most other published reports, as well as lower than the values reported in the PCL-R manual (Hare, 2003) for male prisoners (ranging from .86 to .88). These studies are reviewed in detail in the following section.

"Field Reliability" for PCL Scores

Levenson (2004) examined the stability of PCL-R scores among 69 sex offenders being evaluated for civil commitment who were assessed by (at least) two independent forensic examiners, all of whom were retained by the state. She obtained a multiple rater ICC of .84 for the PCL-R total score, which, according to Boccaccini, Turner, and Murrie (2008), converts to a single-rater ICC value of .72.[6] In addition, Rutherford et al. (1999) reported an ICC value of .60 for the total score in a sample of 200 male methadone patients over a 2-year follow-up period, whereas Tyrer et al. (2005) reported a value of .59 in a small sample of British forensic patients ($N = 15$) over an 11-month follow-up. Although the latter two studies confound interrater reliability with temporal stability because of the time-lag between the first and second assessment, the ostensible immutability of psychopathy (see, e.g., Gacono, 2000; Kernberg, 1998) would seem to suggest that one's level of psychopathic traits—as operationalized by PCL ratings based on lifetime functioning (Hare, 2003)—should not change appreciably over 1 to 2 years.[7] Yet they clearly did in these two studies (see also Burke, Loeber, & Lahey, 2007, for similar results in a clinic-referred sample).

Collectively, the issues noted above raise some concerns about the extent to which the (mostly positive) PCL reliability findings in published research studies and the instrument's manual would generalize to applied settings, particularly adversarial contexts in the United States such as capital murder cases or sexually violent predator civil commitment proceedings. Anecdotally, case examples of highly discrepant scores for criminal defendants who have been assessed with the PCL by "opposing experts" have episodically appeared in the legal literature (for specific case examples, see Edens, 2006; Edens & Petrila, 2006; Edens & Vincent, 2008; Hare, 1998). Fortunately, there has been a recent movement beyond simple anecdotal accounts toward more systematic studies addressing the reliability of PCL scores in adversarial settings.

Murrie, Boccaccini, Johnson, and Janke (2008) examined agreement rates for PCL-R scores from experts retained by opposing sides in Texas Sexually Violent Predator (SVP) civil commitment cases. For the total score, the single-evaluator ICC_1 for absolute agreement across 23 cases was .39, which is markedly lower than the levels of agreement observed for the PCL-R in most research studies. Of perhaps even greater concern, the mean score across state-retained experts (~26) fell at approximately at the 67th percentile for male prisoners (Hare, 2003), whereas the mean score of all defense-retained experts (~18) in these same cases fell at the 32nd percentile. Obviously, mean differences placing the "average" offender in either the upper third or lower third of psychopathy scores (depending on the side retaining the expert) seriously question the reliability and, consequently, validity of these evaluations. Subsequently, Murrie et al. (2009) expanded the initial sample of 23 cases to 35 and reported similar findings (ICC_1 for absolute agreement = .42). In terms of what caused the variability in these scores, 23% of the variance was attributable to the side retaining the examiner, whereas 35% was attributable to other sources of error.

Other research not specific to sexual offender cases has shown some evidence of potential adversarial allegiance in reference to PCL scores, although the effects have been somewhat less pronounced than those reported by Murrie and colleagues. Lloyd, Clark and Forth (2010) reported similar but somewhat weaker differences across PCL scores obtained from Canadian criminal case law reports. Across 15 cases, rater agreement was lower ($ICC_{A,1}$ = .67) than what has been reported in the PCL-R manual (Hare, 2003) though not as low as those obtained by Murrie and colleagues (2008, 2009). Crown-retained expert scores were appreciably higher on average (mean = 27.32) than were those scores derived from defense-retained experts (mean = 24.47).

In addition to potential allegiance effects, field studies have started to examine other factors that might contribute to poor reliability for PCL scores. For example, Boccaccini et al. (2008) examined all civil commitment evaluations in Texas and compared the average PCL-R scores produced across 20 separate state-retained examiners (total N = 321 cases). Very large mean differences were noted across some examiners (e.g., > 10 points) and, using a multilevel linear modeling approach, the authors were able to partition the variance in PCL ratings into their component parts. Although one would hope that the individual who conducts an evaluation would have relatively limited impact on the scores attributed to the defendant, 34% of the variance in PCL scores in this data set was attributable to the examiner conducting the assessment. That is, roughly one-third of an inmate's PCL score was determined by who performed the evaluation rather than anything to do with the inmate's "true" level of psychopathic traits. In addition, 22 of the offenders in this data set were assessed with the PCL on more than one occasion. Agreement for PCL-R total scores was low ($ICC_{A,1}$ = .47), even though evaluators had been retained by the same side (i.e., petitioner; Boccaccini et al., 2008).

Although one might attribute these results to problems with examiners in the particular jurisdiction from which these studies emanated (Texas), other research teams have started reporting similarly modest findings concerning PCL-R interrater reliability, even when the examiners are not retained by opposing counsel in the case at hand. Using data obtained from the Florida Department of Children and Families SVP Program, Miller, Otto, and Kimonis (2010) recently reported preliminary findings regarding the stability of these scores in 313 cases. The ICC for single raters for the total score was .60, with an average score difference across examiners of greater than 5 points.

One limitation of most of the preceding field reliability studies is that they have been restricted to analyses of PCL-R total scores because more detailed scoring information typically has been missing from the archival records used to obtain the data. This is unfortunate because researchers have been unable to examine which components of PCL-R scores might be more or less stable across examiners in applied settings. Whether differences in scores are more attributable to Factor 1 or Factor 2 scores is important for a number of reasons. For example, it has been argued (e.g., DeMatteo & Edens, 2006; Edens et al., 2001, 2009; Edens, Colwell, Desforges, & Fernandez, 2005) that PCL "personality" items (e.g., remorselessness, superficial charm) in particular have the greatest potential for prejudicial impact in criminal justice settings.

It is obvious that some PCL items depend more on subjective judgment (e.g., "pathological" lying) than do others (e.g., many short-term marital relationships) in terms of scoring and that the "personality" items comprising Factor 1 would have greater potential for random error (or perhaps systematic bias) in their scoring than do more behaviorally-based items comprising Factor 2. Individual item ICCs reported by Hare (2003, chapter 5) and Rutherford et al. (1999) support this argument. Even in research reports in which examiners rely on the same file and interview data, ICCs for Factor 1 tend to be somewhat lower than for Factor 2 and the total scores (e.g., ICCs reported in the PCL manual for Factor 2 were .85, .88, and .90, versus .75, .79, and .74 for Factor 1 for three large offender samples).

To our knowledge only two studies have examined the field reliability of the PCL separated out by its factor scores (Edens, Boccaccini, & Johnson, 2010; Miller et al., 2010). In the Edens et al. study, archival data were obtained from a sample of imprisoned sex offenders ($N = 20$) who (1) had been administered the PCL-R and received a total score of ≥ 25 and (2) were then assessed on a second occasion by a different examiner (who, it should be noted, was not blind to the fact that the second assessment was triggered by an elevated score from the first assessment). Unlike earlier studies based on archival samples, factor scores for the PCL-R were available in this particular data set. Intraclass correlations were lower than those typically seen in most research studies for both the total ($ICC_{A,1} = .40$) and Factor 2 ($ICC_{A,1} = .56$) scores—although the total score results were relatively similar to the ICC (.47) reported by Boccaccini et al. (2008). More importantly, Factor 1 scores in this sample were only negligibly related to each other ($ICC_{A,1} = .16$).

Given the relatively limited range of scores in this high scoring sample,[8] analyses also were performed on the total and factor scores to correct for potential range restriction, which might have artificially attenuated the degree of reliability across examiners. Subsequent total and Factor 2 score correlations were somewhat more consistent with published research (corrected Pearson r of .78 and .76, respectively), but Factor 1 continued to demonstrate very poor agreement across examiners (corrected Pearson $r = .24$).

In the only other field study to report data on PCL-R subscales, Miller et al. (2010) were able to obtain these values for the four PCL-R "facets" from approximately 150 SVP cases from their larger sample of 313 cases with total scores from two examiners. The two facets that constitute the "old" Factor 1 displayed poor ICC values: Affective = .39 ($N = 153$), Interpersonal = .48 ($N = 154$). The results for the facets constituting the "old" Factor 2 were somewhat higher: Lifestyle = .56 ($N = 146$), Antisocial = .69 ($N = 150$). This pattern of results is generally consistent with the results reported by Edens et al. (2010) for the older two factor model for the PCL-R, suggesting that these examiners had great difficulty converging on the presence/absence of interpersonal and affective items on the PCL-R in particular.

One other finding from the archival data set analyzed by Edens et al. (2010) warrants brief mention here. In addition to PCL-R scores, self-report measures of interpersonal dominance and warmth were available for 19 of the 20 offenders with two PCL-R scores (Johnson, Edens, & Boccaccini, 2010). Although one might assume that unreliability of Factor 1 scores would be mostly attributable to chance factors or random error, the magnitude of the difference in these scores across examiners was strongly inversely related to both measures of interpersonal style: $r = -.59$ (warmth) and $r = -.52$ (dominance), both $p < .05$. That is, raters tended to agree more on the scoring of (and give higher scores on) Factor 1 when the examinee's self-reported interpersonal style was one of high warmth and high dominance, whereas the raters were much less consistent with each other when the interpersonal style was one of low warmth and low dominance. Although based on a very small sample and in need of replication, these findings raise intriguing hypotheses about potential causes of rater disagreement. For example, high self-reported warmth and dominance among psychopathic offenders may be perceived by both examiners as superficial and manipulative attempts to appear charming and socially adept, whereas the relative absence of these interpersonal characteristics may lead to greater ambiguity across examiners (and less reliable scores) in terms of rating the presence or absence of such traits.

Summary and Concluding Thoughts

Collectively, the results of the studies reviewed in the preceding text raise serious concerns that (1) the interrater reliability statistics from published research studies and the PCL-R manual may not be particularly generalizable to at least some applied settings, particularly adversarial ones, (2) much of the variance in scores (particularly on interpersonal and affective characteristics) reported in applied settings may be attributable to factors other than the examinee's "true" score on the PCL-R, and (3) scores presented in court may be biased specifically toward the side that retained the examiner. Although further replication of these findings clearly is needed, the growing number of studies reporting low interrater reliability in the field[9] suggests that, at a minimum, the forensic and criminal justice systems in which the PCL-R is used should critically examine the basis for any examiner's ratings and not presume a priori that the results are inherently "reliable."

It should be highlighted that concerns about the reliability of psychopathy scales are not limited to interview- and file-review–based methods such as the PCL. The stability of self-report measures of antisocial and psychopathic traits across different contexts is an issue of considerable practical importance as well. For example, experimental research indicates that scores on scales such as the Psychopathic Personality Inventory (PPI; Lilienfeld & Andrews, 1996) can be significantly altered by instructional sets to "fake good" (e.g., Edens, Buffington, Tomicic, & Riley, 2001), raising concerns that individuals motivated to minimize their psychopathic traits may easily do so on such scales. Also, more naturalistic studies suggest that the presence of positive impression management can attenuate the predictive validity of self-report psychopathy scales (e.g., Edens & Ruiz, 2006). The inclusion of structured measures of social desirability within some psychopathy-specific inventories (e.g., PPI) as well as almost all multiscale personality inventories may aid in the identification of such response suppression when it occurs, although more research into this issue is warranted, particularly as it relates to the use of these measures in adversarial contexts.

What might be done to improve the reliability of PCL-R scores in settings in which there is reason to believe it may be less than adequate? Although this is a complex question that likely does not have a simple answer, it is informative to consider other areas of forensic assessment in which the reliability of professional opinions previously have been called into question. For example, evaluations of adjudicative competence ("competence to stand trial") and/or criminal responsibility ("insanity") frequently have been derided as subjective, uninformed, and lacking in legal relevance (e.g., Murrie, Boccaccini, Zapf, Warren, & Henderson, 2008; Nicholson & Norwood, 2000; Skeem & Golding, 1998). Some jurisdictions over the years (e.g., Massachusetts) have moved towards mandatory training or certification programs in an attempt to improve the quality of work performed in these areas (Farkas, DeLeon, & Newman, 1997; Frost, de Camara, & Earl, 2006). Particularly in jurisdictions with SVP statutes in which the PCL-R is routinely used, some form of credentialing or mandated training procedures might result in improved reliability.

Of course, mandatory credentialing or training requirements are much easier to recommend than they are to actually implement, particularly in regard to court cases in which both sides typically have considerable freedom to retain independent expert witnesses. In addition, exactly what constitutes adequate training or "expertise" in the context of using the PCL measures in forensic settings might be a source of considerable disagreement. Although the Darkstone Research Group offers formalized training on the PCL-R and the professional manual (Hare, 2003) provides recommendations concerning qualifications for applied use (e.g., possession of an advanced degree, completion of graduate coursework in psychometrics and psychopathology; see pp. 16–17), it also indicates that Dr. Hare has "no professional or legal authority to determine who can and cannot use the PCL-R, or to provide judgments about the adequacy of specific clinicians and their assessments" (p. 16) and that the Darkstone workshops are not the only method in which examiners can become competent to administer and score the PCL-R. (For a more extensive discussion of some of these issues, see Edens & Petrila, 2006).

Of course, in the context of sex offender risk assessments (as well as other types of risk assessments), one might argue that perhaps the field should simply move away from using inferential, personality-based constructs to inform opinions about risk and focus on more objective risk factors that might eliminate or at least minimize subjective judgment. Recently, Murrie et al. (2009) examined this issue in their data set comparing the reliability of PCL-R scores across forensic examiners in SVP cases. They found that two widely used actuarial tools used in these evaluations, the STATIC-99 (Hanson & Thornton, 1999) and the Minnesota Sex Offender Screening Tool–Revised (MnSOST-R; Epperson, et al., 1998) also demonstrated interreliability coefficients (based on 27 cases) that were weaker than what had been reported in earlier research. In addition, similar to the PCL-R data, the MnSOST-R scores showed evidence of significant adversarial allegiance, with scores provided by state-retained experts being appreciably higher than scores provided by defense experts. There was, however, almost no evidence of adversarial allegiance for the STATIC-99, with only 4% of the variance in scores being attributable to which side retained the expert witness.

As such, there seems to be some evidence that at least some actuarial scales might be less susceptible to systematic forms of rater error than the PCL-R, although field research in this area is in a nascent state. This does not necessarily imply that these scales would perform comparably to or better than the PCL-R in relation to risk assessment, however.[10] We address the issue of the utility of the PCL measures as risk assessment devices, and their utility compared to other approaches, in detail in the following section.

PCL SCORES AND THE PREDICTION OF "BAD OUTCOMES"

Setting aside concerns about field reliability issues for the moment, a second topic of considerable practical significance is the relationship between psychopathy scores and subsequent antisocial or otherwise undesirable behavior. There is a long history of interest in this issue,[11] and some (DeMatteo & Edens, 2006; Edens & Petrila, 2006) have argued that the legal system's interest in psychopathy is driven primarily by its relationship with criminal behavior.

Although there are numerous forms of "bad outcomes" that have been the focus of attention of scholars in this field, such as treatment failure (Skeem, Polaschek, & Manchak, 2009), socially destructive (yet perhaps not overtly criminal) conduct (Hall & Benning, 2006), and dissimulation and malingering (Clark, 1997), in this chapter we focus more specifically on the prediction of behaviors of perhaps greatest interest to the legal field: community and institutional[12] violence and criminal behavior. We first briefly describe global findings from the literature, followed by more extensive reviews of specific controversial topics. Readers should also consult more extensive reviews of these issues, such as DeMatteo et al. (2010), Douglas et al. (2006), Edens and Campbell (2007), Edens, Campbell, and Weir, (2007), Edens et al. (2009), Guy et al. (2005), Leistico et al. (2008), Olver, Stockdale, and Wormith (2009), and Skeem and Cooke (2010).

Psychopathy and Antisocial Conduct: Global Findings

The relationship between the PCL instruments and future antisocial conduct has been the focus of several meta-analyses over the last decade (e.g., Gendreau et al., 2002; Guy et al., 2005; Walters, 2003-a, 2003-b). The most recent and comprehensive of these was recently conducted by Leistico et al. (2008). Negative outcome variables consisted of both institutional misconduct and community recidivism. Antisocial conduct was broadly defined to include general, sexual, and/or violent behavior/aggression. Across 95 nonoverlapping studies ($N = 15,826$ [total score]; $n = 8,653$ [Factor 1]; $n = 8,603$ [Factor 2]) that included retrospective and prospective data, they found that antisocial conduct was moderately related to higher PCL total scores and Factor 2 scores (Hedges' $d = .55$ and $.60$, respectively); weaker (but still significant) effects were noted for Factor 1 scores (Hedges' $d = .38$). A considerable degree of heterogeneity was detected across the results, however, indicating that the "average" effect sizes within the total sample obscured considerable variability in the magnitude of this relationship across individual studies—a finding noted in several earlier meta-analyses as well. That is, some individual studies reported relatively strong relationships, whereas others reported weak or null results, raising questions as to the merits of aggregating across studies with very different findings.

Leistico et al. (2008) conducted a series of moderator analyses to examine whether certain specific sample characteristics influenced the explanatory power of the PCL ratings. Two of the potential moderator variables should be highlighted here, given their applied relevance. First, consistent with some earlier meta-analytic findings in the juvenile literature (Edens et al., 2007) and trends noted in the adult literature (Guy et al., 2005), ethnic status moderated the relationship between PCL scores and outcome criteria. More specifically, samples with greater proportions of Caucasians demonstrated larger effect sizes in the association between PCL Total and Factor 2 scores and antisocial conduct. Second, Leistico et al. reported that effect sizes were *larger* for samples that included greater proportions of women (for Total and Factor 1 scores, in particular). Leistico et al. concluded that this finding contradicted previous results (e.g., Edens et al., 2007) concerning gender and recidivism and argued that PCL measures subsequently should be interpreted more cautiously for males than females.

Such an interpretation seems highly questionable, however, given that Leistico et al. (2008) did not actually analyze effect sizes *separately* by gender, as did Edens et al. in their juvenile meta-analysis in which psychopathy scores for females were for the most part very weakly related to recidivism. The finding of Leistico et al. that the percentage of women across studies was positively related to effect size magnitude could simply reflect an artifact because women tend to receive lower PCL scores and also may be at lower risk to recidivate. As such, combining their data with male offenders in the same analyses might artificially magnify effect sizes that were more a function of gender differences than PCL scores per se.[13]

Leistico et al. (2008) examined several other potential moderator variables, such as the country in which the study was conducted, the type of institutional setting (psychiatric hospital vs. prison), and the type of information used to score the PCL. Given that other recent meta-analyses have addressed several of these same potential moderators in a more nuanced fashion, we will focus on some of their findings in subsequent sections. In addition, several recent studies have been published that were not included in the Leistico et al. results (Cauffman, Kimonis, Dmitrieva, & Monahan, 2009; Edens & Cahill, 2007; Edens, Poythress, Lilienfeld, & Patrick, 2008; McDermott, Edens, Quanbeck, Busse, & Scott, 2008; Vitacco, Rybroek, Rogstad, Yahr, Tomony, & Saewert, 2009; Wormith, Olver, Stevenson, & Girard, 2007). As such, we will briefly review certain key findings from these studies as well.

Community Recidivism: Juveniles

Although the predictive validity of the PCL measures has not been as extensively studied among justice-involved youth relative to adult samples, a growing number of studies have examined the relationship between psychopathy and subsequent offending among youthful offenders. Across 21 nonoverlapping samples of male and female juvenile offenders ($N = 2{,}787$), Edens et al. (2007) recently meta-analyzed prospective recidivism data for the PCL measures. After removing outliers, the effect sizes for the PCL total score for general and violent recidivism were significant, but relatively modest, with weighted mean correlation coefficients of .24 and .25, respectively. The PCL's association with sexual recidivism was nonsignificant ($r_w = .07$). Similar to Leistico et al. (2008), there was a moderate to severe degree of heterogeneity of effect sizes, particularly for violent recidivism. Moderator analyses indicated that the heterogeneity among the effects could be attributed to some extent to the gender and ethnic composition of the sample, as noted earlier in this chapter. That is, multiethnic samples reported weaker effects than did primarily Caucasian samples and effects for samples of female juvenile offenders were weaker than for samples of male juveniles (cf. Leistico et al., 2008).

Recent published studies not included in the preceding meta-analyses have added to the generally heterogeneous collection of effect sizes for juveniles. For example, Edens and Cahill (2007) examined a sample of 75 multiethnic male offenders recruited from a juvenile detention center to prospectively assess general and violent recidivism approximately 10 years subsequent to the administration of the PCL-YV. Many of the youths were current or former gang members, slightly more than half had a history of violent crime, and the sample mean score on the PCL-YV (21.56) was comparable to what has been reported for other institutionalized samples (Forth et al., 2003). Contrary to the results of the only other long-term follow-up recidivism study for juveniles (Gretton et al., 2004), there was no evidence to suggest that the PCL-YV or its subscales could predict any type of community recidivism over this time span. In contrast, Salekin (2009) reported that psychopathy was incrementally predictive of both general and violent recidivism across a 3- to 4-year follow-up period for a sample of 130 multiethnic male and female offenders after controlling for fourteen theoretically relevant variables. As well, the magnitude of the associations between the PCL-YV and its various facets and recidivism were generally similar across criminal outcomes.

Most recently, in the largest published study of psychopathic traits and juvenile recidivism to date, Cauffman et al. (2009) examined the predictive utility of three different psychopathy measures in a sample of 1,170 serious male juvenile offenders over a 3-year period. Of note, only approximately half of this larger sample was included in the recidivism analyses at the 36-month time period for self-reported offending ($N = 641$) and official arrest records ($N = 680$), respectively. The magnitude of the bivariate correlation between PCL-YV scores and self-reported recidivism showed progressive attenuation over successive 6-month follow-up periods (e.g., .32 at 6 months; .21 at 36 months). In terms of subscale performance, unlike Salekin's (2009) findings, Factor 4 (Criminal Behavior) of the PCL-YV almost entirely accounted for the PCL's relationship with subsequent offending. Also, in a variety of multivariate analyses comparing the predictive validity of the three psychopathy measures, the two self-report measures, the Youth Psychopathic Traits Inventory (YPI; Andershed, Gustafson, Kerr, & Stattin, 2002) and the NEO Psychopathy Resemblance Index (NEO-PR-I; Lynam & Widiger, 2007), tended to outperform the PCL-YV, which did not explain any unique variance in outcome measures at 36 months (see Table 5, p. 535).

Institutional Misconduct

The relationship between PCL scores and institutional adjustment has been the focus of considerable research in recent years (see, e.g., Cunningham & Reidy, 2002; Edens, Buffington-Vollum, Keilen, Roskamp, & Anthony, 2005; Guy et al., 2005; Walters, 2003-a, 2003-b), perhaps in some part due to the fairly recent introduction of PCL scores into U.S. capital murder trials in which the long-term prison adjustment of offenders is at issue (Edens et al., 2001).[14] Although Leistico et al. (2008) concluded that there were no significant global differences across institutional and community settings in relation to predictive validity, previous meta-analyses (Guy et al., 2005) suggest that the relationship between PCL scores and institutional violence warrants a more nuanced analysis—particularly in regard to U.S. prisons.

As noted by Guy et al. (2005), studies of institutional misconduct, particularly "violent" misconduct, have suffered from considerable variability in terms of what constitutes the criterion measures of interest. For example, behaviors such as verbal aggression, refusal of medication, and "belligerence" have been characterized as "violent" behavior in some studies (e.g., Hill, Rogers, & Bickford, 1996) that have reported some of the largest predictive validity coefficients for psychopathy measures. Given these (and other) concerns, Guy et al. coded a total of 273 effect sizes derived from correctional, civil psychiatric, and forensic psychiatric facilities to examine the association between PCL measures and a hierarchy of five criterion categories: total/any misconduct, nonaggressive misconduct, general aggression, verbal aggression/property destruction, and physical violence. Mean weighted effect sizes for PCL total scores ranged from .29 (total/any misconduct) to .17 (physical violence). Given the substantial degree of variability across the distribution of effect sizes, Guy et al. (2005) conducted a series of moderator analyses. For example, analyses examining the impact of the country where the data was collected, specifically in terms of prison samples, demonstrated that the effect sizes for physically violent misconduct were trivial in U.S. prison samples ($r_w = .11$) when compared to non-U.S. prison samples ($r_w = .23$).

Subsequent to the completion of the meta-analysis of Guy et al. (2005), more recent studies have reported correspondingly weak effects regarding the prediction of institutional violence (e.g., McDermott, Edens, et al., 2008; Vitacco et al., 2008). It is interesting to note, however, that although Vitacco et al. failed to find a significant relationship with violent behavior broadly construed, they did report a significant relationship between psychopathy scores and *instrumental* institutional aggression. Though intriguing, this finding was based on only 11 individuals in the sample (total $N = 152$) who were classified as engaging in instrumental aggression. Similarly, McDermott, Quanbeck, Busse, Yastro, and Scott (2008), in a follow-up to McDermott, Edens, et al. (2008), reported a modest relationship between PCL-R scores and instrumental inpatient violence, although the base rate of this type of aggression was very low (only 14% of all aggressive acts). Further prospective research examining subtypes of aggressive behavior clearly is warranted, given that most research has not examined this issue (beyond perhaps differentiating verbal and physical aggression).

Despite the relatively small number of studies that have examined the association between psychopathy and adjustment to incarceration and institutionalization among juveniles, Edens and Campbell (2007) conducted a meta-analysis summarizing these findings and the heterogeneity of their effects. Examination of total, aggressive, and physical misconduct (operationalized in a manner similar to Guy et al., 2005) yielded weighted mean correlation coefficients ranging from .25 to .28. Similar to the findings of meta-analytic investigations within adult samples, heterogeneous effect sizes were detected across studies, particularly for both aggressive and physically violent misconduct. Of some concern in regard to the relationship between these scores and violent behavior, several studies failed to report effect sizes, in some instances because there was no "violence" to measure within the facility. Although from a statistical perspective it makes little sense to correlate scales with an outcome measure with little or no variability, from an applied perspective the fact that "no violence" occurs among psychopathic youth in some samples is critically important to consider when attempting to gauge the relationship between PCL scores and institutional violence.

Deconstructing the Psychopathy/Violence Relationship: Some Key Findings

In addition to concerns about the predictive validity of psychopathy scores to relation to certain populations (e.g., non-Caucasians, female offenders, youths) and certain contexts (e.g., U.S. prisons), there has been an increased interest in recent years in examining what particular components of psychopathy, particularly PCL measures, are most strongly related to violence and crime. Such questions bear important theoretical implications concerning role of antisocial conduct in conceptualizations of psychopathy, many of which go beyond the scope of this chapter (see Skeem & Cooke, 2010). Here, we focus primarily on the practical implications of research in this area.

In a recent meta-analysis incorporating 26 studies and 32 effect sizes ($N = 10,555$), Kennealy, Skeem, Walters and Camp (2010) directly compared the incremental ability of PCL Factor 1 and Factor 2 to predict violence, as well as examined whether there was any evidence of interactive effects for these two scales. In terms of the individual effects of each scale, Factor 2 was most strongly predictive of violence: odds ratio = 1.15 (after controlling for variance in Factor 1). However, once Factor 2 variance was controlled, Factor 1 only weakly predicted violence (odds ratio = 1.04). Moreover, there was no evidence of any interaction between the two scales. Also, when the same analytical procedures were repeated with only a subsample of the most methodologically sound studies ($k = 12$), the effect for Factor 2 was essentially unchanged (odds ratio = 1.14), but both the interaction term and Factor 1 were nonsignificant predictors (odds ratio = 1.00, 1.02, respectively; see also Yang, Wong, & Coid, 2010).

Other research has begun to further parse the predictive validity of Factor 2 in relation to antisocial conduct. Walters, Knight, Grann, and Dahle (2008) examined the incremental validity of the four PCL-R/SV facet scores as predictors of both general and violent recidivism, rather than only focusing upon Factors 1 and 2. The follow-up time span ranged from 20 weeks to 10 years. Among six relatively large forensic/correctional samples, Facet 4 (Antisocial) uniquely predicted violent and general recidivism above and beyond the first three facets, whereas the first three facets generally failed to predict recidivism above and beyond Facet 4. Walters et al. (2008) concluded that the antisocial component of the PCL measures is a robust predictor of recidivism, and that none of the "personality" aspects of psychopathy, as measured by the first three facet scores of the PCL, consistently predicted recidivism above and beyond the variance accounted for by Facet 4. Finally, Walters et al. (2008) noted that the demonstrated superiority of Factor 2 in prior studies is more a function of Facet 4 than Facet 3, and that the predictive validity of Facet 4 was due largely to items in Parcel H (Criminality).

More recently, Walters and Heilbrun (2010) replicated many of these findings in two relatively large samples: 216 forensic patients (Sample 1) and 230 prison inmates (Sample 2). This study examined the incremental validity and classification accuracy of the four PCL/PCL-R facet scores and the two Facet 4 parcel scores (General Acting Out and Criminality) as predictors of institutional aggression (combination of both physical and verbal acts) and community recidivism (person crimes only).[15] Similar to the findings of Walters et al. (2008), results from Sample 1 indicated that Facet 4 uniquely predicted the outcome variables, above and beyond the other three Facet scores, whereas Facets 1, 2, and 3 failed to add any unique variance to the outcomes. However, in terms of Sample 2, none of the four Facet scores achieved incremental validity in the prediction of institutional aggression or violence recidivism. Across the two samples, however, Parcel G (General Acting Out), showed incremental validity relative to the four Facet scores and Parcel H (Criminality) for three of the four analyses, with the exception being violent recidivism in Sample 1. In contrast to Parcel H's performance in Walters et al. (2008), Parcel H performed poorly overall, whereas Parcel G achieved significance in all 10 incremental validity and Receiver Operating Characteristic analyses.

Do PCL Measures Add Something Unique to Risk Assessment?

In recent years, researchers have increasingly focused on whether psychopathy measures, particularly the PCL scales, explain a meaningful amount of variance in recidivism and other negative outcomes *beyond* other types of assessment instruments. Gendreau et al. (2002, 2003), for example, argued that PCL scores did not perform as well as a widely used risk scale, the Level of Service Inventory–Revised (LSI-R), in a meta-analysis comparing their predictive validity for general and violent recidivism. More recently, however, Campbell, French, and Gendreau (2009) published a large-scale comparison of the PCL and several widely used risk assessment instruments (including the Level of Service Inventory–Revised [LSI-R], Violence Risk Appraisal Guide [VRAG], and Historical-Clinical-Risk Management 20 [HCR-20]) in relation to institutional violence and violent recidivism. Given the widely overlapping confidence intervals for the mean effects reported across instruments, they concluded that there was very little evidence for the superiority of any one specific measure over the others.[16] Similar arguments have been made by other researchers conducting direct comparisons of risk measures within the same sample (Kroner, Mills, & Reddon, 2005). Interestingly, Kroner et al. created four hybrid risk measures by randomly selecting from the item content of the PCL-R, LSI-R, VRAG, and the General Statistical Information on Recidivism (GSIR). When they compared the hybrid measures in terms of their ability to predict general recidivism, they performed comparably to their respective parent scales—suggesting considerable redundancy in what these measures tap in regard to recidivism.

In the juvenile literature, Edens et al. (2007) reported preliminary analyses indicating that the predictive validity of PCL was comparable to the Youth Level of Service/Case Management Inventory (YLS/CMI) for general (combined $N = 799$) and violent recidivism (combined $N = 727$) across five samples in which both scales were administered to the same youths. Although incremental validity analyses could not be performed based on the extant data, the two scales were highly intercorrelated ($r_w = .77$), suggesting it would be relatively unlikely that either explained much unique variance in recidivism. In a somewhat more extensive meta-analysis, Olver et al. (2009) recently addressed the predictive validity of the PCL, the YLS/CMI and the Structured Assessment of Violence Risk for Youth (SAVRY). Similar to the conclusions of Edens et al. (2007) in regard to PCL-YLS/CMI comparisons, the authors suggested that the three instruments generally seemed to perform comparably in the prediction of general and violent recidivism, with correlations in the .25 to .33 range.

Finally, although not the primary focus of this chapter, various recent studies have suggested that self-report scales tapping psychopathic and antisocial traits may perform comparably to (or in some instances somewhat better than) the PCL measures in the prediction of violence and recidivism (e.g., Cauffman et al., 2009; Edens et al., 2009; Salekin, 2009; Skeem, Miller, Mulvey, Tiemann, & Monahan, 2005; Walters, 2006). These findings need to be interpreted somewhat cautiously, however, in that in most studies self-report measures were completed with assurances of confidentiality. If such measures were administered with examinees clearly understanding that the results might impact their case disposition, it is certainly plausible that socially desirable responding might impact the validity of the results.

Collectively, the results described in the preceding text offer little support for the argument that there is something intrinsic to PCL measures of psychopathy that tap something *beyond* what extant risk assessment instruments already assess in relation to future violence. This should not be surprising, given the earlier results by Walters and colleagues suggesting that it is the relatively nonspecific indicators of prior criminality and violence-proneness embedded in Factor 2 that are primarily responsible for the association between "psychopathy" and future crime and violence.

Summary and Concluding Thoughts

The results of the various meta-analyses that have been conducted indicate that, in the aggregate, there is clear evidence that scores from the PCL measures relate to future violent and criminal behavior. The amount of heterogeneity across studies, however, compels a more nuanced consideration of this literature and deconstructing this relationship further reveals some theoretically interesting and practically important qualifiers, such as relatively limited evidence of predictive validity among adolescent girls and within U.S. institutional settings (e.g., prisons) at this time. Also, the weight of the evidence at present would seem to suggest that it is not anything unique about psychopathic *personality* (i.e., Factor 1) per se that drives the relationship between the PCL scales and violence. Rather, it seems to be the PCL measures' quantification of criminal and antisocial history characteristics (and Parcel G in particular) that are the most salient in identifying those most at risk for violence and recidivism. Such an interpretation comports with the global finding that the PCL measures tend to perform comparably to measures that operationalize similar criminogenic background characteristics, such as the LSI-R (adult and adolescent) measures.

Given the current state of affairs one might reasonably ask: Does it appear that there is any compelling need to rely on "psychopathy" to assess for risk—even among populations for which it is reasonably well validated (e.g., Canadian Caucasian adult males)? One argument might be that, for the most part, PCL scores alone do seem to work about as well as various risk-specific measures. However, if other measures perform comparably *and* they are not tied to either (1) the inflammatory label "psychopath" or (2) the stigmatizing characteristics of Factor 1 (e.g., callousness, egocentricity) that seem at best minimally important to the prediction of future violence and crime (Kennealy et al., 2010; Walters et al., 2008; Walters & Heilbrun, 2010), one could counterargue that a forensic examiner might be hard-pressed to justify the need for PCL scales specifically in real-world contexts. Of course, such a counterargument is predicated in part on the premise that the psychopath label and Factor 1 traits actually are "inflammatory" and "stigmatizing." We turn to this issue in the final section of this chapter.

DO "PSYCHOPATHIC" TRAITS AND LABELS INFLUENCE CLINICAL AND/OR LEGAL DECISION MAKING?

Recently, researchers have begun to focus on the potential influence of psychopathic traits and the associated label "psychopath" on legal decision making. Considering some have questioned the rehabilitation potential of psychopaths (e.g., Harris, Rice, & Cormier, 1994; Rice, Harris, & Cormier, 1992; cf. Skeem, Polaschek, & Manchak, 2009) and the juvenile justice system is largely concerned with offender rehabilitation, it is perhaps not surprising that a large portion of this research has focused on legal players within the juvenile justice system.

The Impact of Characterizing Juveniles as Psychopathic

In the first published study examining the effects of attributing psychopathic traits to a juvenile defendant, Edens, Guy, and Fernandez (2003) presented undergraduates ($N = 360$) with a newspaper account of a capital murder trial and excerpts of testimony from various trial participants. When described as exhibiting prototypical psychopathic traits—though not labeled a "psychopath" per se—36% of participants were supportive of a death sentence for the juvenile defendant. When he was described in more pro-social terms, only 21% supported execution.

Following this initial report, Murrie and colleagues subsequently published a series of studies sampling from professionals involved in the treatment and/or supervision of juvenile offenders. In the first study, Murrie, Cornell, and McCoy (2005) asked juvenile probation officers to develop a hypothetical evaluation based on defendant traits and diagnostic labels. Officers rated a defendant characterized as exhibiting psychopathic traits as more likely to recidivate than a defendant who "meets criteria for psychopathy," regardless of antisocial history. Using a similar methodology, Rockett, Murrie, and Boccaccini (2007) sampled 109 juvenile court clinicians and reported that the psychopathic label had an effect on the clinicians' perception of the defendant's future risk to society when low antisocial behavior was present, but not when high antisocial behavior was present. In addition, clinicians gave higher risk ratings to youths described with psychopathic traits than those without.

Murrie, Boccaccini, McCoy, and Cornell (2007) examined the impact of diagnostic labels, psychopathic traits, and antisocial behavior on juvenile court judges' ($N = 326$) decision making. The defendant's diagnostic criteria had a significant effect on the judges' disposition and perception that the defendant would be violent again in the future. In addition, judges rated a defendant who "meets criteria for psychopathy" as more likely to engage in criminal behavior in the future than a defendant diagnosed with conduct disorder, particularly when the psychopathic defendant also presented with minimal history of antisocial behavior.

Most recently, Boccaccini, Murrie, Clark, and Cornell (2008) systematically manipulated variables concerning antisocial behavior (present vs. minimal), psychopathic personality traits (present vs. absent), and mental health diagnosis (psychopathy, conduct disorder, no diagnosis, or a description that the defendant "is a psychopath"). A survey of 891 jury-pool members indicated that psychopathic traits and antisocial behavior appear to be more influential than diagnostic labels. However, jurors who read testimony about a defendant who "is a psychopath" rated him as more likely to be violent in the future and more deserving of harsher punishment than the defendant who meets criteria for psychopathy. These results are important, in that they suggest some of the effects reported by Murrie and colleagues in their earlier studies might have been even more pronounced had they used the ostensibly more stigmatizing term "psychopath" rather than the more technical phrase "meets criteria for psychopathy."

Other researchers have for the most part replicated the general finding of Murrie and colleagues that psychopathic traits carry considerable potential for stigmatization. For example, using a juvenile probation officer sample, Vidal and Skeem (2007) experimentally manipulated a juvenile offender's psychopathic traits and diagnosis (reported together as either present or absent), child abuse history (presented as either "stable upbringing" or "unstable upbringing") and ethnicity (Caucasian or African American). Probation officers viewed psychopathic youth as difficult cases who are likely to reoffend as adults and unlikely to respond to treatment. Although officers reported being willing to "go the extra mile" for abused youth, they were especially strict about enforcing rules with psychopathic offenders. Also, similar to the results of Murrie et al. (2008), Jones and Cauffman (2008) recently reported that judges ($N = 100$) were significantly influenced by statements that a juvenile was psychopathic. Both the traits and the "psychopath" label impacted judges' perceptions of a youth's future dangerousness. They also had an impact on their opinions about amenability to treatment and the need for restrictive sanctions (also see Chauhan, Burnette, & Repucci, 2007 for additional results regarding effects on judges).

The Impact of Characterizing Adults as Psychopathic

In addition to research focused on psychopathy within the juvenile justice system, studies have examined the impact of these traits within the adult system, specifically in regard to its impact on juror decision making. Edens, Desforges, Fernandez, and Palac (2004) presented 338 college students with a case summary of a capital trial that varied the defendant's diagnosis (psychopathic, psychotic, or none) and risk of future violence (high vs. low). Participants were likely to rate the defendant as dangerous in the psychopathic condition regardless of the high-risk/low-risk condition, although this effect was not found for the no diagnosis or psychotic conditions. Subsequently, Edens, Colwell, Desforges, and Fernandez (2005) replicated this study, again using undergraduates and manipulating the defendant's mental health diagnosis and potential for future violence. Of note, participants were much more supportive of the death penalty when the defendant was presented as psychopathic (60%) as opposed to psychotic (30%) or not mentally impaired (38%).[17]

Recently, Cox, DeMatteo, and Foster (2010) presented participants with vignettes that altered the defendant's mental health diagnosis (psychopath vs. no diagnosis) and likelihood of future violence (high vs. low). Importantly, the defendant's personality traits—which included descriptions of prototypical psychopathic characteristics—were identical in each scenario. As such, any experimental effects would be due specifically to the addition of the label "psychopath" beyond the traits associated with psychopathy. Participants were more likely to impose a death sentence when the defendant was rated as a high risk for future violence, regardless of diagnostic label. Interestingly, the defendant who presented as a high risk for future violence and not labeled a psychopath was sentenced to death at a higher rate than the defendant who was high risk and a psychopath, although this difference was not statistically significant.

In addition to capital cases, experiments have also investigated the effect of psychopathy on decision making in sexually violent predator civil commitment cases (Guy & Edens, 2003, 2006). When presented with one of three types of testimony (clinical opinion, actuarial assessment, and ratings of psychopathy), female undergraduates were more likely to vote for commitment when they were exposed to testimony regarding psychopathy than either clinical opinion or actuarial testimony. Interestingly, this effect was not seen for male participants. Women were consistently more influenced by psychopathy testimony than men across both studies.

One additional finding from some of the earlier mock jury studies is worth briefly noting here. Edens, Davis, Fernandez Smith, and Guy (2012) reexamined data from the control conditions in several earlier studies of capital cases (Edens et al., 2003, 2005; Davis, 2003) to assess the relationship between perceptions of psychopathic traits and support for execution. Although not exposed to expert testimony concerning psychopathy in the control conditions,[18] participants were asked to rate the extent to which they believed the defendant would exhibit psychopathy-like traits after reviewing the case facts. These ratings of how psychopathic they perceived him to be were moderately strongly related to death verdicts (AUC = .69). Moreover, support for execution was attributable almost exclusively to participants' attributions of interpersonal and affective features of psychopathy to the defendant—rather than those traits consistent with PCL Factor 2. As well, in their mock jury study, Guy and Edens (2006) reported a similar effect concerning perceptions of a defendant's psychopathic traits and support for civilly committing him as a sexually violent predator.

In addition to lab-based experimental research, more naturalistic studies have supported the findings that psychopathic traits can influence the decision-making processes within the courtroom. Costanzo and Peterson (1994) reviewed 20 prosecution closing arguments about defendants who were characterized as liars, manipulators, and "cold, remorseless" killers (p. 125). The authors concluded that these prototypic psychopathic characterizations appeared to have an effect on trial outcome. Further evidence comes from the Capital Jury project (Sundby, 1998), which reported that jurors in California used adjectives such as "emotionless," "calculating," and "cocky" to describe defendants whom they sentenced to death.

Summary and Concluding Thoughts

Describing individuals involved in the juvenile and criminal justice systems as "psychopathic" has the potential to have a profound impact on legal decision making. Of course, if the degree of impact is commensurate with the relevance of psychopathy to the case at hand, then there appears to be little reason to be concerned about the potential for biased decision making. For example, a hypothetical 25-year-old white man who has greater difficulty procuring early release from a Canadian prison owing in part to an extremely elevated PCL-R score (particularly based on Factor 2 items) might have little or no grounds for claiming that the effects of the assessment were *prejudicially* impacting his release decision, given the relationship between PCL-R scores and community recidivism (Leistico et al., 2008).[19]

Conversely, we would argue that a hypothetical black defendant who receives a death sentence (at least in part) because a state-retained expert witness labels him at trial as a "psychopath whose callous and egocentric personality traits make him very likely to be highly dangerous even if serving life in prison with no chance of parole" would have a very legitimate claim that the expert was making pejorative assertions that run counter to the published research on psychopathy and prison violence in the U.S. (Edens et al., 2005; Guy et al., 2005). Although the courts are the ultimate arbiters of what constitutes probative and prejudicial information as it relates to expert testimony, we would argue that forensic examiners have an ethical burden to ensure that the content of their assessments and testimony are based on solid empirical footing *and* that they do not unduly stigmatize those individuals who have been evaluated (Edens & Petrila, 2006).

GENERAL SUMMARY

Legal systems around the world seem to show considerable interest in the concept of psychopathy, primarily in regard to questions of violence risk and the potential for future criminal behavior. Survey data suggest that the forensic mental health field has accepted the PCL measures as useful operationalizations of this construct and empirical data clearly indicate that these measures *can* be reliably scored and are, in the aggregate, modestly to moderately related to subsequent violence and crime. As such, should the field conclude that forensic examiners' scores on the PCL-R and its derivatives are "reliable and valid" and focus its attention on other, potentially more theoretically interesting questions that also might bear significant implications for the legal system (e.g., what are the primary mechanisms involved in the etiology and development of psychopathy; what risk and protective factors are the most promising for altering the developmental trajectory of "fledgling" psychopaths; are there clinically meaningful subtypes of psychopathy)? In this chapter we have highlighted what we believe are significant and as of yet unresolved questions regarding the reliability and validity of psychopathy measures, and the PCL instruments in particular, when used to inform legal decision-making. These controversies include the (1) field reliability of scores in applied, and especially adversarial, legal proceedings (e.g., SVP cases), (2) predictive validity of PCL-R scores in relation to specific settings (e.g., capital murder trials) and particular populations (e.g., adolescent girls), (3) relevance of affective and interpersonal features in relation to violence risk assessment, and (4) incremental contribution of the PCL measures beyond extant risk assessment instruments. In addition, as noted at the beginning of this chapter, reliability and validity are not static properties of psychological scales. No instrument therefore is *intrinsically* reliable and valid and the legitimacy of the scores derived from it should be considered on a case-by-case basis.

More generally, we would argue that any broad-based evaluation of the applied uses of an instrument, particularly a clinician rating scale, is as much if not more so a commentary on the field as it is on the instrument itself. When that field provides its services in a highly adversarial and high stakes legal system, no one should be entirely surprised by evidence that some examiners may be misusing or abusing certain tools at their disposal (Edens, 2006; Shuman & Greenberg, 2003)—whether it involves biased ratings that exaggerate or minimize a defendant's psychopathic traits or biased conclusions that exaggerate the relevance of psychopathy in relation to a particular legal issue (DeMatteo & Edens, 2006; Edens, 2001; Edens et al., 2001).

An interesting parallel in the forensic arena at the moment relates to the suddenly prominent role of intelligence tests and measures of adaptive functioning in capital murder cases following the U.S. Supreme Court's ruling in the Atkins case in which it prohibited the execution of the mentally retarded. No one should be shocked by field studies that will almost certainly find that mental retardation diagnoses in these cases are notably less reliable across examiners than would be expected based on ICCs reported in the manuals of widely used measures of intelligence and adaptive functioning (see, e.g., Greenspan, 2009, and the accompanying special issue of *Applied Neuropsychology*, for a review of various concerns about the reliability and validity of diagnoses in these cases).[20]

We do believe that there is at least one glaring difference between assessments of mental retardation and psychopathy in forensic settings: the term "mental retardation," although carrying considerable stigmatization in its own right, does not bring to mind images of Ted Bundy, Charles Manson, and Hannibal Lecter—but the term "psychopath" clearly does (Helfgott, 1997). The final section of this chapter addressed the potential for psychopathic traits and labels attributed to defendants to significantly influence the attitudes and decision making of consumers of forensic evaluations (e.g., judges, probation officers, potential jurors). Being characterized as psychopathic or having psychopathic traits generally does not bode well for the juvenile or adult defendant. Here again, no one should really be surprised by this.

Given the current state of affairs, we would argue that the global bar for justifying the introduction of psychopathy measures into legal cases should be somewhat higher than other psychological instruments that would seem to have less potential for prejudicial impact in forensic contexts. Of course, there is no concrete way to operationalize such a standard and it seems relatively unlikely that the legal system will impose rigorous admissibility guidelines in the context of expert testimony as it relates to risk assessment (see, e.g., Sales & Shuman, 2005)—preferring instead to rely on the adversarial system to ostensibly sort out the "wheat from the chaff" (*Barefoot v. Estelle*, 1983: 901). As such, it is likely to fall primarily on the shoulders of individual examiners to (1) evaluate their competence in relation to the reliable and valid use of the PCL-R and other psychopathy scales and (2) judge whether the potential for stigmatization may outweigh the relevance of the instrument to address the referral question in the case at hand. We hope that the information we have reviewed here will help foster a critical examination of both of these issues for individual examiners and the forensic field more generally.

NOTES

1. Unless otherwise noted, we use "PCL" generically to refer to any of the Hare scales, including the original version, the screening version, and the youth version.
2. It is worth mentioning that the issues addressed here are germane to many other assessment methods used in applied settings (e.g., intelligence tests, projective tests) and relate to distinctions between the "efficacy" versus "effectiveness" of these instruments (e.g., Mash & Hunsley, 2005), similar to the same types of distinctions discussed in the evidence-based treatment literature.
3. For example, Rogers, Vitacco, Jackson, Martin, Collins, and Sewell (2002) administered the PCL-YV to 77 juvenile offenders with instructions to appear socially desirable ("fake good") or socially nonconforming ("fake bad"). Participants presenting themselves as less psychopathic produced scores on average four points below their scores obtained under standard instructions. Also, those presenting as socially nonconforming produced scores that were on average 8 points higher than standard scores.
4. Given their advocacy position, attorneys may very well withhold (or attempt to withhold) from examiners information they think might adversely impact their client's PCL score.
5. Noted earlier, there is also no guarantee that these examiners would necessarily have access to the same file/collateral data.
6. Single-rater values are more relevant to examining the reliability of scales in applied settings such as those in which the PCL-R is being employed.
7. Along these same lines, because the PCL-R was designed to be scored based on lifetime functioning (Hare, 2003), it was never intended to be sensitive to putative changes in psychopathic traits over relatively short periods of time (e.g., as a result of successful treatment).
8. Unfortunately, examinees who initially obtained scores below 25 in this sample were not assessed with the PCL-R a second time.
9. Heilbrun (1992) recommended a reliability coefficient of at least .80 be demonstrated for tests that require significant clinical subjectivity, arguing that the use of a measure that fails to meet this generally accepted standard is questionable and requires significant justification on the part of the clinician.
10. Recently, Boccaccini, Murrie, Caperton, and Hawes (2009) examined the applied predictive validity of these two actuarial scales in a sample of 1,928 male sex offenders from Texas. Neither tool proved to be a strong predictor of sexually violent recidivism, with the MnSOST-R performing no better than chance in predicting both sexually violent recidivism and the combination of violent and sexually violent recidivism.
11. Recall that "*inadequately motivated* antisocial behavior" [emphasis added] was one of Cleckley's (1941) original criteria for diagnosing psychopathy.
12. Though not technically classified as "crimes" in most in most instances, disciplinary infractions in correctional and forensic settings may result

in serious sanctions and penalties, such as loss of gain time, administrative segregation, and other consequences. As such, we believe they are important to address here even though the criteria (particularly nonviolent infractions) are in many ways conceptually distinct from community recidivism.
13. Such an interpretation for the ethnicity finding seems less plausible, however, as there do not appear to be appreciable differences across PCL scores across most ethnic groups (Skeem, Edens, Camp, & Colwell, 2004; McCoy & Edens, 2006; Sullivan & Kosson, 2006).
14. Given that almost all jurisdictions that employ capital punishment now offer life-without-parole as the sentencing alternative to execution, the potential "future dangerousness" of a non-death-row inmate essentially is restricted to a consideration of his behavior while in prison for the remainder of his life.
15. Of note, none of the participants in either of these samples was included in the previous Walters et al. (2008) incremental validity analyses.
16. Although the PCL is included within the item content of the VRAG and HCR-20, some research has demonstrated that when this item is removed (e.g., Douglas, Ogloff, Nicholls, & Grant, 1999) or replaced (Quinsey, Harris, Rice, & Cormier, 2006), predictive validity is not significantly impacted. For example, Quinsey et al. argue that eight indicators of childhood behavior problems can serve as a substitute for the PCL-R in the calculation of the VRAG total score.
17. In terms of death verdicts, there were no significant differences across conditions in the Edens et al. (2004) study. This was because manipulation checks indicated that the majority of participants failed to comprehend the complex sentencing instructions typically used in capital cases (see Wiener, Pritchard, & Westen, 1995, for an extensive discussion of the problem of poor comprehension of judicial instructions in death penalty cases). In the 2005 study, these instructions were simplified to ensure that participants understood the verdict they were recommending for the defendant.
18. The expert testimony indicated that there was no evidence of mental disorder in the control conditions.
19. Assuming, of course, that the obtained score is a reliable indicator of what his true PCL-R score is.
20. Anecdotally, in the *Atkins v. Virginia* (2002) case itself, the defendant was found *not* to be mentally retarded by the jury during his second trial, in which examiners for both sides provided conflicting evidence concerning his diagnostic status. When Atkins' second death sentence was overturned on appeal, the courts noted in particular grave concerns about the reliability and validity of the evaluation performed by the prosecution-retained expert.

REFERENCES

American Educational Research Association, American Psychological Association, & National Council on Measurement in Education. (1999). *Standards for educational and psychological testing* (3rd ed.). Washington, DC: American Educational Research Association.

Andershed, H. A., Gustafson, S. B., Kerr, M., & Stattin, H. (2002). The usefulness of self-reported psychopathy-like traits in the study of antisocial behavior among non-referred adolescents. *European Journal of Personality, 16*(5), 383–402.

Archer, R. P., Buffington-Vollum, J. K., Stredny, R. V., & Handel, R. W. (2006). A survey of psychological test use patterns among forensic psychologists. *Journal of Personality Assessment, 87*, 84–94.

Atkins v. Virginia. (2002). 536 U.S. 304.

Barefoot v. Estelle. (1983). 463 U.S. 880.

Boccaccini, M. T., Murrie, D. C., Clark, J., & Cornell, D. (2008). Describing, diagnosing and naming psychopathy: How do youth psychopathy labels influence jurors? *Behavioral Sciences and the Law, 26*, 487–510.

Boccaccini, M. T., Turner, D. T., & Murrie, D. C. (2008). Do some evaluators report consistently higher or lower scores on the PCL-R? Findings from a statewide sample of sexually violent predator evaluations. *Psychology, Public Policy, and Law, 14*, 262–283.

Burke, J. D., Loeber, R., & Lahey, B. B. (2007). Adolescent conduct disorder and interpersonal callousness as predictors of psychopathy in young adults. *Journal of Clinical Child and Adolescent Psychology, 36*, 334–346.

Campbell, M., French, S., & Gendreau, P. (2009). The prediction of violence in adult offenders: A meta-analytic comparison of instruments and methods of assessment. *Criminal Justice and Behavior, 36*, 567–590.

Campbell, J. S., Pulos, S., Hogan, M., & Murry, F. (2005). Reliability generalization of the Psychopathy Checklist applied in youthful samples. *Educational & Psychological Measurement, 65*, 639–656.

Cauffman, E., Kimonis, E. R., Dmitrieva, J., & Monahan, K. C. (2009). A multi-method assessment of juvenile psychopathy: Comparing the predictive utility of the PCL:YV, YPI, and NEO-PIR. *Psychological Assessment, 21*, 528–542.

Chauhan, P., Reppucci, N., & Burnette, M. (2007). Application and impact of the psychopathy label to juveniles. *International Journal of Forensic Mental Health, 6*, 3–14.

Clark, C. R. (1997). Sociopathy, malingering, and defensiveness. In: R. Rogers (Ed.), *Clinical assessment of malingering and deception* (2nd ed., pp. 68–84). New York: Guilford.

Cleckley, H. (1941). *The mask of sanity*. St. Louis, MO: Mosby.

Costanzo, M., & Peterson, J. (1994). Attorney persuasion in the capital penalty phase: A content analysis of closing arguments. *Journal of Social Issues, 50*, 125–147.

Cox, J., DeMatteo, D. S., & Foster, E. E. (2010). The effect of the Psychopathy Checklist—Revised in capital cases: Mock jurors' responses to the label of psychopathy. *Behavioral Sciences & the Law, 28*, 878–891.

Cunningham, M. D., & Reidy, T. J. (1999). Don't confuse me with the facts: Common errors in violence risk assessment at capital sentencing. *Criminal Justice and Behavior, 26*, 20–43.

Davis, K. M. (2003). *The influence of the label "psychopath" on mock juror decisions in a capital trial.* Master's thesis, Sam Houston State University, Huntsville, TX.

DeMatteo, D., & Edens, J. F. (2006). The role and relevance of the Psychopathy Checklist-Revised in court: A case law survey of U.S. courts (1991–2004). *Psychology, Public Policy, and Law, 12*, 214–241.

DeMatteo, D., Edens, J. F., & Hart, A. B. (2010). The use of measures of psychopathy in violence risk assessment. In: R. K. Otto & K. S. Douglas (Eds.), *Violence risk assessment tools* (pp. 19–40). London: Routledge/Taylor & Francis.

Douglas, K. S., Ogloff, J. R. P., Nicholls, T. L., & Grant, I. (1999). Assessing risk for violence among psychiatric patients: The HCR-20 violence risk assessment scheme and the Psychopathy Checklist: Screening Version. *Journal of Consulting and Clinical Psychology, 67*, 917–930.

Douglas, K. S., Vincent, G., & Edens, J. F. (2006). Risk for criminal recidivism: The role of psychopathy. In: C. Patrick (Ed.), *Handbook of psychopathy* (pp. 533–554). New York: Guilford.

Edens, J. F. (2006). Unresolved controversies concerning psychopathy: Implications for clinical & forensic decision-making. *Professional Psychology: Research and Practice, 37*, 59–65.

Edens, J. F., Boccaccini, M. T., & Johnson, D. W. (2010). Inter-rater reliability of the PCL-R total and factor scores among psychopathic sex offenders: Are personality features more prone to disagreement than behavioral features? *Behavioral Sciences and the Law, 28*, 106–119.

Edens, J. F., Buffington, J. K., Tomicic, T. L., & Riley, B. D. (2001). Effects of positive impression management on the Psychopathic Personality Inventory. *Law and Human Behavior, 25*, 235–256.

Edens, J. F., Buffington-Vollum, J. K., Keilen, A., Roskamp, P., & Anthony, C. (2005). Predictions of future dangerousness in capital murder trials: Is it time to "disinvent the wheel?" *Law and Human Behavior, 29*, 55–86.

Edens, J. F., & Cahill, M. A. (2007). Psychopathy in adolescence and criminal recidivism in young adulthood: Longitudinal results from a multi-ethnic sample of youthful offenders. *Assessment, 14*, 57–64.

Edens, J. F., & Campbell, J. S. (2007). Identifying youths at risk for institutional misconduct: A meta-analytic investigation of the psychopathy checklist measures. *Psychological Services, 4*, 13–27.

Edens, J. F., Campbell, J. S., & Weir, J. M. (2007). Youth psychopathy and criminal recidivism: A meta-analysis of the Psychopathy Checklist measures. *Law and Human Behavior, 31*, 53–75.

Edens, J. F., Colwell, L. H., Desforges, D. M., & Fernandez, K. (2005). The impact of mental health evidence on support for capital punishment: Are defendants labeled psychopathic considered more deserving of death? *Behavioral Sciences and the Law, 23*, 603–625.

Edens, J. F., & Cox, J. (2012). Examining the prevalence, role and impact of evidence regarding antisocial personality, sociopathy and psychopathy in capital cases: A survey of defense team members. *Behavioral Sciences & the Law, 30*(3), 239–255.

Edens, J. F., Davis, K., Fernandez Smith, K., & Guy, L. S. (January 23, 2012). No sympathy for the devil: Attributing psychopathic traits to capital murderers also predicts support for executing them. *Personality Disorders: Theory, Research, and Treatment*, doi:http://dx.doi.org/10.1037/a0026442

Edens, J. F., Desforges, D., Fernandez, K., & Palac, C. (2004). Effects of psychopathy and violence risk testimony on mock juror perceptions of dangerousness in a capital murder trial. *Psychology, Crime & Law, 10*, 393–412.

Edens, J. F., Guy, L. S., & Fernandez, K. (2003). Psychopathic traits predict attitudes toward a juvenile capital murderer. *Behavioral Sciences and the Law, 21*, 807–828.

Edens, J. F., & Petrila, J. (2006). Legal and ethical issues in the assessment and treatment of psychopathy. In: C. Patrick (Ed.), *Handbook of psychopathy* (pp. 573–588). New York: Guilford.

Edens, J. F., Petrila, J., & Buffington-Vollum, J. K. (2001). Psychopathy and the death penalty: Can the Psychopathy Checklist-Revised identify offenders who represent "a continuing threat to society?" *Journal of Psychiatry and Law, 29*, 433–481.

Edens, J. F., Poythress, N. G., Lilienfeld, S. O., & Patrick, C. J. (2008). A prospective comparison of two measures of psychopathy in the prediction of institutional misconduct. *Behavioral Sciences and the Law, 26*, 529–541.

Edens, J. F., & Ruiz, M. A. (2006). On the validity of validity scales: The importance of defensive responding in the prediction of institutional misconduct. *Psychological Assessment, 18*, 220–224.

Edens, J. F., Skeem, J. L., & Kennealy, P. (2009). The Psychopathy Checklist in the courtroom: Consensus and controversies. In: J. L. Skeem, K. S. Douglas, & S. O. Lilienfeld (Eds.), *Psychological science in the courtroom: Consensus and controversy* (pp. 175–201). New York: Guilford.

Edens, J. F., & Vincent, G. M. (2008). Juvenile psychopathy: A clinical construct in need of restraint? *Journal of Forensic Psychology Practice, 8*, 186–197.

Epperson, D. L., Kaul, J. D., Huot, S. J., Hesselton, D., Alexander, W., & Goldman, R. (1998). *Minnesota Sex Offender Screening Tool-Revised (MnSOST-R)*. St. Paul, MN: Minnesota Department of Corrections.

Farkas, G., DeLeon, P., & Newman, R. (1997). Sanity examiner certification: An evolving national agenda. *Professional Psychology: Research and Practice, 28*(1), 73–76.

Forth, A., Kosson, D., & Hare, R. (2003). *Psychopathy Checklist: Youth Version*. Toronto: Multi-Health Systems.

Frost, L. E., deCamara, R. L., & Earl, T. R. (2006). Training, certification, and regulation of forensic evaluators. *Journal of Forensic Psychology Practice, 6*, 77–91.

Gacono, C. B. (2000). Preface. In: C. Gacono (Ed.), *The clinical and forensic assessment of psychopathy: A practitioner's guide* (pp. xv–xxii). Mahwah, NJ: Erlbaum.

Gendreau, P., Goggin, C., & Smith, P. (2002). Is the PCL-R really the "unparalleled" measure of offender risk? A lesson in knowledge cumulation. *Criminal Justice and Behavior, 29*, 397–426.

Gendreau, P., Goggin, C., & Smith, P. (2003). Is the PCL-R really the "unparalleled" measure of offender risk? A lesson in knowledge cumulation: Erratum. *Criminal Justice and Behavior, 30*, 722–724.

Greenspan, S. (2009). Assessment and diagnosis of mental retardation in death penalty cases: Introduction and overview of the special "Atkins" issue. *Applied Neuropsychology, 16*, 89–90.

Gretton, H. M., Hare, R. D., & Catchpole, R. (2004). Psychopathy and offending from adolescence to adulthood: A 10-year follow up. *Journal of Consulting and Clinical Psychology, 72*, 636–645.

Guy, L. S., & Edens, J. F. (2003). Juror decision-making in a mock sexually violent predator trial: Gender differences in the impact of divergent types of expert testimony. *Behavioral Sciences and the Law, 21*, 215–237.

Guy, L. S., & Edens, J. F. (2006). Gender differences in attitudes toward psychopathic sexual offenders. *Behavioral Sciences and the Law, 24*, 65–85.

Guy, L. S., Edens, J. F., Anthony, C., & Douglas, K. (2005). Does psychopathy predict institutional misconduct among adults? A meta-analytic investigation. *Journal of Consulting and Clinical Psychology, 73*, 1056–1064.

Hall, J. R., & Benning, S. D. (2006). The "successful" psychopath: Adaptive and subclinical manifestations of psychopathy in the general population. In: C. Patrick (Ed.), *Handbook of psychopathy* (pp. 459–478). New York: Guilford.

Hanson, R. K., & Thornton, D. (1999). *Static-99: Improving actuarial risk assessments for sex offenders* (User Report 99–02). Ottawa: Department of the Solicitor General of Canada.

Hare, R. D. (1991). *The Hare Psychopathy Checklist-Revised manual*. North Tonawanda, NY: Multi-Health Systems.

Hare, R. D. (1998). The Hare PCL-R: Some issues concerning its use and misuse. *Legal & Criminological Psychology, 3*, 99–119.

Hare, R. D. (2003). *The Hare Psychopathy Checklist-Revised manual* (2nd ed.). North Tonawanda, NY: Multi-Health Systems.

Hare, R. D., & Neumann, C. (2008). Psychopathy as a clinical and empirical construct. *Annual Review of Clinical Psychology, 4*, 217–246.

Harris, G. T., Rice, M. E., & Cormier C. A. (1994). Psychopaths: Is a therapeutic community therapeutic? *Therapeutic Communities, 15*, 283–299.

Heilbrun, K. (1992). The role of psychological testing in forensic assessment. *Law and Human Behavior, 16*, 257–272.

Hill, C. D., Rogers, R., & Bickford, M. E. (1996). Predicting aggressive and socially disruptive behavior in a maximum security forensic psychiatric hospital. *Journal of Forensic Sciences, 41*, 56–59.

Jackson, R., & Hess, D. T. (2007). Evaluation for civil commitment of sex offenders: A survey of experts. *Sexual Abuse: A Journal of Research and Treatment, 19*, 425–448.

Johnson, D., Edens, J. F., & Boccaccini, M. T. (March, 2010). *Unreliability of PCL-R Factor 1 among sexual offenders*. Paper presented at the annual meeting of the American Psychology-Law Society, Vancouver, BC, Canada.

Jones, S., & Cauffman, E. (2008). Juvenile psychopathy and judicial decision making: An empirical analysis of an ethical dilemma. *Behavioral Sciences and the Law, 26*, 151–165.

Kennealy, P. J., Skeem, J. L., Walters, G. D., & Camp, J. (2010). Do core interpersonal and affective traits of PCL-R psychopathy interact with antisocial behavior and disinhibition to predict violence? *Psychological Assessment, 22*, 569–580.

Kernberg, O. (1998). The psychotherapeutic management of psychopathic, narcissistic, and paranoid transferences. In: T. Millon, E. Simonsen, M. Birket-Smith & R.D. Davis (Eds.), *Psychopathy: Antisocial, criminal, and violent behavior* (pp. 372–382). New York: Guilford.

Kroner, D. G., Mills, J. F., & Reddon, J. R. (2005). A coffee can, factor analysis, and prediction of antisocial behavior: The structure of criminal risk. *International Journal of Law and Psychiatry, 28*, 360–374.

Lally, S. J. (2003). What tests are acceptable for use in forensic evaluations? A survey of experts. *Professional Psychology: Research and Practice, 5*, 491–498.

Leistico, A. M., Salekin, R. T., DeCosta, J., & Rogers, R. (2008). A large-scale meta-analysis relating the Hare measures of psychopathy to antisocial conduct. *Law and Human Behavior, 32*, 28–45.

Levenson, J. S. (2004). Reliability of sexually violent predator civil commitment criteria in Florida. *Law and Human Behavior, 28*, 357–368.

Lilienfeld, S. O., & Andrews, B. P. (1996). Development and preliminary validation of a self-report measure of psychopathic personality traits in noncriminal populations. *Journal of Personality Assessment, 66*, 488–524.

Lloyd, C. D., Clark, H. J., & Forth, A. E. (2010). Psychopathy, expert testimony, and indeterminate sentences: Exploring the relationship between psychopathy checklist-revised testimony and trial outcome in Canada. *Legal and Criminological Psychology, 15*, 323–339.

Lynam, D. R., & Widiger, T. A. (2007). Using a general model of personality to understand sex differences in the personality disorders. *Journal of Personality Disorders, 21*, 583–602.

Mash, E. J., & Hunsley, J. (2005). Evidence-based assessment of child and adolescent disorders: Issues and challenges. *Journal of Clinical Child and Adolescent Psychology, 34*, 362–379.

McCoy, W. K., & Edens, J. F. (2006). Do black and white youths differ in levels of psychopathic traits? A meta-analysis of the Psychopathy Checklist measures. *Journal of Consulting and Clinical Psychology, 74*, 386–392.

McDermott, B. E., Edens, J. F., Quanbeck, C. E., Busse, D., & Scott, C. L. (2008). Examining the role of static and dynamic risk factors in the prediction of inpatient violence: Variable-and person-focused analyses. *Law and Human Behavior, 32*, 325–338.

McDermott, B. E., Quanbeck, C. E., Busse, D., Yastro, K., & Scott, C. L. (2008). The accuracy of risk assessment instruments in the prediction of impulsive versus predatory aggression. *Behavioral Sciences and the Law, 26*, 759–777.

Messick, S. (1995). Validity of psychological assessment: Validation of inferences from persons' responses and performances as scientific inquiry into score meaning. *American Psychologist, 50*, 741–749.

Miller, C., Otto, R.K., & Kimonis, E. (March, 2010). *Reliability of PCL-R scoring in sexually violent predator proceedings*. Paper to be presented at the annual meeting of the American Psychology-Law Society, Vancouver, BC, Canada.

Morse, S. J. (2008). Psychopathy and criminal responsibility. *Neuroethics, 1*, 205–212.

Murrie, D. C., Boccaccini, M., Johnson, J. & Janke, C. (2008). Does interrater (dis)agreement on Psychopathy Checklist scores in sexually violent predator trials suggest partisan allegiance in forensic evaluations? *Law and Human Behavior, 32*, 352–362.

Murrie, D. C., Boccaccini, M., McCoy, W., & Cornell, D. (2007). Diagnostic labeling in juvenile court: How do descriptions of psychopathy and conduct disorder influence Judges? *Journal of Clinical Child and Adolescent Psychology, 36*, 228–241.

Murrie, D. C., Boccaccini, M., Turner, D., Meeks, M., Woods, C., & Tussey, C. (2009). Rater (dis)agreement on risk assessment measures in sexually violent predator proceedings: Evidence of adversarial allegiance in forensic evaluation? *Psychology, Public Policy, and Law, 15*, 19–53.

Murrie, D. C., Boccaccini, M., Zapf, P., Warren, J., & Henderson, C. (2008). Clinician variation in findings of competence to stand trial. *Psychology, Public Policy, and Law, 14*, 177–193.

Murrie, D. C., Cornell, D. G., & McCoy, W. K. (2005). Psychopathy, conduct disorder and stigma: Does diagnostic labeling influence juvenile probation officer recommendations? *Law and Human Behavior, 29*, 323–342.

Nicholson, R. A., & Norwood, S. (2000). The quality of forensic psychological assessments, reports, and testimony: Acknowledging the gap between promise and practice. *Law and Human Behavior, 24*, 9–44.

Ogloff, J. R. P., & Lyon, D. R. (1998). Legal issues associated with the concept of psychopathy. In: D.J. Cooke, A.E. Forth, & R.D. Hare (Eds.), *Psychopathy: Theory, research, and implications for society* (pp. 401–422). New York: Kluwer.

Olver, M. E., Stockdale, K., & Wormith, J. S. (2009). Risk assessment with young offenders: A meta-analysis of three assessment measures. *Criminal Justice and Behavior, 36*, 329–353.

Patrick, C. J. (Ed). (2006). *Handbook of psychopathy*. New York: Guilford.

Quinsey, V. L., Harris, G. T., Rice, M. E., & Cormier, C. A. (2006). *Violent offenders: Appraising and managing risk* (2nd ed.). Washington, DC: American Psychological Association.

Rice, M. E., Harris, G. T., & Cormier, C. A. (1992). An evaluation of a maximum security therapeutic community for psychopaths and other mentally disordered offenders. *Law and Human Behavior, 16*, 399–412.

Rockett, J. L., Murrie, D.C., & Boccaccini, M. (2007). Diagnostic labeling in juvenile justice settings: Do psychopathy and conduct disorder findings influence clinicians? *Psychological Services, 4*, 107–122.

Rogers, R., Vitacco, M. J., Jackson, R. L., Martin, M., Collins, M., & Sewell, K. W. (2002). Faking psychopathy? An examination of response styles with antisocial youth. *Journal of Personality Assessment, 78*, 31–46.

Rufino, K., Heinonen, L., Boccaccini, M. T., Murrie, D. C., & Edens, J. F. (March, 2009). *What do PCL rater-agreement coefficients tell us about forensic practice?* Paper presented at the annual meeting of the American Psychology-Law Society, San Antonio, TX.

Rutherford, M., Cacciola, J. S., Alterman, A. I., McKay, J. R., & Cook, T. G. (1999). The 2-year test-retest reliability of the Psychopathy Checklist-Revised in methadone patients. *Assessment, 6*, 285–291.

Salekin, R. (2009). Psychopathy and recidivism from mid-adolescence to young adulthood: Cumulating legal problems and limiting life opportunities. *Journal of Abnormal Psychology, 117*, 386–395.

Sales, B. D., & Shuman, D. W. (2005). Experts in court: Reconciling law, science, and professional knowledge. In *Law and public policy: Psychology and the social sciences* (pp. 43–95). Washington, DC: American Psychological Association.

Schopp, R. F., & Slain, A. J. (2000). Psychopathy, criminal responsibility, and civil commitment as a sexual predator. *Behavioral Sciences and the Law, 18*, 247–274.

Shuman, D. W., & Greenberg, S. A. (2003). The expert witness, the adversary system, and the voice of reason: Reconciling impartiality and advocacy. *Professional Psychology: Research and Practice, 34*, 219–224.

Skeem, J. L., & Cooke, D. J. (2010). Is criminal behavior a central component of psychopathy? Conceptual directions for resolving the debate. *Psychological Assessment, 22*, 433–445.

Skeem, J. L., Edens, J. F., Camp, J., & Colwell, L. H. (2004). Are there racial differences in levels of psychopathy? A meta-analysis. *Law and Human Behavior, 28*, 505–527.

Skeem, J. L., & Golding, S. L. (1998). Community examiners' evaluations of competence to stand trial: Common problems and suggestions for improvement. *Professional Psychology: Research and Practice, 29*, 357–367.

Skeem, J. L., Miller, J. D., Mulvey, E., Tiemann, J., & Monahan, J. (2005). Using a five-factor lens to explore the relation between personality traits and violence in psychiatric patients. *Journal of Consulting and Clinical Psychology, 73*, 454–465.

Skeem, J. L., Polaschek, D. L., & Manchak, S. (2009). Appropriate treatment works, but how? Rehabilitating general, psychopathic, and high-risk offenders. In: J. Skeem, K. Douglas, & S. Lilienfeld (Eds.), *Psychological science in the courtroom: Consensus and controversy* (358–384). New York: Guilford.

Sullivan, E., & Kosson, D. (2006). Ethnic and cultural variations in psychopathy. In C. Patrick (Ed.), *Handbook of psychopathy* (pp. 437–458). New York: Guilford.

Sundby, S. E. (1998). The capital jury and absolution: The intersection of trial strategy, remorse, and the death penalty. *Cornell Law Review, 83*, 1557–1598.

Tyrer, P., Cooper, S., Seivewright, H., Duggan, C., Rao, B., & Hogue, T. (2005). Temporal reliability of psychological assessments for patients in a special hospital with severe personality disorder: A preliminary note. *Criminal Behaviour & Mental Health, 15*, 87–92.

Vidal, S., & Skeem, J.L. (2007). Effect of psychopathy, abuse, and ethnicity on juvenile probation officers' decision-making and supervision strategies. *Law and Human Behavior, 31*, 479–498.

Viljoen, J. L., MacDougall, E. A. M., Gagnon, N. C., & Douglas, K. S. (2010). Psychopathy evidence in legal proceedings involving adolescent offenders. *Psychology, Public Policy, and Law, 16*, 254–283.

Viljoen, J. L., McLachlan, K., & Vincent, G. M. (2010). Assessing violence risk and psychopathy in juvenile and adult offenders: A survey of clinical practices. *Assessment, 17*, 377–395.

Vitacco, M. J., Rybroekvan, G. J., Rogstad, J. E., Yahr, L. E., Tomony, J. D., & Saewert, E. (2009). Predicting short-term institutional aggression in forensic patients: A multi-trait method for understanding subtypes of aggression. *Law and Human Behavior, 33*, 308–319.

Walters, G. D. (2003-a). Predicting criminal justice outcomes with the Psychopathy Checklist and Lifestyle Criminality Screening Form: A meta-analytic comparison. *Behavioral Sciences and the Law, 21*, 89–102.

Walters, G. D. (2003-b). Predicting institutional adjustment and recidivism with the Psychopathy Checklist factor scores: A meta-analysis. *Law and Human Behavior, 27*, 541–558.

Walters, G. D. (2006). Risk-appraisal versus self-report in the prediction of criminal justice outcomes: A meta-analysis. *Criminal Justice and Behavior, 33*, 279–304.

Walters, G. D., & Heilbrun, K. (2010). Violence risk assessment and facet 4 of the Psychopathy Checklist: Predicting institutional and community aggression in two forensic samples. *Assessment, 17*, 259–268.

Walters, G. D., Knight, R. A., Grann, M., & Dahle, K. (2008). Incremental validity of the Psychopathy checklist facet scores: Predicting release outcome in six samples. *Journal of Abnormal Psychology, 117*, 396–405.

Wiener, R. L., Pritchard, C. C., & Weston, M. (1995). Comprehensibility of approved jury instructions in capital murder cases. *Journal of Applied Psychology, 80*, 455–467.

Wormith, J. S., Olver, M. E., Stevenson, H., & Girard, L. (2007). The long-term prediction of offender recidivism using diagnostic, personality, and risk/need approaches to offender assessment. *Psychological Services, 4*, 287–305.

Yang, M., Wong, S. C. P., & Coid, J. (2010). The efficacy of violence prediction: A meta-analytic comparison of nine risk assessment tools. *Psychological Bulletin, 136*, 740–767.

Zinger, I., & Forth, A. (1998). Psychopathy and Canadian criminal proceedings: The potential for human rights abuses. *Canadian Journal of Criminology, 40*, 237–276.

PART EIGHT
RESPONSIBILITY OF PSYCHOPATHS

CHAPTER 15

Criminal Responsibility and Psychopathy: Do Psychopaths Have a Right to Excuse?

Paul Litton

INTRODUCTION

The subjects of this collection are agents who do not merely commit antisocial acts. Psychopaths repetitively manipulate, deceive, leech, exploit, and commit crimes without compunction, guilt, remorse, or empathy for their victims. They fail to hold themselves responsible for their actions and the suffering they cause, while possessing a grandiose sense of self-worth. Undoubtedly, we associate all these characteristics with *increased* blameworthiness. Capital defendants, for example, attempt to show their sentencers that they are remorseful and accept responsibility, knowing that jurors find the remorseless more blameworthy. The coldly calculating criminal who feels nothing for his victims and views them as mere means to his gratification is the epitome of immoral. Accordingly, it might seem easy to answer whether psychopaths are morally responsible for their wrongdoing or should be held criminally responsible for their crimes: their characteristics indicate that they are *especially* blameworthy, not candidates for excuse.

Questions about psychopaths' responsibility status, however, are not easy. Their difficulty becomes apparent when we attempt to articulate general criteria of moral and criminal responsibility and then ask whether psychopaths satisfy those criteria. Responsibility theorists systematically examine the circumstances under which we hold some agents responsible and excuse others and attempt to articulate the general principles implicit in our firm, shared judgments about responsibility. They search for principles that explain and justify why we exempt the insane and children from the moral and legal demands we place on most adults. On one influential and compelling theory, we are committed to the principle that general classes of agents, such as children and the insane, are not morally responsible for wrongdoing insofar as they lack capacity to control their behavior in light of moral considerations (Wallace, 1994). Schoolchildren make choices and may be able to respond to threats and rewards, but, the view argues, we excuse children because they do not yet have a developed capacity to appreciate and respond to moral considerations. If that account plausibly explains and justifies our practice of not holding certain agents responsible, then you see why questions about the psychopath's responsibility arise. Psychopaths reason, but are they able to control their behavior in light of *moral* considerations? Can they appreciate moral considerations, given their significantly impaired capacity to experience empathy and other moral emotions? Of course, as discussed in more detail later, we can question whether moral responsibility does require the capacity to appreciate moral considerations: one might be particularly skeptical of the principle if it does, in fact, imply that psychopaths are blameless. But regardless, these preliminary remarks hopefully provide some sense of why the responsibility status of psychopaths presents a real challenge even though their traits generally indicate increased blameworthiness.

Current law does not recognize psychopathy as a basis for insanity. Psychopathy does not involve psychosis or other mental defect that normally disqualifies an individual for criminal responsibility. According to the M'Naghten test, used in most American jurisdictions,

> [t]o establish a defense on the ground of insanity it must be clearly proved that, at the time of committing the act, the party accused was laboring under such a defect of reason, from disease of the mind, as not to know the nature and quality of the act he was doing, or if he did know it, that he did not know that what he was doing was wrong. (M'Naghten's Case, 1843; quoted in *State v. Johnson*, 1979)

A person was insane at the time of his crime if he did not know (1) the nature and quality of his act or (2) that his act was wrong. The knowledge generally required under both prongs is thin. Based on current law, psychopaths know the nature and quality of their crimes as they are in touch with physical reality and can give an accurate, basic description of their conduct. The second prong requires knowledge of legal or moral wrongfulness (depending on the jurisdiction) for sanity; either way, psychopaths are deemed sane because they generally know what conduct society condemns.

If the knowledge required for sanity were more robust, one could argue that some psychopaths are not responsible under the M'Naghten standard. To illustrate, Antony Duff (1977) argues that psychopaths do not truly know the nature and quality of their harmful acts because they do not understand the moral and emotional aspects of their actions. Duff emphasizes that psychopaths know superficially what matters to others; they have enough insight to con, exploit, and manipulate successfully. However, the psychopath's

> understanding [of what people value] is still deficient: for he cannot see *how* these things can be important, how they can provide reasons for action and judgment; he cannot understand the emotional and moral significance these aspects of life have for others. (Duff, 1977: 193)

These considerations would also be relevant if the second prong required more robust knowledge of right and wrong. Some courts have held that sanity requires *affective* knowledge (Dressler, 2009: 350), which goes beyond knowing what we label "right" and "wrong." Affective knowledge entails the ability to "internalize the enormity of [one's] criminal act [and]...emotionally appreciate its wrongfulness" (Slobogin, 2003: 324).

The case for criminally excusing psychopaths is stronger under a literal interpretation of statutes following the American Law Institute's Model Penal Code (MPC). Even if an individual knows his act is considered "wrong," he may be insane under the MPC if he "lacks substantial capacity to appreciate" the criminality or wrongfulness of his conduct (depending on whether the statute refers to "criminality" or "wrongfulness"). Christopher Slobogin criticizes this test, in part, because under a straightforward application, psychopaths are insane: "[b]y definition, they do not emotionally appreciate [the] wrongfulness" of their crimes (Slobogin, 2003: 324). Despite the literal application, psychopaths are not treated as insane under this standard, perhaps because statutes following the MPC explicitly disqualify as a basis for insanity "an abnormality manifested only by repeated criminal or otherwise anti-social conduct."[1] However, this caveat does not undermine Slobogin's observation given psychopaths' severe emotional impairments: their abnormality is not manifest *only* by their antisocial behavior.

The important question, though, is not whether courts are misapplying current law to psychopathy. The essential issue is whether the criminal law *should* hold psychopaths responsible. Should legislators pass laws that deem psychopathy as an excuse? Alternatively, should the law view psychopathy as a mitigating sentencing factor because it diminishes, but does not eliminate, an individual's moral responsibility? Or is current law correct in holding psychopaths fully responsible for their crimes? We must weigh the reasons for holding psychopaths criminally responsible against those for deeming psychopathy an excusing or mitigating condition.

My initial goal in this chapter is to present a strong prima facie case, consistent with existing standards of criminal responsibility, for legally excusing at least a subset of psychopaths and otherwise treating psychopathy as mitigating deserved punishment. The prima facie case concludes that an incapacity to appreciate and respond to moral considerations provides grounds for excuse or, at least, a mitigating factor. In making that case, I argue that the criminal law itself—the substance of criminal prohibitions—contemplates subjects who are, to some degree, morally competent. These initial arguments also conclude that at least some psychopaths lack moral competence and, thus, that prima facie reason exists to excuse psychopaths.

However, this chapter's second aim, which should be evaluated separately, is to argue that other considerations override the preceding prima facie case. That is, I argue that, overall, we currently have stronger reasons for the criminal law to hold psychopaths fully responsible: the law should treat psychopathy as neither mitigating nor excusing. Pragmatic, consequentialist, and other moral considerations counsel against a "psychopathy defense." I argue that such considerations may prevail because psychopaths are not wronged when treated *as if* they are morally responsible agents, even if they are not.

BIOLOGICAL CAUSES AND RESPONSIBILITY

Before commencing the prima facie case, I will first explain why I will not present certain intuitively tempting arguments for the nonresponsibility of psychopaths. One might argue that psychopathy excuses because psychopaths are not responsible *for having* psychopathy. Some contributors to this volume are actively seeking to explain the causes of psychopathy, and, of course, none posits that the ultimate cause of psychopathy is an individual's uncaused choice to become a remorseless antisocial agent. Rather, researchers investigate whether psychopathy is associated with certain genes or neuroanatomical abnormalities regarding reduced gray matter, the hippocampus, and the amygdala. Psychologists offer different theories to explain why psychopaths did not internalize moral principles as children. This empirical research attempts to discern psychopathy's causes, which are all beyond the control of agents with psychopathy.

However, that psychopathy is caused by forces outside anyone's control does not negate the responsibility of psychopaths per se. If moral responsibility is incompatible with the fact that an agent's behavior or traits have causes outside his control, then none of us would be morally responsible for our conduct or traits. Some responsibility theorists do support that conclusion, namely, that we are not responsible if our actions are caused by forces (such as our genes and upbringing) outside our control. I will assume throughout this chapter, however, that an agent may still be morally responsible for her conduct even if her conduct was causally determined. The idea that deterministic causes are incompatible with our responsibility is rooted in an intuition that finds some support within our practices: the intuition is that an agent is blameless for some act if, at the time, she could not have done otherwise. If typing this paragraph was causally determined to occur by events outside my control such that I could not have done otherwise, then, according to the argument, I am not exercising free will. Even if I chose to type, my choice was causally determined by my brain states, dispositions caused by my genetic makeup, and my upbringing. Powerless to step outside these causal forces, I am not morally responsible for my actions. Responsibility theorists refer to this view as *incompatibilist*: it maintains that causation and moral responsibility are incompatible.

My position is that causation and moral responsibility are compatible. "Compatibilism versus incompatibilism" presents age-old questions and live debates within philosophy; I cannot defend compatibilism here. Rather, I briefly explain the compatibilist position and the fact that the criminal law, itself, embraces a compatibilist view of responsibility.

When we systematically examine our moral and legal practices of holding responsible some agents but not others, we see that the distinction between, say, the insane and responsible agents is not that the latter make uncaused choices while the former are causally determined (Morse, 1998; Wallace, 1994). The difference is that the insane lack a minimally adequate capacity to reason practically. We excuse them because their capacity to grasp and respond to reasons is severely impaired, not because they have some diagnosable disorder or because their genes or neurological states cause their conduct. An underlying biological cause could be relevant to responsibility, but not as a cause per se. Even if our actions are causally determined, some of us can reason and others cannot. A biological cause undermines responsibility only if it causes a rationality impairment: only if it interferes with an agent's ability to consider, assess, and respond intelligibly to reasons. Reasons-responsiveness grounds responsibility.

Regarding the criminal law's tests for responsibility, recall the insanity standards: a person was insane at the time of her crime if she did not know the nature and quality of her act or that her act was wrongful (M'Naghten), or if she lacked substantial capacity to appreciate her act's wrongfulness (MPC). Let us stipulate that all of our actions are causally determined by our brain states and the laws of nature, and the former are causally determined by our genes, our upbringing, and other facts outside our control. It is nonetheless true that we (normal adults) know the nature of our actions and can distinguish and appreciate right from wrong. Deterministic forces may have caused your choice to read this chapter, but you nonetheless know what you are doing and that it is morally permissible according to societal standards. The law is not concerned with whether your actions are caused, but with whether you have certain cognitive capacities.

The fact, then, that psychopathy is caused by forces outside the control of psychopaths is not relevant to deciding whether we should hold them responsible for their antisocial acts. Whether it is appropriate to hold them morally and legally responsible for their conduct depends on their reasoning capacities, the topic to which we now turn.

THE PRIMA FACIE CASE FOR EXCUSING THE PSYCHOPATH

The general capacity for reasons-responsiveness entails more specific capacities and psychological traits, such as the capacities to draw reasonable conclusions from accurate perceptions of reality, to reason logically, to form abstract concepts correctly, to make intelligible judgments about how reasons relate to one another, and to avoid focus on irrelevant facts. Disagreement exists, though, among reasons-responsive theorists regarding another capacity. Some argue that the general capacity to govern oneself through reasoning is necessary but not sufficient for moral responsibility because an agent might have that general capacity yet lack a more specific capacity for *moral* competence. They argue that an agent may be superficially responsible for his conduct insofar as he acts for intelligible reasons; however, being held *morally* responsible and *blamed* for wrongful conduct entails very serious implications, and those serious implications are rendered fair and appropriate only if an agent has the capacity to grasp and appreciate the *moral* reasons he allegedly ignored.

Most responsibility theorists who conclude that psychopaths are not morally responsible agree that moral responsibility requires the capacity to grasp and respond to distinctly moral considerations (Fine & Kennett, 2004; Levy, 2007-b; Morse, 2008; Wallace, 1994). To assess their view we must ask at least two questions: First, should the criteria of moral responsibility include this specific moral capacity? Second, if the criteria should, is psychopathy inconsistent with them? Let us start with the second question of whether psychopaths can grasp and respond to distinctly moral considerations.

Psychopathy and Moral Competence

Notably, the Psychopathy Checklist–Revised (PCL-R) criteria do *not* include "an incapacity to appreciate and respond to moral considerations," although they reference psychopaths' antisocial traits. Moreover, frequent immoralities, by themselves, do not demonstrate an incapacity for moral reasoning. A neo-nazi or mafia boss who perpetrates violence against some individuals but treats others within his group with respect and affection may commit immoral acts frequently. Such an agent's attachment to, and ability to respect, at least some persons—even a limited set—evidences his capacity for moral reasoning. Researchers contrast such moral reasoners with psychopaths, describing the latter as having only superficial relationships (Cleckley, 1988). The extent to which psychopaths can appreciate and respond to moral considerations is an open question.

Knowledge of what we label "right" and "wrong" is not sufficient for an ability to appreciate moral reasons. Young children can recite many basic moral rules, yet their moral capacities are not yet developed. Moral agents can do more than merely mimic moral discourse; they can grasp the varying significance of different human concerns, values, and interests. Having some appreciation of various human concerns allows reasoning to go beyond mere mimicry. It allows agents to understand the varying importance of different moral principles and thus enables them to weigh and negotiate intelligibly competing moral considerations in different situations (Duff, 1977: 195; Wallace, 1994:108).

Regarding the psychopath, Cleckley's remarks parallel our view of young children: "He can repeat the words and say glibly that he understands, [yet] there is no way for him to realize that he does not understand" (Cleckley, 1976: 90, quoted by Hare: 1993, 28) Psychopaths do not embrace moral values—not even warped ones, like mafiosos or neo-nazis—and thus have an impaired capacity to understand the importance we attach to moral principles: "nothing in [their] orbit of awareness...can bridge the gap with comparison" (Cleckley, 1976: 90, quoted by Hare: 1993, 28). As Duff remarks, psychopaths have an impaired capacity to appreciate what matters to people, even though their minimal understanding is sufficient to con and manipulate.

Empirical research arguably supports the claim that psychopaths' ability to gauge the varying importance of human interests is seriously impaired. Normal functioning persons, as young as 39 months and found across cultures, implicitly distinguish *moral* from *conventional* rules, but psychopaths' ability to make the distinction is impaired. As defined by this research, moral rules include prohibitions against rights-violations and causing harm to persons. They bind all persons, existing independently of any legislating authority. Because moral transgressions usually involve rights-violations or harm, they are generally viewed as serious. In contrast, conventional rules, as defined, are not objectively or universally prescriptive, but rather are dependent on some authority's command. For example, subjects agree that it is wrong to wear pajamas to school, but this prohibition is a convention in that it would be permissible if the principal or teacher permitted pajamas. Because conventional transgressions do not involve harm or rights-violations, most adults view them as less serious than moral violations (Blair, Monson, & Frederickson, 2001).

Blair reports that psychopaths "have considerable difficulty" distinguishing moral and conventional norms" (Blair, Mitchell, & Blair, 2005: 58). Displaying their lack of empathy, psychopaths are much less likely to appeal to the welfare of victims when offering a justification for following moral rules (Blair, 1995; Blair, Jones, Clark, & Smith, 1995-a). In at least one study, "the quality of [a subject's] moral/conventional distinction and [his] tendency to make reference to *other's welfare* when justifying moral transgressions were not only correlated with the total PCL-R score but also with individual items of the PCL-R; in particular, with the item 'Lack of Remorse/Guilt'" (Blair et al., 1995-a: 749). Moreover, unlike nonpsychopathic controls, psychopaths judge moral and conventional rules similarly in terms of whether their existence is dependent on a legislating authority.[2] This evidence supports the proposition that psychopaths have a significantly impaired capacity to assess the relative strength of different human concerns, thereby undermining their capacity to reason competently about moral considerations.

One immediate objection could highlight that, in these studies, psychopaths did assess moral transgressions as more serious than conventional transgressions (Blair et al., 1995-a). However, as Blair states, the incarcerated subjects, punished for transgressions we describe as moral, had access to what *we* take to be more serious. It is unlikely that their assessments of seriousness were based on their emotional capacities or internalized principles.

Anecdotal accounts provided by Hare depict agents who either have an astonishingly impaired ability to gauge the relative importance of different human concerns or to anticipate that others will view their moral arguments as "crazy." For example, Hare interviewed a psychopath convicted of rape who claimed that his crimes actually benefitted his victims:

> "The next day I'd get a newspaper and read about a caper I'd pulled—a robbery or rape. There'd be interviews with the victims. They'd get their names in the paper. Women, for example, would say nice things about me—that I was really polite and considerate, very meticulous. I wasn't abusive to them, you understand. Some of them thanked me." (Hare, 1993: 43)

This psychopath knows that people like being mentioned in newspapers and being treated politely and considerately. But if he truly believes that these considerations justify the ways in which he harmed his victims, then we must simply view him as morally "crazy." One might respond by saying that he is obviously lying: he does not believe what he is saying, but rather, as a prison inmate without scruples, is merely trying to paint himself well for self-interested purposes. Even if that were true, we would still have to judge him morally insane for being so out of touch with social reality that he fails to understand that others would view his statement as ridiculous, beyond outrageous.

Again, the psychopaths' moral incompetence is not merely the same as having wrong or warped values. An agent may hold thoroughly immoral beliefs (e.g., racist beliefs) yet understand what it means to take moral considerations seriously (as he holds himself to moral principles with regard to his preferred group). This nonpsychopathic yet immoral agent can have an adequate understanding of moral concepts and familiarity with emotions inextricably connected to moral reasoning. Psychopaths, on the other hand, have deficient moral concepts. When presented with short stories aimed to induce certain emotions, psychopaths and nonpsychopathic controls similarly attributed happiness, sadness, and embarrassment to the protagonists; however, in response to stories of protagonists who harmed a victim, nonpsychopaths expectedly attributed guilt whereas psychopaths attributed happiness or indifference, not guilt (Blair et al., 1995-b). In addition to not feeling guilt, psychopaths do not seem to understand how other people feel when they recognize they have acted wrongly. As Blair and colleagues state, this research "at least indicate[s] that the psychopath has a specific difficulty in understanding this emotion" (Blair et al., 1995-b: 435).

More generally, the psychopath's lack of familiarity with moral emotions supports the view that they "give lip-service to understanding morality, but...do not have moral concepts—or at least they do not have moral concepts that are like the ones that normal people possess" (Prinz, 2006: 32). At least with respect to moral norms that prohibit harming others and that require us to understand how others feel, we may conclude that psychopaths possess a severely impaired capacity to appreciate and respond to moral considerations.[3]

Moral Competence and Criminal Responsibility

Even if all or some psychopaths have a significantly impaired capacity for moral reasoning, we should ask whether psychopaths' rational capacities are otherwise adequate for moral and criminal responsibility. If an agent governs himself by assessing reasons and is otherwise adequately rational (e.g., he reasons logically, is in touch with physical reality, can effectively will the means to his ends, etc.), why demand more specific moral capacities for responsibility and blameworthiness? With regard to criminal punishment, if an agent knows the rules and has adequate rational capacity to avoid incurring punishment, why require more?

R. Jay Wallace argues that moral responsibility requires specific moral capacities because of the burdens associated with being held morally responsible. On his account, being held morally responsible for wrongdoing exposes an agent to possible burdens of blame-related emotions, such as indignation and resentment, and their expression through moral sanctions, "such as avoidance, reproach, scolding, denunciation, remonstration, and (at the limit) punishment" (Wallace, 1994: 54). Because of the potential burdens of being held morally responsible, Wallace argues that it is unfair or unreasonable to hold someone morally responsible unless that agent has the capacity to grasp and apply the moral considerations to which we should respond (Wallace, 1994).[4] In other words, on Wallace's view, it is unfair or unreasonable to blame someone for failing to respond to moral considerations when he has inadequate capacity to do so.

Other compatibilist theorists disagree, arguing that a distinct ability to recognize and respond to moral considerations is not an additional requirement for moral responsibility. The key to the underlying disagreement is an opposing view regarding the interpersonal significance of being held responsible. T. M. Scanlon and others deny that the interpersonal significance of being held responsible should be located in blame-related emotions and their expression through sanctioning behaviors, and thus they deny that fairness requires holding responsible only those who appreciate moral considerations. On this view, blaming judgments respond to the perception that another agent has expressed ill will or inappropriate indifference toward another. Given that an agent's ill will or inappropriate indifference, expressed through conduct, is "in principle" dependent on one's judgments about reasons, blaming judgments involve an implicit demand for the blamed agent to justify her conduct, submit an excuse, apologize, or make amends in some fashion. This implicit demand, which accounts for the special depth of blaming judgments, is not a sanction. It is appropriately directed toward agents who can be asked to justify themselves to others, that is, agents with adequate reasoning capacities, regardless of whether they can appreciate moral considerations.

However, if blame involves an implicit demand for moral justification, excuse, or apology, one might argue that they are directed appropriately only toward those who can grasp and appreciate moral considerations (Wallace, 2006). It would be pointless to demand a justification from someone if he is unable to grasp or appreciate the relevant moral considerations. Of course, circumstances exist wherein it would be pointless to demand a justification even though that agent might still be blameworthy: it could be pointless to demand justification from a devastated parent whose negligence led to his child's serious harm. But in those circumstances, the agent has the moral capacities to respond intelligibly to a demand for justification, excuse, or apology: the surrounding context renders the demand inappropriate. The agent could still be blameworthy because she has the requisite moral capacities. In contrast, if an agent lacks the requisite moral capacities then it would be pointless in *all* circumstances to blame and demand justification, excuse, or apology.

Nonetheless, Scanlon resists moral competence as a requirement for responsibility in light of his account of blame. On his view, to judge someone blameworthy is to claim that "something about [her] attitudes towards others...impairs the relations that others can have with [her]" (Scanlon, 2008: 128). If we do not share common interests or passions, some kinds of relationships may be closed to us, but much remains open: I "can still be a good neighbor, worker, or even a friend" (Scanlon, 2000: 159). But if I hold you responsible for failing to respect my value as a person, blaming you carries interpersonal significance regardless of whether I experience any negative reactive emotion or express my resentment. If you cannot be trusted at all—if it is apparent that you see me only as an instrument to your ends—the implications are more severe in that we cannot have any decent or meaningful relationship.

Why is this account of blame's significance relevant to the criteria of moral responsibility? Scanlon argues that if someone wrongs you and fails to see your value as a person, his disrespectful conduct will have the same implications for your relations regardless of whether he (1) was capable of appreciating moral reasons but ignored them or (2) generally lacked appreciation for moral considerations. The bottom line in either case—whether the agent is a psychopath or not—is that the person does not respect your value: "A person who is unable to see why the fact that his action would injure me should count against it still holds that this *doesn't* count against it" (Scanlon, 2000: 288).

Following Scanlon and writing specifically about psychopaths, Matthew Talbert argues that psychopaths' decisions to manipulate and exploit reflects a normative commitment that nothing about his victims provides any reason to refrain from his harmful conduct. Psychopaths know we believe there is reason not to harm people. He rejects our belief. Given that psychopaths reason, their decisions to harm and exploit reflect their normative commitment that no reason exists to refrain from this behavior. Regardless of whether he can appreciate moral reasons but ignores them, or whether he lacks capacity to appreciate moral considerations, his conduct demonstrates his commitment that nothing about other people provides reason to respect them.

Because psychopaths' wrongful conduct does, allegedly, reflect normative commitments about reasons, it is appropriate to hold them responsible, on this view. Robert Schopp and Andrew Slain emphasize that the psychopaths are not relevantly analogous to other agents who deserve excuse. Imagine an agent with dissociative identity disorder who experiences multiple, distinct states of consciousness. Usually, he is a law-abiding, responsible citizen, unaware of his conduct or the traits he assumes in his altered state of consciousness. In that other state, he becomes psychopathic: he lacks empathy, does not care about any moral principles, and engages in antisocial behavior. Schopp and Slain argue that we would excuse this agent for wrongdoing but not because his psychopathic alter-ego lacks empathy or moral competence. Rather, we would excuse him because when he is in his psychopathic state he lacks access to *his own* "principles, extended interests, and moral emotions" (Schopp & Slain, 2000). In other words, in his altered consciousness, this agent lacks access to himself; *he* is not the bad actor. In contrast, the "normal" psychopath *is* the bad actor. Psychopaths' conduct *does* reflect who they are and their belief that others do not matter.

To make the prima facie case for criminally excusing psychopaths, I must respond to these arguments for holding the psychopath morally responsible. It seems plausible to assume that if psychopaths are morally responsible for breaking a criminal law then the criminal law should hold them responsible.

As a first response, one might argue that at least some psychopaths are less than fully responsible, not because of their specific moral incapacities, but because of more general rational impairments. That is, even if moral competence is not a criterion of moral responsibility, at least some extreme psychopaths are not fully responsible agents because their general rational capacities are impaired relative to those of nonpsychopathic individuals. In the end, I will not take this path for making the prima facie case for a legal excuse, but let us briefly see the argument.

Normal human agents have the capacity to evaluate their desires, to decide whether reason exists to act on them. We do not always act on our strongest desires; we do not merely order our desires by the strength with which we feel them. Normal functioning adults endorse standards and principles in light of which they evaluate the desirability of their first-order desires. We may reject acting on a desire not merely because we have some inconsistent, stronger first-order desire, but because to act on it would be base, cowardly, arrogant, dishonorable, degrading, etc. Our principles constitute our conception of the kind of persons we want to be. Consequently, because we hold ourselves to certain evaluative standards and principles, we experience emotions such as remorse, regret, and shame when we fail to live up to them. These reactive emotions are not only moral emotions. Moreover, they are distinct from frustration and anger: remorse, shame, and guilt entail negative self-evaluation and a deep sense of loss.

Psychopaths do not fit this picture of normal human agency. They seem not only to lack moral standards but also to have a weakened capacity to hold themselves to any evaluate principles and evaluate their desires in light of them. Their moral incapacity is an instance of a more general limitation of their capacity to value, to embrace principles, and thus to understand how normal functioning adults reason. They are immune to shame, remorse, and other indicators of deep self-evaluation. They may get angry or think they did something stupid when they frustrate their ends (by, say, getting caught for a crime); anyone might feel anger or frustration from not being able to satisfy first-order desires. However, such feelings are distinct from shame and other reactive emotions we feel when we violate the principles or higher-order desires we embrace and with which we self-identify. Shame and other reactive attitudes we experience upon violating our deeply held convictions entail a deeper sense of negative self-evaluation.

Hare, in fact, distinguishes psychopaths from other antisocial agents by stating that "[u]nlike most other criminals, psychopaths show no loyalty to groups, codes, or principles, other than to 'look out for number one'" (Hare, 1993: 85). Even in "looking out for number one," psychopaths are not attempting to satisfy principles, for they do not endorse any, according to Hare. The psychopath seeks excitement, pleasure, and power, but seeking such things does not imply holding oneself to standards. Consequently, another psychopathic characteristic is not possessing long-term goals. Psychopaths' lifestyle is "chronically unstable and aimless" (Hare, 1993: 57).

Cleckley, too, describes one of his psychopathic patients as "unfamiliar with the primary facts or data of what might be called personal values and is altogether incapable of understanding such matters" (Cleckley, 1988: 40). Because his patient was unfamiliar with values, Cleckley states that he could not understand how other people are moved: "It cannot be explained to him because there is nothing in his orbit of awareness that can bridge the gap with comparison" (Cleckley, 1988: 40). To object, one might point to psychopaths who describe gay persons as "disgusting" or "low," and conclude that these psychopaths endorse *at least some* moral principles. However, that some psychopaths voice such claims does not show they embrace and hold themselves to principles, moral or otherwise. They just might feel repulsed, say, by the thought of gay sex in the same way most of us find repulsive the idea of eating spiders.

Charles Taylor persuasively writes that an agent without any principles, who could only weigh the relative weights of first-order desires and could not evaluate the worth of her desires,

> would lack the depth to be a potential interlocutor, a potential partner of human communion, be it as a friend, lover, confidant, or whatever. And we cannot see one who could not enter into any of these relations as a normal human subject. (Taylor, 1982: 118)

Arguably, an agent who is unfit for interpersonal relationships "as a normal human subject" is not a member of the moral community, someone whom it is inappropriate to hold responsible.

Researchers attribute unintelligible—irrational—statements to psychopaths, statements that appear tied to a weakened capacity to possess personal values and standards. Hare, for example, describes psychopathic patients who make claims about what they care about, yet their behavior demonstrates that they do not really understand their claims. He describes one psychopathic parent as oblivious to the deep tension between her claim that her child needed her love and the fact that she ignored the child so thoroughly "that the baby was severely malnourished" (Hare, 1993: 63). This strange unawareness is part of what lies behind the psychopath's "mask of sanity": he mimics our behavior and displays of concern without grasping the emotional aspects and real implications of his claims to care (see, e.g., Cleckley, 1988).

Related, Hare describes psychopathic patients who express long-term goals but do not understand them.

> The psychopathic inmate thinking about parole might outline vague plans to become a property tycoon or a lawyer for the poor. One inmate, not particularly literate, managed to copyright the title of a book he was planning to write about himself and was already counting the fortune his bestseller would bring.[5] (Hare, 1993: 39)

If there is never any connection between an agent's sincere claims about long-term goals and his subsequent actions, we question his rationality.

Psychopaths' weakened capacity to value and embrace personal standards is connected to other signs of impaired rationality. Researchers describe their antisocial acts as shockingly inadequately motivated. Robert Hare writes that

> family members, employers, and co-workers typically find themselves standing around asking themselves what happened—jobs are quit,

relationships broken off, plans changed, houses ransacked, people hurt, often for what appears little more than a whim. (Hare, 1993: 58)

They are impulsive and often frustrate their own professed ends, but not necessarily because their rationality is overcome by passion or irresistibly strong urges. Rather, as Jeanette Kennett writes, there is "not much to defeat" (Kennett, 2006: 77) in terms of their capacity to evaluate their desires critically.

Researchers have documented other aspects of psychopaths' impaired rational capacities. For example, they have narrow attention spans, an impaired ability to pay attention to multiple things simultaneously. A narrow attention span impairs an agent's ability to satisfy her ends and to assess whether possible consequences of certain actions are compatible with her ends (Maibom, 2005). Research also shows that psychopaths perseverate reward-seeking behavior when it was previously rewarded even when it is subsequently punished consistently (Maibom, 2008).

Despite these signs of psychopaths' impaired rationality, I will not rest on them the prima facie case for excusing psychopaths. First, that psychopaths' rationality is impaired does not show, by itself, that they are too irrational to be held morally and criminally responsible. Their rationality is impaired but may be adequate for the law's purposes given that they know what conduct is criminalized and are deterrable. Hare, in fact, opines that "psychopaths certainly *know enough* about what they are doing to be held accountable for their actions" (Hare, 1993: 143). Second, the argument rests heavily on anecdotal, rather than experimental, evidence that psychopaths do not possess principles or values. The arguments might apply to some psychopaths but perhaps not others. Though responsibility theorists have highlighted psychopaths' general rationality defects, none has argued for a legal excuse based on those impairments. Arguments in favor of criminally excusing psychopaths rely on their specifically moral impairments (Fine & Kennett, 2004; Morse, 2008).

Therefore, to justify the prima facie case for criminal excuse, we need a different response to the Scanlon/Talbert arguments described earlier. That is, we must provide a different response to the claim that it is appropriate to blame psychopaths because their wrongdoing reflects themselves and their judgment that no reason exists to respect others, regardless of whether they have specific moral capacities. To start, recall that Scanlon rejects the idea that fairness requires adding specific moral capacities to the requirements of moral responsibility because, on his view, being held responsible does not necessarily entail exposure to the burdens of resentment, indignation, and their expression through moral sanctioning behavior. Scanlon may be right with regard to moral responsibility, but our ultimate focus here is on the requirements of *criminal* responsibility—the requirements for fair imposition of punishment—which *does* entail public sanction. Criminal punishment does involve a severe sanction, which includes not only incarceration (or worse), but also publicly expressed moral condemnation. What distinguishes crimes from civil wrongs is that the former "incur a formal and solemn pronouncement of the moral condemnation of the community" (Hart, 1958: 405). Serious moral condemnation is necessarily expressed through criminal conviction and punishment. Because the condemnation is expressed and attached to severe sanctions, fairness is relevant: it is fair to impose such public moral condemnation upon someone only if he has adequate capacity to appreciate and respond to the moral considerations to which he has failed to respond, according to the public condemnation.[6]

We can connect the fairness of criminal punishment and the capacity to appreciate and respond to moral reasons more explicitly.[7] First, recognize that the criminal law imposes extremely harsh penalties on persons. For such penalties to be justified, we must have adequate opportunity to avoid incurring punishment by choosing legal activities. The fact that the law requires a voluntary choice to violate a criminal prohibition contributes to the justification of criminal sanctions. If someone had adequate opportunity to avoid punishment by choosing appropriately, his grounds for complaining about being punished are considerably weaker than a complaint by another punished individual who lacked adequate opportunity to choose appropriately. However, that someone has chosen an act that constitutes a crime does not, by itself, justify severe criminal sanctions. The conditions under which someone chooses to violate the law matter to whether the person had fair or adequate opportunity to choose appropriately. For example, if someone chooses under duress to violate the law, that agent lacked fair opportunity to avoid incurring sanction. Thus, the excuse of duress provides a safeguard against incurring criminal punishment. Societal and economic conditions are also relevant to a person's opportunity to avoid violating the law. If, because of economic conditions, it is more rational for citizens to violate the criminal law than to live within it, the state does not have the right to punish even those who voluntarily break the law. Minimally decent economic conditions also provide a necessary safeguard against incurring punishment by helping to provide a fair opportunity to persons to choose legal activities.

That we socialize our children—that they internalize moral principles—provides another safeguard that helps justify the severe sanctions of the criminal law. Most of us avoid violating the criminal law not because we fear punishment, but because we have internalized moral principles and rarely, if ever, seriously consider committing a crime. Mostly, we avoid criminal activity for moral reasons; the threat of punishment adds extra, self-interested reasons for avoiding crime. Successful socialization, then, enhances one's opportunity to choose law-abiding activities and, as such, contributes to the justification of imposing harsh punishment on those who voluntarily choose to violate the law.

Therefore, the agent who cannot appreciate and respond to moral considerations—an agent who has not been successfully socialized—does not benefit from an important safeguard that helps justify the harsh sanctions of the criminal law. As Stephen Morse puts it, such an agent lacks access to the "most rational reasons to behave well," (Morse, 2008: 208) and, as such, his opportunity to avoid criminal activity is not enhanced by socialization. That agent only has access to self-interested reasons to avoid criminal punishment. For an agent who has access only to self-interested reasons, a decision to break the law may represent a failure to appropriately weigh self-interested reasons, but it would not also represent a failure to respond to moral considerations. Perhaps imposing negative consequences on that agent is justified: that the agent had an opportunity to choose appropriately on self-interested grounds may be sufficient to justify at least some incarceration. At minimum, however, we see that relative to others who have been provided the safeguard of successful socialization, it is more difficult to justify the harsh sanctions of criminal law on those who could not be socialized. These considerations support the conclusion that in the eyes of the law, agents who were not—because they could not be—socialized are, at the least, less deserving of punishment than those who were successfully socialized. We would reach the same conclusion about someone who could have been socialized through normal methods but was, say, raised by wolves. Like the psychopath, but based on a different cause, that agent did not have the chance to internalize basic moral rules and dispositions, and accordingly it would be more difficult to justify imposing harsh criminal sanctions on him should he violate the law. The basic idea is that persons who have access to moral considerations, in addition to self-interest, have a superior ability to choose wisely and avoid incurring criminal punishment, and, as such, have weaker grounds to complain should the state impose punishment for their criminal decisions.

To conclude this section, I have argued that criminal responsibility criteria *should* include the capacity to appreciate and respond to moral reasons. Criminal punishment necessarily entails severe, public moral condemnation, and it is unfair to impose such condemnation on agents without capacity to respond to the moral reasons to avoid criminal activity. Moreover, lacking that specific moral capacity should at least be viewed as a mitigating factor because it is more difficult to justify the full extent of the law's harsh sanctions for persons who were not socialized to some minimally adequate degree.

The Criminal Law Implicitly Assumes Moral Competence

This section turns away from justifications for the criminal law to the law itself. The question here is whether the law itself already presumes a position on whether criminal responsibility should require the capacity to appreciate and respond to moral reasons. We may search for the answer to our inquiry that best fits existing criminal law as a whole. Does the criminal law contemplate subjects who are, at least, minimally competent in moral reasoning, or does the law only presume subjects capable of following black letter commands? Having discussed insanity standards in the introduction, I turn to other criminal law doctrines. I argue that their substance implicitly contemplates subjects who have the capacity to grasp and apply moral considerations.

First, I assume that "a crucial aim of the criminal law is to guide and thus to prevent certain actions" (Morse, 2004: 367). In providing punishment for certain actions the criminal law commands citizens. The directives correspond with the substantive crimes: Unless you wish to incur punishment, do not intentionally or knowingly cause the death of another person;[8] do not intentionally cause physical injury to another person;[9] do not knowingly have sexual contact with another person if you know or have reasonable cause to believe the contact is offensive to that person;[10] do not willfully start a fire with the purpose of destroying someone's building;[11] and so forth. These directives represent some examples of the law's commands.

These particular examples alone do not make clear why the criminal law implicitly assumes moral competence. They do not seem to require complex or even simple moral reasoning. To comply, do not intentionally or knowingly kill anyone. However, the criminal law is rife with commands not as simple which do require moral reasoning.

The aforementioned examples represent crimes for which the state must prove a defendant acted intentionally or knowingly. However, some crimes involve risky conduct performed *recklessly* or even with *criminal negligence*. In many states a person commits involuntary manslaughter by causing another's death recklessly[12] or with criminal negligence.[13] What does it mean to act recklessly? In most jurisdictions, recklessness implies the actor "disregarded a *substantial* and *unjustifiable* risk of which he was aware" (Dressler, 2009: 135). A person acts with criminal negligence when his conduct represents a gross deviation from the standard of care a reasonable person would have taken under the circumstances, given that the actor took a *substantial* and *unjustifiable* risk of which he *should* have been aware (Dressler, 2009). The law directs us to perceive and avoid certain substantial and unjustifiable risks. To comply with such commands, an agent must be able to distinguish justifiable from unjustifiable risks, a task that requires moral reasoning.

In addition to forbidding certain acts, the criminal law provides other directives associated with affirmative defenses, and these directives contemplate agents who can morally reason competently. United States common law and many state statutes recognize the necessity or choice-of-evils defense. Conduct is justified under necessity, even if it violates the letter of a prohibition, if violating the letter of the law was necessary to avoid an imminent harm that would have been greater than the harm caused by the otherwise criminal act. Translated as a directive, this defense instructs that we may avoid punishment if we choose the lesser of two evils. This directive requires more than the ability to follow simple commands. It requires the capacity to distinguish the relative degree of distinct evils. To use the directive successfully, one must be able to gauge the varying importance of human interests to determine which evil is greater. In fact, some state statutes that codify the necessity defense specifically reference moral reasoning, requiring the avoided harm to be "of such gravity that, according to ordinary standards of intelligence and morality, the desirability and urgency of avoiding such injury clearly outweigh the desirability of avoiding the injury sought to be prevented by the statute [defining the offense]."[14] As discussed, this capacity to gauge intelligibly the importance of different human concerns may be lacking in the most severe psychopaths: recall, for example, the psychopath who claimed that his victims benefitted from their names appearing in the newspaper and that this "benefit" mattered in assessing the overall harm he inflicted upon them.

Let us consider some objections. First, one might argue that necessity is rarely ever a relevant possibility. Avoiding punishment is easy and does not require moral reasoning: simply avoid performing prohibited acts. In response, though, this objection does not undermine my argument that the substance of criminal law doctrine generally contemplates agents who can reason morally. Moreover, although necessity is almost never relevant, in recognizing a necessity defense, the law contemplates agents who can distinguish circumstances under which necessity is and is not a plausible justification. Our judgment that necessity is almost never relevant is based on our ability to engage competently with morally relevant facts.

Second, an objector might claim that the law does not require subjects to reason competently about morality; the necessity defense, for example, does not require that someone choose the actual lesser evil, but directs him to choose the evil lesser in the community's judgment (via the jury or legislature).[15] A successful defendant only had to anticipate what conventional morals required, and psychopaths know the content of conventional morality, just as they know our basic criminal prohibitions.

In response, even to anticipate the community's moral appraisal of a particular complex situation, competent moral reasoning is required. An agent possesses basic knowledge of conventional morality when able to recite our moral rules and principles and to recognize paradigm examples of violations. Children above a certain age have this basic knowledge. But to anticipate and grasp how the community would appraise a morally complex set of facts, one must engage competently in moral reasoning: an agent would have to think about how others would reason by identifying the relevant moral principles, weighing evidence relevant to those principles, and balancing competing principles and interests. Note that we do not have explicit, verbally formulated rules for recognizing the moral principles salient to a particular situation, for weighing relevant evidence, or for balancing principles and interests. Perhaps the nonexistence of such explicit rules plays a role in explaining the psychopath's diminished capacity for moral reasoning.

To summarize the prima facie case, criminal responsibility should require some capacity for moral competence. It is unfair for society to express severe moral condemnation toward agents who lack capacity to appreciate and respond to moral considerations. Moreover, it is more difficult to justify the criminal law's harsh sanction for agents who have access only to self-interested reasons compared to agents who benefitted from socialization. Finally, the criminal law, itself, contemplates subjects who have some minimal ability to reason morally. The prima facie case concludes that psychopaths with warped moral concepts, unfamiliar with empathy and moral emotions, and unable to gauge the relative importance of different human interests should be exempted from punishment and consequently, in appropriate cases, exposed to civil confinement.

COUNTERING THE PRIMA FACIE CASE: HOLDING PSYCHOPATHS CRIMINALLY RESPONSIBLE

The prima facie case is based on a non-consequentialist moral principle: the law should not punish agents who lack adequate moral capacities regardless of whether societal welfare, as a whole, would be greater by treating psychopaths as responsible for their crimes. It is wrong, for example, to punish someone who is insane under current law, regardless of any good consequences (such as deterrence) that would result from punishing her: to punish an insane person is to violate his rights.

The counter to the prima facie case, however, does appeal to pragmatic, consequentialist, as well as other moral reasons. To proceed then, I must explain why psychopaths' criminal responsibility status may be settled by consequentialist reasoning when a non-consequentialist moral principle supports excuse. My explanation is that psychopaths are not wronged by being treated *as if* they are morally responsible even if they are not, in fact, morally responsible and deserving of punishment. Psychopaths, themselves, have greater interest in being treated as responsible agents; they are better off not being excused. If psychopaths are not wronged by being treated as responsible, then we are morally free to consult pragmatic, consequentialist, and other considerations to decide whether to hold them criminally responsible.

No Right to Criminal Excuse

One might argue that psychopaths are not wronged when treated as if they are responsible because they cannot be wronged at all. Put differently, one might argue that psychopaths lack a right to be excused because they do not have any moral rights. Jeffrie Murphy famously argues that the existence of a system of rules, implying rights and imposing correlating obligations, rests upon its subjects' capacity and willingness to treat each other respectfully, even when doing so conflicts with their desires. Because the psychopath cannot participate in a system requiring such reciprocity, he "violates a condition" for the existence of obligations and rights (Murphy, 1972: 291).[16] As such, he "is in no position to claim rights for himself" (Murphy, 1972: 291). Interestingly, Murphy emphasizes that his conclusion explains why psychopaths may be *excused*: "punishment may be regarded as a *right*" (Murphy, 1972: 291). However, Murphy also states that psychopaths need not be excused, presumably because they lack the right to be excused, as well. In fact, Murphy argues that the law should treat psychopaths as responsible (Murphy, 1972).

Murphy's conclusion that psychopaths lack rights is controversial.[17] Without assessing his view, I instead will assume psychopaths have rights (e.g., the right not to be killed unjustly) and can be wronged. I will, though, argue that they are not wronged if treated as if they are morally responsible. Stated differently, I argue that even if they have rights generally, they lack the specific right to be criminally excused for failing the requirements of moral responsibility. I begin with an interest theory of rights: rights protect important interests. More specifically, rights protect interests that are of sufficient importance to one's well-being that they ground a duty on others (Oberdiek, 2009; Raz, 1986). With that starting point, we must identify the interest that grounds the right of agents without requisite moral capacities to be exempted from punishment and then ask whether psychopaths possess that interest.

Let's first rule out a potential candidate for the relevant interest: the interest in avoiding confinement and remaining at liberty. Regardless of whether psychopaths are treated as responsible or not, they will not avoid confinement and remain free when caught committing crimes. A defendant who successfully raises the insanity defense does not walk from the courtroom freely. In some jurisdictions he is automatically committed to a psychiatric facility, and in other states the court sends him to a mental health facility to assess whether he should be committed civilly (Dressler, 2009). Exposure to indefinite civil commitment represents a significant threat to liberty, and thus the interest in living freely cannot ground the nonresponsible person's right to be excused.

If a defendant will lose his liberty through punishment or civil commitment, then presumably the interest underlying the right to be excused must be associated with an important difference between punishment and civil confinement. In what way does punishment, but not civil confinement, significantly impair one's well-being, given that both entail the loss of liberty?

The prima facie case for excusing psychopaths relied on the fact that punishment necessarily expresses public condemnation; but now that fact is essential to rebutting the prima facie case. Punishment, but not civil confinement, carries the community's expression of moral blame. Henry M. Hart stated a half-century ago that criminal penalties are distinct from civil remedies due to the community condemnation that accompanies the former (Hart, 1958: 405). As Joel Feinberg argues, punishment has a "symbolic significance" found in its expression of blame-related attitudes held by the punishing authority or those it represents (Feinberg, 1970). If punishment's association with blame distinguishes it from civil commitment, we may conclude that the interest underlying the right to be excused from criminal sanction is the interest in not being morally blamed. If the right to excuse protects each individual's interest in avoiding moral blame, then our next question is the following: Does being morally blamed by individuals or by the community negatively impact the well-being of a person with psychopathy? Do psychopaths have an interest in not being blamed—justly or unjustly—independent from their interest in not being confined? If they do not, then they do not have the interest protected by the right to be criminally excused for failing to have adequate capacities.

To assess blame's impact on a psychopath's well-being, we may start with whether being blamed causes a negative subjective experience for psychopaths. One reason blame is uncomfortable for its target is because it can induce unpleasant guilt. If you express blame to me for wronging you, I may reevaluate my actions, conclude that I have wronged you, and subsequently feel bad. This guilt-inducing aspect of expressions of blame, though, are inconsequential to the psychopath who is unfamiliar with guilt. He notoriously does not take responsibility for his actions, a prerequisite for feeling guilt. Being blamed does not induce the psychopath to accept moral criticism, feel guilt, and decide to treat others more respectfully. Insofar as feeling bad is relevant to one's well-being, being blamed does not negatively impact the psychopath's well-being.

Moreover, from an objective standpoint, being blamed does not negatively impact psychopaths' relations with others.[18] If a friend blames you for betraying her trust, her resentment does not capture the full significance of her blame. Your relation of friendship is impaired (Scanlon, 2008). Perhaps she will no longer confide in you, value your time together, or consider you a friend. As T. M. Scanlon persuasively argues, even in the case of strangers, blame indicates the impairment of relations. If, for example, someone does not care whether his actions are justifiable to others—if he shows no concern for others' well-being—our blame is not simply our feeling of resentment. Blame implies the modification of the normal attitudes we possess with regard to others (Scanlon, 2008). Although we may be ready to interact with others in certain, friendly ways, we will not be so ready with this agent. Entailing the impairment of relations, blame signifies "rifts [in] the social fabric[]" (Wallace, 2002, reprinted in Wallace, 2006: 444). I presume that being blamed does negatively impact an individual's well-being, whether she cares about being blamed or not, because it is good to live in unity with others, engaged in relationships of mutual respect and concern.

However, being the target of community blame, expressed through criminal conviction, does not negatively impact the psychopath's interests (thus obviating the need for a right to be excused from the blame expressed through punishment). As researchers describe the psychopath, he does not participate in relationships of mutual concern; he does not know and cannot comprehend the good of living in unity with others or in participating in meaningful interpersonal relationships. A rift already exists in the social fabric between normal functioning adults and the psychopath in virtue of the psychopath's characteristics. In other words, being blamed by others does not impair the psychopath's relationships with others; he already is deprived of meaningful or even decent relationships with others because of his psychopathic traits. Of course, he does not want the community to blame him through criminal conviction, but that is because prison is an obstacle to satisfying his desires, no different than civil commitment. The point is that being blamed does not diminish the goodness of his relationships with others because he lacks good relationships (regardless of whether he knows what a good relationship is). Psychopaths view others as mere instruments to their ends.

One might wonder whether infants represent an obstacle to my argument. The objection would be that infants do not feel badly about being blamed and yet it would be wrong to blame an infant for, say, causing a mother pain while nursing because the infant is not responsible. In other words, the objection contends, no one thinks the infant should be held responsible—the infant has a moral right to excuse—even though the infant, like the psychopath, lacks negative feelings about being blamed.

This objection, though, misunderstands the argument. Subjective experience is only *one* aspect of an agent's objective well-being. The preceding argument is that being blamed by the community does not negatively impact the psychopath's well-being because being blamed *neither* causes the psychopath a negative subjective experience *nor* does it objectively impair his relations with others. The psychopath does not participate in good relationships based on mutual respect and concern. Blame does not impair his relations; they already are terribly deficient.

Let us return now to the parent–infant relationship. Normative standards governing this relationship require the parent to perform certain acts and have particular feelings and attitudes for the infant. To say that the parent truly blames the infant would be for the parent to see their relationship as impaired, making it appropriate for the parent to alter her intentions and attitudes toward the infant (Scanlon, 2008). Beyond the fact that it makes no sense to think the infant violated any norm, the infant certainly has an interest in his or her parent maintaining loving and caring attitudes. The infant has a very significant interest in not being morally blamed.

A second objection might be that to deny psychopaths the right to excuse is morally problematic because it denies humanity to psychopaths. As Samuel Pillsbury points out in this volume, the "history of declaring groups inhuman is an ugly one," and, the objection contends, my argument dehumanizes the psychopath.

To the contrary, concerns about dehumanization *support* my overall conclusion, which is that the law should treat psychopaths as if they are morally responsible even if the prima facie argument for nonresponsibility is strong. To treat psychopaths as nonresponsible—to conclude that they lack the right to be *held responsible*, as opposed to the right to excuse—is much more worrisome in terms of dehumanization. The concern that I share here with Pillsbury is not original: Murphy concludes that psychopaths are nonpersons but argues that it is too dangerous for the law to declare them as such (Murphy, 1972). The philosophical argument of the prima facie case certainly seems plausible if not strong; but even I worry that it does not provide the certainty that could justify treating psychopaths as nonresponsible and civilly confining them indefinitely.

To summarize this subsection, being blamed does not decrease the psychopath's well-being, whether by producing unpleasant feelings for him or by impairing the objective quality of his relationships with others. His interest in avoiding criminal punishment is comparable to his interest in avoiding civil confinement. In fact, psychopaths may have greater interest in being treated as criminally responsible. Excusing psychopaths exposes them to indefinite civil commitment, which in many cases could be longer than a defined prison term. If blame does not impair psychopaths' well-being *and* they would be better off treated as responsible, then they lack the interests that ground a right to excuse based on incapacity. Because they lack the right to excuse, we may decide whether the law should hold psychopaths criminally responsible by appeal to pragmatic, consequentialist, and other moral considerations.

Pragmatic, Consequentialist, and Other Moral Considerations Against Excusing Psychopaths

In the first subsection that follows, I describe some pragmatic considerations regarding the practical implementation of any "psychopath excuse." These considerations pose serious challenges for implementing such an excuse. Perhaps it will be possible to meet them as we learn more about psychopathy through research. However, in the second subsection, I discuss other consequentialist and moral reasons that strongly speak against enacting a psychopathy excuse and for maintaining the status quo policy of holding psychopaths criminally responsible.

Challenges to Implementing an Excuse

The first pragmatic concern regards how the system would identify defendants who should be excused from the set of those diagnosed with psychopathy. A psychopath defense raises questions about the standard factfinders would use to assess a defendant's claim, how factfinders would make that assessment, and about the proper role for mental health experts. Under current insanity law, an expert's psychiatric diagnosis should not be important: the standard is not medical. Rather, as Stephen Morse urges, experts should provide "full, rich, clinical descriptions of [a defendant's] thoughts, feelings, and actions" at the time of the crime and leave to the factfinder whether the defendant was too "crazy" to be held responsible (Morse, 1985). Experts should provide neither diagnoses nor their own legal conclusions because they are irrelevant and mislead jurors about the nature of their decision. Verdicts should not rest heavily on a battle of expert diagnoses and legal conclusions, but on jurors' assessment of whether the defendant's capacity to respond to good reasons was too impaired (e.g., because of delusions of physical reality or an inadequate ability to reason logically) to be held responsible. Of course, M'Naghten and the ALI standard are not mathematical algorithms: people disagree on who is and is not insane under them, and the system produces errors. Nevertheless, the questions jurors must answer in applying insanity standards do require, as Morse points out, "common-sense judgments" (Morse, 1985: 822).

One significant concern about implementing a psychopathy defense is that, in practice, each claim would be settled by a battle of opposing expert diagnoses. The reason is that as of now, the best evidence that an antisocial agent lacks adequate and moral capacities seems to be the best diagnostic instrument. A psychopathy defense would be based on the idea that some antisocial agents lack, whereas others possess, an adequate capacity to appreciate and respond to moral considerations How would jurors decide the camp into which a defendant falls? In other words, how would jurors decide whether a defendant lacks capacity to appreciate and respond to moral reasons or, instead, possesses that capacity but failed to respond to moral reasons at the time of the crime? Recall that the prima facie case for nonresponsibility appeals to data regarding psychopaths *in general*. Factfinders, however, would decide each case individually. Could jurors determine whether a particular antisocial agent lacked capacity for empathy or rather just failed to exercise his capacity? For normal insanity cases, juries focus on the defendant's thoughts, feelings, and behavior at the time of the crime. To answer the "psychopathy defense" question, jurors need to know much more about the defendant. They would heavily rely on experts' application of the best available diagnostic instrument. In other words, if a defendant were to raise a psychopathy defense, the best evidence that he is nonresponsible would be a diagnosis that he is a psychopath according to the accepted instrument. But that fact sets up a battle of opposing expert diagnoses, which is especially problematic because experts' application of the PCL-R is often biased by potential legal implications (Murrie, Boccaccini, Johnson, & Janke, 2008). Moreover, with so much riding on an expert's diagnosis, we would have to be concerned about serious differences in experts' abilities and training. Some research shows that experts have testified despite inadequate training, inappropriate use of the PCL-R, and reaching diagnoses with insufficient evidence (e.g., Zinger & Forth, 1998).

The prospect of relying on the PCL-R or other diagnostic tool raises further questions and practical obstacles. Psychopathy seems to come in degrees. Mental health experts generally conclude that a subject scoring 30 or more on the PCL-R has psychopathy. The PCL-R, however, was not designed for moral or legal purposes.[19] A score of thirty does not necessarily imply that the subject's moral capacities are insufficient for criminal responsibility. The prima facie case for excusing psychopaths, based on moral incapacities, may apply only to primary psychopaths, who are defined predominantly by their Factor-1 traits, including unfamiliarity with guilt, as opposed to secondary psychopaths, who do experience guilt and otherwise possess primarily Factor-2 antisocial traits (Blair et al., 1995-b). That possibility raises many questions: Must one be diagnosed as a primary psychopath, with a score greater than 30, in order to be excused, or must he only have a certain score on Factor-1? What score on Factor-1 represents the degree of moral incapacity insufficient for moral and criminal responsibility? Would all primary psychopaths with an entire PCL-R score greater than 30 qualify for excuse? If not, does an increase in PCL-R score indicate greater moral incompetence, and if so, at what point is an agent's moral incompetence too great for criminal responsibility?

Perhaps future psychopathy research will help us answer these questions and provide some support for implementing a psychopathy defense that avoids a mere battle of opposing diagnoses. First, recall that the prima facie case for nonresponsibility appealed to the poor performance by psychopaths, generally, on the moral/conventional task. Some psychopaths perform this task better than others and may have adequate capacity to grasp moral considerations. Maybe, as Neil Levy suggests, the moral/conventional task will play a more significant role in distinguishing individuals who are currently grouped together (Levy, 2007-a). Furthermore, it might be that particular items on the PCL-R (such as lack of empathy, lack of guilt, and failure to accept responsibility) are linked more closely than other items (such as criminal versatility and promiscuity) with an inadequate understanding or appreciation of moral principles. Possible future findings might help us develop sensible standards for distinguishing psychopaths whose moral capacities are adequate for moral blame from those whose capacities are not.

Moreover, in light of such research, a psychopathy defense might not necessarily rest on a battle of opposing expert diagnoses. Experts could convey a defendant's performance on the moral/conventional task or other tests relevant to assessing moral understanding, or discuss the defendant's capacity for feeling guilt or remorse. In light of such testimony, factfinders could ask whether the defendant lacks substantial capacity to appreciate the moral wrongfulness of his conduct. We would still have to overcome concerns about testimony from experts inadequately trained or biased by potential legal implications, but we could possibly avoid a battle of expert diagnoses.

Nevertheless, it is still difficult to imagine jurors being able to decide confidently whether a defendant lacked substantial capacity to empathize or just failed to empathize on the particular occasion. It is even more difficult to imagine jurors agreeing that a defendant should not be punished based on his lack of empathy, remorse, or capacity to distinguish morals from conventions. Undoubtedly many citizens will view an incapacity for empathy and guilt as an aggravating fact, not grounds for excuse. It is especially difficult to imagine jurors agreeing to excuse a psychopath in jurisdictions that do not inform jurors of the consequences of excuse (i.e., mandatory or possible civil commitment): insanity jury instructions in most jurisdictions do not inform jurors of the consequences (LaFave, 2010). Of course, if a jurisdiction changed its law to provide for a psychopathy defense, it could require courts to inform jurors of the consequences of excuse. Moreover, perhaps a defendant could waive his right to a jury and hope for a judge open to the defense.

Reasons to Hold Psychopaths Criminally Responsible

Even if the practical challenges discussed in the preceding text could be met in the future, other reasons speak very strongly against implementing a psychopathy excuse. Significant moral and other costs would accompany a psychopath defense. First, antagonism toward the insanity defense, even for defendants who are uncontroversially insane, already exists in some jurisdictions, not to mention the handful of state legislatures that have abolished it.[20] If legislators passed a "psychopath defense," or if judges read current standards to imply that psychopaths are blameless, public antagonism toward excusing people may renew or increase. Because psychopaths' violence is instrumental and they know what the law prohibits, it is next to impossible to imagine public support for the idea that psychopaths are blameless; a backlash against laws excusing them could threaten nonresponsibility standards more generally. The absence of an available insanity defense for significantly irrational individuals represents a grave injustice and must not extend to more jurisdictions.

A related moral cost regards the victims of psychopaths. Failing to blame a wrongdoer can be blameworthy itself because such a failure may disrespect the value of the victim. Of course, if psychopaths are not responsible agents, then excusing them does not actually disrespect their victims. Excusing someone under current insanity standards shows no disrespect to victims even if some victims continue to blame the insane offender. However, again, society, including crime victims, will not accept the idea that psychopaths are blameless. The philosophical arguments of the prima facie case are controversial among those familiar with the underlying considerations; such arguments, reduced to and warped through media soundbites, will not change society's general sense that psychopaths are bad, not mad. Most victims will feel disrespected by the law for not blaming their assailants. This consideration would not carry weight if psychopaths, like persons who are insane under current standards, were wronged if they are morally blamed; but they are not. As such, it morally matters that most victims will blame psychopaths and therefore feel disrespected by the law if it deems psychopaths blameless.

The dilution of the criminal law's message represents another moral cost, given the sheer number of antisocial defendants who could contest their responsibility. "The law wants to reinforce societal assumptions that most of us are morally accountable actors but [a psychopathy excuse] would permit most criminal defendants to challenge that expectation of accountability" (Arenella, 1992). As a society, we do not want to associate callousness, refusal to accept responsibility, remorselessness, pathological lying, and impulsivity with nonresponsibility and blamelessness. Current insanity laws do not undermine the law's message. Individuals, consciously or unconsciously, do not think that if their reasoning were just more disordered and irrational—if their grip on reality were just more tenuous—they could avoid holding themselves responsible for their choices. A psychopathy defense, though, could have an effect on people's attitudes toward their own responsibility. The law does not want to convey that the more remorseless, cold, and impulsive you are, the less responsible you are for your actions and attitudes. To be clear, as a normative matter, endorsing the view that psychopaths are not responsible *should not* convey the message that acting coldly, guiltlessly, impulsively, and so forth, is excusing or mitigating. Nonetheless and more importantly, many people will interpret the message as such.

A psychopathy excuse could carry serious financial costs and a burden on the system. A significant percentage of defendants would be able to contest their responsibility, even if unable to raise the defense successfully, given the number of offenders with antisocial personality disorder. Some studies "suggest that between 50 and 80 percent of US inmates reach criteria for ASPD" (Blair et al., 2005: 129). Regarding the more narrow diagnosis of psychopathy, other studies conclude that 15% to 25% of prison inmates meet the PCL-R criteria (Hare, 1996). Morse rightfully points out that defendants charged with minor crimes would not raise the defense, preferring a minimal prison term to indefinite civil commitment (Morse, 2008). But nevertheless, the number of defendants raising the defense—perhaps those facing the prospect of moderately serious to severe punishment—could be very significant and costly to the judicial system. This concern might be amplified if the law, in deeming psychopaths nonresponsible for their conduct, authorized the state to pursue post-incarceration civil confinement for individuals diagnosed as psychopaths prior to prison release. Even defendants charged with moderately serious crimes might try to avoid criminal punishment if they will nonetheless face the prospect of civil commitment following completion of their sentences. Some might feel they will be able to manipulate their way to release from civil confinement.

Civilly confining psychopaths would, itself, be very expensive. It would seem to require special segregated units to keep psychopaths away from individuals with other mental illnesses and disorders who would be especially vulnerable to psychopaths' exploitation and manipulation. Moreover, excused psychopaths would be subjected to extremely long periods of civil commitment. Effective treatment does not exist yet. Worse, in at least one study, among psychopaths with the most severe emotional impairments, recidivism rates were actually higher for those who received mental health treatment (Blair et al., 2005). The rules governing commitment may allow for release at certain ages because psychopaths' antisocial conduct eventually decreases (Blair et al., 2005); but otherwise their confinement would be lengthy and costly. Perhaps rendering psychopaths eligible for indefinite civil commitment would reduce crime by confining psychopaths for periods longer than prison sentences. But this enticement represents too much of a danger to civil liberties. With so many antisocial agents falling into a gray area—as to whether they are incapable of appreciating moral considerations or just ignore them—and with so much riding on possible battles of experts, it is too dangerous to allow the state the power to pursue indefinite civil commitment for so many individuals.[21]

CONCLUSION

The prima facie case for excusing psychopaths is based on their severely impaired capacity to appreciate and respond to moral considerations. In the end, it is their moral incapacity that undermines any moral imperative to excuse them. Whether the criminal law should hold them responsible may be determined by pragmatic, consequentialist, and other moral concerns. Undoubtedly I have missed many such considerations and hope future research addresses whether the pragmatic and civil liberty concerns raised about a psychopathy defense can be addressed and whether societal welfare would be maximized by blaming or excusing individuals with psychopathy.

ACKNOWLEDGMENTS

For very helpful comments and conversations about this chapter, I thank Walter Sinnott-Armstrong, Samuel Pillsbury, Jim Tabery, Michelle Madden Dempsey, Michael Corrado, Scott Lilienfeld, Stephen Morse, and John Oberdiek. I am also grateful to the Villanova Law School faculty for a very helpful discussion.

NOTES

1. American Law Institute, Model Penal Code, § 4.01(2) (1985); see also Ala. Code 1975 § 13A-3-1(b); West's Arkansas Code Annotated § 5-2-312(b); West's Colorado Revised Statutes Annotated § 16-8-101(2); Connecticut General Statutes Annotated § 53a-13(c).
2. In contrast to expectations, studied psychopaths classified both moral and convention rules as authority independent; however, Blair argues that the incarcerated psychopathic subjects "would be motivated to show how they had learned the rules of society." Not being able to make the distinction, they chose to classify all rules as moral (Blair, 1995:, 23).
3. For more on moral judgments by psychopaths, see Chapter 7 by Schaich Borg and Sinnott-Armstrong.
4. As another example, Susan Wolf argues that judgments of blame have a distinct significance absent from other evaluative judgments in that they target "the moral quality of the individual herself...[in a] seemingly more serious way" (Wolf, 1990: 40–41). Accordingly, she argues, an agent does not deserve blame if he is incapable of acting in light of true moral values.
5. This passage appears here to show the disconnect between this psychopath's voiced plans and his actual actions. However, with respect to the previous argument regarding the possession of principles, one might argue that this subject's statement shows that he does, indeed, hold himself to some evaluative principles if he thinks the book and earning money will be good. Again, though, deeming some things as "good" does not demonstrate possession of higher-order desires or holding oneself to principles. An agent with mere-first-order desires and a weakened capacity for higher-order desires can think money is a good thing because it permits him to satisfy first-order desires. He may also simply have a first-order desire for power and money, which he thinks his book will bring.
6. Fine and Kennett (2004: 34) also argue that if rehabilitation is a justifying aim of punishment, specific moral capacities are required: "The goal of reform involves seeing the offender as redeemable, as having certain (though perhaps underdeveloped) moral capacities, as being within the reach of moral address" A criminal will be able to grasp punishment's moral message only if he has an adequate level of moral understanding.
7. The basis for the next three paragraphs can be found in Scanlon (2000). See also Litton (2005).
8. See, for example, Hawaii's Revised Statutes § 707–701.5, defining murder in the second degree as intentionally or knowingly causing the death of another person.
9. See, for example, Ala. Code 1975 §13A-6-21(a)(1), defining one way that someone can commit assault in the second degree: "A person commits

the crime of assault in the second degree if the person does any of the following: (1) With intent to cause physical injury to another person, he or she causes physical injury to any person."
10. See, for example, North Dakota Century Code 12.1-20-07(1), defining one manner of sexual assault: "A person who knowingly has sexual contact with another person, or who causes another person to have sexual contact with that person, is guilty of an offense if: a. That person knows or has reasonable cause to believe that the contact is offensive to the other person."
11. See, for example, New Mexico Statutes Annotated 1978, § 30-17-5(A)(1): "Arson consists of a person maliciously or willfully starting a fire or causing an explosion with the purpose of destroying or damaging: (1) a building, occupied structure or property of another person...."
12. See, for example, Vernon's Annotated Missouri Statutes 565.024 (1) (defining involuntary manslaughter in the first degree).
13. See, for example, Vernon's Annotated Missouri Statutes 565.024(3) (defining involuntary manslaughter in the second degree).
14. West's Delaware Code Annotated § 463; West's Colorado Revised Statutes Annotated § 18–1–702(1); Vernon's Annotated Missouri Statutes 563.026(1).
15. I am grateful to Michael Corrado for raising this objection and for a very helpful discussion.
16. Presumably, the psychopath's lack of potential to respect others is also relevant here, given that young children also cannot treat others respectfully but nonetheless possess rights.
17. Scanlon (2008), for example, asserts that even agents with no moral regard for others retain basic moral rights, such as the right not to be killed.
18. My remarks on the significance of blame follow T. M. Scanlon's (2008) compelling account of blame. .
19. Samuel Pillsbury emphasizes the same point in his essay at page 300.
20. See, for example, Idaho Code § 18–207 (2004); Kan. Stat. Ann. § 22–3220 (1995) (stating that evidence of mental disease or defect is not a defense to a crime unless it shows the defendant lacked the requisite mental state); Utah Code Ann. § 76-2-305 (2003).
21. Jeffrie Murphy makes a similar point (1972: 296).

REFERENCES

Arenella, P. (1992). Convicting the morally blameless: Reassessing the relationship between legal and moral accountability. *UCLA Law Review*, 39, 1511, 1599.

Blair, R. J. R. (1995). A cognitive developmental approach to morality: Investigating the psychopath. *Cognition*, 57, 1–29.

Blair, R. J. R., Jones, L., Clark, F., & Smith, M. (1995-a). Is the psychopath "morally insane?" *Personality and Individual Differences*, 19, 741–752.

Blair, R. J. R., Mitchell, D., & Blair, K. (2005). *The psychopath: Emotion and the brain*. Malden, MA: Blackwell.

Blair, R. J. R., Monson, J., & Frederickson, N. (2001). Moral reasoning and conduct problems in children with emotional and behavioural difficulties. *Personality and Individual Differences*, 31, 799–811.

Blair, R. J. R., Sellars, C., Strickland, R., Clark, F., Williams, A. O., & Smith, M. (1995-b). Emotion attributions in the psychopath. *Personality and Individual Differences*, 19, 431–437.

Cleckley, H. (1988). *The mask of sanity: An attempt to clarify some issues about the so-called psychopathic personality* (5th ed.). St. Louis, MO: Mosby.

Dressler, J. (2009). *Cases and materials on criminal law* (5th ed.). St. Paul, MN: West.

Duff, A. (1977). Psychopathy and moral understanding. *American Philosophical Quarterly*, 14, 189–200, 191.

Duff, A., & Garland, D., Eds. (1994). *A reader on punishment*. Oxford: Oxford University Press.

Feinberg, J. (1970). The expressive function of punishment. In: *Doing and deserving: Essays in the theory of responsibility* (pp. 95–118). Princeton, NJ: Princeton University Press.

Fine, C., & Kennett, J. (2004). Mental impairment, moral understanding and criminal responsibility: Psychopathy and the purposes of punishment. *International Journal of Law and Psychiatry*, 27, 425–443.

Hare, R. D. (1993). *Without conscience: The disturbing world of the psychopaths among us*. New York: Guilford.

Hart, H. M., Jr. (1958). The aims of the criminal law. *Law and Contemporary Problems*, 23, 401–441.

Kennett, J. (2006). Do psychopaths threaten moral rationalism? *Philosophical Explorations*, 9, 69, 77.

LaFave, W. R. (2010). *Criminal law* (5th ed.). St. Paul, MN: West.

Levy, N. (2007a). Norms, conventions, and psychopaths. *Philosophy, Psychiatry, & Psychology, 14*, 163–170.

Levy, N. (2007b). The responsibility of the psychopath revisited. *Philosophy, Psychiatry, & Psychology, 14*, 128–138.

Litton, P. (2005). The 'abuse excuse' in capital sentencing trials: Is it relevant to responsibility, punishment, or neither? *American Criminal Law Review, 42*, 1027–1072.

Maibom, H. (2005). Moral unreason: The case of psychopathy. *Mind & Language, 20*, 237–257.

Maibom, H. (2008). The mad, the bad, the psychopath. *Neuroethics, 1*, 167–184.

M'Naghten's Case, 8 Eng. Rep. 718, 722 (1843). Quoted in *State v. Johnson*, 399 A.2d 469, 472 (R.I. 1979).

Murrie, D. C., Boccaccini, M. T., Johnson, J. T., & Janke, C. (2008). Does interrater (dis)agreement on psychopathy checklist scores in sexually violent predator trials suggest partisan allegiance in forensic evaluations? *Law and Human Behavior, 32*, 352–362.

Morse, S. J. (1985). Excusing the crazy: The insanity defense reconsidered. *SouthernCalifornia Law Review, 58*, 777–836.

Morse, S. J. (1998). Excusing and the new excuse defense: A legal and conceptual review. *Crime and Justice, 23*, 329–406.

Morse, S. J. (2004). Reasons, results, and criminal responsibility. *Illinois Law Review, 2004*, 363– 444, 367.

Morse, S. J. (2008). Psychopathy and criminal responsibility. *Neuroethics, 1*, 205–212.

Murphy, J. G. (1972). Moral death: A Kantian essay on psychopathy. *Ethics, 82*, 284–298.

Oberdiek, J. (2009). Towards a right against risking. *Law & Philosophy, 28*, 367–392.

Prinz, J. (2006, March). The emotional basis of moral judgments. *Philosophical Explorations, 9*, 29–43.

Raz, J. (1986). *The morality of freedom*. Oxford: Oxford University Press.

Scanlon, T. M. (2000). *What we owe to each other*. Cambridge, MA: Harvard University Press.

Scanlon, T. M. (2008). *Moral dimensions: Permissibility, meaning, blame*. Cambridge, MA: Harvard University Press.

Schopp, R. F., & Slain, A. J. (2000). Psychopathy, criminal responsibility, and civil commitment as a sexual predator. *Behavioral Sciences and the Law, 18*, 247–274.

Taylor, C. (1982). Responsibility for self. In: G. Watson, *Free will*. Oxford: Oxford University Press.

Wallace, R. J. (1994). *Responsibility and the moral sentiments*. Cambridge, MA: Harvard University Press.

Wallace, R. J. (2002). Scanlon's contractualism. *Ethics, 112*, 429–470.

Wallace, R. J. (2006). *Normativity and the will*. Oxford: Oxford University Press.

Wolf, S. (1990). *Freedom within reason*. New York: Oxford University Press.

Zinger, I., & Forth, A. E. (1998). Psychopathy and Canadian criminal proceedings: The potential for human rights abuses. *Canadian Journal of Criminology, 40*, 237–276.

CHAPTER 16

Why Psychopaths Are Responsible

Samuel H. Pillsbury

The gap between the academic and the "real" world is a staple of modern policy discourse. Critics claim that because professors live in an ivory tower, disconnected from reality, their analysis of problems is misguided and their solutions portend disaster.

Oh, but it's not true! academics protest. We might not mind taking up residence in an ivory tower—sounds quite pleasant, actually—but in fact we fight the same traffic, pay the same taxes, and run the same risks of crime and terrorism and environmental catastrophe as do others in society. If we see the world differently than the lay public, it is not because we do not know the world—quite the reverse. By virtue of careful study and rigorous analysis, we actually know it better. We see what others miss.

As an academic myself, I have a stake in this debate—that "we" just mentioned includes me—and yet there are times when I find myself taking the lay position against the academic, when I too wonder what world my colleagues inhabit. An example is the criminal responsibility of psychopaths. The "real" world's view—the public's view—is clearly expressed in current criminal law, which holds that psychopathy provides neither excuse from liability nor mitigation of punishment. If considered at all in a criminal case, psychopathy operates to worsen the defendant's situation, serving as an aggravating factor in determining sentence (Edens & Petrila, 2007; Ells, 2005). Similarly, in popular culture, psychopaths personify evil. They practically constitute the class of Hollywood bad guys, explaining, in part, the prevalence of serial killers on television and on movie screens. In sum, the public views psychopaths as the worst of the worst, the scariest of all, the most deserving of all subjects of punishment.

By contrast, consider the academic perspective on the psychopath. Although far from uniform, a number of academics in both philosophy and law have recently argued that psychopaths may not be moral agents, suggesting they should be excused from criminal responsibility. Indeed, most philosophers and legal academics who have written on the subject in recent years have expressed serious doubts about the moral responsibility of the psychopath (Arenella, 1992; Duff, 1986; Fischer & Ravizza, 1998; Fischette, 2004; Horder, 1993; Maya Mei-Tal, 2004; Morse, 2008; Shoemaker, 2007; Wallace, 1994; cf. Maibom, 2008). Persons that behavioral science labels psychopaths appear to have pervasive and profound emotional deficits; they appear incapable of feeling empathy for others. As a result, many argue that they either cannot understand moral reasons or cannot appreciate their special significance in human affairs and therefore cannot be moved by such reasons. For one or both reasons, many argue that psychopaths lack moral capacity and do not deserve blame for their otherwise immoral acts. They may be dangerous and society may protect itself against their predations just as it does threats from animals or natural forces, but without the capacity for fellow feeling, psychopaths lack a prerequisite for moral and criminal responsibility. How can they deserve punishment if they cannot comprehend the wrong that they do?

The gulf between worldviews could not be more dramatic: one side sees evil; the other, pathology. One side seeks moral condemnation and the other an excuse from blame.

Although an academic, I support the public's—and the criminal law's—view that psychopathy should not excuse a person from full criminal responsibility. I agree with the lay view that the criminal acts of psychopaths help define evil-doing. I cannot in a single book chapter provide a complete argument to support this position, because to do so would require nothing less than a complete theory of criminal responsibility. I can explain, though, why the academic and public perspectives on the psychopath diverge so widely today and present basic arguments for why the law's current view of criminal responsibility should be preferred.

The argument in favor of psychopath responsibility is simply stated. At its core, criminal law condemns harmful acts committed for reasons that demonstrate basic disregard for the value of other human beings. Psychopaths frequently act in harmful ways that demonstrate just such disregard. When their acts of disregard meet the definitions of criminal offenses, they should be held to account like any other person. Criminal liability here depends on what I call conduct relationality: how persons interact in society. Criminal liability depends on the social meaning of human interactions, not on the moral or emotional resources of the individual harm doer. The psychopath's apparent inability to feel for others or appreciate moral principle does not affect the social meaning of his conduct. In social terms—meaning conduct as perceived and experienced by victim and society at large—a rape is equally vicious whether committed by a psychopath or a nonpsychopath. In both cases the perpetrator acts in a way that demonstrates profound disregard for the victim's value, for his or her sexual integrity. Criminal liability currently, and I believe properly, depends on the social/moral meaning of the interaction between perpetrator and victim, a meaning that does not depend on the perpetrator's ability to appreciate wrong.

Moral capacity proponents argue that conduct demonstrating disregard is not sufficient for moral responsibility because it does not take account of the moral and emotional capacities of the perpetrator. Their contention is that criminal liability must rest in part on an assessment of the individual's moral capacities. One such capacity may be the ability to recognize the special significance of moral rules, which psychopaths lack in significant measure. Others contend that moral responsibility requires the ability to be moved by moral reasons, a quality of the will, and that this is what psychopaths critically lack.[1] Proponents of both approaches agree that because the psychopath appears to lack the capacity for empathy and remorse he or she is not a full moral agent and cannot be fully blamed. As a result, psychopaths perhaps should be entirely excused from criminal liability. Or perhaps their punishment should be mitigated (e.g., Glannon, 2008). Or perhaps we should distinguish between moral and criminal responsibility, and for a variety of essentially practical reasons, continue to hold them criminally liable even in the absence of moral responsibility (Haji, 1998).[2]

The issue of the psychopath's criminal responsibility turns, I believe, on whether criminal liability requires only judgments about conduct demonstrating disregard, or whether a judgment of the person in the form of his or her moral capacities is also required. To explore this question, we begin with the behavioral science of psychopathy. A brief review presents its basic findings, and familiar tensions between science and law. Next we take up a particular form of moral capacity theory, known as reason-responsiveness, which promises to bridge the gap between behavioral science and moral responsibility. I argue that such a reason-responsiveness excuse cannot be reconciled with the practical and political demands of the criminal law, which justify limiting guilt to judgment of bad conduct. This argument is then developed in a particular case, the perpetration of the Columbine massacre by Eric Harris, an individual later identified as psychopathic.

BEHAVIORAL SCIENCE AND CRIMINAL LAW: THE PROBLEM OF PSYCHOPATHY

The difficulties in moving from behavioral science concepts to criminal law norms are well known. Especially with respect to mental illness and insanity, many have written about the challenge of translating scientific concepts into legal terms and vice versa. As a result, the discussion of this problem here will be brief, limited to the basic differences between the fields that are important to the responsibility of psychopaths.

To put the matter simply, behavioral science seeks to identify and explain particular behavioral dispositions[3] according to their physical causes in the environment and genetics. Criminal law by contrast seeks to judge an individual's responsibility for choosing to act in a prohibited fashion, according to the person's reasons for acting. Two sets of distinctions between the fields are particularly important for our purposes: (1) the difference between focusing on behavioral dispositions and particular acts and (2) the difference between analysis of physical causes and moral judgment.

Psychopathy is a behavioral science concept. In its study, scientists have documented and sought to explain a distinctive pattern of human behavior found in a relatively small number of identifiable individuals. These individuals have distinctive patterns of thought, feeling, and action that many scientists say can be reliably identified through interviews and analysis of life histories. Psychopathy is generally identified by distinctive personality traits, exploitative interpersonal relations, and antisocial conduct. Psychopaths are characterized by superficial charm, egocentricity, and lack of guilt or shame at moral transgressions or any form of hurting others. They are often highly skilled at deception. In terms of life history, the psychopath typically has committed numerous antisocial acts, often beginning in early childhood. Usually this history will include significant proven criminality, but not always (Hare, 1996; Patrick, 2007). Often the psychopath's personal behavior is highly impulsive, with little attention given to its likely consequences.

Researchers of psychopathy seek to explain and document the condition: who suffers from it and to what extent. They look at how the condition arises, whether it changes over time, and how it functions in physiological, cognitive, and neuroscientific senses. A significant concern is its etiology: its causes in environmental influences and genetics. Recent evidence suggests a strong role played by heredity here (see Blair, Mitchell, & Blair, 2005; Harris, Skilling, & Rice, 2001). Finally, behavioral scientists research the possibilities of treatment (Hare, 1996).[4] In all of these endeavors, scientific researchers focus on psychopathy's physical causes: explaining its origins, dynamics, and providing clues to its possible treatment.[5]

With this as introduction, we turn to the first of our important distinctions between behavioral science and law: the difference between a focus on behavioral dispositions in psychopathy and individual acts in criminal law.

A person is labeled a psychopath according to a checklist of factors developed from extensive study of other persons; the currently accepted standard is known as the Psychopathy Checklist–Revised (PCL-R) (Hare, 1996). A PCL-R score of 30 is often seen as the distinguishing mark of psychopathy, but researchers say that for some purposes the mark may be lowered to 25 (Hare, 1996). There is considerable debate in the field about whether psychopathy is better understood as a dimensional concept, meaning that it identifies certain traits that are shared, to a greater or lesser extent by a significant number of persons, or whether it represents a categorical type, identifying a discrete set of individuals, essentially different from all others.[6]

The critical point for our purposes is that a psychopathy determination establishes the individual's disposition to certain kinds of thoughts, feelings, and actions. Psychopaths are distinguished by their patterns of exploitative interactions with others and a restricted emotional life. These combine to indicate a person who lacks positive feeling for other people and has little personal regard for moral rules. Thus to be classified as a psychopath is to be classified as a person who lacks significant empathy for others, and who is disposed to treat others—to use them—without regard for their welfare.[7]

The study of psychopathy is similar to other efforts in behavioral science to identify types of persons according to social misconduct. Social and behavioral scientists have sometimes sought to identify behavioral types specific to certain kinds of criminal offenses, such as sex offenders, perpetrators of domestic violence, or victims of domestic violence. Although the scientific reliability of resulting labels may vary considerably—and here psychopathy may well be in a class by itself—all these efforts involve the development of a human typology. The scientist seeks to explain why persons of this type think, feel, and act as they do. It explains their characteristic dispositions to involvement in crime as perpetrators or victims. When the typology focuses on social dysfunction, as with groups defined partly by criminal behavior, the typology often includes an assumption that the typed individual is in some respect (socially, genetically, or otherwise) abnormal or defective.[8]

This kind of scientific typing fits the kind of judgment of the person that we commonly make in our everyday lives, and that the law must engage in on occasion: it categorizes the person according to disposition to certain forms of socially dangerous conduct. Each of us must decide upon meeting another whether that person is someone with whom we want to interact, indeed whether the person is someone with whom it is safe to interact. What kind of person is this? What is his or her character? Decisions about hiring and firing, inviting to social occasions, and electing to office all involve this kind of nature-of-the-person judgment. Determinations about criminal liability, about guilt and innocence, depend on somewhat different questions, however.

Unlike the science of psychopathy, criminal liability does not involve an assessment of an individual's general disposition to certain thoughts, feelings, or even actions; instead, guilt depends on the assessment of particular past acts. In the criminal law, judges and juries must decide whether the defendant is guilty of a crime by committing certain conduct. Although prior patterns of behavior are sometimes relevant to that determination, the focus is on whether the individual, *in this particular instance*, violated the criminal law. Generally speaking, the criminal law punishes according to the character of the criminal act, and not the character of the defendant. Most commentators, including most moral capacity proponents who have addressed the issue, agree that criminal law should not judge the person, but the crime (Moore, 1990, 1997; Morse, 2004; Murphy, 2004).[9] Especially when the question is guilt or innocence, American criminal law focuses fundamentally on what a person has done rather than who the person is.[10]

To sum up the discussion so far: psychopathy uses scientific means to identify a particular type of person. It provides valuable information about a distinctive set of behavioral dispositions that may be useful to many decisions, including sentencing in criminal cases. It tells us who the psychopath is and what we may expect from him or her generally and in some cases it may provide insight into culpability for particular acts, by providing information on why the accused acted as he or she did. The categorization says nothing directly about guilt, however. As a behavioral typology, psychopathy cannot directly inform criminal liability judgments, unless we decide that liability should depend on an individual's general disposition to particular thoughts, feelings, or action.

This brings us to our second distinction between science and law, involving the distinction between causal explanation and moral judgment. Briefly stated, behavioral science seeks causal explanations for why humans behave the way they do; in moral and criminal responsibility we pass judgment on chosen conduct. Behavioral science concerns physical cause; criminal responsibility focuses on choice.

Because the distinctions between explanation and judgment, cause and choice, are familiar to modern legal discourse, they do not require much elaboration here. The obvious point is that law requires a moral judgment about choice, which behavioral science cannot render because of its focus on physical causation. The two fields speak different languages: choice and cause. Nor can these be reconciled as merely different stages of the same analytic process: objective explanation followed by normative assessment. Scientific explanation and moral judgment involve different subjects and methodologies. In fact, behavioral science and criminal law are distinctive fields, each with its distinctive subjects, aims, and methods, each with its own distinct areas of interest and disinterest. Neglect these differences and we will soon become confused.

FROM MORAL PHILOSOPHY TO CRIMINAL LAW: REASON-RESPONSIVENESS AS AN EXCUSE FROM LIABILITY

To connect behavioral science learning about psychopathy with criminal law doctrine we must build a linguistic and conceptual bridge. A natural candidate for this work is moral philosophy: it respects scientific explanations, yet it aims to make moral judgments about human action based on choice, as does the criminal law. Indeed most of the writing on the moral responsibility of psychopaths in recent years has been done either by moral philosophers or law professors who work with philosophic concepts and methods.[11] Contemporary moral philosophers and those working within the modern philosophic tradition seek to identify the essential traits of moral agents, a quest that naturally leads to a concern with essential moral capacities. An influential form of what may be called moral capacity theory is reason-responsiveness. Because of its acceptance by some of the most respected scholars in moral and legal responsibility, it is my focus here (e.g., Duff, 2007; Morse, 2008; Wallace, 1994).

The theory of reason-responsiveness holds that to be morally responsible, an individual must be able to respond to moral reasons. It sees individual rationality as the core of moral responsibility, and defines that rationality in moral terms. In short, an individual who cannot respond to moral reasons is not morally rational (Glannon, 2008; Litton, 2008; Morse, 2008; see also Fischette, 2004).[12] This makes the psychopath a prime candidate for a responsibility excuse. Because psychopaths apparently cannot appreciate the pain caused by hurting others, it appears they cannot internalize moral standards. They do not see the difference between moral and conventional rules—between, for example, talking out of turn in class and stealing another's purse—suggesting that they cannot subject their own conduct to moral critique. Given these incapacities, the psychopath lacks vital aspects of practical reason with respect to moral rules. Not being personally able to grasp the most basic of moral propositions, they cannot engage in a meaningful moral dialogue. Therefore they cannot deserve punishment.[13]

Before taking up the particulars of reason-responsiveness and its potential application to criminal liability, a few notes about its approach. Reason-responsiveness proponents, like others using a moral capacity approach to moral responsibility, frame the issue of the psychopath's responsibility in terms of moral agency. Moral agency is seen as comprised of the capacities that a person needs to engage in a meaningful discussion about wrongdoing and personal accountability, to engage in the dialogue of morality (e.g., Litton, 2008; Shoemaker, 2007).[14] The decision to focus on moral agency centers attention on the chooser and his or her capacities. The inquiry includes the individual's conduct, but the critical issue is the individual's capacity to make moral choices. To answer this question we need to know not just what the person has done, but who the person *is*, in terms of moral capacity. This lines up well with the behavioral science of psychopathy, which as we have seen, focuses on the characteristic dispositions of the individual psychopath: who the person is according to a scientific typology.

Determining the fundamentals of moral agency has great appeal for deserved punishment, because this project promises to make criminal judgment fair to the person judged. It promises a system in which wrongdoers can be blamed for wrongful actions *according to their own* abilities. A person with the capacities of a moral agent—here meaning reason-responsiveness—who chooses to do wrong, deserves blame. The judge determining this person's responsibility should have no moral qualms in rendering harsh judgment on the wrongdoer, should suffer no pangs of conscience for moral condemnation, because punishment is deserved, merited. Given that the individual had the rational capacity to see and appreciate moral wrong, punishment is fair. Notice that here the promise of fairness rests explicitly on the accused's moral capacities. Without determining the defendant's moral capacity, punishment might not be deserved; it might represent blame for choosing to do an act that the defendant could not recognize as morally wrong.

Finally, note that the critical punishment question for most reason-responsiveness—and indeed most moral capacity—theorists, is that of deserved punishment. Deterrence is not explicitly considered by contemporary proponents. In fact, deterrence would not seem to present the same conceptual challenge to punishment of the psychopath as does retribution.[15]

REASON-RESPONSIVENESS AND CRIMINAL LIABILITY

Who does and who does not deserve punishment under reason-responsiveness? Proponents generally agree that a person without a demonstrated capacity for rationality should not be held morally or legally responsible (e.g., Morse, 1998). In their view this explains a central excuse from responsibility in criminal law: the insanity defense. Such persons, displaying fundamental irrationality, simply cannot choose rationally, and so should not be held responsible.[16]

A more difficult question under the reason-responsiveness approach is whether lack of volitional capacity should excuse. What if the individual is unable to refrain from wrongful conduct that she realizes is wrong? How should we handle the individual who knows she should not steal, or sexually touch, or use drugs, but nevertheless cannot refrain from doing so? Putting aside psychopathy for the moment, the current legal-philosophical consensus seems to be against including volitional deficits in moral capacity. Few in academics argue today that the addict or alcoholic lacks the capacity for moral agency because of addiction.[17] The same likely applies to compulsive conditions such as kleptomania or certain paraphilias that are associated with criminality.[18] Such sufferers are usually presumed to be morally capable by virtue of their rational capacities. They can see the wrongness of their actions, or can at least see what makes them wrong, even if they cannot refrain from committing them.

Psychopathy has been viewed differently. As the science of psychopathy has developed, the view of psychopaths as fundamentally deficient moral choosers has become more prevalent. Because psychopaths have such limited capacity for empathy, they cannot appreciate the force of morality. Without the basic feelings of remorse that represent the foundation of all morality they do not and perhaps cannot understand why it is wrong to hurt others, at least not in the way that makes moral wrongs special. As a result, a number of scholars have suggested that psychopaths should not be morally blamed and therefore do not deserve punishment, even for intentionally harmful conduct (e.g., Duff, 1986; Morse, 2008).[19] Some commentators concede that there may be practical problems with implementing this view in law (Litton, 2008), but the basic thrust of most moral capacity writing is that criminal law is on shaky moral ground in holding psychopaths fully responsible.

As applied to criminal liability, the reason-responsiveness approach depends on two critical propositions: (1) that in criminal litigation we can reliably determine the reason-responsiveness of an accused; and (2) that reason-responsiveness is necessary for justified punishment. I will dispute both points. The larger argument that I will develop is that reason-responsiveness sees individual responsibility as requiring a form of person judgment in the assessment of assessing individual moral capacities, while I believe that criminal liability should rest exclusively on conduct judgment: proven instances of intentional misconduct. This argument depends significantly on the nature of criminal law, on its distinctive function and aims.[20]

To return to the bridge building metaphor used before, we started our responsibility inquiry on the grounds of behavioral science, and soon learned that we needed to build a linguistic and conceptual connection to determinations of criminal liability. Moral capacity in the form of reason-responsiveness has been offered as a possible bridge between these realms. To test the connection, we need to look more closely at the basic work and principles of criminal law. We must examine the legal grounds on which the other end of this span must rest.

CRIMINAL LIABILITY AND CONDUCT JUDGMENT: A FUNCTIONAL APPROACH TO THE CRIMINAL LAW

Logically, the foundations of criminal law might be found in moral philosophy. After all, at least in its most important provisions, criminal law represents one form of moral responsibility. Thus we might begin any consideration of criminal responsibility with foundational concerns in moral philosophy. Here I choose a different starting place, however: the function of criminal law. I suggest we look at what the criminal law does and must do to succeed. This is its true bedrock.

In the United States, the aim of the criminal law is to do justice according to the core values of American citizens. This means that criminal law has both a moral and political dimension.[21] The criminal law must resolve particular disputes—accusations of serious individual wrongdoing—according to legal principles that accord with the most important values of the population. Criminal law expresses and defends baseline moral principles, setting out what conduct is fundamentally prohibited in society.[22] Its condemnations must carry moral weight with the public to have full effect. This means that the criminal law must reflect a general public ethos, not just that of certain groups or individuals. Over the long term, criminal law must maintain the public's respect to succeed. To this extent, its work is political as well as moral.[23]

What values are expressed and defended in criminal law depends significantly on our conception of democracy. Criminal law in a modern secular democracy is limited because in such a democracy, government's powers are limited. Thus the line that criminal law draws between public moral precepts that the state can enforce by punishment, and others left to individuals, groups, and other institutions, depends on liberal ideas about the proper relationship between government and citizen. The state's authority to punish, to exert direct force against citizens, helps define the extent of citizen freedom. That is why questions about the justification of punishment involve political as well as moral theory.[24]

The political function of the criminal law justifies its conduct focus: its attention to the observable physical conduct of persons within the society. It focuses on the social/moral meaning of human interaction, seeking to determine whether particular conduct violates essential social norms. It helps explain a prerequisite of most modern criminal law: proof of wrongful conduct that harms others.[25] Wrongful thoughts, expressions, or bad character in general should not be the concern of criminal law, because in a liberal democracy, these are not the concern of government. Criminal law judges the social/moral character of the accused's conduct, based on the reasons for which the accused acted. The reasons for the conduct matter, because they are critical to determining whether the conduct violated fundamental public norms. It's the character of the conduct that matters, not that of the individual.

I recognize that this account of criminal law is more assertion than proof. Criminal law can be and has been described in quite different responsibility terms, terms far more amenable to moral capacity conceptions of responsibility such as reason-responsiveness. My aim so far has just been to present the conduct-limitation argument in outline, so as to introduce basic themes. Next I take up two basic questions about reason-responsiveness as applied to criminal liability: (1) its susceptibility to proof and (2) the necessity of person judgment via moral capacity for justified punishment.

PROOF

The behavioral science of psychopathy presents empirical evidence of individuals who lack significant capacity for reason-responsiveness. Behavioral scientists have developed sophisticated measuring instruments to identify psychopathy, and advances in psychology and neuroscience may supply new measuring methods in the future. This suggests that the most important question concerning psychopathic responsibility is normative rather than empirical. If reason-responsiveness is to become a legal requirement, though, the lawyer will have some hard questions—at least the prosecuting attorney will. Exactly how will this lack of reason-responsiveness be determined? Even if we can establish that some individuals lack reason-responsiveness, or suffer some significant diminution in it, how can we reliably establish the defendant's relevant capacity or incapacity at the time of the crime?

Proof of facts is at the heart of criminal adjudication. Most of criminal adjudication concerns fact-finding, both procedural and substantive. Facts may be stubborn things, as John Adams famously declared, but they can also be messy, complex, and elusive.[26] Fact-finding in U.S. criminal law is an especially challenging task due to the elaborate rules of evidence and constitutional procedure that govern the admissibility and interpretation of facts in jury trials. The privilege against self-incrimination, the fourth amendment's prohibition on unreasonable searches and seizures, restrictions on hearsay and character evidence, and the due process requirement of proof beyond a reasonable doubt are some of the restrictions and qualifications placed on factual proof. Even without these, there is the challenge of gathering reliable evidence initially: finding witnesses, recording their accounts, recovering physical evidence, and performing accurate analysis upon it.

Given the importance and difficulty of proof, it's not surprising that criminal law normally sets relatively simple questions for decision makers: What did the accused do, and why? These questions usually can be answered from third-party sources: persons or physical evidence that describe the accused's conduct and so provide basic information that the decision maker can analyze to determine mens rea and other aspects of legal culpability. Can reason-responsiveness be proven in similar fashion? I have serious doubts.

What would constitute proof or disproof of a lack of capacity to appreciate moral reasons? Given that in our ordinary lives, we frequently talk about individual human capacities, we might think that such proof should not pose too big a problem. If we could not discern others' relevant capacities, we would not bother discussing the physical, psychological and, yes, moral capacities of others, as we so often do. The criminal law places special demands on language and evidence, however. We often find that concepts adequate for everyday discourse do not suffice in the courtroom.

To explore the proof of moral capacity, let us begin by considering capacity in its simplest and most verifiable form. Consider machines and their capacities. We can reliably measure the memory and processing capacities of a microchip or the torque, horsepower, and miles per gallon of a motor vehicle. We can produce relatively accurate and observer-independent measurements of particular machine capacities. (By observer-independent I mean simply that we can measure with instruments whose readings do not depend on the idiosyncrasies of the instrument operator.)

What about measuring the physical capacities of humans? We certainly have reliable methods and instruments for measuring human physical *performance.* For example, medical professionals can measure lung and heart function. Electronic timers can measure a swimmer's speed in the pool or a runner on the race track. A problem is lurking here if our concern is with *capacity* and not just particular performance, however. Human physical capacities are variable, depending in significant measure on training, effort, and expectation. For example, for years, many believed that the human body simply was not capable of running a sub–4-minute mile. Certainly no one had ever been recorded running that fast. Yet today, world-class milers regularly break the 4-minute mark. The expectation of better performance has changed our concept of human capacity. This states a basic truism in sports and in many other forms of human endeavor. Neither individuals nor groups necessarily know their actual physical limits; changing expectations for physical performance can change our understanding of physical capacity. Surely, though, this is not an insurmountable problem in measurement. We can at least establish a basic range of physical capacities for most individuals and thus set outer boundaries on human physical capacities. I agree; such rough measures are feasible. Recall, though, that the real test is yet to come, with the measurement not of physical but of moral capacity.

Unlike with physical capacity, we have no observer-independent measuring instruments for moral resources, no timers, EKGs, or heart monitors. Although brain scanning technology has considerable promise, its documentation of brain states cannot quantify *moral* capacity: it does not measure what a person can and cannot understand about moral reasons.

The most reliable data we have about a person's capacity to perceive, appreciate, and act upon certain stimuli in a given situation is what the person has done in similar situations in the past. We can document the individual's past patterns of behavior and compare these with the behavior of others similarly situated. From these data, we may be able to predict future behavior. This is the stuff of behavioral science, not moral or criminal responsibility, however. For moral or criminal responsibility we must assess *moral* capacity. Moral capacity cannot depend just on past patterns of action, because that would eliminate evaluation of choice in determining moral responsibility (Litton, 2008).

The same point—the difficulty in distinguishing behavioral prediction from moral judgment—can be made by comparing social science explanations of criminality with principles of criminal responsibility. A number of criminologists in recent years have focused on lack of self-control as key to criminality (Gottfiedson & Hirschi, 1990).[27] Persons who commit crimes tend to be impulsive, acting in harmful ways without considering the consequences. But impulsivity, by itself, is not an excuse from responsibility. This would confuse explanation with excuse.

We also know that the qualities needed to restrain impulsivity—self-discipline and self-control—are often fostered by parents, teachers, and others through their enforced expectations. Consistently applying negative consequences to negative (here impulsive) actions can change behavior. Similarly, we know that persons who previously seemed incapable of abiding by certain social rules can, in different settings, subject to different inducements and expectations, develop new capacities. In sum, individual capacities and social expectations for the individual are connected.

Current evidence indicates that psychopaths have little appreciation for the significance of moral reasons.[28] It may be that they cannot grasp the significance of moral reasons, but how are we to know? For psychopaths, like others, what the individual *can* do depends in part on what others *expect* of him, the standards to which he is held. In the case of the psychopath, the most important such standards are likely set by criminal law.[29] We encounter a basic tension between capacity determination and expectation setting in criminal law. The assessment of reason-responsiveness depends on separating these functions, on the notion that we can set general expectations for conduct in the rules of criminal law and also excuse certain persons from those expectations because of their individual moral capacities. The difficulty is that we know expectations can powerfully influence capacity. It's hard to see how any law or decision maker can consistently reconcile these competing demands.

The tension between capacity and expectation presents more than a practical problem for determining the legal responsibility of an alleged psychopath. It also introduces a major conceptual issue: the normative status of the psychopath's moral motivation. Assume that we can prove that the psychopath lacks any significant appreciation of morals and therefore any internalized reason to behave morally. As a result, she lacks moral motivation. Normally we consider lack of moral motivation to be blameworthy. Not caring about the basic value of another is cause for condemnation. It is the essence of cruelty. With the psychopath, though, her lack of moral motivation may be traced to a lack of moral comprehension, to an inability to respond to moral reasons. As a result, perhaps lack of moral motivation here supports an *excuse* from responsibility. This brings us to our central question: should conduct judgment be enough for criminal liability, or is person judgment in the form of moral capacity analysis required as well?

My answer centers on the kind of rationality needed for criminal liability. I argue that the psychopath acts for rational reasons that challenge basic social values and that this kind of rationality is all that should be required for criminal responsibility.

THE RATIONALITY OF THE PSYCHOPATH

Psychopaths commit harmful actions for readily comprehensible reasons, reasons that directly challenge our commitment to universal human value. Psychopaths act out a basic philosophy that might be called me-now. One of the most primitive imaginable, this philosophy is most commonly seen in the actions of young children, but it is hardly restricted to youth. All humans feel its pull and are susceptible, at least occasionally, to its influence.[30]

What is a me-now philosophy? It has three essential elements: (1) only I count; (2) only the most basic animal drives matter; and (3) short-term desires trump long-term considerations. Me-now is hopelessly inadequate as a true moral philosophy because it is by definition incapable of universalizability. Moral responsibility surely cannot turn on the individual's powers of philosophical reasoning, however. That we may philosophize about morality does not make morality synonymous with philosophy.

The psychopath puts himself first. He is the only one who truly matters. He is a god.[31] No wonder he is mystified at the idea that he should care about others. How do they matter to him? This selfishness constitutes a moral stance because it sets normative priorities in relations with others. The anecdotal literature on psychopaths indicates that some are essentially oblivious to social/moral standards. Basic moral standards simply do not seem to have personal significance for them. Other psychopaths appear positively motivated to violate social/moral standards. They enjoy breaking important (which is to say, moral) rules and getting away with it. This is revealed in one of the most distinctive traits of psychopaths, the persistent and skilled use of deception. We see psychopaths lying both instrumentally, to gain immediate selfish ends, and also for the sheer pleasure of fooling others.

The psychopath's desires are base: a desire for power over others, for material goods (especially money), for the sensual pleasures of sex and violence, drink and drugs. The psychopath's commitment to the base and shunning of love, respect, loyalty, and friendship, to name a few of the higher goods, is radical and therefore extraordinary, but like his selfishness, the tendency itself is common.

Finally, the psychopath typically privileges now over later. Impulsivity is a common characteristic, as we have seen. What I want, I want *now*. In some psychopaths this trait is particularly pronounced and leads to counterproductive behaviors, where the failure to engage in the most basic consideration of longer-term consequences defeats the psychopath's immediate aims. Other psychopaths seem to display considerable planning ability to achieve their aims, and yet even these aims are relatively short term. Once more, this trait in psychopaths is unusual in degree, but not in nature. We have seen that impulsivity is common to criminal offenders generally. It is also characteristic of the young, but can be found, commonly, in persons of all ages. Valuing immediate desires over long-term goods is a deeply human tendency.

What does all this say about the rationality of psychopaths? Previously I noted that some reason-responsiveness advocates contend that psychopaths are irrational in their noncomprehension of morals (Litton, 2008; Morse, 2008). They do not comprehend a fundamental aspect of human life, the importance of not hurting others, which is said to preclude their ability to engage in an important form of practical reasoning. This begs the question of why such understanding must be included in the rationality needed for moral responsibility, however. Psychopaths generally are not tormented souls. They are neither troubled by a conscience nor by the lack of a conscience. They may lead pointless (to us) lives, but their main complaint seems to be not with their (to us) defective makeup, but with a defective (to them) world that will not give them what they want, now. Their conduct in pursuit of a me-now philosophy is readily comprehensible to all humans. It has its own primitive rationality.[32]

The actions of psychopaths should now appear in a different light. They make sense. They make sense not only to psychopaths, but in an ugly way, to society as well. In committing harmful actions that disregard the basic value of humans, psychopaths act out a me-now philosophy that directly challenges social/moral principles; in short, they commit criminal acts that merit punishment.

JUSTIFYING THE CONDUCT LIMITATION IN CRIMINAL LIABILITY

Still, reason-responsiveness proponents may say: none of this addresses the basic fairness problem.

The selfishness of psychopaths is explained by their inability to internalize moral rules. They do not "get it" emotionally when it comes to wrongful conduct; indeed, it appears that emotionally they cannot "get it." This seems to preclude any meaningful moral dialogue. To analogize to other perceptual abilities, how can we discuss music with someone who can only hear noise, or color with someone who is color blind?[33] Although powerful, the analogy is inapposite to criminal liability, I believe. It works only if we are judging the moral person (according to personal capacities) and not moral performance. In criminal liability, we judge performance. To excuse for moral capacity in criminal law would be like excusing a musician for being tone deaf, or a painter for being color blind. We judge moral performance in criminal liability, to demonstrate that abiding by fundamental moral principle is essential to society and indeed to humanity.

Person judgment, in the form of assessment of an individual's dispositions or emotional resources, is critical to many normative judgments. It is at the heart of judging another's character. It is involved, to some extent, in criminal sentencing. But criminal liability should not turn on this kind of person judgment. To explain, we need to return to the idea that moral responsibility involves a moral dialogue.

As applied to criminal liability, we might envision this dialogue as a verbal exchange about wrongdoing between judge and accused, between Law and Defendant. We have seen that reason-responsiveness proponents ask whether the accused has the ability to participate in this dialogue, whether he or she can meaningfully understand the principles of wrongdoing under consideration. But what kind of dialogue does criminal liability actually entail? Who is addressed, and how?

Criminal law speaks to the offender *and* the victim of crime *and* society at large. Society is an important audience for, and judge of, all legal decisions (Binder, 2008). Emile Durkheim famously claimed: "punishment is above all intended to have its effect upon honest people" (Durkheim, 1984: 62–63). Because the dialogue of criminal responsibility involves the public as well as the decision-maker and offender, responsibility determinations necessarily look forward as well as back. The requirement that criminal law maintain the general public's respect means that it, like the general public, normally takes a pluralist view of punishment principles.[34] Retribution matters, but so does deterrence. If offenders do not have the internal moral motivations required for law obedience, then the law will supply its own, external motivations in the form of punishment. Thus the fact that psychopaths may be deterred by the prospect of punishment (albeit with great difficulty), and others may be deterred by the example of their punishment, helps justify that punishment, regardless of the psychopath's inability to appreciate moral principle.[35] This brings us to another potentially distinctive feature of criminal law: its use of force.

In moral philosophy, the dialogue of responsibility is usually imagined as a kind of conversation between wrongdoer and moral authority. The conversation conducted by the criminal law is similar, except that it expresses norms not just with words, but with force. We hope that the conviction and punishment of the offender will prove morally persuasive, but frankly we're not holding our breath. We don't bring the accused into court to effect their moral conversion. If legal judgments of guilt and deserved punishment do not convince the wrongdoer of his or her wrongdoing, then force alone must bear the law's social/moral message. What you did—killing, robbing, attacking, raping, stealing—that, you cannot do with impunity. That is unacceptable and will be punished.

Regardless of what the offender believes morally, or can grasp morally, the law by force declares the unacceptability, the basic social wrongness, of criminal conduct. In this dialogue, the offender may remain unrepentant. He may persist in his moral obtuseness. He may lack the moral resources required at that time to comprehend wrongdoing. Such moral thickness is hardly unusual among offenders. Again what counts in law is chosen conduct: action that demonstrates fundamental disregard for the value of others. This is the target of criminal law.

Finally, as Paul Litton has noted, there are serious practical problems with a legal regime that excuses psychopaths according to reason-responsiveness, or any other form of moral incapacity. These are in addition to the proof concerns mentioned earlier. To meet its basic moral and political aims, criminal law must establish uniform conduct expectations for members of society. Individualizing responsibility according to varying moral and emotional resources would severely undercut that project. Excusing psychopaths who commit wrongs, including some of the very worst wrongs, because of their individual emotional and moral incapacities—or providing lesser punishment because of diminished such capacities—would undermine the expectation that *all* persons must refrain from acts of disregard. The effect would be especially severe when we consider estimates that psychopaths make up perhaps 20% of our current prison populations.[36]

GETTING DOWN TO CASES: ERIC HARRIS AND COLUMBINE

One of the great weaknesses of an otherwise impressive and rapidly growing literature on the responsibility of psychopaths is the lack of consideration of specific criminal cases. The portrait of psychopaths is usually drawn from the behavioral science literature, with particular traits illustrated by particularly vivid individual examples. But the construct of psychopathy permits significant trait variation, meaning that this method may provide a misleading picture of the full range of dispositions contained within the psychopathy category. For example, some may be highly impulsive, living essentially without a plan. Others may display considerable planning abilities with respect to their own exploitative projects.[37] More serious for our purposes, though, is the failure to identify particular harmful conduct committed by psychopaths. To test our understanding of and commitment to criminal responsibility, we need to look at concrete examples of harmful actions by psychopaths that demonstrate basic disregard for the welfare of others. This is conduct that criminal law now condemns, but that reason-responsiveness might excuse or mitigate.[38] To remedy this deficiency, I will discuss a recent, infamous case, in which there is considerable evidence that one of the perpetrators was a psychopath.[39]

On April 20, 1999, Eric Harris and Dylan Kleybold arrived at their high school on a killing mission (Cullen, 2009). Armed with two shotguns, a semiautomatic rifle, two semiautomatic handguns with large magazines, numerous homemade pipe bombs, and two large propane gas cylinder bombs, the two killed 13 and injured 21 others. They then killed themselves. But deadly as this toll was, it was far less than they had sought. If the two propane bombs in the high school cafeteria had gone off during lunch as planned, the explosions would have brought down the structure, killing hundreds. Then the two planned to shoot any survivors who fled the scene. They had set car bombs to kill those who managed to reach the apparent safety of the parking lot.

Eric Harris was the leader of this homicidal enterprise. He kept a journal of his thoughts and plans, and a website. He wanted the world to know who he was and why he did what he did. A senior in high school, he was an angry young man who believed himself intellectually superior to most, perhaps all, around him. "I am GOD compared to some of these un-existable zombies" (Cullen, 2009: 182).[40] On his website he declared war on all humankind, writing "i don't care if I live or die in the shootout, all I want to do is kill and injure as many of you pricks as I can" (Cullen, 2009: 216). Despite a great many media stories after the shootings that attributed the killings to Goth culture or resentment of athletes or other popular students, it now appears that the motive for the killings, at least for Harris, who was an otherwise successful and popular student, was hatred or disdain for fellow human beings.[41]

Harris was a highly accomplished liar, who managed to deceive his parents, teachers, a probation officer, and friends about his nature and his plans. He took great pleasure in his ability to deceive, seeing it as proof of his superiority. The journalist David Cullen, who has written the most extensive account of Columbine, argues that Harris demonstrated what psychologist Paul Ekman called "duping delight" (Cullen, 2009: 241). Harris was an intelligent young man, at least in the cognitive sense. On several occasions he demonstrated a significant facility for formal moral reasoning.[42] As for his feelings for others, these are harder to evaluate. He apparently did care, *to some extent*, for other persons. Harris warned one fellow student away from school prior to the attacks. "Brooks, I like you now. Get out of here. Go home" (Brown & Merritt, 2002: 4). In his journal, Harris described the need to discipline himself to avoid feelings of empathy in order to carry out his homicidal plan. "I have a goal to destroy as much as possible, so I must not be sidetracked by my feelings of sympathy, mercy, or any of that" (Cullen, 2009: 276). In his final notes to his parents, he apologized to his mother, recognized her care and thoughtfulness, and said his father was great. He described trying to keep a distance from them as the date of the attack approached. "I don't want to spend any more time with them I wish they were out of town so I didn't have to look at them and bond more" (Cullen, 2009: 326).[43] In his final videotaped message, Eric Harris said: "Yeah...everyone I love, I'm really sorry about all this. I know my mom and dad will be just like...just fucking shocked beyond belief. I'm sorry, all right. I can't help it" (Cullen, 2009: 348). Nevertheless, a complete analysis of the incident and perpetrators by a psychologist-FBI agent and a psychiatrist concluded that Eric Harris was a psychopath. He killed "to demonstrate his superiority and to enjoy it" (Cullen, 2009: 236–243, 239). The assessment of psychopathy was made on the basis of Harris's life history, which included previous criminality, his demonstrated commitment to deception, his conception of his own superiority, his superficial charm, and lack of any significant concern for the pain he caused other human beings.

Assuming that Harris was a psychopath, or at least would score high on the psychopathic checklist, then what does that say about the relationship between psychopathy and criminal responsibility? As discussed in our behavioral science review, the categorization by itself cannot determine moral or legal responsibility because the categorization explains physical cause; it does not judge choice. Assuming that reason-responsiveness would have to be proven independent of the psychopath categorization, how might that be done? For example, what are we to do with Harris's occasional expressions of empathy for certain persons such as his parents? Does this indicate sufficient grasp of moral reasons with respect to his actual victims, or did such persons remain beyond his limited empathic, and therefore moral, capacity? More fundamentally, were Harris's expressions of empathy sincere? Did they indicate actual feelings for others, or did they represent mimicking of moral behavior, a classic sign of psychopathy? Researchers have long noted the way in which psychopaths can fool others about their normality by their ability to pretend concern for others. Any feeling for others that psychopaths display is actually superficial and done for self-interested reasons, researchers claim. Given this, if reason-responsiveness is required for conviction, then presumably we must litigate in the criminal courtroom the sincerity of expressions of compassion that defendant may have made at different points in his life. But note the peculiarity of this endeavor, especially considering the public purposes of criminal law. Do we really want a criminal law that excuses Eric Harris from criminal liability *if* he was pervasively callous throughout his life, but holds him responsible if at least once prior to the crime he truly felt for another human being? It's hard to imagine such a law maintaining the public's respect for long.[44]

The key to Eric Harris's criminal and moral responsibility comes from his own declared intentions, I believe. In his attacks at Columbine High School he meant to, and did engage law-abiding society in a moral dialogue. Harris saw himself as acting on a public moral stage. He understood social concepts of right and wrong, and rejected them. A significant measure of his motivation, both in his efforts to deceive others and in his criminal offenses, was to demonstrate his superiority to and therefore rejection of conventional moral limits.[45] It's true that he appeared to lack feeling for others in a way that makes him seem strange. Perhaps he did lack a basic capacity for fellow feeling. But does that change the social/moral status of his conduct?

Harris was a moral actor, *in the social sense*, whose conduct represented, and was meant to represent, a radical rejection of basic social/moral norms.[46] Regardless of how he came to be the person who rejected standard morality, he did rationally reject it and that rejection, when expressed in prohibited conduct, presents a powerful challenge to basic social/moral norms.[47]

CRIMINAL LIABILITY: A QUESTION OF HUMANITY

Imagine three different outcomes for a criminal prosecution, assuming that Eric Harris survived the massacre.

First, following conviction, a court excoriates Eric Harris for the cruelty of his murders, observing that in these actions he demonstrated a casual and nearly complete disregard for the value of human life. Such reprehensible conduct demands condemnation by a severe sentence, the court holds.

Second, a court decides that Harris did not completely lack capacity for reason-responsiveness, but because he had such diminished capacities in this regard, he deserves less punishment. His psychopathy does not excuse him from liability, but mitigates his punishment.

Third, a judge following a hearing declares that Harris is a psychopath, incapable of feeling for others and therefore lacking the ability to respond to moral reasons, an essential requirement of moral and legal responsibility. Under a civil commitment scheme, Harris is ordered confined to a secure treatment facility until such time as his psychopathic condition might be alleviated (Morse, 2008: 211–212). Given the lack of any effective treatment for psychopathy, such confinement might well be permanent.

Of these outcomes, the mitigation alternative seems the least plausible as a matter of law. In reason-responsiveness terms, mitigation might make sense, by linking a judgment of partial moral deficit with a determination of partial legal responsibility. But consider this outcome in terms of the criminal law's public (read here political) mission. Such mitigation would provide for punishment disproportionate to the severity of the criminal conduct—the highly calculated and entirely impersonal murder of multiple people in a high school—*and* a diminished prospect for protection of the public from future violence. Given its lack of legal appeal, by virtue of undercutting essential public functions of the criminal law, I focus my remaining attention on the all-or-nothing options: full excuse or full responsibility.

In both the first and third instances (full excuse and full responsibility respectively) the court's ruling would take Harris's violence seriously by providing society with significant protection from its repetition.[48] Both legal conclusions could be justified on the facts presented. So how do we choose between them? Perhaps the choice does not really even matter, except to theoretically inclined legal academics. The choice is more than academic, however. The choice between scenarios turns on a basic moral question: Are psychopaths human?

Finding a person nonresponsible, not because of the reasons for their conduct, but because of their individual capacities, constitutes a judgment on their natures, what I have called a person judgment. It judges their status as members of the community, indeed as human beings. In this sense, holding psychopaths criminal responsible declares their full humanity; involuntary commitment denies it (Murphy, 1972).

Throughout history, select groups of persons have been denied full human status because of their perceived intellectual, physical or moral shortcomings. Given that psychopaths apparently lack the emotional capacities needed to love and be loved, qualities vital to good moral conduct and essential to civilized life, they appear inhuman. In this way we see that moral capacity is not just another trait which a person has or does not. It is fundamental to society and therefore fundamental to humanity. Seen this way, the earlier analogies to the colorblind or tone deaf appear seriously mistaken. Human society does not require color or music; it does require morality. Subjecting psychopaths to indefinite civil confinement rather than punishment is consistent with deciding that, in moral terms, they possess essentially animal natures. Lacking any true conception of social/moral obligation, they are not fully human.

The history of declaring groups inhuman is an ugly one, and should raise many questions about how any legal categorization based on psychopathy might be used, and abused. But my present concern here is not with the welfare of "them" but with "us." What would declaring psychopaths alien mean for our own conception of humanity?

Excluding psychopaths from humanity would reshape the species in accord with our preferred self-image, our image of humans as humane. We would like to think of human beings as fundamentally empathetic, capable of concern for others and remorse at causing others' suffering. But is this an accurate depiction of the human world? Conscienceless deception and manipulation, selfishness and egocentricity, the pursuit of power, sex, and money at all costs, a disregard for or even delight in the suffering of others, the casual use of violence to dominate and inflict pain—these are *inhuman* traits? We would like to think so, but surely we know better.

PSYCHOPATHY AS A TEST

Psychopaths have an extraordinary ability to fool others. Even experienced researchers and savvy law enforcement officers have been duped by psychopaths (Hare, 1993: 14). Psychopaths often display a genius for short-term deception and manipulation that allows them to gain money, sexual pleasure, and power in ways that often seem incomprehensible in retrospect. But their genius is readily explicable, being based on their often intense focus on other persons' value systems, which they see as key to successful manipulation and exploitation. They are frequently brilliant at playing on others' hopes and fears.[49]

In researching and writing this chapter I have had the occasional sense that the same dynamic might somehow be at work in the academic consideration of psychopath responsibility. It is almost as if psychopaths identified the enormous value that the modern academic places on rational choice and then used this knowledge to construct a plausible excuse from all responsibility.

I realize this is, literally, nonsense. Psychopaths are not engaged in this debate (to our knowledge), and though psychopaths consistently deny responsibility for particular wrongs, they normally do not assert a lack of moral agency. Instead, they frequently believe themselves worthy of praise (a form of individual responsibility) for many actions and traits.[50] Yet the notion that the worse their moral deficits, the better their argument for an excuse, has a weird, psychopathic ring to it. It appears that in the presence of psychopaths, what otherwise seems plain becomes doubtful, for academics no less than others.[51]

Psychopaths may be seen as our psychological and moral testers. They test our cognitive and emotional intelligence with misrepresentations and manipulations; they test our moral principles with deeds of disregard. They also test our conception of criminal responsibility.

The example of the psychopath shows, I believe, that criminal liability is, and should be, limited to conduct judgment: assessing what persons do to and with others in society, rather than judging the person in the form of his or her moral capacities. The criminal law's condemnation of conduct is a fundamentally social judgment concerning social/moral standards. The offender is punished as the author of the conduct, but criminal blameworthiness depends on the meaning of that conduct to all concerned in the society. It is not a Godlike judgment (rendered by that modern deity, Science) of the individual according to how his or her moral capacities were utilized.

This conception of criminal liability has major implications for the question of justification, for the fairness of punishment. Holding psychopaths responsible as we do, illustrates that criminal punishment is justified, is fair, according to a universal standard of expected conduct, not an individual standard of choice-making according to certain emotional or moral capacities.

Finally, the conduct conception of criminal liability emphasizes the limits of public fact-finding in criminal law. Limiting determinations of guilt to analysis of wrongful conduct and not moral capacity, means that we judge based on what we can see of the accused's conduct and what it means to us, not what he or she may have experienced (or not) in its choosing.

All of this I think is consistent with the basic rules of criminal liability today, but that is not to say that this is how most people actually understand the work of criminal law. Speaking now in the most general sense of "we," we generally presume, despite considerable evidence to the contrary, that those who commit criminal acts had sufficient moral and emotional resources to do better. We do this in part because the emotional temptation to judge persons is enormous. It is so much simpler and so much more satisfying to judge people Good and Bad, Criminal and Law-Abider, than it is to value all persons, while condemning their particular wrongful acts. We feel much safer—and in some respects may be safer—making person judgments about dispositions to criminality rather than limited judgments about particular acts. Most fundamentally, we would like to rest our condemnation of the offender on the morally secure ground that, in their situation, we would have chosen better. The science of psychopathy suggests otherwise.

If we blame psychopaths for their depredations regardless of their emotional and moral shortcomings, then we must acknowledge that criminal liability does not depend on the person's moral resources. We must acknowledge that blameworthiness does not necessarily represent an assessment of what the person emotionally and morally understood or could have understood, but just an assessment of what the person did, and why. The only question is whether the accused violated the minimum conduct standard that we set for all persons in society. We are left with a thinner conception of criminal responsibility than previously, but also I believe one that is more honest and that, overall, might be less prone to penal excess. That, however, is very much another story (e.g., Pillsbury, 2002, 2008).

CONCLUSION

It's one of the treasured myths of modernity, and one particularly treasured by academics, that increased knowledge is always good. Each of us engaged in the life of the mind can list manifold ways in which our own contributions to human knowledge promise improvements in human life. Knowledge by itself is not a good, however. Sometimes increased knowledge can lead to confusion, because we misread the significance of new insights.

Science teaches us more, practically every day, about the human species, and especially about the mind and behavior. We can see further into the human organism, especially into cognitive and emotional processes, than ever before. The study of psychopathy is a startling example of science's improved abilities to identify, quantify, and explain particular patterns of especially problematic human behavior. We seem to be on the verge of a true science of antisocial behavior, or at least one critical subset of it.[52]

It's natural to believe that new insights into antisocial behavior will have important consequences for moral and criminal responsibility.[53] Indeed the science of psychopathy powerfully challenges our understanding of responsibility, because while the actions of psychopaths appear blameworthy, psychopaths themselves seem to lack the feelings needed to comprehend wrongdoing. We would like to believe that responsibility requires a free and fully rational, meaning fully comprehended, choice to do wrong. That cannot be true if psychopaths are held responsible, however.

I have titled this chapter "Why Psychopaths Are Responsible" because it represents a defense of current law, which holds that psychopaths *are* fully responsible. The use of the passive voice here, though, might suggest that responsibility is a natural state to be discovered by the decision maker. It might suggest that an accused either is or is not responsible according to certain objectively ascertainable facts about the world. Given my emphasis on criminal liability as enforcing general social/moral expectations for conduct, perhaps the better title would be: "Why We Should Hold Psychopaths Responsible." The shift from passive to active voice would signal the importance of society's role, our role, in setting standards to be enforced by criminal liability. In moral and criminal responsibility we observe and set fundamental expectations for each other. Moral responsibility is a human creation and not a natural feature of the world. I believe it is our greatest creation. Deciding whether psychopaths are responsible is our responsibility. We have a choice.

We can excuse psychopaths *if we wish*. Such an excuse could be justified under a particular view of responsibility and the rationality it requires. Such an excuse might even be reconciled with the needs of public safety, if paired with an expansion of involuntary civil commitment. Nevertheless I think it would be a serious mistake. To excuse psychopaths would undercut the basic expectation that all persons in society should refrain from cruel and exploitive conduct, from harmful acts that demonstrate basic disregard for the welfare of others. It would diminish our commitment to basic social/moral expectations that we should defend in all of our responsibility practices.

NOTES

1. For an excellent overview of these different positions, see Litton (2008). Litton argues that a close look at the behavioral science literature reveals that psychopaths suffer extensive deficits in practical reasoning, meaning that they fail both conceptions of moral agency. See also Fischette (2004).
2. See Chapter 15 by Litton.
3. I define behavioral dispositions broadly to include cognitive processes and affective states as well as actions.
4. To date, treatment efforts have not been very successful. There is evidence that treatment that works for nonpsychopaths may actually worsen psychopathic behavior. Hare (1996).
5. The study of physical cause is distinct from analysis of whether an individual is responsible for a result that has multiple contributors, which is called causation in law.
6. Just as many individuals have depressive tendencies, but would not be diagnosed as clinically depressed, so many individuals may display psychopathic tendencies without ever being categorized as a psychopath. See Litton (2008). On the other hand, some researchers hold that psychopathy is better understood as a category distinction, in which persons so labeled are fundamentally different from others in society. See Hare (1996) and Harris (2001).
7. Psychopathy therefore involves more than just social dysfunction. By comparison, autism also involves an apparent inability to connect emotionally with other persons, but without the aggression and exploitation associated with psychopathy. See Blair, Mitchell, and Blair (2008); Maibom (2008); and Wauhop (2009).
8. For an insightful commentary on the normative significance of scientific terminology and language, see Reimer (2008).
9. There are dissenters who argue that character is assessed by criminal law, perhaps especially in the excuses. See, for example, Huigens (2002); Kahan (1997), arguing for character assessment rationale for mistake of law doctrine. Compare Gardner (1998), arguing that excuses from criminal liability involve conduct generally characteristic of what we expect of law-abiding persons, rather than conduct that is uncharacteristic of the individual. Moral capacity analysis represents a kind of character theory, in that it analyzes a person's dispositions and abilities to think, feel, or act in ways that disrespect others, and these represent a critical aspect of character. See Moore (1997: 564–565), describing character in terms of dispositional states. Nevertheless I use the term person judgment here instead of character judgment to avoid confusion with the judgment of a person's overall good or bad character, a quite different endeavor potentially from what moral capacity analysis involves.
10. At sentencing, criminal histories and demonstrated dispositions to criminality can be very important, meaning that here the criminal law does consider character to some extent.
11. Not exclusively, of course. See, for example, see Blair (2008).
12. Litton argues that psychopaths suffer from deep rationality deficits; they have no evaluative standards, which means they have a reduced capacity for rational self-governance; thus they are not fully morally responsible under both reason-responsiveness theories and theories that focus on a deficit in the will.
13. We might also distinguish between moral and legal responsibility. See, foe example, Litton (2008: 387–391). Many commentators focus exclusively on the demands of moral responsibility without taking up its consequences for law. In this chapter I assume that we should not find a psychopath criminally responsible without also determining that he or she is morally responsible. See Schopp and Slain (2000), noting the contradiction between current American criminal law and civil commitment law on the question of the psychopath's ability to exercise self-control.
14. Litton argues that the psychopath's lack of evaluative standards for his own conduct "renders

much of his behavior unintelligible to us." See also Sasso (2009).
15. For psychopaths' ability to respond to penal and other external stimuli, see Morse (2008: 210); Glannon (2008: 164); Schopp and Slain (2000: 250–251).

   Under deterrence, the critical question is whether punishment of psychopaths meets utilitarian standards, in which the disutility of punishment's pain is outweighed by the utility of crime reduction. Specific deterrence certainly presents significant challenges, because psychopaths appear essentially immune from the social shame that deters many from crime commission. Some theorize that fearlessness is key to the condition itself. Nevertheless, psychopaths do appear to calculate their behavior, seeking to minimize the prospects of detection and therefore punishment, and maximize the opportunities for successful exploitation. This suggests that they may be deterrable, though with great difficulty. Supporting this point, some behavioral scientists believe that there are large numbers of so-called successful psychopaths, who do not engage in extended criminal careers, but live largely within legal bounds. See generally, Hall and Benning (2007) and Lykken (1995). Such psychopaths may not be inclined to crime, perhaps because they can obtain what they want largely without the risks involved in criminal activity. The latter explanation might support a specific deterrence rationale for psychopaths generally, because it suggests that psychopaths are—to varying degrees—sensitive to the threat of punishment. Finally, there is strong evidence that many psychopaths "age out" of criminality, suggesting that they may modify their behavior so as to avoid future punishment. See Hare (1996). This pattern also suggests the efficacy of incapacitation for psychopaths during their early adulthood.

   Meanwhile the example of their punishment will provide general deterrence, as crimes by psychopaths often represent among the worst of their types for the remainder of society.
16. The other main responsibility defense, infancy, also has a potential rationality-capacity explanation: the young lack the practical reasoning capacities necessary for being a fully rational chooser. See Vincent (2008); Morse (2008). Although I offer this only as a preliminary speculation, there might be another way that the infancy excuse might be explained, and that is as a lack of experience of life. Because young children are so inexperienced in all senses, any choice they make to violate basic moral/social rules does not represent the same challenge to social/moral principle as when committed by someone older and therefore more life-experienced.
17. On addiction, see Morse (2006); Wallace (1999); see also Fingarette (1988); Husak (1999); and Moore (1997)In the law of insanity, the strong trend is to eliminate an excuse for lack of volition, See *United States v. Lyons* (1984) (en banc); 18 U.S.C. sec. 17(a). Even where such a claim is legally recognized, a successful insanity defense almost always involves a defendant with a serious rationality problem as well as volitional difficulties. See Morse (1999: 203).
18. With respect to compulsive sex offenders, see Morse (2002), arguing that if any excuse is recognized it should turn on irrationality, not offender loss of self-control. But see sex predator civil commitment laws, which presume a lack of self-control (volition) as a predicate for indefinite commitment, and in which psychopathy is seen as a mental disease that involves such a lack of volition, for example, *Kansas v. Hendricks* (1997), finding state's scheme for civil commitment of violent sex offenders constitutional; *Kansas v. Crane* (2002), holding that proof of deficit in volitional control is needed for civil commitment of sex offenders; and Schopp and Slain (2000).
19. Though Morse argues the excuse should be limited to extreme psychopaths.
20. Cane (2003) argues that moral philosophy may sometimes learn from law rather than always serving as law's teacher.
21. On the importance of political theory to criminal law, see Binder (2008) and Fletcher (1998). On the essentially political nature of modern criminal law and its consequences, see Dripps (2009) and Schopp and Slain (2000).
22. Expressive theories of punishment are often associated with the French sociologist Emile Durkheim. For an introduction, see Sasso (2009: 1202–1204). See also Feinberg (1970).
23. Obviously the criminal law should have no political function in the partisan or electoral sense. The adjudicative processes of criminal justice must be insulated from the surging to and fro of general public opinion.
24. For example, the reluctance to require legal duties to aid strangers (Kleinig, 1976) and the

significance of results independent of mens rea (Binder, 2008).
25. Crimes that do not involve such concrete harm to others, such as conspiracy, or crimes of consensual illegality (drugs, prostitution) are often controversial for just this reason.
26. "Facts are stubborn things and whatever may be our wishes, our inclinations, or the dictates of our passion, they cannot alter the state of facts and evidence." Argument in Defense of the Soldiers in the Boston Massacre Trial, December 1770.
27. See also DeLisi (2009), arguing that the traits of psychopathy provide a template for explaining all criminal behavior.
28. Their deficit in this regard is not complete, however. See Maibom (2008).
29. In general, we know that psychopaths are not moved by shame or other moral or social sanctions, but their conduct may be affected by the concrete consequences of penal confinement. See note 15 supra (on deterrence).
30. If you have any doubts, try driving in bad traffic when you are very tired, cranky, or otherwise indisposed.
31. As Heidi Maibom (2008: 180) puts it: "He does not think that nobody has any intrinsic value; he just thinks that *only he* has such value."
32. Psychopathy may even be, in an evolutionary sense, adaptive. As long as limited to a small number of persons, it may provide individuals with an evolutionary advantage in reproduction. Psychopathic traits of fearlessness and ruthless pursuit of selfish aims may also be important to success in some noncriminal endeavors. See Lykken (1995: 116), referring to a "talent for psychopathy"; Maibom (2008: 177); and Reimer (2008: 187–192); Maibom (2008: 177).
33. See Hare (2003: 28), quoting Hervey Cleckly on the psychopath's lack of appreciation for beauty or love or goodness: "It is as if he were color blind, despite his sharp intelligence, to this aspect of human existence." See also Morse (2008: 209).
34. As a result, although the criminal law may aspire to a coherent and consistent theory of liability and punishment, this will never be achieved. This does not mean that the effort to achieve coherence and consistency, to construct a principled basis for criminal law, should be abandoned. It remains enormously important. But coherence and consistency are not in criminal law the ultimate values that they are in philosophy. A stable criminal law can be based on a mix of rival perspectives and justifications. See Dripps (2009). From a functional perspective, what matters is that the criminal law resolves criminal disputes to the long-term satisfaction of a great majority of the public.
35. See note 15 supra (deterrence).
36. Hare (1996: 43–44) noting prevalence of psychopathy among sex offenders. Recidivism rates and especially violent recidivism rates are much higher for psychopaths (Harris, G. T., Rice, M. E., & Cormier, C. A. (1991). A possible counter example is insanity. Reason-responsiveness proponents often argue that insanity depends on rational incapacity (Morse, 1999). If true, this gives some ground for believing that the criminal law could handle a moral incapacity excuse. Although controversial, the insanity defense has been a part of our criminal law for centuries. Its assessment is difficult and verdicts rendered often unpopular, but the excuse remains viable in almost all jurisdictions. Perhaps this shows that we could also solve any practical problems posed by an excuse for the psychopath's rational incapacity. I contest the premise of this argument though, that insanity is based on rational incapacity. I would argue that the insane are excused from responsibility because their reasons for action, and therefore their actions, are fundamentally irrational. Perceiving a different reality than the rest of us, their actions make no moral sense; therefore, punishment makes no moral sense as a response. Insanity is about crazy reasons for action, not rational incapacity.
37. Behavioral scientists and moral capacity proponents alike often emphasize the impulsivity of psychopaths, their disinclination and perhaps incapacity for anticipating the consequences of conduct. If this were a prerequisite for psychopathy, however, it would preclude that categorization for the most feared category of criminals who are generally presumed to be psychopaths: serial killers. Such individuals often do engage in significant planning of their offenses. See Hare (1993: 3–4, 114–115), listing by name serial killers who were likely psychopaths; and Lykken (1995: 115, 143).
38. Moral capacity proponents do sometimes provide factual detail about psychopaths and their conduct, but the facts presented are usually directed to particular personality traits or

aspects of personal histories deemed relevant to responsibility. Rarely are the full details of acts of violence or other wrongful conduct provided. As I have argued throughout, such conduct is the primary concern of criminal law. Without concrete examples of that conduct, an important part of our social/moral resources—the way in which particular actions trigger responses based on our own values and experiences—is excluded.

39. A brief note about the choice of this case. As we will see, although Eric Harris undoubtedly displayed psychopathic traits, it is not clear how high he would score on the PCL-R. If only those who demonstrate the most severe lack of reason-responsiveness should be excused, then Harris might not qualify. Thus it might be that Eric Harris simply does not support the case against reason-responsiveness in criminal law. Nevertheless, I think the case is useful to provide a concrete example of psychopathic, me-now philosophy in action. There is also a larger point which is necessarily speculative. For myself, I do not believe that an individual who had a longer criminal record, was socially less successful, more impulsive, less empathetic, and overall more clueless about morals—and therefore scored higher on the PCL and perhaps lacked more reason-responsiveness—who committed the same acts, would present significantly different responsibility issues. For a comparison to such an offender, see Trillin (2009: 32).

40. In his journal which he called "The Book of God" Harris wrote: "I feel like God. I am higher than almost anyone in the floating world in terms of universal intelligence" (Cullen, 2009: 234).

41. See generally Cullen (2009). When Harris wrote about going to college, he fantasized about doing great violence to a female student. "I want to tear a throat out with my teeth like a pop can. I want to grab some weak little freshman and just tear them apart like a fucking wolf, strangle them, squish their head, rip off their jaw, break their arms in half, show them who is god" (Cullen, 2009: 294). Kleybold has been described as a depressive (Cullen, 2009: 187–188).

42. He articulated sophisticated moral reasoning in a high school essay (Cullen, 2009: 265). In another essay he insightfully expressed how another might feel when victimized by a property crime (Cullen, 2009: 258), describing a letter Harris wrote to the victim of a car burglary that he and Kleybold have committed, a letter written as part of his probation. Nevertheless, Harris elsewhere railed at the victim for his stupidity in leaving his vehicle in a vulnerable place (Cullen, 2009: 260).

43. For a similar account of empathy insufficient to dissuade from violence, see Maibom (2008: 172), quoting the serial killer and presumed psychopath Ted Bundy.

44. As we will see shortly, the same may also be said of the possibility of mitigating punishment according to diminished reason-responsiveness.

45. But see Litton (2008), in which Litton argues that psychopaths have essentially no moral standards, good or bad. If true, perhaps Harris was not a psychopath. But the rejection of standard morality, when expressed in harmful conduct, undertaken for selfish reasons, represents taking a moral position. Immorality and amorality are equally challenging to social/moral values. To be responsible, the wrongdoer's reasoning need not meet any criteria for living a good life; his reasons just must not involve a fundamental misunderstanding of reality.

46. Consistent with most others who are labeled psychopathic, Harris never claimed a lack of moral agency. That is, he never said that he was incapable of moral conduct. He just did not care to follow moral rules.

47. On the similarity of ordinary "criminal thinking" to that of psychopaths, see Samenow (2002).

48. A civil confinement scheme might, overall, provide more protection from psychopathic violence than does the criminal law. Of course, providing protection from future harms is not the criminal law's only concern.

49. This distinguishes psychopaths from the autistic, who also display significant deficits in social learning capacities and might be the basis for an argument that psychopaths have sufficient moral capacity for responsibility even given a severely diminished experience of empathy (Maibom, 2008: 170 n.6).

50. Given the likelihood of a civil commitment alternative to replace criminal responsibility, psychopaths might be worse off overall, were they to lose their status as responsible individuals. See Litton (2008: 391).

51. In fact, education and other traditional measures of intelligence may be a hindrance

in dealing with psychopaths. Robert Hare describes a psychopath who used his stare to powerful effect on others, including a prosecutor. When he turned it on a police officer, the unfazed officer told him: "That bullshit only works on intelligent people" (Hare, 1993: 210).

52. See Matt DeLisi (2009), arguing that the behavioral science concept of psychopathy provides the essential basis for a theory of criminology.

53. It is worth noting that those calling for changes in the criminal responsibility of psychopaths are generally legal academics and philosophers, not behavioral scientists. This is unsurprising given that behavioral scientists pursue causal explanation and not responsibility judgments. Perhaps the leading contemporary researcher on psychopathy, Robert Hare, supports the responsibility of psychopaths. "In my opinion, psychopaths certainly know enough about what they are doing to be held accountable for their actions" (Hare, 1993: 143). But see Seabrook (2008) (quoting Jean Decety, a social neuroscientist: "what neuroscience is showing us is that a great many crimes are committed out of compulsion—the offenders couldn't help it. Once that is clear, and science proves it, what will the justice system do?")

REFERENCES

Arenella, P. (1992). Convicting the morally blameless: Reassessing the relationship between legal and moral accountability. *UCLA Law Review*, *39*, 1511.

Binder, G. (2008). Victims and the significance of causing crime. *Pace Law Review*, *28*, 713.

Blair, R. J. R. (2008). The cognitive neuroscience of psychopathy and implications for judgments of responsibility. *Neuroethics*, *1*, 149.

Blair, R. J. R., Mitchell, D., & Blair, K. (2005). *The psychopath: Emotion and the brain*. Malden, MA: Blackwell.

Brown, B., & Merritt, R. (2002). *No easy answers: The truth behind death at Columbine*. New York: Lantern Books.

Cane, P. (2003). *Responsibility in law and morality*. Oxford: Hart.

Cullen, D. (2009). *Columbine*. New York: Twelve.

DeLisi, M. (2009). Psychopathy is the unified theory of crime. *Youth Violence & Juvenile Justice*, *7*, 256.

Dripps, D. (2009) Rehabilitating Bentham's theory of excuses. *Texas Tech Law Review*, *42*, 383.

Duff, R. A. (1986). *Trials and punishments*. Cambridge, MA: Cambridge University Press.

Duff, R. A. (2007). *Answering for crime: Responsibility and liability in the criminal law*. Oxford: Hart.

Durkheim, E. (1984). The Division of Labor in Society (pp. 62–63). New York: Free Press.

Edens, J. F., & Petrila, J. (2007). Legal and ethical issues in the assessment and treatment of psychopathy. In: C. J. Patrick (Ed.), *Handbook of psychopathy* (pp. 573–588). New York: Guilford.

Ells, L. (2005). Juvenile psychopathy: The hollow promise of prediction. *Columbia Law Review*, *105*, 158, 178–185.

Feinberg, J. (1970). *Doing and deserving: Essays in the theory of responsibility*. Princeton, NJ: Princeton University Press.

Fingarette, H. (1988). *Heavy drinking: The myth of alcoholism as a disease*. Berkeley: University of California Press.

Fischer, M., & Ravizza, M. (1998). *Responsibility and control: A theory of moral responsibility*. New York: Cambridge University Press.

Fischette, C. (2004). Psychopathy and responsibility. *Virginia Law Review*, *90*, 1423.

Fischette, C. (2007). Psycho moral community. *Ethics*, *118*, 70, 77–85.

Fletcher, G. P. (1998). The fall and rise of criminal theory. *Buffalo Criminal Law Review*, *1*, 275, 287–294.

Gardner, J. (1998). The gist of excuses. *Buffalo Criminal Law Review*, *1*, 575.

Glannon, W. (2008). Moral responsibility and the psychopath. *Neuroethics*, *1*, 158.

Gottfredson, M. R., & Hirschi, T. (1990). *A general theory of crime*. Stanford, CA: Stanford University Press.

Haji, I. (1998). On psychopaths and culpability. *Law & Philosophy*, *17*, 117, 138–139.

Hall, J. P., & Benning, S. D. (2007). The "successful" psychopath: Adaptive and subclinical manifestations of psychopathy in the general population. In: C. J. Patrick (Ed.), *Handbook of psychopathy* (459–480). New York: Guilford.

Hare, R. D. (1993). *Without conscience: The disturbing world of the psychopaths among us*. New York: Guilford.

Hare, R. D. (1996). Psychopathy: A clinical construct whose time has come. *Criminal Justice & Behavior*, *23*, 25.

Harris, G. T., Rice, M. E., & Cormier, C. A. (1991). Psychopathy and violent recidivism. *Law & Human Behavior*, *15*, 625.

Harris, G. T., Skilling, T. A., & Rice, M. E. (2001). The construct of psychopathy. *Crime & Justice, 28*, 197, 214–216.

Horder, J. (1993). Pleading involuntary lack of capacity. *Cambridge Law Journal, 52*, 298, 302–304.

Huigens, K. (2002). Homicide in aretaic terms. *Buffalo Criminal Law Review, 6*, 97.

Husak, D. N. (1999). Addiction and criminal liability. *Law & Philosophy, 18*, 655.

Kahan, D. M. (1997). Ignorance of the law *is* an excuse—but only for the virtuous. *Michigan Law Review, 96*, 127.

Kansas v. Crane, 534 U.S. 407 (2002).

Kansas v. Hendricks, 521 U.S. 346 (1997).

Kleinig, J. (1976). Good Samaritanism. *Philosophy & Public Affairs, 5*, 382.

Litton, P. (2008). Responsibility status of the psychopath: On moral reasoning and rational self-governance. *Rutgers Law Journal, 39*, 349

Lykken, D. T. (1995). *The antisocial personalities,* The Psychopath: An Introduction to the Genus (p. 115), A Theory of Primary Psychopathy (p. 143). New York: Psychology Press.

Maibom, H. L. (2008). The mad, the bad, and the psychopath. *Neuroethics, 1*, 167.

Mei-Tal, M. (2004). The criminal responsibility of psychopathic offenders. *Israel Law Review, 36*, 103.

Moore, M. (1990). Choice, character, and excuse. *Social Philosophy & Policy, 7*, 29.

Moore, M. (1997). *Placing blame: A general theory of the criminal law*. New York: Oxford University Press.

Morse, S. J. (1999). Crazy reasons. *Journal of Contemporary Legal Issues, 10*, 189, 193–203.

Morse, S. J. (2002). Uncontrollable urges and irrational people. *Virginia Law Review, 88*, 1025.

Morse, S. J. (2004). Reasons, results and criminal responsibility. *University of Illinois Law Review, 363*, 374–377.

Morse, S. J. (2006). Addiction, genetics, and criminal responsibility. *Law and Contemporary Problems, 69*, 165.

Morse, S. J. (2008). Reasons, psychopathy and criminal responsibility. *Neuroethics, 1*, 205.

Murphy, J. G. (1972). Moral death: A Kantian essay on psychopathy. *Ethics, 82*, 284.

Murphy, J. G. (2004). *Getting even: Forgiveness and its limits,* Christianity and Criminal Punishment (pp. 106–108). New York: Oxford University Press.

Pillsbury, S. H. (2002). A problem in emotive due process: California's Three Strikes Law. *Buffalo Criminal Law Review, 6*, 483.

Pillsbury, S. H. (2008). Learning from forgiveness. *Criminal Justice Ethics, 28*, 135.

Reimer, M. (2008). Psychopathy without (the language of) disorder. *Neuroethics, 1*, 185.

Samenow, S. E. (2002). Understanding the criminal mind: A phenomenological approach. *Journal of Psychiatry & Law, 29*, 275.

Sasso, P. (2009). Criminal responsibility in the age of "mind-reading." *American Criminal Law Review, 46*, 1191, 1203–1205.

Schopp, R. F., & Slain, A. J. (2000). Psychopathy, criminal responsibility, and civil commitment as a sexual predator. *Behavioral Science & Law, 18*, 247.

Seabrook, J. (November 10, 2008). Suffering souls: The search for the roots of psychopathy. *The New Yorker*.

Shoemaker, D. (2007). Moral address, moral responsibility, and the boundaries of the moral community. *Ethics, 118*, 70, 73.

Trillin, C. (July 27, 2009). Annals of Crime: At the train bridge. *The New Yorker*.

United States v. Lyons, 739 F.2d 994 (5th Cir. 1984).

Vincent, N. A. (2008). Responsibility, dysfunction and capacity. *Neuroethics, 1*, 199, 201.

Wallace, R. J. (1994). *Responsibility and the moral sentiments*. Cambridge, MA: Harvard University Press.

Wallace, R. J. (1999). Addiction as defect of the will: Philosophical reflections. *Law & Philosophy, 18*, 621, 652–654.

Wauhop, B. (2009). Mindblindness: Three nations approach the special case of the criminally accused individual with Asperger's syndrome. *Pennsylvania State International Law Review, 27*, 959.

PART NINE
DETENTION OF PSYCHOPATHS

CHAPTER 17

Preventive Detention of Psychopaths and Dangerous Offenders

Stephen J. Morse

INTRODUCTION

Dangerous offenders are a threat to any society, and all civilized societies will be motivated to try to reduce the danger such offenders present. The questions for the law are how to do that effectively and with respect for the rights of the offender.

This chapter begins by defining three classes of nonpsychotic offenders who seem to present the greatest danger to public safety and how the law responds to them: psychopaths, people with antisocial personality disorder, and recidivist offenders who seem to suffer from no identifiable mental abnormality that is a risk factor for criminal behavior. Then it considers the law's justification for preventive detention generally, and argues that Anglo-American, and especially American, practice is dominated by what I term "desert-disease jurisprudence" (Morse, 1999). Next, the chapter turns in detail to primarily the dominant American approaches to reducing the danger these people present. It argues that current doctrines and practices are not effective or unduly threaten the rights of offenders. Two constitutionally acceptable alternatives for psychopaths are then canvassed: holding psychopaths nonresponsible and an analog to quasi-criminal commitments imposed on so-called mentally abnormal sexual predators. I argue that the former is unworkable and the latter has potentially grave dangers. In short, there is no optimal solution to the danger these types of offenders present that is consistent with protection of civil liberties. Then the chapter briefly considers the possibility of genuinely pure preventive detention based on dangerousness alone and untethered from the current desert-disease limits

THREE CLASSES OF OFFENDERS AND LAW'S CURRENT RESPONSE TO THEM

Psychopathy is a condition characterized by emotional traits, such as lack of empathy, conscience, and concern for others, and by conduct abnormalities, such as repetitive antisocial behavior. It is estimated that 25% of convicts serving prison terms have psychopathy, which is a substantial risk factor for crime. Psychopathy is also a major risk factor for antisocial conduct among those suffering from other mental disorders. There is considerable controversy about whether psychopathy is a mental disorder, but the dominant position is that it is a personality disorder and that its signs and symptoms are pathological. At the least, psychopaths lack psychological attributes that seem central to successful, cooperative life. At present, psychopaths are considered criminally responsible, psychopathy is not considered a mitigating condition for sentencing, and psychopathy is not a sufficient mental abnormality to qualify for ordinary civil commitment.

Psychopathy must be distinguished from Antisocial Personality Disorder (APD), which, unlike psychopathy, is a diagnostic category included in the American Psychiatric Association's authoritative *Diagnostic and Statistical Manual of Mental Disorders*, 4th Edition-Text Revision (American Psychiatric Association, 2000). All but two of the criteria for APD are repetitive antisocial behaviors, and the psychological criteria, lack of remorse and impulsivity, are not necessary to make the diagnosis. About 40% to 60% of prisoners have APD and there is substantial overlap with psychopathy. Despite inclusion of APD in *DSM–IV*, there is also great controversy about whether as defined it should be considered a mental disorder. People with APD are considered criminally responsible, the disorder is not a basis for mitigation in sentencing, and the disorder does not qualify for ordinary involuntary civil commitment.

Some recidivist dangerous offenders are neither psychopaths nor suffering from APD. There are various causes that predispose them to be at enhanced risk for offending, including genetic or psychologically abnormal variables. Unlike psychopaths or people with APD, there is no hypothesized mental abnormality, however, that would justify attribution of a diagnostic category. In short, this is a diverse category of dangerous but otherwise normal people. Not surprisingly, there is no question about their criminal responsibility, mitigation is not warranted unless they also have some independent mitigating condition, and such people cannot be civilly committed.

THE LOGIC OF PREVENTION

The basic logic of prevention is quite straightforward. To some unknown degree, human beings must live in cooperative societies to survive. Such societies are viable only if, also to some unknown degree, members forbear from putting each other unreasonably at risk. Within the limits of viability, how much risk is unreasonable is a normative, moral and political question, but all societies that survive surely place limits on risk and will act to prevent danger from those for whom socialization has apparently failed. This story can be told in the crudest evolutionary biological terms, but in more politically and philosophically sophisticated societies, these necessary preventive practices are the subject of rich theoretical analysis, usually from a consequential or a rights perspective. Broadly speaking, two types of stories, each rooted in a theory of the person, inform these perspectives: "Good bacteria, bad bacteria" and "Taking people seriously."

Let us begin with the short form of the former, which I abbreviate because it is rejected by all but the most extravagantly hard-nosed consequentialists. According to this account, we could treat each other like bacteria. Some bacteria that inhabit our gastrointestinal system, our gut, are crucial to the smooth operation of the system. They are the good bacteria. We try to enhance their survival and do nothing to inhibit their growth. On occasion, alas, our guts are invaded by bacteria that interfere with the proper operation of the system, causing various unseemly ailments and, in extreme cases, death. These are the bad bacteria. We try to prevent these critters from entering our gut in sufficient numbers to overwhelm the body's natural defenses, and if the natural defenses fail, we try with various techniques, such as antibiotics, to kill the offensive, bad bacteria.

Now, despite the potential of various bacteria to confer benefits and harms, as the case may be, and despite our consequential, substantial efforts to deal rationally with these bacteria, no one holds either kind of bacteria responsible for smooth or rocky gastrointestinal functioning and we would not dream of either praising or blaming bacteria. Similarly, we would not dream of considering antibiotic treatment a means of punishing the bad bacteria. We treat bacteria purely as objects, and never as potentially responsible subjects, as potential moral agents. We could, by analogy, simply treat each other like bacteria, as potentially beneficial or harmful objects, and act accordingly. This conception of people, much beloved by eliminative materialists, would support a purely predictive and preventive scheme of social organization, in which the emotional and societal response to the organism could be entirely independent of the moral goodness or badness of the person's conduct. Indeed, any other regime may appear founded on irrational dreams about our privileged place in the natural order.[1] It would be a regime of utterly strict liability. We do not at present have the emotional repertoire or the predictive and therapeutic technology to institute this vision very precisely or effectively, but this is a technoquibble. In principle, I suppose, it is a possible form of social organization. Indeed, in some senses we might all be "safer" and, to some, social life might appear more rational if we acted accordingly.

The alternative, dominant story, "Taking people seriously," is familiar. It admits that, like bacteria, human beings are part of the physical universe and subject to the laws of that universe, but it also insists that, as far as we know, we are the only creatures on earth capable of acting fully for reasons and self-consciously. Only human beings are genuinely reason-responsive and live in societies that are in part governed by behavior-guiding norms. We are the only creatures to whom the questions "Why did you do that?" and "How *should* we behave?" are properly addressed, and only human beings hurt and kill each other in response to the answers to such questions. As a consequence of this view of ourselves, human beings typically have developed rich sets of interpersonal, social attitudes, practices, and institutions, including those that deal with the risk we present to each other. Among these are the practice of holding others morally responsible, which includes moral expectations, attitudes about deserved praise and blame, and practices and institutions that express those attitudes, such as reward and punishment.

The concern with justifying and protecting liberty associated with "taking people seriously" is deeply rooted in the conception of rational personhood I have sketched. Only human beings self-consciously and intentionally decide how they should live; only human beings have projects that are essential to living a good life. Only human beings have expectations of each other and require justification for interference in each other's lives that will prevent the pursuit of projects and seeking the good. If liberty is unjustifiably deprived, a good life is impossible. Some would attempt to collapse the two accounts, claiming that many of our seemingly retrospective, nonconsequential practices, such as holding others responsible, can in fact be justified by a fully prospective, consequential theory (Dennett, 1984). This account recognizes that evolution has designed us to be intentional, self-conscious creatures, but practices such as holding others responsible are, allegedly, simply stimuli that increase the probability of safe (good) behavior and decrease the probability of dangerous (bad) behavior. No one, in other words, is "really" responsible. In the words of H. L. A. Hart (1968), it is an "economy of threats." The economy-of-threats approach does not successfully explain our practices, however, and suffers from defects of its own. Nothing in this approach would prohibit blaming and punishing innocent people if doing so would maximize the good. This is a familiar criticism, but one that has no answer if it is unjust to punish the innocent, as virtually all theories of justice, except the most unflinchingly consequentialist, hold.

Second, as Jay Wallace points out, the economy-of-threats approach fails to explain our practices, because it omits the central attitudinal aspect of blaming (Wallace, 1994). To hold an agent responsible and to blame that agent is not simply a behavioral disposition, whose purpose is the maximization of some future good. Blaming fundamentally expresses retrospective disapproval. Even if it has the good consequence of decreasing future harmdoing, our current practice is undeniably focused in large measure on past events.[2] In sum, many of our most important moral and political concepts depend on taking people seriously as people, as practical reasoners and potentially moral agents.

The desire to be safe ultimately conflicts with and complements the desire to be free. People who live in constant terror of dangerous neighbors do not feel free or cannot enjoy their freedom, even if their society is politically liberal. But achieving the safety that makes freedom possible inevitably requires substantial infringement on the liberty of dangerous agents.

As a consequence of taking people seriously as people, as potential moral agents, however, we believe that it is crucial to cabin the potentially broad power of the state to provide protection by depriving people of liberty. Thus our polity has imposed two fundamental legal limits on the state's power to intervene: The agent must be dangerous because he or she is suffering from a disease (especially a mental disorder) or because the agent is a criminal who deserves punishment.[3] For the purposes of this chapter, the notion of desert I am employing is simply the traditional retributive conclusion that if an offender's behavior satisfies the elements of a charged offense and no justification or excuse obtains, then the offender is culpable and deserves the ensuing blame and punishment. To avoid confusion, I should add that the two generic excusing or nonresponsibility conditions are lack of rational capacity and lack of control or compulsion. Lack of "free will" and "causation" are not legal excusing conditions.

For people who are dangerous because they are disordered or because they are too young to "know better," the usual presumption in favor of maximum liberty yields. Because the agent is not rational or not fully rational, the person's choice about how to live demands less respect, and he or she is not morally responsible for his or her dangerousness. The person can therefore be treated more "objectively," like the rest of the world's dangerous but nonresponsible instrumentalities, ranging from hurricanes to microbes to wild beasts. Although all human beings deserve to be treated with dignity and respect in virtue of their humanity, agents incapable of rationality do not actually have to cause harm to justify nonpunitive intervention. We can take preemptive precautions, including broad preventive detention, with nonresponsible agents based on an estimate of the risk they present. Justified on consequential grounds, such deprivation will be acceptable if the conditions of deprivation are both humane and no more stringent than is necessary to reduce the risk of harm. Such deprivations are forms of greater or lesser quarantine and may include "treatment," but in theory they are not punishment and they should never have a punitive justification or effect.[4]

Virtually all criminals are rational, responsible agents, however, and according to the dominant story, the deprivation imposed on them, punishment, is premised on considerations of desert. No agent should be punished without desert for wrongdoing, which exists only if the agent culpably caused or attempted prohibited harm. The threat of punishment for a culpable violation of the criminal law is itself arguably a form of preventive infringement on liberty, but it is an ordinary, "base-rate" infringement that requires no special justification. After all, no one has a right to harm other people unjustifiably. In our society the punishment for virtually all serious crimes, and thus for dangerous criminals, is incapacitation, which is preventive during the term of imprisonment. But criminals must actually have culpably caused or attempted harm to warrant the intervention of punishment. We cannot detain them unless they deserve it and desert requires *wrongdoing*. In the interest of liberty, we leave potentially dangerous people free to pursue their projects until they actually offend, even if their future wrongdoing is quite certain. Indeed, we are willing to take great risks in the name of liberty.

In sum, both the criminal and the medical/psychological systems of behavior control require a justification in addition to public safety—desert for wrongdoing or nonresponsibility (based on disease)—to justify the extraordinary liberty infringements that these systems impose. The normative basis of this system of desert-disease jurisprudence is that it enhances liberty and autonomy by leaving people free to pursue their projects unless they responsibly commit a crime and unless through no fault of their own they are nonresponsibly dangerous.

The story about crime and disease is, of course, not so simple. Do we really believe that responsible, dangerous agents have a right to be at liberty when their potential harmdoing is serious and quite certain? In theory we do, and "gaps" (Schulhofer, 1996) between the disease and crime justifications for intervention remain. For example, imagine a young male prisoner about to be released from his third incarceration for armed robbery who boasts that he will immediately do it again upon release and will kill any victim who might potentially identify him. There is nothing at present the law can do to prevent this. We can only hope that the armed robber has some sort of conversion experience. But such cases properly cause grave social and legal concern, and, in fact, the law insistently seeks to fill the desert-disease gap with both civil and criminal preemptive remedies. It either adopts more harshly incapacitative criminal incarceration or expands the non-responsibility criterion and thus also expands the categories of people who can be civilly detained even if they do not deserve punishment. Whether these "gap-filling" measures are efficacious and just are the issues the law must address.

EXPANDING DESERT-BASED DETENTION

This section considers the two dominant types of desert-based preventive detention approaches: increasing sentence length and recidivist offender enhancements. The focus is on the American experience, but English and Canadian approaches are on occasion briefly examined for comparison. Although in principle both look promising as a means differentially to preventively incapacitate particularly dangerous offenders, neither is promising in fact for theoretical or practical reasons.

Sentencing

For much of the 20th century, American sentencing schemes strongly favored indeterminate sentencing with very wide ranges for most serious crimes.[5] Although in theory and to some degree in practice, desert sets a cap to the permissible ranges, the justification for this practice was largely a rehabilitation model. Deciding when to release the prisoner would depend on the professional judgment of the correctional officials and parole boards concerning the prisoner's rehabilitation progress. At the maximum term, the prisoner would of course have to be released even if he had not been rehabilitated and was still dangerous, but the maximum was typically quite long. Thus, lengthy incapacitation was possible and many otherwise unrehabilitated prisoners would simply "age out" of future violent conduct. Those who were rehabilitated and needed no further incarceration to protect society could be released. Because the three classes of offenders under consideration were all criminally responsible, indeterminate sentencing was a potentially effective means to keep these offenders incarcerated for long periods.

Although the indeterminate sentencing model was coherent, in practice it was subject to many severe problems that led to its demise in many jurisdictions. First, it led to unprincipled and arbitrary discretion. Parole officials had virtually unreviewable authority to make release decisions but they had neither a coherent conceptual approach to such problems nor accurate predictive validity about who would be dangerous if released. The alleged rehabilitative programs available in the prisons were either paltry or unvalidated. In a word, the officials were "flying blind" and this produced differential treatment that could not possibly be justified and was widely believed to be racially biased. In addition, many believed that the sentences actually served in general and the maximum terms in particular bore insufficient relation to the punishment that offenders deserved.

In the late 1960s and early 1970s the critiques of indeterminate sentencing reached an unusually bipartisan crescendo. Critics and politicians from across the political spectrum called for a new regime that would tie an offender's punishment to his just deserts and the legislatures responded. Simultaneously, and in response to concerns about arbitrary discretion and unjustifiably unequal treatment, legislatures also began to adopt determinate sentences with relatively limited ranges for most crimes and sentencing guidelines that limited judicial sentencing discretion to various degrees. Although the new regime also faces substantial criticisms, it has been adopted in a substantial number of jurisdictions, including federal criminal jurisdiction, and it exerts influence on those jurisdictions that still retain more indeterminate sentencing.

The just deserts/determinate sentencing system lacks the resources differentially to preventively detain particularly dangerous offenders, but incapacitation can be achieved simply by increasing the sentences for serious crimes and insuring through so-called "truth in sentencing" that few offenders are released early. Escalating sentence length has indeed been the American response. Although dangerous offenders are incapacitated for lengthy terms, the problem is that many nondangerous offenders are incapacitated as well. This is not an effective means of differentially incapacitating the most dangerous classes of offenders.

Recidivist Offender Enhancements

This type of enhancement of desert-based detention simply increases a multiple offender's sentence beyond the range normally imposed for the crime based on the offender's criminal history. Jurisdictions vary substantially concerning which prior crimes should suffice as a basis for enhancement and how much enhancement is justified. In principle, however, this type of penal program can work effectively to preventively incapacitate a specific, targeted group of offenders who pose a particular threat to society. The questions for the law, of course, are whether these controversial enhancements are fair and efficacious. As one commentator says of such programs generally, there is "no easy way out" (Lippke, 2008; McAlinden, 2001). Rights and social safety are inevitably in conflict.

English law for most of the 20th century allowed such enhancements for offenders deemed especially dangerous, but judges were loath to impose them because they seemed too much like double punishment for the same crime (McAlinden, 2001). Various substitutes for enhancement, such as increased use of life imprisonment and some limited enhancements, have resulted since the early 1990s (McAlinden, 2001). England's parliamentary system ensures that such schemes will be upheld if duly passed by Parliament, so the politics of the issue is crucial and there is little sympathy for repeat, dangerous offenders.

The United States Supreme Court has considered on numerous occasions whether such enhancements violate the 8th Amendment prohibition of cruel and unusual punishments, but there was no clear general answer until *Ewing v. California* (2003; Duff, 1998).[6] *Ewing* considered the constitutionality of the State of California's so-called "three strikes and you're out" law, which imposed a minimum term of 25 years for any defendant convicted of a third felony. Defendant Gary Albert Ewing, who was 36 years old at the time, shoplifted three golf clubs worth just over $1,000 from a shop. He had numerous prior convictions, including one armed robbery that did not result in injury and for which he served 6 years of a 9-year term in state prison. The "three strikes" law was imposed and Ewing received a sentence of at least 25 years although the felony that triggered the enhancement, grand larceny, carries generally light penalties.

A plurality opinion upheld the constitutionality of the law, holding that the proportionality requirement of the 8th Amendment was strictly limited when applied to a legislatively mandated term of years in prison. Only extreme sentences that are grossly disproportionate violate the 8th Amendment, the Court held, and this was to be determined in light of great deference that should be granted to legislative decisions about what penal justifications to adopt and about what terms of years was warranted. The court was unable to fashion a retributive justification for enhancements, but deemed it sufficient that they could easily be justified on the consequential grounds of general prevention and incapacitation. In his concurrence, Justice Scalia, who believes that constitutional proportionality analysis does not apply at all to terms of years, pointed out that proportionality would impose a coherent limit on prison terms only if it was justified retributively. In short, legislatures are free to impose draconian enhancements and thus preventively to detain dangerous offenders for far longer than the triggering offense alone would permit.

Just deserts is the moral shoal upon which retributive enhancements founder. By definition, the offender has already been punished for his prior offenses as much as the state was willing and able to do so. In metaphorical terms, the "slate has been wiped clean" for the prior offenses and the triggering offense does not alone warrant the enhancement. Most criminal justice theorists and commentators in the United States recognize the important limit desert places on just punishment, so there have been many attempts to justify recidivist enhancements retributively E.g., (Walen, 2011; Yaffe, 2011). It would go beyond the purposes of this chapter to canvass all the attempts. It suffices to say, however, that there is general agreement that most fail and even those that appear somewhat promising, such as Antony Duff's proposal that some courses of criminal conduct indicate such a complete rejection of respect for society that "banishment" by enhancements is just (Duff, 1998), are highly controversial or undeveloped.

It is especially difficult to justify such enhancements retributively if one believes that criminal punishment should respond to what the criminal did and not to who the criminal is. Repetitive offending certainly shows that the agent has antisocial dispositions and has done far more than his fair share of criminal harmdoing. Nonetheless, it is not a crime to have a criminal predisposition or criminogenic character and people do not deserve punishment for their characters. Persistent offenders have also received substantially more punishment than less repetitive criminals for the disproportionate amount of crime the former commit. Recidivism does not make the last crime worse or more culpable in itself than if it had been the agent's first offense. It simply indicates that the agent is a worse and more dangerous person, but again, it is not a punishable crime to be a bad, dangerous agent. Defenders of a retributive justification for such enhanced punishment schemes are an extremely rare species precisely because it is so difficult, and perhaps impossible, to provide an adequate retributive justification for enhanced punishment.[7]

Suppose, however, that like the Supreme Court, one accepts the adequacy of purely consequential justifications for recidivist enhancements. At that point, the problem of efficacy arises, especially because lengthy imprisonment is immensely expensive. Additional imprisonment is costly in itself and prevents the imprisonment of more people unless the state is willing to build more prison cells to hold all the new offenders who also need to be imprisoned. The question is whether we can accurately predict who will seriously reoffend, even among multiple offenders, and how good are the data about the general preventive efficacy of these enhancements? In any case, note that they can only be effective as a form of incapacitation if the dangerous offender meets the triggering criteria.

Although prior conviction does increase the risk of future offending because past behavior is the best predictor of future behavior, we often have little precise information with which accurately to estimate in individual cases how much the risk is increased or how much general prevention will be increased by imposing recidivist enhancements. Recidivism interacts with other variables to create a complex prediction equation. There is so much crime in the United States and we have so much data that in principle we might develop an adequately accurate prediction tool based on recidivism that would permit rational enhancement on incapacitative and deterrent grounds. Nonetheless, present use of recidivism is too empirically blunderbuss to be fair.

For example, the 25 years to life enhanced sentence that Gary Albert Ewing received was vastly more than could be rationally justified on incapacitative or deterrent grounds. Now, Ewing was no choir boy. He had a long history of criminal offenses, and even though his triggering offense was quite minor, it is legitimate to conclude that he presented a continuing danger to the public. A lengthy prison term for prior burglary and robbery had failed specifically to deter him. Apparently, only incapacitation could prevent him from reoffending. Moreover, although he was not deterred by the possibility of lengthy sentences, others might well be. Still, even if some enhancement were warranted, we have no sensible idea how much. One might argue that Ewing or others subject to enhancements "assumed the risk" by offending, again knowing that they were so subject, but no citizen should be asked to assume such an irrational risk of State infliction of pain.

Enhanced sentencing based on incapacitative needs also raises the prediction problem generally. The difficulties attending accurate prediction of future behavior, especially low base rate behavior, such as very serious crime, present a serious moral and practical problem for the legitimacy of using predictions of future dangerousness to enhance criminal sentences. An enormous amount has been written about this issue, so I shall cover only the minimum necessary for the wider purpose addressed. To begin, however, note that using recidivism to enhance sentences at least uses a factor involving intentional criminal wrongdoing for which the agent was responsible. In contrast, demographic or personality characteristics, for example, are factors for which the defendant is not responsible. Also, third-time offenders tend to be older and have often reached an age when they will soon "age out" of high risk for criminal offending.

It is a truism of behavioral science that statistical or mechanical prediction and highly structured clinical decision making based on empirically validated risk factors are more accurate than pure clinical prediction, but despite advances in the database that have improved the cookbook (e.g., Monahan et al., 2001; Skeem & Monahan, 2011), highly accurate prediction by any method eludes us in all but the most obvious cases. Nonetheless, and although the limitations of prediction are well recognized, our fear of danger permits prediction of dangerousness to ground sentencing length (and various forms of quasi-criminal civil confinement). The inevitable errors are morally and practically costly, however. The most common error, a false positive, is especially problematic because it results in unnecessary massive and expensive deprivation of liberty. Such erroneous deprivations of liberty may have general deterrent effects, but only at the expense of confining an offender who would not recidivate. Antony Duff tries to avoid this problem by claiming that dangerousness is not a prediction of future conduct but is instead a present state assessment of predisposition to reoffending (Duff, 1998). Thus, there is no genuine actuarial problem. We do not think a person has a predisposition, however, unless we also think that there is some substantial probability that it will produce action. If no such probability exists, we think that there is no predisposition or that it has eroded.

Despite the prediction problems, the gatekeepers, such as judges or parole boards, are conservative because false negatives tend to be more politically costly even though they are infrequent. The costs of unnecessary imprisonment of false positives are not before the public eye, we are not sure that they are false positives because those predicted to re-offend are confined, and convicts seldom have much sympathy from the public. On the other hand, if grave harm is done by a formerly imprisoned person who has been released although a longer prison term was possible, the public is outraged. The incentive structure predisposes decision makers in cases involving danger to overpredict and thus to imprison longer than is necessary. In the case of recidivist enhancements, the imposition of which is optional, there are strong incentives to apply them.

Empirical problems may be solved by research. Increasing knowledge may allow us accurately to estimate the need for enhanced sentences based on recidivism and on other factors, such as sex and age, that we know are related to the risk of reoffending.[8] Reliable information may permit rational enhancement and will decrease the especially worrisome false-positive rate, but it might also lead to extremely lengthy confinement unless the technology of treatment increased simultaneously and we were willing to release offenders serving enhanced sentences prior to the termination of the enhanced term. Moreover, even effective treatments would raise serious moral and legal issues if they were intrusive and had likely and serious side effects. Administering treatment involuntarily also implicates important constitutional liberty interests recognized by the Supreme Court (e.g., *Washington v. Harper*, 1990; *United States v. Sell*, 2003). On the other hand, few confined people would probably refuse safe treatments that would help them reduce their violence potential and thus gain their release. At present, however, this whole line of thought is speculative because the data are not yet available and we do not have safe, effective methods that can reduce the risk of future offending. The problem with predictive accuracy and treatment efficacy will apply to all preventive regimes, including those that seem most theoretically satisfying and fair.

In conclusion, recidivist offender enhancements would in principle be effective to incapacitate our three dangerous classes of offenders for lengthy periods, but these enhancements violate retributive constraints on just punishment, suffer from empirical problems concerning prediction and treatment, and are enormously costly. Such programs are politically popular, but they are probably not good criminal justice policy at present.

EXPANDING DISEASE-BASED DETENTION

The three forms of disease-based detention this section considers are, straightforward, traditional involuntary civil commitment; commitment following acquittal by reason of insanity; and an analog to the quasi-criminal commitment of mentally abnormal sexually violent predators. I conclude that the former two are not useful means of accomplishing preventive detention under the current legal regime, but the latter is. Nonetheless, I conclude that quasi-criminal commitment is a frightening method of social control and that society is better off acquitting psychopaths by reason of insanity and then committing them.

Traditional Involuntary Civil Commitment

Once again in principle, involuntary civil commitment might be an effective means to preventively detain dangerous offenders, but this method fails because it does not satisfy the disease constraint and because contemporary involuntary commitment is aimed not at long-term incapacitation but at short-treatment and return to the community.

Our three classes of dangerous offenders are all legally responsible and not subject to involuntary commitment because they do not suffer from a sufficient mental disorder. Even if they satisfy the dangerousness standard for commitment, neither psychopathy nor APD qualifies for the mental disorder criterion for commitment and in some jurisdictions these conditions are specifically excluded by statute. Psychopaths were previously not considered proper subjects for commitment in England because they were thought not treatable, but the Mental Health Act of 2007 makes clear that psychopaths can be committed. The problem, however, is that long-term incapacitation is not contemplated. Even if some members of our classes of offenders suffered from a comorbid disorder that would more traditionally qualify for involuntary commitment, this would not be true of most of them and the resultant commitments would once again be brief. The comorbid disorder would be treated to re-compensate or stabilize the person and then he would be released.

Finally, traditional civil hospitals do not prefer to admit dangerous offenders. Such patients place severe demands on the security and safety of the institution and are more like management problems than proper patients (Appelbaum, 2005). Such criticisms fueled proposals to remove dangerousness to others as a criterion for involuntary commitment and to respond to people whose primary problem is the risk of violence in the criminal justice system.

Traditional Insanity Acquittal Followed by Commitment

Legal insanity is the only current American doctrine that instantiates an excuse for lack of rational or control capacity. Forty-six states and the federal criminal law contain some version of the excuse. The most common is a so-called "cognitive" test that adopts some variant of the traditional English rule derived from M'Naghten's Case (1843). That test holds that a person will be excused if he was acting under such a defect of reason arising from disease of the mind that (1) he did not know the nature and quality of the act that he was doing, or, (2) if he did know it, he did not know that what he was doing was wrong. A minority of American jurisdictions have adopted a "control" test in addition to a cognitive test. These tests excuse if, as a result of mental disorder, the defendant was acting under an irresistible or uncontrollable impulse or had lost the ability to choose the right conduct.[9] No jurisdiction has adopted a control test as its sole test for legal insanity. A minority of jurisdictions have adopted the American Law Institute's Model Penal Code test, which includes both a cognitive and control test. It excuses a defendant, if as a result of mental disease of defect, the defendant lacks substantial capacity either to appreciate the criminality [or, in the alternative, the wrongfulness] of his conduct or to conform his conduct to the requirements of the law.[10]

Cognitive tests provide a distinct folk psychological mechanism for excuse or mitigation, including the inability to attend to the proper considerations for guiding conduct in a specific context and the inability to use those considerations actually to guide conduct. An agent who lacks these abilities for any nonculpable reason has a rationality defect. Such explanations make sense of the common sense claim that a defendant could not control himself. Indeed, I have no problem calling this standard a control standard as long as one understands clearly that the problem that undermines self-control is a cognitive defect and not some overwhelming force or the like. The rationality standard is a genuine and limiting condition of nonresponsibility rather than a metaphoric ground. It can be applied workably and fairly and leaves room for moral, political, and legal debate about the appropriate limits on responsibility.

If we consider the legal and moral standards of responsibility, it is clear that the capacity for rationality is the primary criterion. Only lack of rational capacity can explain the diverse conditions that undermine responsibility, including, among others, infancy, mental disorder, dementia, and extreme stress or fatigue. Reflection on the concept of the person that law and morality employ and on the nature of law and morality suggest that the capacity for rationality must be the central criterion. What distinguishes human beings from the rest of the natural world is that we are endowed with the capacity for reason, the capacity to use moral and instrumental reasons to guide our conduct. Law would be powerless to achieve its primary goal of regulating human interaction if it did not operate through the practical reason of the agents it addresses and if agents were not capable of rationally understanding the rules and their application under the circumstances in which the agent acts (Shapiro, 2000). The central reason why an agent might not be able to be guided by moral and legal expectations is that the agent was not capable of being guided by reason. It is sufficient if the agent retained the capacity for rationality even if the capacity was not exercised on the occasion.

It should be apparent from the potential capaciousness of the language of all the insanity defense tests quoted in the preceding text that there is a reasonable argument for including psychopathy as a potential predicate for an insanity plea but almost none to include APD. Nonetheless, American law either explicitly by statute or by judicial interpretation excludes psychopathy as the basis for an insanity defense.

A further provision of the Model Penal Code is instructive. Referring to its insanity defense test quoted previously, it says that, "the terms 'mental disease or defect' do not include an abnormality manifested only by repeated criminal or otherwise antisocial conduct."[11] According to the strict language of this provision, neither psychopaths nor those with APD are excluded because neither is manifested *only* by repeated criminal or otherwise antisocial conduct.[12] Psychopathy, as clinically described (Cleckley, 1976) and as measured by the Hare Psychopathy Checklist–Revised (PCL-R; Hare, 2003), includes many psychological criteria and is not manifested solely by repetitive criminal or antisocial behavior. APD is a more arguable case. Its criteria do include two psychological variables—lack of remorse and impulsiveness—but neither criterion needs to be present to make the diagnosis and all the other criteria are repetitive criminal or antisocial behaviors. Moreover, the psychological criteria for APD are vague and thus manipulable. One could argue that APD is not excluded because the diagnostic criteria include psychological criteria. On the other hand, because it is not a necessary criterion, perhaps APD should be excluded. A third possibility is that APD would not be excluded only in those cases in which lack of remorse or impulsiveness was one of the diagnostic criteria used. Despite the logic, however, this influential Model Penal Code provision has been interpreted to exclude psychopathy, and, a fortiori, APD as a sufficient mental abnormality to satisfy the insanity defense test. Despite United States law's exclusion of psychopathy as the basis for an insanity defense, the argument for excusing psychopaths, or some of them, from crimes of moral turpitude is that they are not rational because they lack the strongest reasons for complying with the law, such as understanding that what they are doing is wrong and empathic understanding of their victim's plight (Litton, 2008; Morse, 2008). Most people can use empathy, conscience, understanding of the reason underlying a criminal law's prohibition, and prudential reasons to guide behavior. Psychopaths can be guided only by strictly prudential, entirely egoistic reasons not to be caught and punished. In other words, they cannot grasp or be guided by the good reasons not to offend, which could be expressed either as a cognitive or control defect. In addition, according to the same argument, psychopaths with lesser psychopathy should qualify for mitigation, which is considered virtually entirely at sentencing in the United States.

In response, most advocates for continuing the exclusion of psychopathy as a basis for the insanity defense argue that it is sufficient for criminal responsibility if psychopaths can reason prudentially about their own self-interest (e.g., Maibom, 2008; Pillsbury,1992; Glannon, 2008, holds an intermediate position). First, psychopathy does not prevent agents from acting as the law defines action, nor does it prevent psychopaths from forming prohibited mental states. A psychopath who kills another human being intentionally is fully prima facie criminally responsible. Further, psychopaths are not excused because they do possess many rational capacities. They usually know the facts and are generally in touch with reality, they understand that there are rules and consequences for violating them, which they treat as a "pricing" system, and they feel pleasure and pain, the anticipation of which can potentially guide their conduct. This is a relatively thin, prudential conception of rational capacity, but the law deems it sufficient to justify punishment on desert and deterrence grounds. Finally, psychopaths do not suffer from lack of self-control as it is traditionally understood. They do not act in response to desires or impulses that are subjectively experienced as overwhelming, uncontrollable, or irresistible. Proponents of holding psychopaths responsible argue that there is no need to excuse according to either a desert or deterrence justification for punishment. In short, the law should view the psychopath as bad and not as mad.

Those who wish to hold psychopaths responsible also point out that a criminal defendant with major mental illness, such as schizophrenia, also does not qualify for the insanity defense if the defendant knew that what he was doing was criminal or wrong. Consequently, there is even less reason to excuse the psychopath who knows the rules, as virtually all do, because the psychopath is in touch with reality. Moreover, even if a defendant with major mental disorder knows technically that what she is doing is against the law, she may not, unlike the psychopath, even retain substantial prudential reasoning ability. In contrast, the psychopath always knows that the rule applies to him. A defendant who is grossly out of touch with reality and delusionally believes that she is doing the right thing, such as God's will, is paradoxically the mirror image of the psychopath. Her general capacity for moral reasoning remains intact, but her psychotic reasons for action undermine the potential of the rules to guide her prudentially.

Suppose one accepts on normative grounds, as so many do, that the capacity for prudential reasoning is sufficient for criminal responsibility. There remains one final argument for excusing at least extreme psychopaths based on their lack of even prudential reasoning ability. According to one plausible but controversial, broad characterization of psychopathy, most ably advanced by Paul Litton (2008), psychopaths are not rational at all because they lack any evaluative standards to assess and guide their conduct. They do not even possess evaluative standards related to the pursuit of excitement and pleasure. Psychopaths are like Frankfurt's concept of the "wanton." They do not feel regret, remorse, shame, and guilt, feelings that are typically experienced in reaction to our failure to meet the standards we have set for ourselves. They may feel frustration and anger if they fail to get what they want, but these are not reactive emotions. Such frustration or anger does not entail negative self-evaluation. Moreover, severe psychopaths are out of touch with ordinary social reality. They say that they have goals, but act in ways inconsistent with understanding what having and achieving a goal entails. They do not consistently follow life plans and are impulsive. Litton concludes that "it is not surprising that agents with a very weak capacity of internalizing standards act on unevaluated whims and impulses" (Litton, 2008: 382). Much of their conduct appears unintelligible because we cannot imagine what good reason would motivate it. In brief, psychopaths have a generally diminished capacity for rational self-governance that is not limited to the sphere of morality. They cannot even reason prudentially. Again, it is possible that future research may convince legislatures or courts to accept such an understanding of some psychopaths and to extend the insanity defense to them, but this is not the current law, even for such extreme cases.

Suppose the insanity defense were extended to psychopaths. In all jurisdictions, commitment of people who have been acquitted by reason of insanity is automatic in one form or another, although subject to periodic review. Because these commitments are triggered ab initio by proof beyond a reasonable doubt of a criminal offense—if the defendant is able to cast doubt on the prima facie case there is no need to raise the insanity defense—they are considered distinguishable from ordinary civil commitment, which does not require any criminal act for incarceration. Thus, they may impose more onerous conditions on the person committed. The Supreme Court held in *Jones v. United States* (1983) that both mandatory initial commitment and indefinite confinement of such people is constitutional. The reasoning behind the holding was the common sense view that it is presumed that mental illness and dangerousness continue and that public safety requires commitment if the person remains mentally disordered and dangerous. In a later opinion, the Court made clear that an insanity acquittee had to be released from commitment if the person either was no longer mentally disordered or no longer dangerous (*Foucha v. Louisiana*, 1992). The reasoning for the latter decision was straightforward. If the person were no longer suffering from a disorder, there was no disease justification for preventive detention. If the person were no longer dangerous, even if still suffering from disorder, the public safety rationale for preventive detention was not satisfied. But as *Jones* made clear, if mental disorder and dangerousness continued, there was no constitutional limit on the length of these commitments. In short, post-insanity acquittal commitment would seem an excellent means to incapacitate psychopathic offenders.

Despite the initial attractiveness of this solution to the danger psychopaths present, there is nevertheless a major practical objection. The insanity defense cannot be imposed on a competent defendant who does not wish to raise it (e.g.,

United States v. Marble, 1991), and virtually no psychopath would then raise the insanity defense. At present, there is no effective treatment for adult psychopaths, so any psychopath acquitted by reason of insanity would be facing a lifelong commitment to an essentially prison-like facility. In contrast, except for crimes carrying the possibility of the death penalty or life without possibility of parole, the defendant would be much more sensible to arrange a favorable plea bargain for a lesser terms of years or to face conviction and imprisonment for the maximum term the law permits for the crime charged, which would be shorter than the potential commitment. Even in cases involving the potential for life without possibility of parole, a plea bargain to a lesser charge or sentence would be preferable. Moreover, a conviction and the imposition of life without the possibility of parole might be successfully appealed, whereas the only hope for release from an indefinite involuntary commitment would be the discovery of a successful treatment for psychopathy or the hope that the hospital would release the psychopath when he would clearly have "aged out" of dangerousness.

In short, even if American law came to the conclusion that psychopaths should be excused, few psychopaths would be willing to accept such "lenient" treatment and we would still have to rely on a pure criminal justice response.

Finally, the potential use of post-insanity acquittal would not apply to offenders with APD or other dangerous offenders. There simply is no credible argument that such offenders are not criminally responsible unless they also suffer from some comorbid disorder that does negate responsibility.

Quasi-Criminal Commitment

Another means reventively to detain our three classes of offenders would be by a law analogous to the quasi-criminal commitments of so-called "mentally abnormal sexually violent predators." These laws, unlike commitment of a person following acquittal by reason of insanity, permit the conviction and punishment of the person for an offense *and* potentially indefinite civil confinement. They are a strange hybrid of desert-disease jurisprudence, but the ultimate rationale for the commitment is an expansion of disease jurisprudence. They are apparently the most promising means under current law to accomplish preventive detention, so let us consider them in substantial detail to understand how they are justified and what problems they raise.

The History and Criteria: Sexual Predator Commitments

Sexual predators fall into the gap between criminal and civil confinement. They are routinely held fully responsible and blameworthy for their behavior because they almost always retain substantial capacity for rationality, they remain entirely in touch with reality, and they know the applicable moral and legal rules. Consequently, even if their sexual violence is in part caused by a mental abnormality, they do not meet the usual standards for an insanity defense.[13] For the same reason, they do not meet the usual nonresponsibility standards for civil commitment and retain the competence to make rational decisions about treatment. Moreover, as we have seen, in most cases in which civil commitment is justified, most states no longer maintain routine indefinite involuntary civil commitment but instead tend to limit the permissible length of commitment.

To fill the gap, Kansas and a substantial minority of other states have adopted a form of indefinite involuntary civil commitment that applies to "sexually violent predators who have a mental abnormality or personality disorder."[14] Kansas defined a sexually violent predator similarly to other states that have adopted such legislation:

> "any person who has been convicted of or charged with a sexually violent offense and who suffers from a mental abnormality or personality disorder which makes the person likely to engage in repeat acts of sexual violence." (Kan. Stat. Ann. § 59-29a02(a))

In turn, "mental abnormality" is defined as:

> a "congenital or acquired condition affecting the emotional or volitional capacity which predisposes the person to commit sexually violent offenses in a degree constituting such person a menace to the health and safety of others." (Kan. Stat. Ann. § 59-29a02(b))

The statute did not define "personality disorder." As I have indicated elsewhere (Morse, 2002) and suggest below, the definition of "mental abnormality" is entirely tautological and not a definition of abnormality at all—it is a definition of the causes of all behavior—but, astonishingly, all nine members of the Court found it acceptable.

The state may impose this form of civil commitment not only when a person has been charged with or convicted of a sexual offense, but also after an alleged predator has completed a prison term for precisely the type of sexually violent conduct that provides part of the basis for commitment. Commitment is for an indefinite period, and thus potentially for life, although an annual review of the validity of the commitment is required.

In *Kansas v. Hendricks* (1997), the Supreme Court rejected a substantive due process challenge to the constitutionality of the Kansas statute. The majority's primary rationale was that the Kansas criteria were similar to civil commitment criteria that the Court had long approved and that the purpose of the commitment was not punitive (*Kansas v. Hendricks*, 1997). The Court emphasized that legislative judgments were entitled to great deference and that states were free to use any terminology they wished and did not need to use the specific nomenclature of any professional group, such as psychiatrists (*Kansas v. Hendricks*, 1997). Thus, Kansas was permitted to make "mental abnormality," which is *not* a recognized diagnostic term in psychiatry or psychology, a predicate for commitment. Personality disorder is a traditional diagnostic category class, but states are free to define this class differently from psychiatric or psychological standards.

The Court properly looked beyond labels, however, to determine what potentially justifiable ground for civil commitment the criterion represented. In this case, civil commitment was justified because the mental abnormality or personality disorder criterion limited confinement

> ...to those who suffer from a volitional impairment rendering them dangerous beyond their control. The Kansas Act...requires a finding of future dangerousness, and then links that

finding to the existence of a "mental abnormality" or "personality disorder" that makes it difficult, if not impossible, for the person to control his dangerous behavior.... The precommitment requirement of a "mental abnormality" or "personality disorder"...narrows the class of persons eligible for confinement to those who are unable to control their dangerousness. (*Kansas v. Hendricks*, 1997: 358, internal citations omitted)

Thus, loss of control was apparently the crucial nonresponsibility condition, although it was not listed in the Kansas criteria and the state did not need to prove lack of control before imposing commitment. Indeed, this was precisely the type of problem allegedly exhibited by Hendricks, who had a history of multiple convictions for sexual molestation of children and who described himself as having uncontrollable urges to molest children when he was stressed. According to Hendricks, only death could prevent him from acting on those urges when they occured.

In *Kansas v. Crane* (2002), the Supreme Court was asked to decide "[w]hether the Fourteenth Amendment's Due Process Clause requires a state to prove that a sexually violent predator 'cannot control' his criminal sexual behavior before the State can civilly commit him for residential care and treatment."[15] *Crane* thus addressed precisely the question *Hendricks* left unanswered and presented a watershed opportunity for the Supreme Court to clarify both the nonresponsibility condition that justifies civil commitment of sexually violent predators and the constitutional limits on preventive detention.

Justice Stephen Breyer's majority opinion rejected pure preventive civil detention based on dangerousness alone and held that substantive due process required "proof of serious difficulty in controlling behavior" as an independent predicate for the civil commitment of mentally abnormal sexual predators (*Kansas v. Crane*, 2002). Although the Court constitutionalized the lack of control standard, it rejected the argument that the lack had to be "total or complete" because such a standard was unworkable (*Kansas v. Crane*, 2002). The Court reiterated that both the mental abnormality or personality disorder criterion and a lack of control criterion (*Kansas v. Crane*, 2002) were necessary to narrow the class of persons eligible for confinement. These strict eligibility requirements prevent such commitments from becoming mechanisms for retribution or deterrence, which are justifications for criminal punishment but not for civil commitment (*Kansas v. Crane*, 2002). The Court noted that, in *Hendricks*, the presence of an undeniably serious mental disorder that created a "special and serious lack of ability to control behavior" was crucial to justify the civil nature of the commitment (*Kansas v. Crane*, 2002).

Defining the quantum of lack of control necessary to justify these onerous civil commitments thus assumes supreme constitutional importance, but the *Crane* opinion provides little guidance. The relevant language is worth quoting in full:

In recognizing that [lack of control is required], we did not give to the phrase "lack of control" a particularly narrow or technical meaning. And we recognize that in cases where lack of control is at issue, "inability to control behavior" will not be demonstrable with mathematical precision. It is enough to say that there must be proof of serious difficulty in controlling behavior. And this, when viewed in light of such features of the case as the nature of the psychiatric diagnosis, and the severity of the mental abnormality itself, must be sufficient to distinguish the dangerous sexual offender whose serious mental illness, abnormality, or disorder subjects him to civil commitment from the dangerous but typical recidivist convicted in an ordinary criminal case. (*Kansas v. Crane*, 2002: 413)

The Court characterized this language as a description of the inability to control behavior in a "general sense" (*Kansas v. Crane*, 2002: 414).

The Court recognized that this is not a precise constitutional standard, but asserted that constitutional safeguards of liberty in mental health law "are not always best enforced through precise bright-line rules" (*Kansas v. Crane*, 2002: 413). The Court defended this assertion with two arguments. First, states have considerable discretion to define the mental abnormalities and personality disorders that are predicates for civil commitment. Second, psychiatry, which informs but does not control mental health law determinations, is "ever-advancing," and its "distinctions do not seek precisely to mirror those of the law" (*Kansas v. Crane*, 2002: 413–14). Consequently, Justice Breyer concluded, the Court has provided constitutional guidance in the area of mental health law "by proceeding deliberately and contextually, elaborating generally stated constitutional standards and objectives as specific circumstances require" (*Kansas v. Crane*, 2002: 414). Finally, the Court implied, but did not decide, that the Constitution does not require that a serious control problem must be caused by a volitional impairment. The Court suggested that an emotional or cognitive impairment that caused a sufficient control problem would also pass constitutional muster (*Kansas v. Crane*, 2002: 414–415). To summarize, the disease rationale for these commitments was furnished by the requirements that the offender suffer from a personality disorder or "mental abnormality," that the offender has serious difficulty control himself, and that the mental impairment must "cause" the person to have serious control difficulty. As a result of these three requirements, the offender is allegedly not fully responsible for his sexually violent comment and comes squarely within the realm of disease jurisprudence.

Expansion of Sexual Predator Commitments to Violent Offenders

To see why this form of commitment is a potentially promising model for the preventive detention of our three classes of offenders, imagine the following hypothetical changes to the definitions of a predator and of mental abnormality. Recall that the Court indicated that it will be very deferential to legislative judgments in this area. Assume that the legislature has announced, as it did about sexual predators, that our classes of dangerous offenders create a very acute need for social protection that requires special legislation. Then, it defines a "dangerous predator" and "mental abnormality" as follows, in each case using the definitions in sex predator legislation, but simply removing any reference to sex and leaving references to violence:

> ...any person who has been convicted of or charged with a violent offense and who suffers from a mental abnormality or personality disorder which makes the person likely to engage in repeat acts of violence.
>
> ...a congenital or acquired condition affecting the emotional or volitional capacity which predisposes the person to commit violent offenses in a degree constituting such person a menace to the health and safety of others.

In principle, the new statute could cover *all* three classes of offenders because there would be no trouble concluding that most within each class suffer from a mental abnormality or personality disorder. The only question would be whether they have serious difficulty controlling themselves or some other emotional or cognitive impairment resulting from the mental abnormality or personality disorder.

There are grave difficulties, however, with every aspect of the sexual predator commitments that would apply equally to any extension to commitment of dangerous offenders more generally. They employ (1) a problematic differential responsibility standard compared to criminal responsibility; (2) an empty mental abnormality definition; and (3) a vague, unoperationalized nonresponsibility criterion.

The state can of course employ different responsibility standards in different contexts. When indefinite confinement is at stake, however, the state has a heavy duty to justify the lesser standard for responsibility that is the foundation for indefinite commitment. Should not the state be more "forgiving" when it is blaming and punishing for crime than when it is imposing involuntary commitment for people who at the moment cannot be punished for any offense? No state or commentator has yet provided an adequate justification for the distinction. If a potential predator is insufficiently responsible to be left at liberty until he commits another offense, why should he be held criminally responsible for such an offense in the first place? After all, offenders held responsible enough to warrant fully the state's most severe infliction—the imposition of criminal blame and punishment—are now being committed at the end of a prison term justified by *desert* because they are not responsible for precisely the same type of behavior for which they were convicted and punished.

Assuming that there is an adequate justification for the difference, there are problems with all three criteria that support the disease justification. The justifiable purpose of the personality disorder and mental abnormality criterion is to identify those dangerous offenders who are not responsible, but neither criterion will serve the purpose.

"Personality disorder" is a recognized category of psychiatric diagnoses (American Psychiatric Association, 2000: 685), but people with personality disorders rarely suffer on that basis alone from the types of psychotic cognition or extremely severe mood problems that are the standard touchstones for a finding of nonresponsibility. Most are perfectly in touch with reality, their instrumental rationality is intact, and they have adequate knowledge of the applicable moral and legal rules that apply to their conduct.[16] Although their abnormalities might make it harder for them to behave well, they seldom manifest the grave problems that might satisfy an insanity defense or even warrant a commonsense excuse on the ground that the person cannot "help" himself or herself. Even if interpreted to exclude less severe defects, the term would still be overinclusive as a predicate for genuine nonresponsibility in the case of most people who fit within our three classes of violent offenders.

"Mental abnormality," as Kansas defines it, may be constitutionally acceptable after *Hendricks*, but it cannot possibly satisfy the nonresponsibility condition in fact because it would apply to every person who is potentially violent, whether or not the person's conduct warranted a recognized diagnosis. The "mental abnormality" criterion is obscure, circular, and mostly incoherent.

The definition states that a person is abnormal if any genetically inherited/prenatally acquired (congenital) or environmental (acquired through life experience) variable that affects the person's emotional or volitional ability predisposes the person to engage in violent crime. It is not clear what is meant by "emotional" or "volitional" ability. Neither word is a term of art or a technical term in the behavioral or philosophical literature. The former has a common sense, intuitive meaning. In contrast, the concept of volition is extraordinarily vexed.[17] If it refers to the ability of an agent to execute his or her intention, psychopaths have no volitional disability whatsoever (Fingarette & Finagarette Hasse, 1979). If it refers to states of desiring or wanting, it is redundant with the requirement of a "predisposition" to criminal violence. Predisposing cognitive variables were evidently excluded from Kansas's definition, probably because cognitive problems are rarely factors in sexual abnormalities. *Crane* appears to recognize, however, the possibility that cognitive impairments would suffice and a statute could be rewritten or interpreted to include them. If a cognitive factor were included and did seem relevant, as in the case of manifest delusions about what the offender was doing, standard nonresponsibility conditions, such as gross irrationality, would obtain.

Assume that we have clear understanding of the meaning of emotional, volitional, and cognitive abilities. What else would predispose any agent to any conduct—criminal and noncriminal, normal and abnormal—if not biological and environmental variables that affect the agent's emotional, volitional, and cognitive abilities? In other words, the definition is simply a partial, generic description of the causation of all behavior, and it is not a limiting definition of abnormality. All behavior is (partially) caused by emotional, volitional, and cognitive abilities that have themselves been caused by congenital and acquired characteristics. The condition that makes violent predators mentally abnormal— congenital or acquired causes of a predisposition—applies to all behavior and is thus vacuous. It certainly cannot explain why the inevitable presence of congenital and acquired causes of a predisposition means that the agent cannot control and is not responsible for action that expresses the predisposition. Indeed, according to this criterion, no one would ever be responsible for any conduct.

The revised criterion is entirely dependent on the requirement of a specific predisposition to commit violent offenses to limit the definition to violent predators, but it is not a definition of mental abnormality even in the case of violent people. If any agent who has a predisposition to commit violent offenses is mentally abnormal, as the revised definition implies, then the definition of mental abnormality is circular, and abnormality does not independently provide even part of the necessary causal link. The definition presupposes what it is trying to explain. Moreover, such a circular definition collapses the clichéd but important distinction between "badness" and "madness," which is precisely the distinction the definition is meant to achieve to justify civil rather than criminal commitment.

Despite the glaring flaws in this crucial criterion for the disease justification, the Supreme Court has upheld its constitutionality. Let us therefore see how it might apply to our three classes of offenders. The emotional predisposing capacities might straightforwardly describe abnormalities psychopaths exhibit. One reason they may be predisposed to crime is that they have no concern or empathy, which can plausibly be construed as emotional capacities. If cognitive capacities were considered, it is clear that psychopaths are not psychotically out of touch with reality. But if one were to interpret their empathy and conscience impairments as producing cognitive impairments in their practical reasoning about rights infringements, then the conclusion that a cognitive abnormality is present might be warranted. Note again, however, that this interpretation would be inconsistent with criminal responsibility standards. Moreover, it is not clear that people with APD and other dangerous offenders have such emotional and cognitive impairments, but the emptiness of the definition would probably permit diagnosing these people as suffering from "mental abnormality."

Recall that a criterion for these commitments is that there must be a causal link between the mental impairment and serious difficulty controlling oneself or some other non-responsibility condition. When mental abnormality is causally related to legally relevant behavior, such as violent, future conduct, two effects are possible: the abnormality may simply play a predisposing causal role, and the abnormality may undermine the agent's responsibility for the legally relevant behavior. Consider first how a mental abnormality may operate as a predisposing cause of behavior. A mental abnormality does not cause legally relevant bodily movements to become mere biophysical mechanisms, such as a neuromuscular spasm. Abnormal thoughts, desires, perceptions, and the like are not simply irresistible mechanical causes of further conduct, even if, ultimately, biophysical explanations can be given for them (and for normal thoughts, desires, and perceptions). Rather, such abnormalities create irrational reasons for action or compromise the agent's general capacity for rationality or self-control. A mental abnormality thus sometimes plays a causal role by affecting the agent's practical reasoning that leads to the legally relevant behavior. If such irrationality had not existed, the legally relevant behavior would have been less likely to occur. A mental abnormality is not a necessary cause of legally relevant behavior—and it is virtually never sufficient—but it may be a strongly predisposing cause. As we have seen, however, mental abnormalities that impair cognition are likely to play this role only for psychopaths among our three classes of offenders if the defects of psychopaths are interpreted to satisfy the cognitive impairment criterion. Consequently, the "serious difficulty controlling" oneself criterion may assume the paramount role for the other two and perhaps for psychopathy as well.

Lack of self-control was at the heart of the disease justification for quasi-criminal commitment in *Hendricks* and *Crane* so it is crucial to understand what it means and how successfully we can assess it. The rationale for an independent control test is that some agents allegedly do not have rationality defects and therefore cannot satisfy cognitive tests, but they nonetheless cannot control their conduct and therefore are not responsible for the behavior they cannot control. The question for quasi-criminal commitment law is whether an independent control test for excuse or mitigation is conceptually sound and practically feasible. I suggest that at present control tests are poorly conceptualized and cannot be adequately assessed. Thus they are poor predicates for disease jurisprudence within any doctrine, and not just for quasi-criminal commitment.

Four false starts or distractions bedevil clear thinking about control tests: (1) the belief that allegedly uncontrollable behavior is not action;

(2) the belief that behavior must be out of control if it is the sign or symptom of a disease; (3) the belief that the metaphysical argument about free will and responsibility has any relevance to the criminal law problem of whether a control test is necessary; and (4) the belief that causation at any level of causal explanation, including abnormal causation, is per se an excusing condition or the equivalent of compulsion.

Control test cases uniformly involve human action and not mechanism. If the agent's conduct is a literal mechanism, such as a reflex, or if it is performed in a state of substantially clouded or divided consciousness, then the defendant does not act at all and there is no need for a control test. Cases in which a control test seems necessary often have the feature that the relevant conduct, such as sexually touching children and seeking and using drugs, are allegedly symptoms of a disease, but that does not mean that the defendant is not acting. Touching childfen and seeking and using drugs are quintessentially intentional human actions and at least potentially subject to the control of reason.

Conduct is not per se out of control simply because it is the symptom of an alleged disorder. Most signs and symptoms of diseases are literally mechanisms and not human action. Once the disease process begins, one cannot stop it only by intentionally deciding to end it. In contrast, the signs and symptoms for which a control test is allegedly necessary are per se human actions, and simply refraining from acting in the objectionable way is sufficient to end the sign of the disease. If actions that are signs and symptoms of a disease are to be excused because they are involuntary, involuntariness or compulsion must be independently demonstrated to avoid begging the question.

Control tests have nothing to do with free will understood as contra-causal freedom or agent origination. All criminal law responsibility doctrines are compatible with the truth of determinism (Morse, 2007). Control problems must be demonstrated independently of the external, metaphysical debate about free will and responsibility because doctrines of excuse are internal to law. Moreover, if some behavior is randomly caused or the product of indeterminacy, this would not be a secure foundation for responsibility or nonresponsibility. Even if it were, there is no reason to believe that random or indeterminate causation plays a greater role in supposed control test cases.

Causation of behavior is not per se an excusing condition, and it is not the equivalent of compulsion or involuntariness. To believe otherwise is to make "the fundamental psycholegal error" (Morse, 1994). In a causal universe that is massively regular, that satisfies what philosopher Galen Strawson (1989: 12) terms "the realism constraint," all behavior is presumably caused by necessary and sufficient conditions. If causation were per se an excuse or the equivalent of compulsion, then no one could ever be responsible for any behavior. Causation is not the equivalent of compulsion because the non-literal compulsion that control tests address is normative. It applies only to some defendants. All behavior is caused, but only some behavior is compelled. The external critique of all responsibility practices based on universal causation does not explain or improve understanding of positive law.

Even if the causal process is considered "abnormal," it does not follow that the caused behavior cannot be controlled. For example, the dominant biological theory of addiction hypothesizes that persistent use of rewarding substances usurps the brain's normal mechanisms of reward. Even so, lack of control must be proven independently by showing how this account indicates lack of control.

Lack of control must be explained and understood in the terms of folk psychology. Folk psychology refers to the theory of explaining behavior that treats mental states, such as desires, beliefs, intentions, plans, and reasons, as genuinely causal and that treats people as agents who can potentially be guided by reason, who are potentially reason responsive. It is the law's implicit theory of action because all legal criteria presuppose folk psychology. Evidence concerning action, disease or disorder mechanisms, and causation may be relevant to the proof of whether a control problem exists, but the definition of and the criteria for a control problem must be folk psychological. To claim that folk psychology is "wrong" or unscientific is an external attack on all current conceptions of law. Such critiques should be addressed directly and should not be smuggled in partially through a control test.

An adequate, independent folk psychological account of loss of control must fulfill at least five criteria. First, it must be a capacity account. Otherwise, simple failure to exercise the capacity for self-control the agent possesses would be sufficient for excuse, which would be a morally and legally indefensible result. Second, the account must be distinguishable from weakness of will, which is considered a moral failure. Drawing this distinction will be difficult because the definition of weakness of will is fraught. Third, loss of control must be a continuum capacity. It is virtually inconceivable that control capacity would be all-or-none. Fourth, the capacity should be applicable in an ordinary environment broadly conceived. An agent's ability to restrain himself if extraordinary restraining influences were present does not entail that he can control himself under ordinary circumstances. Fifth, the criteria must be folk psychological because the law is resolutely folk psychological.

Virtually all proposed loss of control theories already meet all or most of these criteria, except perhaps the second and last. Finally, a nonconceptual criterion is that the capacity must be practically subject to reasonably objective application.

Let us begin with the phenomenology. Suppose that a person has a powerful desire to do something that it is unwise, immoral, or illegal. That is, the agent really, really, really wants to do something wrong, such as violently attacking another or touching a child sexually. Desires, whether "normal" or "abnormal," may be strong or weak, persistent or sudden. It is of course easier, in the colloquial sense, to behave wisely, morally, and legally if an agent does not have suddenly arising, strong desires. Moreover, failure to satisfy strong desires can cause very unpleasant feelings, such as tension and anxiety. The agent's instrumental practical reason may seem unimpaired when powerful desires arise and virtually all agents who yield to strong and even sudden or surprising desires to behave unwisely, immorally, or illegally fully recognize that yielding is wrong. What does it mean to say that an agent "can't help it" when the agent yields?

Scientific discoveries about behavior often furnish mechanistic causes, but the problem of control remains because causation per se, at any level of causation and whether or not it is "normal," is not an excusing condition or the equivalent of compulsion. Humans clearly have "stop" folk psychological processes that are influenced by mechanistic causes. Successful human interaction would otherwise be impossible. Nevertheless, we still need an adequate, independent folk psychological account of why the psychopath or other dangerous offenders have trouble controlling themselves.

A common approach is to conceive of loss of control as motivational compulsion, as occurring when a desire has too much motivational force to be resisted under ordinary circumstances. The analogy is to overwhelming physical force, but rather than being compelled by external force majeure, the agent is compelled by his own "overpowering" desires. Some desires are stronger than others, but desires are not like external physical forces that physically overwhelm the agent's ability to resist. If this were true, the claim would be no action. The agent who loses control nonetheless acts.

There are not "forces" of desire. Physical forces can bypass intentionality and assent; desires cannot (Watson, 1999). It is more likely that strong desires redirect rather than bypass intentionality. Resisting the desire causes the agent so much effort and discomfort that resisting is not worth the effort, even though it is possible, so the agent collaborates with the desire. This account also fails to distinguish between strong desires because all strong desires appear to be sources of loss of control. Focusing on "abnormal" desires will not solve the problem because "normal" desires may be equally strong and abnormal desires may be weak. Moreover, once the desire is considered resistible with effort, how is this case different from weakness of will? The motivational compulsion account of loss of control leaves all the important issues unresolved.

Another theory hypothesizes that compulsion arises from a conflict between first-order desires, what the agent wants to do now, and second-order desires, the desires that the agent reflectively has about what he should want. Conflict between first- and second-order desires may make it more difficult to avoid acting on one's first-order desires, but this theory has weaknesses and why it is a theory of compulsion is unclear. The observation that an agent is in conflict does not mean that the agent cannot control his conduct unless there is an account of why that conflict produces lack of control.

A promising approach to control difficulty is based on "reason-responsiveness." If an agent cannot be persuaded, actually or hypothetically, by good reasons to avoid acting, or if he cannot bring those reasons to bear, then the agent probably cannot control himself. The reasons must be ordinary reasons or the criteria would be too demanding. A gun at the head would constitute an extraordinary reason. If the agent can control himself in such circumstances, it would not follow that the agent could control himself in ordinary circumstances.

Although this account is subject to objections and difficulty distinguishing weakness of will, it is intuitively appealing because it does not suffer from the dis-analogy to physical force and because it provides a common sense folk psychological process for loss of control. Nonetheless, to the extent it is valid, it is a rationality account. The capacity to grasp and be guided by good reason is the heart of normative rationality. Once again, this might be an attractive characterization of the deficits that psychopaths have.

A final theory for an independent self-control failure is the analogy to the two-party excuse of duress, but we do not excuse in these cases because the agent had a volitional or control problem. The agent's reasoning is intact, and his will operated effectively to save him from the threat. We excuse the agent because he faced a dreadfully hard choice for which he is not responsible, and we could not fairly expect him not to yield. The agent faced with the threat of frustration of strong internal desires is essentially claiming an "internal duress" excuse. Such accounts may seem plausible for "disorders of desire," such as addiction and the paraphilias, but it does not seem applicable to psychopaths and the other types of offenders under consideration in this chapter.

Proponents of an independent control test have not yet provided a persuasive folk psychological account independent of a rationality problem. In addition, control tests suffer from the defect that I have termed "the lure of mechanism," the tendency to analogize allegedly out of control agents to literal mechanisms. Sophisticated proponents do not do this, but many academic lawyers, practitioners, and mental health experts do. The usual basis is the mistaken belief that if behavior is caused, the agent could not have acted otherwise. Control tests inadvertently fuel this pernicious problem because they mask the difference between the folk psychological sense of loss of control and the metaphysical question of whether determinism or universal causation undermines all deontological responsibility.

Control tests also raise difficult practical problems. Even the American Psychiatric Association (1983) supported the movement to abolish control tests for legal insanity on the ground that it was impossible to evaluate lack of control objectively. Recall that in *Crane*, Justice Breyer provided a typically thin and seemingly commonsense test.

> …we did not give to the phrase 'lack of control' a particularly narrow or technical meaning. And we recognize that in cases where lack of control is at issue, 'inability to control behavior' will not be demonstrable with mathematical precision. It is enough to say that there must be proof of serious difficulty in controlling behavior.(*Kansas v. Crane*, 2002: 413)

It would have been harder for Justice Breyer to do better because there is essentially nothing to say that is not conclusory or circular. (See Pierson, 2011, for a review and analysis of how state courts have interpreted Justice Breyer's test.) There is no consensual scientific definition or measure of lack of control. Nor is there yet an adequate folk psychological process that has been identified as normatively justifiable for legal purposes.

Justice Breyer's vague and unhelpful "serious difficulty" control criterion was the wrong test. How would a fact finder know if the defendant had serious difficulty controlling himself except on the bases of the defendant's self-report and observations of the defendant's seemingly self-destructive conduct? Justice Scalia's dissent observed that the test would give trial judges "not a clue" about how to charge juries. Justice Scalia speculated that the majority offered no further elaboration because "elaboration...which passes the laugh test is impossible" (*Kansas v. Crane*, 2002: 423). Justice Scalia wondered whether the test was a quantitative measure of loss of control capacity or of how frequently the inability to control arises. In the alternative, he questioned whether the standard was "adverbial," a descriptive characterization of the inability to control one's penchant for sexual violence. The adverbs he used as examples were "appreciably," "moderately," "substantially," and "almost totally" (*Kansas v. Crane*, 2002: 423–424). Justice Scalia's common sense criticism of the test was apt. To date, advocates of an independent control test have not demonstrated the ability to identify "can't" versus "won't."

We do talk colloquially about and appear to have an everyday understanding of loss of control, but we do not, in fact, have a good understanding. Moreover, successful human interaction does not depend on successfully assessing control capacity. Even when we appear to be making common sense, ordinary judgments of lack of self-control, the psychological process is unspecified. If it were analyzed, rationality impairment would appear.

If one examines most cases of alleged "loss of control," they raise claims that, for some reason, the agent could not "think straight" or bring reason to bear. The "control" language used in *Crane* and in other cases and statutes is metaphorical and better understood in terms of rationality defects. Human beings control themselves by using their reason. "Stop" mechanisms are primarily cognitive. If agents cannot use their reason, it is difficult to behave properly and why some people seem "out of control." I suggest again that this is the best understanding of why psychopaths may have difficulty "controlling themselves:" They do not have access to empathy, concern, and conscience that give agents the normatively best and empirically most motivating reasons not to harm others.

The abnormality, causal link, and serious control difficulty criteria are not adequate nonresponsibility standards. They cannot conceivably limit quasi-criminal commitment only to those mentally abnormal potential violent predators who cannot control themselves and thus are not responsible for their potential violence. Using such criteria, virtually every predator would be both convictable and committable. This would be unjust.

Even if we could limit the class of offenders who somehow properly met the quasi-criminal commitment criteria, we would still have the problems of prediction and treatment that all preventive detention justifications present. Although psychopathy is a serious risk factor for crime and enhances the probability of recidivism, there will still be large numbers of false positives, adequate treatments do not exist, and gatekeepers will tend to be conservative. The result will be lengthy and often lifelong commitments for people who might be released earlier and lengthy commitments for some offenders who might not be dangerous at all. And unlike recidivist enhancements, which suffer from the same defects but are at least based on intentional wrongs for which the offender is culpable, the ground for preventive detention in this case—psychopathy or some other disease criterion—is a disorder, an attribute of the person for which the offender is not responsible. Many psychopaths committed quasi-criminally might spend the rest of their lives unnecessarily in institutions at immense cost to them personally and at immense fiscal and moral cost to society.

PURE PREVENTIVE DETENTION

Suppose the law were to abandon desert-disease jurisprudence wholly or in part in order to protect society more adequately. What would result? It is beyond the scope of this chapter to explore this speculation in great detail, but a brief sketch will be instructive.

To abandon desert-disease jurisprudence altogether would be a radical change in the relation between the state and the individual. Notions of autonomy and of genuine positive and negative desert and their consequences—now a bulwark of liberty and a core of interpersonal relations—would be abandoned for a pure prediction-prevention, "screen and intervene" system of social control. Prior criminal or other antisocial conduct would no longer be necessary as a precondition for state intervention unless such conduct was necessary for reasonably accurate prediction. But that is simply an empirical question. If it were not necessary, pure prediction would suffice. What degree of intervention the state would be warranted in imposing would be an open question. Citizens might still retain some rights against particularly intrusive treatments or the like, but what would prevent the state from imposing indefinite confinement upon those who are predictively dangerous, cannot be controlled in the community, and can otherwise be treated to avoid incapacitation only by extremely intrusive methods that virtually anyone would rather avoid? What would prevent the state from wholesale screening of children for violent propensity, followed by interventions for those apparently at risk. We already permit mandatory vaccination of children to protect them and others from infectious diseases. Moreover, once the state begins to treat people more like objects and less like people, the structure and meaning of rights will surely and substantially change. Let us assume, then, that the state will not abandon desert-disease jurisprudence entirely.

The law already allows pure preventive detention in a wide variety of contexts, but they are virtually all limited in temporal scope and borrow from desert-disease jurisprudence to some extent. Denial of bail on grounds of dangerousness is an excellent example (*United States v. Salerno*, 1987), but the accused is brought to trial relatively quickly and the incarceration is justified by probable cause to believe that the accused has culpably committed a criminal offense. Suppose, however, that society could identify an identifiable class of people, say those with certain recidivist records or patterns of childhood misconduct, for whom a subset would be predicted with great accuracy to commit very violent acts unless they were incapacitated or otherwise treated. Would society be justified in screening people in that class and purely preventively intervening for those predicted to be violent, even if the person was responsible and no criminal punishment was justified at the time?

The argument in favor would be that although desert-disease jurisprudence protects liberty and autonomy interests that we cherish and that are constitutionally protected, no individual right, including those protected specifically by the constitution, is absolute. Any might yield in the face of a sufficiently compelling state interest. Preventing serious violence is certainly a compelling state interest. I suspect that the political and constitutional acceptability of screening of limited classes followed by preventive intervention would depend entirely on the accuracy of the screening and predictive methods. If the classes screened were strictly limited by clear criteria that were not racially or class-biased, the predictions were exceptionally accurate—very few false positives—and especially if there were non-incapacitative, nonintrusive successful interventions possible, I expect that such screening and preventive detention would be upheld.

CONCLUSION: PREVENTIVE DETENTION AND FUTURE SCIENCE

Despite vast amounts of research, the ability accurately to predict serious violence over considerable time periods is limited at present. Current technologies are improving, especially with mechanical or highly structured clinical methods of prediction, but we are far from reaching the levels of sensitivity that could justify widespread pure preventive detention. As we get better, the positive result will be that there will be fewer false positives and unnecessary loss of liberty using traditional desert-disease jurisprudence. But if we ever reach the exceptional levels of accuracy described in the last section, increased abandonment of desert-disease jurisprudence could result. Although predictability does not mean nonresponsibility, the lure of accurate social engineering may be irresistible. Some important protections for liberty and autonomy may hang by a technological thread. For now, however, desert-disease jurisprudence remains the template for thinking about how society may protect itself consistent with human rights. The cost of such protection, however, is reduced public safety. Such trade-offs are inevitable in a free society, which believes that liberty is worth substantial costs. The best hope for the future is that we can discover rehabilitation techniques that will lower the risk of violent offending in our three classes of offenders.

ACKNOWLEDGMENTS

I thank Ed Greenlee for his invaluable help.

NOTES

1. For an attempt to argue this point, see Wilson (1998).
2. Finally, the economy of threats approach makes the world entirely too "safe for determinism." The determinist anxieties that seem inevitably to arise cannot be banished so easily, without doing violence to our conceptual concerns. A full, satisfying account of responsibility and blaming, paradoxically, should be subject to anxieties about determinism.
3. I have explored the civil–criminal distinction as a basis for confinement elsewhere and will therefore provide only the briefest sketch here. See Morse (1996, 1999, 2002).
4. The nonpunitive characterization of such interventions often justifies lesser procedural protections for the potential subject. See, for example, Allen v. Illinois (1986). (Fifth Amendment guarantee against compelled self-incrimination does not apply in a proceeding to determine whether a person is a "sexually dangerous person" because the proceeding is not "criminal").
5. For the English experience, see Ashworth (2000).
6. Compare, for example, Rummel v. Estelle (1980), upholding life imprisonment with possibility of parole for a three-time offender whose third crime was larceny of a small amount by false pretenses, with Solem v. Helm (1983), rejecting life imprisonment without the possibility of parole for a seven-time offender whose seventh offense was uttering a no account check for a small amount of money.
7. See Davis (1992), noting the scarcity of retributive justifications and offering an account of the "special advantage" recidivists receive by reoffending.
8. There is a strong argument, based on our history of discrimination and negative stereotyping, that race is the only variable that might have validated predictive validity that nevertheless should not be used (Monahan, 2006).
9. For example, Parsons v. State (Ala. 1887). In his great history of the English criminal law, Sir James Fitzjames Stephen advocated for a control test in England (Stephen, 1883). Careful examination of his argument discloses, however, that the test was in fact a cognitive or rationality test.
10. American Law Institute, Model Penal Code §4.01 (1).
11. American Law Institute, Model Penal Code §4.01(2).
12. This point was recognized by at least one United States court (United States v. Currens, 1961: 774).
13. Consider the remarks of Justice Owen Dixon of Australia in King v. Porter (1933:187):

    [A] great number of people who come into a Criminal Court are abnormal. They would not be there if they were the normal type of average everyday people. Many of them are very peculiar in their dispositions and peculiarly tempered. That is markedly the case in sexual offenes [sic]. Nevertheless, they are mentally quite able to appreciate what they are doing and quite able to appreciate the threatened punishment of the law and the wrongness of their acts, and they are held in check by the prospect of punishment.

14. Kan. Stat. Ann. §§ 59-29aO1-59-29a20 (Supp. 2000). Kansas amended its statute after Kansas v. Hendricks was decided. The version of the statute considered by the Court, which can be found at Kan. Stat. Ann. § 59-29aO1-59-29a15 (1994), applied to mentally abnormal sexual predators.
15. Petition for a Writ of Certiorari to the Supreme Court of Kansas at i, Kansas v. Crane (2002, No. 00–957).
16. In many cases, the conduct that is the basis for the diagnosis does not per se cause the person distress. For example, an agent whose conduct warrants the diagnosis of Antisocial Personality Disorder may be distressed by the reactions of the police, creditors, and others, but the conduct itself might not be distressing. Moreover, the degree of distress or impairment such disorders cause is very much a function of the particular social, moral, and legal regime in which the person lives, which once again suggests the highly value-relative nature of the judgment of disorder in these cases.
17. The meaning of volition is controversial in philosophy and psychology. See Moore (1993), providing the most extensive discussion of volition in the legal literature, criticizing the view that volitions are desires, and arguing that a volition is an intention to execute a basic action.

REFERENCES

Allen v. Illinois, 478 U.S. 464 (1986).
American Law Institute (1985). *Model Penal Code*. Philadelphia, PA: Author.
American Psychiatric Association (2000). *Diagnostic and statistical manual of mental disorders* (4th ed., text rev.). Washington, DC: Author.
American Psychiatric Association Insanity Defense Working Group (1983).American Psychiatric Association statement on the insanity defense. *American Journal of Psychiatry, 140,* 681–688.
Appelbaum, P. S. (2005). Dangerous severe personality disorders: England's experiment in using psychiatry for public protection. *Psychiatric Services, 56,* 397–399.
Ashworth, A. (2000). *Sentencing and criminal justice* (5th ed.). Cambridge: Cambridge University Press.
Cleckley, H. (1976). *The mask of sanity: An attempt to clarify some issues about the so-called psychopathic personality* (5th ed.). St. Louis, MO: Mosby.
Davis, M. (1992). *To make the punishment fit the crime: Essays in the theory of criminal justice*. Boulder, CO: Westview.
Dennett, D. (1984). *Elbow room: The varieties of free will worth wanting*. Cambridge, MA: MIT Press.
Duff, A. (1998). Dangerousness and citizenship. In A. Ashworth & M. Wasik (Eds.), *Fundamentals of sentencing theory: Essays in honour of Andrew von Hirsch* (pp. 141–163). Oxford: Oxford University Press.
Ewing v. California, 538 U.S. 11 (2003).
Fingarette, H., & Hasse, A. F. (1979). *Mental disabilities and criminal responsibility*. Berkeley: University of California Press.
Foucha v. Louisiana, 504 U.S. 71 (1992).
Glannon, W. (2008). Moral responsibility and the psychopath. *Neuroethics, 1,* 158–166.
Hare, R. D. (2003). *Manual for the psychopathy checklist–revised* (2d ed.). Toronto, Ontario, Canada: Multi-Health Systems.
Hart, H. L. A. (1968). *Punishment and responsibility: Essays in the philosophy of law*. New York: Oxford University Press.
Jones v. United States, 463 U.S. 354 (1983).
Kansas Statutes Annotated § 59-29a01-59-29a15 (1994).
Kansas Statutes Annotated §§ 59-29a01-59-29a20 (Supp. 2000).
Kansas Statutes Annotated § 59-29a02(a) (2000 Cum.Supp.).
Kansas Statutes Annotated § 59-29a02(b) (2000 Cum.Supp.).
Kansas v. Crane, 534 U.S. 407 (2002).
Kansas v. Hendricks, 521 U.S. 346 (1997).
King v. Porter, 55 C.L.R. 182 (1933).
Lippke, R. (2008). No easy way out: Dangerous offenders and preventive detention. *Law and Philosophy, 27,* 383–414.
Litton, P. (2008). Responsibility status of the psychopath: On moral reasoning and rational self-governance. *Rutgers Law Journal, 39,* 349–392.
Maibom, H. (2008). The mad, the bad, and the psychopath. *Neuroethics, 1,* 167–184.
McAlinden, A-M. (2001). Indeterminate sentences for the severely personality disordered. *Criminal Law Review,* 108–123.
M'Naghten's Case, 8 Eng. Rep. 718 (H.L. 1843).
Monahan, J. (2006). A jurisprudence of risk assessment: Forecasting harm among prisoners, predators, and patients. *Virginia Law Review, 92,* 391–435.
Monahan, J., et al. (2001). *Rethinking risk assessment: The MacArthur Study of Mental Disorder and Violence*. New York: Oxford University Press.
Moore, M. S. (1993). *Act and crime: The philosophy of action and its implications for criminal law*. New York: Oxford University Press.
Morse, S. J. (1994). Culpability and control. *University of Pennsylvania Law Review, 142,* 1587–1660.
Morse, S. J. (1996). Blame and danger: An essay on preventive detention. *Boston University Law Review, 76,* 113–155.
Morse, S. J. (1999). Neither desert nor disease. *Legal Theory, 5,* 265–309.
Morse, S. J. (2002). Uncontrollable urges and irrational people. *Virginia Law Review, 88,* 1025–1078.
Morse, S. J. (2007). The non-problem of free will in forensic psychiatry and psychology. *Behavioral Sciences & the Law, 25,* 203–220.
Morse, S. J. (2008). Psychopathy and criminal responsibility. *Neuroethics, 1,* 205–212.
Parsons v. State, 2 So. 854 (Ala. 1887).
Petition for a Writ of Certiorari to the Supreme Court of Kansas, *Kansas v. Crane* (2002, No. 00–957).
Pierson, J. (2011). Construing *Crane*: Examining how state courts have applied its lack-of-control standard. *University of Pennsylvania Law Review, 160,* 1527–1560.
Pillsbury, S. H. (1992). The meaning of deserved punishment: an essay on choice, character, and responsibility. *Indiana Law Journal, 67,* 719–752.

Rummel v. Estelle, 445 U.S. 263 (1980).

Schulhofer, S. J. (1996). Two systems of social protection: Comments on the civil-criminal distinction, with particular reference to sexually violent predator laws. *Contemporary Legal Issues, 7*, 69–96.

Shapiro, S. J., (2000). Law, morality and the guidance of conduct. *Legal Theory,6*, 127–170.

Skeem, J. L. & Monahan, J. (2011). Current directions in violence risk assessment. *Current Directions in Psychological Science, 20*, 38–42.

Solem v. Helm, 463 U.S. 277 (1983).

Stephen, J. F. (1883). *A history of the criminal law of England* (Vol. 2). London: Macmillan and Company.

Strawson G. (1989). Consciousness, free will, and the unimportance of determinism. *Inquiry, 32*, 3–27.

United States v. Currens, 290 F.2d 751 (3d Cir. 1961).

United States v. Marble, 940 F. 2d 1543 (D.C. Cir. 1991).

United States v. Salerno, 481 U.S. 739 (1987).

United States v. Sell, 539 U.S. 166 (2003).

Walen, A. (2011). A punitive precondition for preventive detention: Lost status as a foundation for lost immunity. *San Diego Law Review, 48*, 1229–1272.

Wallace, R. J. (1994). *Responsibility and the moral sentiments.* Cambridge, MA: Harvard University Press.

Washington v. Harper, 494 U.S. 211 (1990).

Watson, G. (1999). Disordered appetites: Addiction, compulsion and dependence. In: J. Elster (Ed.), *Addiction: Entries and exits* (pp. 3–28). New York: Russell Sage Foundation.

Wilson, E. O. (1998). *Consilience.* New York: Knopf.

Yaffe, G. (2011). Prevention and imminence, pre-punishment an actuality. *San Diego Law Review, 48*, 1205–1228.

CHAPTER 18

Some Notes on Preventive Detention and Psychopathy

Michael Louis Corrado

By "preventive detention" I mean indefinite detention that is aimed at preventing future crimes by the person detained, and not intended to punish for past crimes. Commitment of the mentally ill, when designed to prevent harm to themselves or others, is a form of preventive detention; they are not being detained as punishment for past crimes, though past crimes may serve as evidence of the dangerousness that accounts for their detention. "Pure" preventive detention is detention of those who are not legally insane, detention inflicted for the purpose of preventing harm to themselves or others and not justified by prior criminal behavior (though again prior criminal behavior may serve as evidence of dangerousness). Indefinite detention of sexual predators, after they have been adjudged legally sane and have served a prison term for their crimes, would be an example of pure preventive detention.

THE JUSTIFICATION OF PREVENTIVE DETENTION

Generally

Philosophy students often ask to write papers on preventive detention. I usually discourage them, because my own efforts in this area, an area I have revisited a number of times, have been so disappointing (Corrado, 1996-a, 1996-b, 1999, 2005-b). Richard Lippke for the most part summed up everything we know about it in a recent paper, the discouraging title of which was "No Easy Way Out" (Lippke, 2008). There is no generally recognized theory of preventive detention to compare to retributive and consequentialist theories of punishment. Yet if a theory is needed anywhere, a justifying theory, it is here. For one thing, preventive detention, just like punishment, takes away freedom, and a liberal society does not do that sort of thing without justification. For another, a theory of justification carries with it its own limits, and can tell us how far we can go. That is the role that retribution or fairness is supposed to play in the theory of punishment.

We do not snatch theories out of thin air, of course. Our theories owe as much to our intuitive judgments as our final judgments owe to our theories. We begin with some deeply held beliefs, and posit a theory to account for them; from then on it is a matter of give and take between theory and belief. So, for example, theories of punishment work with an established institution of punishment and with our intuitions about what is appropriate punishment and what is not. If the theory does not match the intuitions about the institution, then either the theory must be rejected or some of the intuitions must be changed.

The problem with preventive detention, of course, is that there is no established and generally accepted institution, and our intuitions are weak. There is commitment of the legally insane: That's an established institution, and our intuitions about it are relatively strong. Even there, of course, it is difficult to say precisely what justifies the detention. We want to take those who might be dangerous off the street, but that is only one side of justification. Where do we find the limits? What plays the role, in the commitment of the mentally ill, that desert plays in punishment? There is still no generally accepted theory to point to.

But when we leave the field of those acknowledged to be legally insane and enter the domain of preventive detention generally, the case is much worse. As for our intuitions about the matter, there is simply no consensus. I find among my students little agreement even about whether those who are legally insane should be preventively detained or punished, and of course pure preventive detention, detention of the legally sane on grounds of dangerousness alone, is even more controversial. It would mark some progress in the area we are discussing today if we could agree that the psychopath fits the legal definition of insanity, but there is controversy there. Richard Lippke, for example, in his excellent paper, states: "I take it to be fairly well-established that psychopaths are not appropriate subjects of retributive legal punishment" (Lippke, 2008: 391, n. 16; see also Morse, 2002). On the other hand, Richard Bonnie assumes, as many people do, that psychopaths should not be considered legally insane: In defending his version of the insanity test, Bonnie says that it might be "theoretically accurate to say that a free-standing 'appreciation' test [of insanity], not predicated on a psychotic condition, might include psychopaths." But he finds that outcome to be unacceptable, and he would exclude it by limiting the insanity defense to psychotic conditions; for though the psychopath *may* be psychotic, he need not be (Bonnie, 1998.[1] The important thing to notice is that both think that the assumption they are working from is more or less obvious.

And what of the "institution" of preventive detention? I think it is fair to say that there is no such generally recognized institution. Even the most well-established form of preventive detention, the commitment of the insane, has begun to lose ground. Of five states that have recently tried to eliminate the insanity defense, four have succeeded, and there is no indication that the Supreme Court will find such a defense to be an entrenched part of our tradition.

Although the insanity defense survives in most states, at least for the present, there is even less consensus about whether we should limit indefinite detention for the purposes of criminal justice to insanity acquittees. In Kansas and other states, legislation provides for preventive detention of sexual predators who are presumed not to be mentally ill, but are considered to be undeterrable. The Supreme Court has insisted that the state must demonstrate that those who are to be detained in this way lack control over their behavior, which only adds to the confusion.

The sex predator legislation creates a class of persons who, due to a psychological disorder, cannot control their dangerous behavior but, at the same time, are not legally insane. If they were insane, as we might think someone who because of a psychological condition cannot control his behavior is insane, then they could not be punished, and in fact under this legislation they *are* punished before they are subjected to preventive detention. But if they were sane, we might think, then they could not be preventively detained. It might seem obvious that because they are either sane or insane, they can be punished or detained, but not both; the legislation is incoherent. But it may be, instead, that it is our basic suppositions that will have to be abandoned, our suppositions that only the sane may be punished and, in particular, that only the insane may be detained.

Preventive Detention of Psychopaths

It is not clear whether a consensus will be reached soon as to what constitutes the essential features of psychopathy—not the observable traits associated with psychopathy, but the deep defining features—but I take it that when we refer to the psychopath we are referring to someone who does not respond to moral considerations, but who would satisfy the requirements of sanity in most states. Theoretically that makes the psychopath a wonderful test for any theory of preventive detention; but it also makes the question how to handle the psychopath, who might be quite dangerous, a pressing problem.[2]

Without any firm intuitions to hold onto, and without any generally accepted institution to guide us, we remain adrift. In the hopes of bringing us closer to shore, I look at two very different ways of attacking this question of the justification of the preventive detention of the psychopath. It may be that in this narrow area some progress is possible. I consider the work of four different scholars who have waded into these waters:, Stephen Morse, Jeffrie Murphy, Christopher Slobogin, and Paul Litton.[3]

The first approach starts from the supposition that preventive detention should be reserved for those who are not responsible for what they do. Those who are sane and have committed a crime may only be punished, and may not be preventively detained. Those who are not responsible, on the other hand, and who are dangerous may be preventively detained. On this line of thought, there is, or rather should be, no such thing as pure indefinite preventive detention: preventive detention is reserved for those who are not responsible. The question then becomes whether the psychopath is responsible for what he does. I call this the responsibility approach.

The second approach is to assume that preventive detention should be extended to those who will not or who cannot respect the rights of others, whether or not they are responsible for what they do. To be punished for a definite period of time is a right, and those who do not recognize rights should not themselves be granted any rights. On this view, punishment and preventive detention are not mutually exclusive. There is no reason why a person without rights might not be both punished and subsequently detained. Indeed, there is no reason, as the most famous of those who have held this view has said, why we might not even subject the psychopath "(when the case is hopeless) to painless extermination" (Murphy, 1972).[4] Those who take this approach sometimes talk about responsibility, but it is clear that for them determinations of responsibility follow from determinations about the susceptibility to preventive detention, and not the other way around. I call this the rights approach.

Both approaches aim at the justification of preventive detention, and both assume that there are limits to justified detention. As we will see, however, the implications of the two approaches are very different. One last note by way of introduction: this discussion of preventive detention should make clear that the choice is not between punishing psychopaths for their crimes and letting them go free. There is a common fear that lawyers will attempt to use neuroscience to "get their clients off." We can see this attitude in an important recent paper on the relationship between neuroscience and law:

> To span the divide between the way neuroscientists describe mental states and the way the law applies them, we must develop a set of rules for evaluating when a defendant's neurological profile meets or fails a particular legal requirement. *Too liberal and the guilty run free; too strict, and the ill and innocent suffer imprisonment or death.* (Aharoni, Funk, Sinnott-Armstrong,& Gazzaniga, 2008; emphasis added)

But at least in the case of the psychopath a legal finding of nonresponsibility will not result in the offender running free. It will result in indefinite detention; psychopathy would fall, along with psychosis, into the realm of legal insanity.

THE RESPONSIBILITY APPROACH

Morse: Preventive Detention and Responsibility.

Let's begin with the responsibility approach, which,, is the approach to preventive detention adopted in many of the papers of Stephen Morse (2002). According to this approach, preventive detention should be limited to those who are not criminally responsible. I will assume that criminal responsibility is not a purely consequentialist notion, and that criminal responsibility will depend in part upon moral responsibility. According to this approach, an agent may be indefinitely detained on grounds of dangerousness if and only if he is not morally responsible for what he does.[5] If he *is* morally responsible, then the only control the state may exercise over him, for the purpose of controlling crime, is by means of the deterrent effect of the threat of punishment. The imposition of punishment is both justified by and limited by the notion of responsibility. Conversely, the imposition of preventive detention is limited, and perhaps justified, by the nonresponsibility of the person detained.[6]

The psychopath presents a problem for this approach. The psychopath, if he exists, does not respond to moral reasons for acting. Is he, then, morally responsible for what he does, or not? The law itself appears to be of two minds about this. In most jurisdictions that have an insanity defense, a mental disorder that substantially interferes with the ability to understand or appreciate that his act is wrong is a sufficient condition for finding an agent legally insane.[7] In Nevada the Supreme Court struck down legislation that would have abolished the insanity defense on the ground that the ability to understand the moral wrongness of the act is an element of every crime (*Finger v. State,* 2001).[8] If the psychopath does not understand that his act is wrong, then by the very terms of the law he should not be considered responsible for his criminal acts.

On the other hand, there is also reason to think that the law would consider the psychopath responsible. In practice the psychopath is not treated as legally insane, and I am not sure that this position, represented by the quote from Professor Bonnie, is not the dominant point of view among legal theorists. Although Bonnie has been the main advocate of an expanded understanding of the knowledge of wrongness requirement in the insanity defense, he would explicitly exclude the psychopath from its benefit (Bonnie, 2003: 60–61).

Litton: Responsibility and the Psychopath

The answer, on the responsibility approach, will depend upon just what role responsiveness to moral reasons plays in our definition of responsibility. There is a wonderful exploration of this question in a recent article by Paul Litton (2008). Much of what follows in this section comes from reflecting on what he has to say about psychopathy in that article.

The fundamental question is whether moral responsiveness is an *element* of responsibility. Litton's answer to that question is in the affirmative: Moral responsiveness is an element of responsibility, so that the psychopath, defined as one who is not morally responsive to the rights of others, is not responsible for his criminal behavior. Litton would distinguish between the person who does not understand moral concerns and the person who, though he understands moral concerns, is not motivated by them. The first is the psychopath; the second, says Litton, is the sociopath. The first comes into the world without a capacity to understand moral reasons. The second, the one for whom moral reasons have no motivational force, has had his responsiveness to moral reasons beaten out of him, either physically or psychologically.

The importance of the distinction for Litton is this: The person who does not understand moral considerations, the psychopath, is not fully rational, because an understanding of moral considerations is a necessary condition of rationality. Therefore on any view the psychopath is not responsible for what he does, because rationality is by general consensus an element of responsibility. The person who understands but is not motivated by moral considerations, the sociopath, may or may not be fully rational, but in any case he is not a psychopath. As Litton describes the psychopath, he is incapable of planning or, when he does plan, of carrying out his plans. He interferes with his own aims; he is impulsive; he is generally unsuccessful in pursuing his own interests. it is precisely because he is unresponsive to evaluative factors that he is so ineffective.

The argument is as follows:

1. One who, through some sort of mental defect, is not responsible for his behavior may not be punished, but may be preventively detained.
2. Psychopaths lack the ability to understand moral concerns.
3. The psychopath's inability to understand moral concerns is a mental defect that interferes with his ability to reason practically, which is essential for responsibility.
4. Therefore the psychopath is not responsible, and may be preventively detained but may not be punished.[9]

For Litton, then, as with all responsibility theorists, the right to detain depends upon the absence of responsibility and is inconsistent with the right to punish. Where Litton differs from some responsibility theorists is this: He believes that responsibility depends upon the ability to respond to moral concerns. If the psychopath is indeed incapable of responding to moral concerns, then a responsibility theorist who accepted Litton's assumption would detain but not punish the psychopath. It is possible to be a responsibility theorist without accepting that particular premise.

Conclusions

How should we judge the moral responsiveness requirement for criminal responsibility? I do not think a good case can be made for it. I would confine the rationality required for responsibility to purely prudential forms of reasoning. Why should the ability to reason prudentially not be enough for criminal responsibility? It is sufficient if the actor is able to respond to any reason for acting, even something as raw and blunt as the criminal law. The law is aimed at those who can respond, but it is not aimed only at those who can respond to moral concerns.

That appears to be the view of Kant, whose philosophy of law was not generally very forgiving. According to Kant, it must be possible to organize a state even for a race of devils. He says:

> The problem of organizing a state, however hard it may seem, can be solved even for a race of devils, if only they are intelligent. The problem is: 'Given a multitude of rational beings requiring universal laws for their preservation, but *each of whom is secretly inclined to exempt himself from them,* to establish a constitution in such a way that, although their private intentions conflict, they check each other, with the result that their public conduct is the same as if they had no such intentions.'
>
> A problem like this must be capable of solution; it does not require that we know how to attain the moral improvement of men but only that we should know the mechanism of nature in order to use it on men, organizing the conflict of the hostile intentions present in a people in such a way that they must compel themselves to submit to coercive laws. Thus a state of peace is established in which laws have force. (Immanuel Kant, *To Perpetual Peace*)

If the law were aimed only at those capable of moral reasoning, then the devils he speaks of would be outside the reach of the law, and consequently candidates for preventive detention.

My take on the matter then is this: If the psychopath is as described, a person who does not respond to moral considerations, whether because he does not understand *or* because he lacks motivation, but who is fully capable of prudential reasoning and able to control his behavior, then he is responsible for what he does, and if we were to accept the responsibility approach to the matter, it would follow that he may not be preventively detained.

But this is perhaps the point at which to raise the questions I mentioned earlier about the nature of the psychopath—and here, I think, is where neuroscience enters the picture: Is he in fact fully capable of prudential reasoning, and is he in control of his behavior? As to the first question, it seems to me that we are given two very different pictures of what psychopathy entails. One is of a monster, a cold-blooded criminal, a careful calculator, who feels no remorse and has no empathy, who is capable of fooling all but the most experienced investigator. The second is the picture of a schlemiel, or perhaps a "schlemiel-schlemozzle," who is incapable of carrying out plans, who steps on his own toes, who defeats his own purposes.[10] If this latter is the correct picture of the psychopath, then indeed it may be the case that the psychopath is not responsible for what he does. He may be incapable of prudential reasoning; more would have to be said. But even if it turns out that he is not responsible, it will not be because he is incapable of moral feelings; it will be precisely because he is incapable of prudential reasoning. The monster, the person who acts deliberately and cleverly and utterly without moral sentiment, if there is such a person, should be accountable for what he does; therefore it does not seem to me that the capacity to respond to moral concerns should be an element either of prudential reasoning or responsibility.

The other point has to do with control. It is generally assumed that the psychopath is in control of his behavior, to the same extent that the normal person is. If that is not true, then again I would agree that the psychopath is not responsible for what he does, because it seems to me that rationality is not, in the traditional view of things, the only condition of responsibility. Control is also required. Here again recent neuroscientific work would seem to have something to say about responsibility: If it can be demonstrated that the psychological traits associated with psychopathy stem from a lack of the apparatus of control, then there is a prima facie case to be made that those traits show a lack of responsibility. On the responsibility approach, that would qualify the psychopath for detention and would rule out punishment.

In sum, if his lack of moral responsiveness is the only disability the psychopath suffers from, then I believe that under traditional views of responsibility he is responsible for his behavior. On the other hand, if I did believe that the psychopath was not responsible for what he did, several things would follow on this first approach. First, the state would be entitled to preventively detain him, but would not be entitled to impose punishment, either by itself or in conjunction with preventive detention. For on this approach, the state may punish only those who are responsible for what they do. If we are persuaded that the responsibility approach is the right one, and that psychopaths are not responsible, then it might make sense to expand the insanity defense to include psychopaths.

Second, the state would not, by reason of the psychopath's lack of responsibility and the state's consequent obligation to prevent him from harming others, be released from its obligations to the psychopath himself. The psychopath, like the psychotic individual, would become a ward of the state, and where treatment was possible the state would not be released from its obligation to treat him, to offer him any other benefits not inconsistent with his condition, and to make him as comfortable as possible. Finally, I would argue for a right to compensation (Corrado, 1996-b: 813–814; Schoeman, 1979).

To repeat my own view of these points, if the psychopath's only deficit is in moral responsiveness, then under traditional accounts of responsibility he must be held responsible. Under the responsibility approach to preventive detention that means that he may not be preventively detained. I now look at the rights approach has to say about this.

A Footnote on Responsibility and Control

We must be careful not to confuse control with metaphysical or contracausal freedom; we know enough about the operations of the brain to know that even if human beings do not have an unmoved mover's power to originate action, we may still distinguish between actions they have control over and those that they do not (Aharoni, 2008). My own view is that both the power to originate and the power to control are requirements for responsibility. I assume, in fact, that there is no power of origination, and therefore my own view is that there is no such thing as moral responsibility or criminal desert. I will say a word more about this at the end of the chapter. For present purposes it is enough to repeat that origination and control are separate faculties, and that both are required for responsibility and desert in the traditional view.[11] Neuroscience is not likely to answer the question about origination definitively any time soon, but it appears to have a pretty good handle on the machinery of control.

THE RIGHTS APPROACH

Murphy: Dignity and Preventive Detention

According to the rights approach, what justifies preventive detention is that the offender does not have the right *not* to be detained. It is different from the responsibility approach in many ways, not least of which is that the agent himself may be responsible for his actions, and may nevertheless have lost the right not to be detained.

What does this approach have to say about the psychopath? I want to use for this discussion the answer that Jeffrie Murphy gave, in one of the classic philosophical papers on psychopathy. The answer, for Murphy, lay in the reciprocal nature of rights: We have rights only to the extent that we are prepared to recognize the rights of others. The psychopath does not recognize the rights of others, and therefore he has no rights of his own.

Murphy argued that the psychopath could be detained indefinitely (and even exterminated in certain cases). He did not argue the converse, however; he did not argue that the psychopath *had to be* detained rather than punished,[12] and I think that it is fair to infer from his argument that the psychopath might be punished if it served our purposes, and that the legislator who provided for both punishment and indefinite detention of sexual predators has not violated any moral principle, provided that the sexual predator is a psychopath. The reason we may, in theory, subject the psychopath to detention, treatment, even execution, is that he lacks what Murphy calls "dignity." And he lacks dignity precisely because he does not respond to the rights and interests of others.

The argument goes as follows:

1. The psychopath does not care about the rights of others; he fails to "care about his own moral responsibilities [toward others],...[or] to accept them even if he recognizes them" (Murphy, 1972: 295).
2. Therefore the psychopath is in no position to claim any rights "on grounds of moral merit or desert" (Murphy, 1972: 294).

This step is warranted by the reciprocal nature of rights. I do not have rights to make any claims under a particular institution if I am not prepared to respect the rights of others under that institution; I do not have any rights at all if I am not prepared to respect any rights.[13]

3. Therefore though we can act wrongly toward psychopaths, we cannot wrong them. "They can be injured, but they can be done no moral injury" (Murphy, 1972: 294). Our moral response to them is to be on a par with our moral response to animals.
4. Thus, "[w]hen an individual has been diagnosed a psychopath, his 'rights' may be suspended and he may be subjected to involuntary indefinite preventive detention and therapy and perhaps even (if his case is hopeless) to painless extermination" (Murphy, 1972: 296).

On Murphy's view, responsibility plays a role in determining how we are to treat the psychopath. But responsibility follows from rights, and lack of responsibility follows from lack of rights, and not the other way around. The psychopath is not to be held morally responsible for what he does precisely because he cannot claim any rights, and being held responsible is a right. Neither is the psychopath entitled to moral respect, and for the same reason.

Murphy's discussion is harsh.[14] It seems clear that he thinks of the psychopath as someone who is at fault for his behavior, and not someone who, like those with mental disease, cannot be blamed for what he does. He says:

> [T]he psychopath, by his failure to care about his own moral responsibilities, his failure to accept them even if he recognizes them, becomes morally dead—an animal rather than a person....The man who, when told that his conduct harms the rights of others, sincerely responds "Who cares?" is hardly in any position to demand that others recognize and respect any rights that he might want to claim. (Murphy, 1972: 295)

When we say that the psychopath does not deserve to be punished, then, what we mean is that he is not *worthy* of being punished.

Given Murphy's appraisal of the psychopath, a natural question to ask would be: Must the dangerous person be *unable* to respond to the interests of others in order to lose all his rights? Or is it sufficient that he *refuses* to respond to the interests of others? In other words, must he be a psychopath, or will the hardened criminal also lose his right to be punished? Notice that the assumption that the psychopath is *unable* to respond to moral concerns appears to play no role in the argument. Why should the fact that someone is unable to take moral considerations into account, rather than just unwilling to, matter? As far as the premises of the argument are concerned, they would seem to apply as well to someone who simply chose not to respond to moral concerns.

Slobogin: Undeterrability and Preventive Detention

In fact, we seem to find this version of the theory suggested in the work of Christopher Slobogin.

From this point of view, it does not matter whether the offender cannot respect the rights of others, or simply will not respect the rights of others. Slobogin considers enemy "combatants under orders to kill, terrorists willing to die for their cause, and extremely impulsive individuals like sex offenders who are willing to commit their crimes even with a policeman at their elbow" to fall within an exception to Hegel's right to punishment.

> Unlike people with serious mental illness, these latter individuals can be considered autonomous. But they do not deserve the right to punishment because they exercise their autonomy in the wrong direction even when faced with death or a significant punishment. (Slobogin, 2000: 74)

Although he describes all such individuals as irrational, what he means by that is simply that they disregard "society's most significant prohibitions." They are part of the "universe of individuals who are 'undeterrable,' i.e., those who are unaffected by the prospect of criminal punishment or significant harm."

> An alternative regime of liberty deprivation is justifiable for these individuals because the dictates of the criminal justice system have little or no impact on them. (Slobogin, 2000: 74)

Slobogin's net is cast pretty wide, and if, indeed, we cannot distinguish between the person who cannot respond to the rights and interests of others and the person who simply will not recognize that right, then this position is a *reductio ad absurdum* of the rights approach. For from this point of view preventive detention is appropriate for anyone who is "undeterrable." Slobogin has already given us the beginning of a list of those who may qualify as undeterrable: the combatant under orders to kill; the terrorist willing to die for his cause; impulsive sex offenders; hardened criminals of all sorts.[15]

Conclusions

Although Murphy does not take up this point about the distinction between those who cannot and those who can but will not, I would argue that the theory has plausibility only if applies only to those who cannot respect the rights of others. This will be true on the one hand of offenders who cannot control their behavior, and on the other of offenders who are incapable of understanding the moral concerns of others. In restricting the theory in this way, I am raising questions that were also questions for the responsibility theorist: Is control a necessary condition of responsibility and thus punishment? Is moral responsiveness a necessary condition? But even with the rights approach restricted in this way, the difference between the two approaches is great, as is the gap between the rights approach and our current understanding of responsibility and punishment.

Putting it in the best light and assuming it applies only to those who *cannot* respect the rights of others, the rights approach has these consequences: First of all, there are no safeguards within the theory to protect the rights of the psychopath because, of course, the psychopath has no rights. In particular, the psychopath has no right not to be detained, or even exterminated. On the other hand neither does he have a right not to be punished, if it suits our purposes to punish him. It is therefore possible to design legislation which will both punish the psychopath for what he has done and subject him to subsequent preventive detention if he remains dangerous. That means that it would justify the legislation which, in Kansas and elsewhere, has developed precisely that sort of regime for sexual offenders; it would justify indefinite sentences within the sentencing system for those who refuse to be deterred; and it would justify guilty-but-mentally-ill pleas (resulting in punishment followed by indefinite detention) for those criminals who cannot control their behavior.

Nor do I see why treatment would be *required* on this approach, rather than simple warehousing. Certainly if it were for the good of society, to save expense or more completely establish the security of the community, the state *could* impose treatment on the psychopath; but only in the interest of the community, and never because required by the rights of the psychopath.

Finally, no compensation would be required for the infringement of his right to freedom, because again he has no rights. This approach has nothing to do with what the state owes the psychopath, because it owes him nothing. Being the sort of human being he is, he is entitled to nothing, and that's what justifies his detention. On the responsibility approach, on the other hand, it is precisely the fact that the psychopath is not responsible—if in fact he isn't—that imposes an obligation toward him on the state, an obligation to protect him from retaliation and other forms of harm, to treat him if possible, and to compensate him for the freedom that it has taken away for the safety of the community. These obligations are part of the justification for indefinite detention.

CONCLUSION

These are two very different ways of approaching the psychopath and the question of preventive detention. Given my conclusion that the psychopath is responsible for what he does, he would be punished but not detained under the first theory, but under the second he would be detained and perhaps punished as well.

Which of the two theories is more plausible? If the rights theory is to have any plausibility in the present context, then, as I said, it must be limited to those who cannot appreciate the wrongness of their actions, and must not apply to those who simply will not acknowledge their moral responsibilities. But there is nothing about the theory that would seem to justify such a restriction. The basis of the judgment that an actor is lacking in dignity, that he has no rights that he can claim, is that he has failed to respect the rights of others; that he has said "Who cares?" Why should our treatment of the one who cannot help himself in this matter be more harsh than our treatment of the one who can help himself but will not? It is not clear to me that the theory can avoid the conclusion that Slobogin would draw.

In any case, I do not see any safeguards for the psychopath on this rights approach. Indeed, some additional assumptions will have to added to the theory just to block the conclusion that the psychopath was at the mercy of anyone who wanted to abuse him. How are we to avoid the conclusion that the fate of the psychopath should be the fate of the outlaw: that anyone might do what they choose to him; that anyone might detain him; that anyone might kill him for good reason? Murphy does not address this point, but I think it is something that must be addressed. Here is one possible way of dealing with it: The psychopath, having no rights of his own, must become the property of the state, protected from injury by others. However feeble our intuitions about preventive detention, one thing that is clear is that we will not countenance random abuse of a human being, even one entirely lacking in rights.

Even given that particular assumption, however, he may be preventively detained by the state where it suits our purposes, and even exterminated if his case is hopeless; and that, I believe, robs the approach of whatever plausibility it might otherwise have. And finally there is a kind of paradoxicality: Take the most favorable version of the theory, limiting it to those who cannot respect the rights of others. On this theory those who cannot respect the rights of others may be exterminated if hopeless, while one who can but will not must be imprisoned for a fixed term. Our sense of fairness rebels at the thought.

Therefore of the two the responsibility approach is the preferable approach; and therefore it seems to me that the psychopath may not be detained but must be punished for his crimes.

PHILOSOPHICAL AFTERWORD

Everything I have argued here is based upon certain assumptions, most importantly the assumptions that human beings are sometimes responsible for their behavior, that the traditional analysis of responsibility in terms of control and understanding are correct, and that the fact that human beings are responsible entitles us to punish them when they commit crimes. It seems reasonable to make these assumptions for the sake of the debate we are engaged in here. However, they are not assumptions I share.

The justification of punishment depends upon the notions of criminal responsibility and retribution, and retribution depends upon desert: We are justified in punishing an offender only if he deserves punishment. Although a great deal of energy has been expended trying to show that desert and retribution are compatible with the scientific view of man—I mean the view that the actions of human beings are just like other events in the natural universe, subject to cause and chance just as other events are—I believe that all such attempts fail. My view of these things may be characterized as "hard incompatibilism," a label suggested by Derk Pereboom (2006).[16]

The difficult question for this point of view is how to replace punishment in the social control of crime. I do not have an answer to that question,[17] but this much is clear: however we end up using the power of the state to control crime, certain distinctions that exist in our current system will have to be more or less maintained, though not exactly as they are. There is a difference between those who are found legally sane and those who are found legally insane under the present system, and it stands to reason that when we give up the notion of punishment we will not be able to lose that difference entirely. The criminal law recognizes a difference between the adolescent and the adult, and that distinction will have to find some place in the new system. Though the law does not currently recognize it, I would argue that there is a significant difference between the addict and the nonaddict, and whatever new methods of control are developed will have to respect that difference. The project for the legal theorist who accepts the hard incompatibilist point of view (and who rejects the "illusion" view of Saul Smilansky) is to draw the outlines of such a new system.

Should the psychopathic be distinguished from the nonpsychopathic personality? Assuming that we will not be inclined to subject criminals who are rational and can control their behavior to indefinite detention except under very special circumstances, is there any reason to make an exception of the psychopath? It is perhaps worth remarking that our increasing knowledge of the operations of the brain parallels an increasing willingness to use indefinite preventive detention in the law. As I have already observed, a number of states permit both the punishment and the preventive detention of sexual predators, utterly confounding the traditional notion of punishment; a number of states have created a plea of guilty-but-mentally-ill which permits detention of offenders not found to be legally insane.

My point is not to suggest that there is causation in either direction. The increased use of preventive detention is due to purely political motives. My point is that any movement we make in the direction of undermining the notion of punishment must be extremely cautious. Once the wall of responsibility has fallen, there will be little to block the advance of preventive detention into areas in which it is both unnecessary and dangerous, even from an incompatibilist point of view.

Although I believe that there is no room for responsibility in the scientific view of man, I believe that there is room for moral rights, a point I will not try to argue here. This is not to adopt anything like the rights approach to preventive detention of the psychopath. Just as there are rights, there are certain limited ways of losing rights. Injuring another or the property of another may take away part of your right not to be made an example of by the state. Lack of capacity for control or rationality, taken together with a tendency to violence, may deprive you of your right to liberty for as long as you remain unable to control your behavior or understand the nature of it. If an actor retains his ability to control and understand the consequences of his behavior, he may be made an example of by the state, but he should not be subject to preventive detention; that is, he does not lose the right to liberty for an indefinite period of time. I would argue that on this view, just as on the traditional view of human responsibility, the psychopath should not be subject to indefinite detention unless he is unable to control or unable to understand the consequences of his behavior. His inability to process moral concerns does not deprive him of the ability to understand the consequences of his behavior in the important sense that controlling crimes requires.

The claim that is sometimes made, that neuroscience and law simply speak different languages, and that neuroscience is unlikely to have anything to say about law and in particular about preventive detention, is based in part on legitimate questions like the question whether in fact the responsibility approach is the correct approach, questions that do not depend on scientific knowledge for their answer. But it is also based in part (it seems to me) on the supposition that an inability to control behavior and an inability to originate action are both irrelevant to responsibility. That is just not so. Where science can demonstrate a lack of control, then responsibility (in the traditional understanding of the term) is missing; and should it be true eventually that science can rule out the power of origination, then science will have ruled out responsibility in the traditional sense. The optimistic belief that compatibilism would provide us a way around this difficulty, that it would allow us to maintain a belief in responsibility and desert in spite of the advances of science, has for the most part been crushed.

NOTES

1. See also Scanlon (1998: 284), where he says of the psychopath: "If he commits these crimes because he does not place any value on other people's lives or interests, what clearer grounds could one have for saying that he is a bad person and behaves wrongly?" He may be blamed for his behavior, according to Scanlon, and to the extent that moral responsibility is sufficient for criminal responsibility he may be held responsible and punished.
2. I'm aware that the problem of psychopathy is a problem of more or less, and not a black-and-white problem. For simplicity's sake I will ignore that fact in this chapter and assume that we are dealing only with the most pronounced cases of psychopathy.
3. I should also mention the work of the philosopher Ferdinand Schoeman, whose work on quarantine and preventive detention got me interested in the problem of justifying preventive detention. See Schoeman (1979).
4. Christopher Slobogin has also advanced this view. See in particular his contribution, "Defending Preventive Detention," to *Criminal Law Conversations*.
5. As these terms are commonly used, moral responsibility is not coextensive with criminal responsibility. Moral responsibility is a necessary condition of criminal responsibility, but the converse is not true. Nevertheless I believe that (again, as these terms are commonly used) there may be preventive detention if and only if there is a lack of moral responsibility (or, to be more precise, lack of a capacity for moral responsibility). Even though, for various reasons, there may be cases in which there is moral responsibility but no criminal responsibility, in those cases preventive detention is not justified on this "responsibility" approach.
6. "Perhaps justified": Perhaps the facts of non-responsibility and dangerousness override the offender's right to liberty. I am not completely happy with arguments of that sort.
7. The three most common versions of the insanity defense are the M'Naghten defense, the Model Penal Code defense, and the appreciation-or-wrongfulness approach found in federal legislation. All require an ability to understand or appreciate that the act was wrong. See Corrado (2009).
8. *Finger v. State*, 27 P.3d 66 (Nev. 2001).
9. We would reach the same result, of course, if moral responsiveness were not an element of rationality, but were required all the same for responsibility.
10. On the difference between the schlemiel and the schlemozzle, see Eric Berne (1964).
11. On the distinction between origination and control, I follow roughly the lines that Saul Smilansky takes in *Free Will and Illusion* (2002). The assumption that origination is required for moral responsibility is the traditional mark of the incompatibilist; and the incompatibilist who accepts the scientific view of man will deny that there is such a thing as origination, and she will deny the state's right to impose punishment (understood as retribution). But to deny origination is not to deny control: Even the incompatibilist who accepts the scientific view of man must make room, in whatever system of control succeeds punishment, for the distinction between the sane and the insane that is made in our present law. Philosophically that distinction may be based upon rationality or upon control, understood as it is presently understood by compatibilists. And scientifically it may be based upon the work of neuroscientists. But it should be free of the notions of desert and responsibility.
12. Paul Litton makes this point.
13. I presume the qualifier here, "on grounds of moral merit or desert," is to distinguish the rights Murphy is talking about from the rights we find in utilitarian theories of rights, on which theories the psychopath might indeed have rights—as indeed a piece of wood might have rights.

14. This is a view that Murphy no longer holds. See Murphy (2007: 423–453; the footnote occurs on pp. 425–426).
15. We may add to this list the saint. People we consider saints need not be passive and long-suffering; often they will be active and aggressive in the cause of what they believe to be right. Often they are willing to break the law to make their point, and are not easily deterred (Corrado, 2005-a: 74–75).
16. See also Greene and Cohen (2004). But see Morse (2008).
17. However, see Corrado (2001 and forthcoming).

REFERENCES

Aharoni, E., Funk C., Sinnott-Armstrong, W., & Gazzaniga, M. (2008). Can neurological evidence help courts assess criminal responsibility? *Annals of the NY Academy of Sciences, 1124,* 145.

Berne, E. (1964). *Games people play.* New York: Ballantine.

Bonnie, B. (2003). Why 'appreciation of wrongfulness' is a morally preferable standard for the insanity defense. In: Proceedings of Texas Society of Psychiatric Physicians, *The Affirmative Defense of Insanity in Texas,* 50, 60.

Corrado, M. (1996-a). Punishment, quarantine, and preventive detention. *Criminal Justice, 15, Ethics* 3.

Corrado, M. (1996-b). Punishment and the wild beast of prey: The problem of preventive detention. *Journal of Criminal Law and Criminology, 86,* 778.

Corrado, M. (1999). Preventive detention. In: *The philosophy of law: An encyclopedia.* New York: Garland.

Corrado, M. (2001). Abolition of punishment. *Suffolk Law Review, 35,* 257.

Corrado, M. (2005-a). Responsibility and control. *Hofstra Law Review, 34,* 59, 74–75.

Corrado, M. (2005-b). Sex offenders, unlawful combatants, and preventive detention. *North Carolina Law Review, 84,* 77.

Corrado, M. (2009). The case for a purely volitional insanity defense. *Texas Tech Law Review, 42,* 481.

Corrado, M. (forthcoming). Why do we resist hard determinism. In: T. Nadelhoffer (ed.), *The future of rehabilitation and punishment.* New York: Oxford University Press.

Greene, J., & Cohen, J. (2004). For the law, neuroscience changes nothing and everything. *Philosophical Transactions of the Royal Society B: Biological Sciences, 359,* 1775, 1778.

Kant, I. *To perpetual peace: A philosophical sketch* (Humphrey trans. 1983), p. 124. Indianapolis, IN: Hackett.

Lippke, R. (2008). No easy way out: Dangerous offenders and preventive detention. *Law and Philosophy, 27,* 383.

Litton, P. (2008). The responsibility status of the psychopath. *Rutgers Law Journal, 39,* 349.

Morse, S. (2002). Psychopathy. In: Dressler, J. (Ed.), *The encyclopedia of crime and justice.* New York: Macmillan.

Murphy, J. (1972). Moral death: A Kantian essay on psychopathy. *Ethics, 82,* 284.

Murphy, J. G. (2007). Remorse, apology, and mercy. *Ohio State Journal of Criminal Law, 4,* 423–453.

Pereboom, D. (2006). *Living without free will.* New York: Cambridge University Press.

Morse, S. (2008). Determinism and the death of folk psychology. *Minnesota Journal of Law, Science and Technology, 9,* 1.

Scanlon, T. (1998). *What we owe to each other.* Cambridge, MA: Belknap Press of Harvard University Press.

Schoeman, F. D. (1979). On incapacitating the dangerous. *American Philosophical Quarterly, 16,* 27.

Slobogin, C. (2009). Defending preventive detention. In: *Criminal Law Conversations* 67, 74.

Smilansky, S. (2002). *Free will and illusion.* New York: Oxford University Press.

CHAPTER 19

Psychopathy and Sentencing

Erik Luna

INTRODUCTION

A recent cartoon offered the following verdict form to depict the reality of decision-making in criminal cases:

> We, the jury, find:
> [ ] The defendant guilty beyond a reasonable doubt.
> [ ] The defendant not guilty.
> [X] The defendant may or may not be guilty, but looks like one scary mother-fu****, so, guilty.
>
> Foreperson

Experienced defense attorneys would readily admit the kernel of truth in this faux verdict, but other criminal justice actors (especially prosecutors and judges) might only acknowledge as much in moments of unguarded honesty. Defendants who provoke dread and revulsion—by their words and deeds, their affect and appearance, their criminal and psychiatric history, and so on—are more likely to be convicted and receive harsh sentences, even if the basis for fear is legally irrelevant to issues of responsibility or, worse yet, where it might indicate grounds for excuse or mitigation. No matter how strong the defense case, there is little chance that a jury would acquit someone such as Jeffrey Dahmer, the cannibalistic serial killer. At trial, Dahmer was judged legally sane despite evidence of mental disease and acts so disturbing that they seemed to scream out the word "crazy" (O'Meara, 2004). In fact, some people expressed genuine pleasure when Dahmer was later bludgeoned to death in prison. "I'm happy and very excited that the monster is finally dead," said the sister of one murder victim. "The Devil is gone" (*Chicago Sun-Times*, 1994).

Dahmer ranks among the most infamous "psychopaths" in American history, joined by the likes of David Berkowitz, Ted Bundy, John Gacy, and Richard Ramirez. To be sure, the label of popular culture does not always square with the medico-scientific diagnosis of *psychopathy*. This disorder is marked by, among other things: repetitive antisocial behavior and criminal versatility; pathological lying, conning, and manipulation; impulsivity, carelessness, and the need for stimulation; a grandiose sense of self-worth and failure to accept responsibility for one's actions; and a shallow affect and the lack of empathy, remorse, or other psychological aspects associated with the concept of conscience (Hare, 2003). Yet psychopaths do not appear to suffer from delusional or irrational thinking, nor do they tend to be nervous or neurotic. As Hervey Cleckley observed in his pioneering work on the disorder, the psychopath may come across as charming, a person with "a technical appearance of sanity, often one of high intellectual capabilities," someone who may even achieve some success in the professional world (Cleckley, 1988).

Precisely because they maintain the facade of normality—a *mask of sanity*, as Cleckley described it—psychopaths present a theoretical and practical challenge for a criminal justice system. "[T]heir unspeakable acts, their grotesque sexual fantasies, and their fascination with power,

(© 2009 Courtoons and David E. Mills)

torture, and death severely test the bounds of sanity," writes Robert Hare, perhaps the most prominent researcher in the field today. "Such morally incomprehensible behavior, exhibited by a seemingly normal person, leaves us feeling bewildered and helpless" (Hare, 1993: 5). Under one gloss, the psychopath's antisocial conduct stems from a cognitive and affective ailment that should excuse him from criminal responsibility. But a different interpretive account sees the psychopath as acting out of cold, self-interested calculations rather than impulse or coercion, thus justifying full liability for his crimes.

Today, there appears to be some agreement about psychopathy and excuse, although the bottom line diverges between philosophical tracts and legal doctrine. Most contemporary theorists believe that psychopaths should be excused from criminal responsibility (but still subject to civil commitment). In contrast, modern penal law rejects out of hand any argument for psychopathy as grounds for acquittal (Morse, 2008-a, 2008-b). The divergence is striking and worthy of further consideration by scholars and practitioners alike. But assuming a psychopathic defendant is deemed legally responsible and found guilty of criminal conduct, an important question remains: Should a diagnosis of psychopathy serve to increase or decrease his sentence? This is not a marginal issue, prompted only by Dahmer-esque characters. Estimates suggest that 15% to 25% of all prison inmates may meet the criteria for psychopathy, with studies finding that psychopaths could constitute, for instance, a substantial proportion of all rapists and cop killers (e.g., Hare, 2002; Kiehl & Hoffman, 2011; Schopp, Scalora, & Pearce, 1999).

Theoretical arguments can be compiled on either side of the ledger, mitigation versus aggravation, derived in large part from the debate regarding psychopathy as excuse. Current doctrine also offers plausible claims for both severity and leniency in sentencing. As is true at the guilt stage, however, a defendant rarely benefits at sentencing from the label of psychopath. This issue is presented most starkly in capital cases. Death penalty jurisprudence calls for the dispassionate weighing of aggravating and mitigating factors, but the actual decision can come down to gut feelings of fear and disgust. To rephrase the above cartoon: Theory and doctrine may or may not support capital punishment in a given case—but the psychopath cuts a scary, even monstrous figure, so execute him.

This chapter does not consider in any depth whether psychopathy can serve as an excuse from criminal responsibility. Nor will it evaluate official responses to the psychopathic offender outside of the criminal justice system, such as civil commitment. These issues are dealt with elsewhere in the present volume.[1] The chapter also accepts (but will later critique) the prevailing convention among legal actors and some mental health experts to use interchangeably the terms *psychopathy, sociopathy,* and *antisocial personality disorder* (American Psychiatric Association, 2000; Edens, Petrila, & Buffington-Vollum, 2001; *Graham v. Collins* (1993); Hare, 1993, 2002; Litton, 2008; Lykken, 1995; Ogloff & Lyon, 1998; *Roper v. Simmons* (2005); Wolfson, 2008; Zinger & Forth, 1998).[2] Most of all, the following assumes that the psychopathic offender can be and has been found guilty of a crime, and as a result, punishment is to be doled out. The question here is the appropriate sentence, or at least whether psychopathy should be a reason to increase or decrease punishment.

The first section provides a brief overview of the justifications for punishment and lists several propositions about psychopathy for analytical purposes. The second and third sections describe theoretical arguments for psychopathy as mitigation and aggravation at sentencing. The fourth section considers noncapital sentencing doctrine, describing the various punishment schemes and how psychopathy might fit within them. The fifth section reviews the actual practice of noncapital sentencing and the use of psychopathy to increase or decrease punishment. The sixth section describes the contemporary history of capital punishment and the structure of modern death penalty statutes, followed by a review of the U.S. Supreme Court's jurisprudence on mitigating and aggravating factors. The seventh section examines the Supreme Court case most relevant to psychopathy, *Barefoot v. Estelle*, which considered the constitutionality of future dangerousness as an aggravating factor in capital sentencing. The eighth section discusses some issues related to punishment and psychopathy, concentrating on the right to counsel and the obligation to investigate potentially mitigating evidence. Finally, the ninth section offers some reflections on the use and abuse of psychopathy at sentencing.

PUNISHMENT THEORY AND PSYCHOPATHY: A FRAMEWORK

Justifications for punishment can be roughly divided into two broad categories: consequentialist theories and non-consequentialist theories.[3] The former are forward-looking, motivated by the future consequences of punishment; the latter are backward-looking, typically concerned with the defendant's blameworthiness for his past acts.

The leading consequentialist theory, utilitarianism, seeks to maximize the social benefit (utility) of punishment while minimizing its costs (e.g., Dressler, 2009; Vitiello, 1993; Zimring & Hawkins, 1973, 1995). Toward this end, punishment may deter the offender and other members of society from committing future crimes (specific and general deterrence); it may provide the offender treatment and other services to curtail his commission of further crimes (rehabilitation); and it may confine or otherwise debilitate the offender, thereby rendering him unable to commit crime (incapacitation). Appropriate punishment could be gauged by the sentence in the particular case that produces the greatest utility (act utilitarianism) or instead the sentencing rule across all cases that generates the best consequences (rule utilitarianism). In general, utilitarian punishment seeks to reduce the quantity and severity of crime, but it may also take into consideration other concerns—maintaining the perception of fairness and institutional legitimacy, for example, and upholding communal values and expressing condemnation for antisocial conduct—with the goal of increasing aggregate utility throughout society.

The best known non-consequentialist rationale for the criminal sanction, deontological retributivism retributivism, often conceptualizes punishment as "just deserts."[4] Under this theory, criminals should be punished because they deserve it and regardless of social consequences. Blameworthiness may be seen as a function of an offender's actions, his accompanying mental state, and (arguably) the harm he has caused (Robinson, 1994). Wrongdoing is a predicate for punishment, which must be proportionate to the gravity of the offense (Dressler, 2009; von Hirsch & Ashworth, 2005). One's mental state and the quality of his actions may demonstrate a greater level of depravity that justifies increased punishment. But offenders should not receive harsher sentences based upon predictions of future events, such as the likelihood of recidivism (Dressler, 2009; Fletcher, 2007; von Hirsch & Ashworth, 2005).

Moreover, culpability is often coupled with "free will," one of the most contested concepts in moral and legal theory.[5] At times, judges, attorneys, scholars, and mental health professionals refer to free will in their assessments of criminal responsibility, but presumably they do not mean it in the metaphysical sense of an agent able to act uncaused by anything other than himself.[6] For present purposes, any confusion may be avoided by replacing free will with (or redefining it as) the capacity for rational decision-making and the absence of compulsion. As used here, rational decision-making means the ability to reflect and decide what to do in particular circumstances guided by moral and legal standards and a sufficient grasp of factual and social reality.[7] The two criteria, rationality and non-coercion, are fundamental to issues of criminal liability and excuse, and they can impact judgments about sentencing.

The preceding account glosses over many details and dodges all sorts of issues, including the proper measure of utility, the varieties of retributive theory and principles, and the compatibility of responsibility and determinism. But the description suffices for present purposes. It is also necessary to lay out some basic propositions about the functional impairments, etiology, and social effects of psychopathy. Although founded upon scientific studies and professional experiences, each contention is open to disagreement about methodologies, observations, inferences, conclusions, and so on, as well as debate over relevance and application in legal contexts (Blair, Mitchell, & Blair, 2005; Felthous & Henning Saß,

2008; Hare, 1993; Hare, Clark, Grann, & Thornton, 2000; Patrick, 2006; Raine & Sanmartin, 2001). Nonetheless, these propositions appear throughout the literature and help inform the analysis of theory, doctrine, and practice.

- Psychopathy is associated with neurobiological dysfunction, including abnormalities in brain regions (e.g., amygdala) and neurotransmission (e.g., serotonin activity) critical to cognition and emotion.[8] The disorder appears to have a significant genetic component, with twin studies supporting the inheritability of the core personality traits of psychopathy.[9] Environmental factors, such as poor parental supervision and low family income, also seem to play an important part.
- Psychopaths experience reduced physiological reaction to fear, as well as decreased recognition of and physiological reaction to distress in others (Newman, Curtin, Bertsch, & Baskin-Sommers, 2010). Likewise, they appear to suffer a defect in the normal emotional reactions that accompany anticipated outcomes, which can impede the selection of beneficial options.
- Psychopaths demonstrate deficiencies in emotional learning, such as the ability to learn a desirable response to a stimulus based upon rewards and punishments. In addition, they experience trouble with long-term planning and goal achievement. Psychopaths may also suffer from flawed moral development, as demonstrated by a difficulty in differentiating between acts that harm another person or infringe upon his rights versus acts that merely violate social customs.[10]
- Psychopaths have many rational capacities. In general, they are in touch with reality, have a full understanding of the facts before them, and do not suffer from hallucinations or delusions. Psychopaths may also recognize the existence of rules and penalties for their violation. They engage in instrumental reasoning and action, guided by their own conception of pleasure and pain rather than irresistible impulses or overwhelming desires.
- Psychopathy, like many other mental disorders, is a matter of degree. The severity of cognitive and affective dysfunction may vary across individuals, creating a continuum instead of an absolute measure (Krueger, Markon, Patrick, & Iacono, 2005). Moreover, psychopaths often suffer from additional disorders (e.g., substance abuse), making comorbidity a frequent and confounding phenomenon.
- Psychopaths commit crimes at a disproportionate rate, with illegal behavior beginning at early age and continuing across their lifespan. Psychopathy is a strong predictor of recidivism, including crimes of violence.[11] Psychopathic offenders are several times more likely to commit violent crimes in the future; in fact, most reoffend within the relevant observation period. What is more, there is no known effective treatment for adult psychopathy.[12] Some treatments have even been shown to *increase* recidivism among psychopaths (Hare, 2000).

PSYCHOPATHY AS MITIGATION

Given the foregoing, a number of theoretical arguments might support psychopathy as a mitigating circumstance.[13] One writer depicted the disorder as an issue of compulsion, where "the psychopath is unable to weigh alternative actions and consider which actions among the alternatives is the most appropriate" (Nordenfelt, 2007: 182). In some sense, the psychopath may be able to see other courses of action, but his impulsivity and inability to weigh the options in any meaningful sense leaves him with a single option—the one that most fulfills his desires. However, the best arguments would probably avoid claims of coercion and instead would invoke deficiencies in reasoning and impairments in other-regarding emotions such as empathy and remorse.

Psychopathy may be characterized as a decision-making disorder that obstructs prudential rationality in long-term planning and an appreciation of the effect of one's choices over time (Glannon, 2008). The psychopath may make instrumental choices to achieve immediate goals, but he misses a crucial component of fully developed judgment: the ability to plan for the future or foresee the possible consequences of his conduct. He may also lack the mental faculties to appreciate behavioral expectations and standards. Some scholars frame the issue as the ability to distinguish between types of norms, with psychopaths having considerable difficulty drawing the line between acts that flaunt social conventions (e.g., cross-dressing) and those that violate a moral rule (e.g., prohibition on rape). The difficulty in understanding the wrongfulness of an act such as theft or sexual assault—unable to fully appreciate it as a violation of another's rights rather than, say, a mere breach of social etiquette—may render the psychopath less than fully responsible as a moral agent (Fine & Kennett, 2004; Levy, 2007).

Many theorists focus on the emotional deficit of the psychopath and the resulting cognitive impairment in his reasoning. Of particular importance is the ability to grasp the moral quality of one's actions, especially the effect they have on other people—to appreciate fear and anxiety, to feel empathy for the suffering of others, and to experience remorse if the actor himself is the source of distress. A psychopath suffers a deficit in these emotions, and as a result, he has no standards, no ability to reflect upon the morality of his actions, and no capacity to engage in meaningful moral dialogue. He is not a rebel, Antony Duff notes, someone "who rejects more conventional values and emotions in the light of some favoured conception of the good" (Duff, 1977: 189). Instead, he has no values at all. Without this dimension in his own life, the psychopath cannot understand the emotions and values in the lives of others and is therefore incapable of appreciating the emotional and moral aspects of his actions toward them. He stands outside of the community, unable to participate in common life with other members of society.

In a series of scholarly works, Stephen Morse argues that empathy, remorse, and other feelings associated with conscience are essential for the just attribution of blame, given that these emotions offer the most forceful reasons to abide by proscriptions of law and morality (Morse, 2002-a, 2008-b). The lack of conscience—the inability to empathize with those who would suffer from an actor's conduct and to be guided by the reasons it provides against such action—renders the psychopath "morally insane," to use the historic term (Prichard, 1835). Likewise, Jeffrie Murphy draws upon Kant's phrase in referring to psychopaths as "morally dead" (Murphy, 1972). The psychopath has no conscience; he does not recognize moral obligations; he has no respect for the rights of others; and for these reasons, he has no rights or duties of his own. He is not a moral agent, or at least not a fully responsible one, making him a less fitting subject of punishment.

Other sophisticated philosophical theories, such as the reasons-responsiveness model of John Martin Fischer and Mark Ravizza, have reached similar conclusions. Intelligent animals, young children, and psychopaths "exhibit a certain pattern of responsiveness to reason," but that is not enough for moral responsibility (Fischer & Ravizza, 1998: 76). Psychopaths may recognize that rights and duties exist, yet they do not understand that someone else's rights and interests may create a sufficient reason for him to act (or not act). "This sort of individual is not appropriately receptive to reasons," Fisher and Ravizza conclude, "and thus is not a morally responsible agent" (Fischer & Ravizza, 1998: 79). Another model employs the notion of self-governance. Psychopaths cannot critically evaluate themselves, Paul Litton argues, and they may be deprived of evaluative standards more generally (Litton, 2008). In a sense, a psychopath is out of touch with reality, averring goals and then acting at odds with the very idea of a *goal* as a concept and as a project. The incapacity for rational self-governance may thereby diminish the psychopathic offender's moral responsibility.

Many of these theories might be depicted as using *a priori* argument about the nature of rational thought and the types of reasons that may be taken into account (Fischette, 2004). They seem roughly consistent with deontological retributivism and its requirement of rational and uncoerced decision-making (Morse, 2001, 2002-a, 2007). But the arguments can also be more experimental in nature, converging on widely held sentiments (e.g., empathy) that lie at the core of society's conventions and behavioral expectations, which help define what it means to hold ourselves and others responsible for action. Without the ability to feel the force of societal proscriptions, the psychopath cannot understand the significance of legal and moral commands or be motivated by them. In light of societal sentiments and views of punishment, it could be unfair to hold the psychopath criminally responsible.

This line of analysis might draw upon Peter Strawson's (1992) famous work on reactive attitudes. Moral responsibility is not a matter of metaphysical principles but instead a function of the attitudes we take and the emotions we feel in our interpersonal relationships, constituted in a community and creating expectations of all members. The practice of blaming, assessing responsibility, and assigning punishment involve an evaluation of the attitudes expressed by other's actions. A moral agent may be excused or considered less blameworthy because his actions are not reflections of antisocial attitudes (e.g., "he didn't mean it," "he didn't understand the facts," "he wasn't himself," etc.). However, a different response may be appropriate when the harm is committed by a psychopath: He was himself; he appreciated the facts; and he did it intentionally. Unlike a person with full moral agency, a psychopath is seen as mentally "abnormal" or "deranged." People's reactions toward him may reasonably include repulsion, fear, and pity—but not resentment, forgiveness, or anger—because in the end, he cannot be argued or reasoned with. The psychopath is not a member of the community and falls outside of the society's blaming practices (Arenella, 1992; Glannon, 2008; Haji, 2003, 2008; Vargas & Nichols, 2007; Wallace, 1994).

In scholarship, the above claims are typically offered as an excuse from criminal responsibility, not as a reason to mitigate punishment. Still, these arguments might support sentencing reductions for culpable but psychopathic defendants. As a matter of doctrine and practice, a rationale for mitigation may take the form of unsuccessful excuses, such as a mental disorder that does not amount to legal insanity. An offender's diminished capacity for rational decision-making would seem to be particularly relevant here.[14] As discussed, moral and legal responsibility typically require some level of rationality, where an individual's decisions are founded upon reasons "at least 'in the ballpark' as contenders for being correct" (Fischer & Ravizza, 1993) and a capacity to reflect upon such reasons. An otherwise sane and legally responsible offender may still suffer from a rationality-impeding disorder that affected his reasoning about the crime at issue, making him less culpable for the offense and justifying a reduction in punishment.[15] For legal decision makers, the case for psychopathy as mitigation might be strengthened by advancements in the natural sciences, especially neuroimaging and genetic studies that provide evidence of brain abnormalities in psychopaths and the heritability of the disorder. These works would seem to suggest a causal link between psychopathy and antisocial conduct, with the disorder potentially hampering even minimally rational decision-making.

Considerations of retributive desert, especially proportionality, offer the most obvious grounds for mitigation by incorporating into sentencing calibrations a psychopath's diminished capacity for rational decision-making. Consequentialist theories might also call for reduced punishment, not only because of the ineffectiveness of specific deterrence and rehabilitation, but also due to the rule utilitarian benefit from the perception of fair principles of punishment, including sentencing consistent with societal views of relative blameworthiness. Moreover, psychopathy as mitigation may express society's belief that the criminally responsible but disordered defendant is not a *full* member of the community and does not provoke the same negative sentiments as other offenders.

PSYCHOPATHY AS AGGRAVATION

To some extent, the arguments against mitigation and for increased punishment are the mirror images of the preceding analysis. Although Professor Hare has been a chief proponent of psychopathy as a meaningful diagnosis, he made clear that

> psychopaths do meet current legal and psychiatric standards for sanity. They understand the rules of society and the conventional meanings of right and wrong. They are capable of controlling their behavior, and they are aware of the potential consequences of their acts. (Hare, 1993: 143)

According to Hare, "psychopaths certainly know enough about what they are doing to be held accountable for their actions." The traits first catalogued by Cleckley and refined by Hare and others could even be described as a checklist for human *evil* rather than aspects of a mental disorder. As has been repeatedly noted in the literature, evil is neither an excuse to liability nor a basis for reduced punishment. To the contrary, psychopathy is typically viewed in court as an aggravating factor, not a mitigating one. "This is the way it should be, in my view," Hare opines (Hare, 2002: 205). Psychopaths are calculating "intraspecies predators," who maintain the appearance of sanity, even substantial intellect, yet can harm others without compunction.

Professor Hare is not alone in the belief that psychopathy does not impede basic cognitive abilities or limit responsibility for conduct. Critics of excuse and mitigation arguments emphasize that the disorder has little if any effect on the capacity for instrumental reasoning. Psychopaths tend to grasp reality and understand what they are doing as a practical matter. As a question of law, psychopaths recognize that there are rules and consequences for any violations. They are not compelled or coerced, at least as these concepts are commonly understood. Psychopaths suffer neither irresistible impulses from the internal world of the mind nor the duress of an external agent. Their behavior can demonstrate control and consideration of different courses of action, such as selecting vulnerable victims, waiting for opportune moments, and taking steps to minimize apprehension (Elliott, 1992; Glannon, 2008; Hare, 1993). The lack of repentance in spite of the capacity for instrumental reasoning may provide cause for greater punishment. To reach this type of offender, societal condemnation must be communicated in terms of his self-interest, that is, through the imposition of adverse consequences (Primoratz, 1989).

As for categorization, critics would reject Fisher and Ravizza's grouping of psychopaths with small children and smart animals. Children are still developing their thinking skills, and by definition, animals are outside the ambit of human systems of criminal justice. In contrast, the psychopath's cognition has reached its full development (whatever that may be), and he meets the basic prerequisites for criminal responsibility. Indeed, it is not clear the neurobiological psychopath can be meaningfully distinguished from the "acculturated psychopath." The fervent racist, the fanatical terrorist, and even the crooked Wall Street financier may have many of the same characteristics as the psychopath of organic origins—lack of empathy, for instance, and indifference toward the rights and interests of others—but there is little doubt that the acculturated psychopath is fully responsible for his actions (Pillsbury, 1992).

Moreover, rationality may not be amenable to the fine distinctions drawn by some scholarship, and whatever nuances exist might be irrelevant to issues of criminal justice. The divide between conventional and moral transgressions can be overstated, as laws and customs often embody standards of morality. If a psychopath recognizes conventional rules, and these rules correspond to moral proscriptions, there may be no justification for excusing the psychopath or mitigating his punishment (Glannon, 2008; Haji, 2003, 2008; Vargas & Nichols, 2007). The fact that a psychopath fails to follow either type of rule may simply show that he lacks the motivation to do so, which itself implies nothing about the capacity to follow rules and avoid wrongdoing. By comparison, the average commuter may not be motivated to abide by the speed limit at all times, but that does not mean that he is *incapable* of driving as required by law. The same kind of analysis might apply to issues of long-term planning: Psychopaths may have trouble making plans and setting goals for the future, but difficulty is not equivalent to incapacity. Again, by comparison, some people may find it hard to save money for the proverbial rainy day, but this says nothing about whether they could do so. In addition, psychopaths do seem to have an evaluative standard of sorts, premised on their desire for immediate pleasure and excitement. The standard may not be espoused by society, but it is not incomprehensible to the average citizen.

In this light, excusing the psychopath or mitigating his punishment would follow the "model of the circular process by which mental abnormality is inferred from anti-social behavior, while anti-social behavior is explained by mental abnormality" (Wootton, 1959: 250). To date, the most impressive scientific work in this area can only demonstrate a correlation with antisocial behavior. Even if an association is so tight as to imply a cause-and-effect relationship, causation by itself is not an excusing condition. After all, each and every act is in some sense caused (Morse, 2000-b, 2007). Claims of psychopathy as excuse or mitigation are further complicated by the intricate relationship among biological and environmental influences. The former factors (e.g., genetic diseases) are often characterized as innate and seem to carry greater weight than personal experiences in any assessment of responsibility. As a theoretical and doctrinal matter, some environmental factors (e.g., low family income) may be wholly rejected as bases for excuse or mitigation.[16] Moreover, psychopathy appears to be a matter of degree, where many psychopaths avoid criminal (or at least violent) behavior, and those who do not may only suffer from mild brain dysfunction (Edens & Petrila, 2001; Glannon, 2008; Hare, 1993).

In sum, these arguments suggest that a psychopath may be sufficiently rational to be treated as a moral agent who merits his just deserts, or that his capacity for self-interested reasoning makes him a proper subject for deterrence. When psychopathy is at issue in sentencing, other theoretical considerations come into play, especially rehabilitation and incapacitation. If there is no effective regimen to prevent or reduce future wrongdoing (i.e., rehabilitation does not work), the only option may be to segregate the psychopathic offender from society (i.e., incapacitation) and thereby prevent further crimes. Psychopathy—as a seemingly static, untreatable disorder that heightens the risk of reoffending—thus provides a reasonable justification for increased punishment. Even Professor Murphy, who argued that psychopaths cannot be held morally responsible, acknowledged that such individuals might be deprived of rights, indefinitely detained, and, if their cases are hopeless, subjected to "painless extermination" (Murphy, 1972).[17] In contrast, reductions in punishment for guilty but disordered defendants may be rooted in vague sympathies, concealed by the language of "diminished capacity" and the like, rather than an assessment of morally relevant criteria (Morse, 1984-a, 1984-b; Hare, 1993). By the time of sentencing, insanity claims have already been rejected, and if mental disease is pertinent at all, it might offer a reason for increased punishment to protect society.

Furthermore, it is extremely unlikely that the public would favor reduced punishment for psychopaths (Arenella, 1992; Litton, 2008; Reznek, 1997). If anything, an assessment of socio-political costs and benefits may point in the opposite direction. The popularity of anti-recidivist statutes and animosity toward the insanity defense suggest that the public would demand longer sentences for psychopaths. Along these lines, Samuel Pillsbury argues that the law's social and political role requires an understanding of criminal responsibility that focuses on acts and mental states rather than abstract reasoning capacity (Pillsbury, 2009). Current doctrine rejects psychopathy as an excuse to criminal liability, and as Professor Pillsbury notes, there have been no political efforts in support of psychopathy as excuse or mitigation (Pillsbury, 2009: 159). This makes sense to critics, given that psychopaths epitomize the greatest challenge to social and moral norms: individuals who rationally pursue their own interests based on a clear understanding of reality, harming or killing others without empathy, remorse, or concern for moral and legal standards. Unlike the distorted world of the paranoid schizophrenic, the psychopath abides by an ignoble but certainly comprehensible credo: "Only I count" (Pillsbury, 2009: 159). Not only does his behavior merit the denouncement of conviction, but the psychopath's lack of empathy and remorse and his continuing threat to society might call for enhanced punishment.

SENTENCING DOCTRINE AND PSYCHOPATHY

For the most part, the foregoing arguments were pitched at a high level, grounded in scientific claims and moral theory, often propounding a thick conception of rationality. However, contemporary criminal law and its principles of responsibility are built upon a type of shallow philosophy and folk psychology. Terms such as *act*, *intent*, and *voluntary* are defined by law in a relatively commonsense fashion (Morse, 2007). Under homicide doctrine, for instance, the defendant's bodily movements (e.g., pulling a trigger) must result from a conscious decision to kill, engage in a lethal risk, or participate in certain independent crimes, but the capacity for moral reflection may be immaterial to legal responsibility.[18] Psychopaths, as individuals in touch with reality and capable of deliberate action, can be guilty as a matter of law.

This does not mean that principles of legal and moral responsibility *should* diverge.[19] In fact, plausible *legal* arguments for psychopathy as an excuse might exist under contemporary doctrine.[20] Of particular importance here is the more receptive nature of sentencing provisions, in terms of both mitigation and aggravation. Whereas the liability stage of a criminal trial is framed as a dichotomous judgment, punishment decisions may incorporate considerations that were either irrelevant to guilt or fell below the relevant legal threshold. Conceivably psychopathy-related evidence that was excluded or rejected on the issue of responsibility could impact sentencing determinations consistent with the theories of punishment.

In today's criminal justice systems, no single theory stands as the justifying aim of punishment. Sentencing laws tend to list both retributive notions of just punishment and the utilitarian goals of deterrence, incapacitation, and rehabilitation. The enabling act of the current federal sentencing scheme is a case in point, calling upon courts to impose sentences that "reflect the seriousness of the offense" and "provide just punishment"—in other words, retribution—and that serve the utilitarian objectives of "adequate deterrence to criminal conduct," protecting "the public from further crimes of the defendant," and providing the offender "needed educational or vocational training, medical care, or other correctional treatment."[21] However, modern systems often lack an established methodology to deal with potential conflicts among justifications.[22] Part of the problem is the buffet approach taken by legislators, who may choose the rationale that feels right at the moment or include a bit of everything to avoid political discord, all but ensuring justificatory tensions in the actual practice of punishment.

When the issue is psychopathy, most sentencing schemes hold out the possibility of conflict between utilitarian concerns of public safety and retributive notions of rationality, often codified as the aggravating factor of future criminality and the mitigating factor of diminished capacity. This is true regardless of where a system falls on the continuum of "indeterminate" sentencing (i.e., open-ended terms or broad ranges of punishment) versus "determinate" sentencing (i.e., fixed terms or narrow ranges of punishment). Utah maintains one of the last truly indeterminate sentencing schemes, where all first-degree felonies are punishable by incarceration for 5 years to life, for example, and second-degree felonies receive an indeterminate prison sentence of 1 to 15 years. The ultimate release date is set by the Utah Board of Pardons and Parole, which uses a set of guidelines to decide whether an inmate should be paroled or kept in prison. Typical considerations include the gravity of the inmate's crime of conviction, any prior offenses, and his supervision history. The board will also take into consideration various aggravating and mitigating factors, including those relevant to a diagnosis of psychopathy, namely, whether a defendant presents a serious threat of future violence or, conversely, he has substantial grounds to excuse his behavior.[23]

California adopted one of the nation's first determinate sentencing laws.[24] If a defendant is convicted of a felony offense and sentenced to prison, the court may impose one of three sentences—a triad of lower, middle, and upper terms—with, for instance, voluntary manslaughter subject to imprisonment for 3, 6, or 12 years. By law, the court is required to impose the middle term unless it finds aggravating or mitigating circumstances listed in California court rules. Although the state's sentencing reform statute contains language suggestive of retribution, the court rules incorporate both consequentialist and non-consequentialist considerations in choosing among the sentencing triad for a specific crime.

As was true with Utah's indeterminate scheme, California's sentencing approach might allow psychopathy to be treated as a mitigating or aggravating factor, depending on whether the defendant's disorder is considered a mental condition that significantly reduced his culpability for the crime or instead serves as an indicator that he presents a serious danger to society.[25]

As a final example, the federal government employs an elaborate system of guidelines for sentencing in U.S. District Courts. In setting punishment, the judge must determine which of forty-three categories applies to the crime at issue ("offense level"), and which of six categories applies to the defendant's prior record ("criminal history"). The court can then consult a two-dimensional grid that fixes the range of punishment based on the defendant's criminal history and offense level (Luna, 2005). The range may be adjusted by aggravating and mitigating circumstances, referred to as upward and downward "departures" (or "variances"). On the one hand, psychopathic offender might receive an upward departure if the relevant range substantially under-represents the likelihood that he will commit future crimes.[26] On the other hand, a psychopath might argue that his disorder significantly reduced his mental capacity and contributed substantially to the commission of the offense, thereby justifying a downward departure.[27]

SENTENCING PRACTICE AND PSYCHOPATHY

As the foregoing suggests, most jurisdictions hold out the possibility that psychopathy could serve as a basis to either increase or decrease punishment. In fact, one might assume that the disorder has always had great significance in the practice of sentencing. Human history and popular culture are replete with stories of psychopathic characters and their misdeeds, providing a lively topic for philosophers, researchers, and legal scholars (Kiehl & Hoffman, 2011; Maughs, 1941; Millon, Simonsen, Birket-Smith, & Davis, 2003; Werlinder, 1978). But a scientifically grounded account of the disorder, one that provides a forensically relevant instrument in court, is relatively new.

The first modern-day clinical description of psychopath was offered in Dr. Cleckley's 1941 book, which still remains an important reference for researchers. The work inspired countless studies and served as a starting point for further definitions, including instruments such as Professor Hare's "Psychopathy Checklist." In its revised form (PCL-R), the checklist is a twenty-item rating scale composed of both interpersonal/affective traits (e.g., glibness) and impulsive/antisocial behaviors (e.g., juvenile delinquency) (Hare, 2003; Kiehl & Hoffman, 2011). Each item is scored on a 3-point scale (0–1–2) based on a file review and semi-structured interview, producing a range from 0 to 40. The typical nonoffender scores 5 or less on the PCL-R, whereas criminals average around 22. In general, psychopathy is indicated by a score of 30 or more (Hare, 1996; Kiehl & Hoffman, 2011). Various other tools exist, but the PCL-R is considered by many to be the most reliable and valid measure of a psychopathic disorder (Kiehl & Hoffman, 2011; Morse, 2002-a).

Professor Hare's checklist is increasingly employed in penological contexts and admitted in court, to the point that a high PCL-R score has become a virtual euphemism for psychopathy (DeMatteo & Edens, 2006; Walsh & Walsh, 2006). In particular, instruments such as the PCL-R are used to determine whether soon-to-be-released inmates fit within the statutory category of "sexually violent predator." This classification allows an offender to be civilly committed after the completion of his criminal sentence. As will be discussed in the subsequent parts, perhaps the most controversial use of a psychopathy evaluation is during the sentencing phase of capital trials, where risk assessment tools can play a major part in life-or-death judgments.[28] To date, however, the issue has not generated a substantial body of case law in noncapital sentencing, possibly due to the predominance of plea bargaining, the lack of sufficient resources and time for psychiatric evaluation, or a general unawareness in the legal profession about psychopathy as a disorder.

When the subject has come up in published opinions, the perception (if not diagnosis) of psychopathy has usually been viewed as an aggravating factor (*Boyd v. Brunsman*, 2012; *People v. Garza*, 2012; *People v. Nichols*, 2012). In a 1991 Montana case, *State v. Evans*, a psychologist testified at sentencing that the defendant could be a "sexual psychopath." The trial judge then doled out an 80-year prison term for aggravated burglary and assault, along the way describing the defendant as "shrewd, self-serving, deceitful, extremely dangerous, a high risk for escape and for future crimes" (*State v. Evans*, 1991: 519–520) Likewise, a 2001 Wisconsin case, *State v. Bare*, involved a defendant who exposed himself near children and underwent a battery of psychological tests, including the PCL-R, in preparation for sentencing. He ultimately received the maximum punishment based on the trial court's judgment on the "key issue," the likelihood of reoffending (*State v. Bare*, 2002).[29]

Sentencing decisions over the past decade are mostly in accord, viewing psychopathy as a basis for increased punishment. In a 2007 federal case from Illinois, *United States v. Torres*, a drug dealer was diagnosed by two psychologists as suffering from, among other things, antisocial personality disorder. At sentencing, the trial judge categorized the defendant's disorder as an aggravating factor (*United States v. Torres*, 2007). In a 2009 Texas opinion, an appellate court upheld a 60-year sentence for a juvenile sex offender charged as an adult. The punishment was justified in part by testimony from a clinical psychologist that the defendant's behavior was suggestive of later development of antisocial personality disorder, "a lifelong condition of someone who doesn't have any respect of authority, rules, or remorse for their actions or behaviors" (*McNichols v. State*, 2009: *5).[30] And in 2010, a federal trial court in Virginia sentenced a prison inmate for, among other things, mailing a letter that threatened to kill President Obama and to rape and kill his wife and daughters. In imposing enhanced punishment, the court relied upon the defendant's prior diagnosis of antisocial personality disorder, as well as a prison psychiatrist's previous conclusion that the defendant was "a dangerous psychopath" who "knows exactly what he is doing" (*United States v. Coates*, 2010: *2–3).

On occasion, criminal justice actors appear to "diagnose" defendants as psychopaths in the absence of expert testimony. In *McPhail v. Renico*, a Michigan state judge characterized the defendant as a "social psychopath" without any supporting scientific evidence. In ordering a sentence of 20 to 40 years for armed robbery, the judge stated the following: "I believe that some day that he will murder or kill some individual simply to further his own economical interests in stealing property and committing armed robberies. When I see a person like that blossoming in his early 20s, as far as I'm concerned, the only thing I can do is protect the citizens of the State of Michigan" (*McPhail v. Renico*, 2006: 652). The defendant would later object to his being labeled a psychopath without expert evidence, but a reviewing court concluded that the judge's statement was not a diagnosis *per se*, but instead an expression of his belief that the defendant was violent and unlikely to be rehabilitated.[31]

In contrast, there are almost no reported decisions where psychopathy has actually served as a mitigating factor in noncapital sentencing. One possible exception is a 2007 federal case from Nebraska, *United States v. McNeil*, where the defendant was convicted of selling crack cocaine and subject to a punishment range of 30 years to life. Prior to trial, the defendant underwent four psychiatric evaluations, which indicated delusional disorder, cannabis dependence, and a combination of schizotypal and antisocial personality disorders. Given the combination of mental health problems and the defendant's bizarre behavior at trial—including McNeil telling the jury that he was good drug dealer and one they could trust with their prospective drug purchases—the trial court reduced the sentence to 10 years, concluding that the defendant's reduced mental capacity to comprehend the wrongfulness of his behavior had contributed to the commission of the crime.

A few reported cases have regarded psychopathy as a *potential* mitigating factor but have refused to decrease punishment, either because the evidence failed to demonstrate significantly diminished mental capacities or it did not reduce the defendants' responsibility for their respective crimes (*People v. Boreham*,

2003; *United States v. Chiscihlly*, 1994; *United States v. Motto*, 1999; *U.S. v. Clark*, 2009). After acknowledging that antisocial personality disorder could be mitigation under federal sentencing law, one trial court noted with derision that almost any defendant could qualify for this disorder and make a claim for reduced punishment. "Criminals as a class suffer from 'antisocial personality disorders,'" the judge opined. "All may therefore claim [the law's] grace." (*United States v. Motto*, 1999: 575).[32] This kind of cynicism may help explain the reluctance of criminal justice actors to use or accept psychopathy as mitigation.

CAPITAL SENTENCING AND PSYCHOPATHY

As the preceding demonstrates, punishment theory and doctrine might view psychopathy as a basis to either increase or decrease an offender's term of imprisonment. When the issue is raised in court, however, the disorder tends to serve as a rationale for a longer prison term. The same dynamic applies to the most severe sentence of all: the death penalty. According to the Supreme Court, capital punishment must serve legitimate purposes, such as retribution, general deterrence, and incapacitation (*Gregg v. Georgia*, 1976). In this context, psychopathy is a "two-edged sword" (*Penry v. Lynaugh*, 1989).[33] It may diminish a defendant's blameworthiness for his crime and thus function as mitigation in capital sentencing. At the same time, psychopathy may indicate that the defendant presents a threat of future violence, a factor that supports the death penalty. In practice, this latter consideration dominates capital punishment, making psychopathy grounds for execution.

There are literally hundreds of death penalty cases that either directly or indirectly concern issues of psychopathy, or at least involve defendants who have been diagnosed as such. In one recent lower-court case, for example, a capital defendant's "severe psychopathic madness" implicated questions of his competence to stand trial, his interest in acting as his own lawyer, the implications of his disruptive behavior in court, and the impeachment of mental health experts. The case even raised arguments about the arbitrariness of the death penalty when "the lives of other equally guilty psychopaths are spared (*United States v. Gabrion*, 2011: 323).[34] Nonetheless, the following only provides an overview of the Supreme Court's jurisprudence in this area, beginning with the history and framework of capital punishment in America today.

Basic Background

Four decades ago in *Furman v. Georgia* (1972), the U.S. Supreme Court struck down death penalty schemes across the nation as violating the Eighth Amendment ban on cruel and unusual punishment. The case began with a terse, unsigned decision, followed by separate opinions written by every member of the Court. Each Justice offered slightly different arguments, but several expressed concerns about the arbitrary and capricious nature of capital punishment. For instance, Justice Potter famously analogized a death sentence "to being struck by lightning" (*Furman v. Georgia*, 1972: 309). In a dissenting opinion, however, Chief Justice Warren Burger suggested that legislatures could create new laws that would pass constitutional muster by providing standards to guide juries in the capital sentencing process. And that is exactly what happened in the ensuing years: Lawmakers around the nation enacted death penalty statutes aimed at remedying the unrestricted discretion that appeared to animate key opinions in *Furman*.

In 1976, the Supreme Court upheld the so-called "guided discretion" approach to capital punishment that had been adopted in a number of jurisdictions. The Court's principal decision, *Gregg v. Georgia* (1976), concluded that such schemes sufficiently minimized the risk of arbitrary decision-making by creating a distinct sentencing proceeding after a defendant has been found guilty of a capital crime. During the penalty phase, jurors would hear additional evidence and arguments about whether the death penalty should be imposed, guided by specific aggravating and mitigating circumstances (Carter, Kreitzberg, & Howe, 2008; Coyne & Entzeroth, 2006; Kaplan, Weisberg, & Binder, 2008).

The *Gregg* opinion referenced with approval the Model Penal Code's listing of capital sentencing factors, an approach that was subsequently implemented across the nation. For example, Wyoming provides a relatively long list of aggravating and mitigating circumstances for juries to consider during their deliberations. Among the mitigating factors is a defendant's substantially impaired capacity to appreciate the criminality of his actions or to conform his conduct to the requirements of the law. Arguably, a psychopathy-based mitigation claim could be made under this factor, as well as through a general catch-all provision allowing evidence on any fact or circumstance of the defendant's character that might mitigate his culpability.[35] Indeed, all death penalty schemes—either by statute or, as discussed below, judicial precedent—might permit a defendant to raise psychopathy as mitigation.

In turn, the Wyoming statute lists as an aggravating circumstance that the "defendant poses a substantial and continuing threat of future dangerousness or is likely to commit continued acts of criminal violence."[36] This factor, generically known as *future dangerousness*, provides the most obvious hook for psychopathy as a basis for execution. Future dangerousness is an explicit factor in several other jurisdictions, including the relatively lethal states of Oklahoma and Virginia, and the wholesale leader in capital punishment, Texas, where it serves as the sole aggravating factor.[37] In practice, nearly all capital punishment states allow testimony on future dangerousness in one form or another, and it has played an explicit role in at least half of all executions in the modern era (Blume & Garvey, 2001; Dorland & Krauss, 2005; Shapiro, 2008). With this structure in mind, let's turn to a few of the Supreme Court's post-*Furman* cases that have some relevance to the issue of psychopathy.

Constitutionality of Aggravating and Mitigating Factors

In a companion case to *Gregg*, the Justices examined the capital sentencing procedures in Texas, with particular emphasis on the aggravating factor of future dangerousness. That case, *Jurek v. Texas* (1976), upheld the state's scheme and rejected the claim that it is impossible to accurately predict behavior. In his opinion for the Court, Justice John Paul Stevens noted that the criminal justice system is pervaded by this type of forecasting, from judgments on pretrial release to decisions about parole eligibility. The constitutional requirements are met, Stevens concluded, so long as the jury was able to consider all relevant information on the issue of the defendant's future dangerousness.[38]

Along the way, the *Jurek* opinion referenced with apparent approval a Texas appellate case decided earlier in 1976. In *Smith v. State*, the defendant was found to be a future danger and sentenced to death, based in part on a psychiatrist's judgment that the defendant was a sociopath and would commit further crimes of violence.[39] As an aside, the psychiatrist in *Smith* was James Grigson, a man who would come to be known as "Dr. Death" because of his resolute, oftentimes dramatic testimony that the defendants he examined were psychopaths who would inevitably kill again. In total, Grigson was an expert witness in 167 capital cases, and, as will be seen below, he figured prominently in modern death penalty jurisprudence.

Six years after *Jurek*, the Supreme Court considered the subject of mitigation in *Eddings v. Oklahoma* (1982). The case was chock full of death penalty issues, including the defendant's youth. Sixteen-year-old Monty Lee Eddings was tried as an adult and convicted of the capital murder of an Oklahoma Highway Patrol Officer. During the penalty phase, the state alleged Eddings would commit acts of violence in the future and constituted a continuing threat to society. In mitigation, the defense presented evidence of Eddings's poor upbringing and physical abuse, testimony regarding his developmental age, and most relevant for present purposes, a psychologist's opinion that Eddings had a sociopathic or antisocial personality. Pursuant to the Oklahoma scheme, the trial judge found all of the aggravating factors to be true, including his own assessment that Eddings posed a continuing threat of violence. The defendant's age was deemed a mitigating factor, but the judge refused to consider any of the other mitigation evidence. The state appellate court affirmed the sentence of death, pointing out that although Eddings suffered from a personality disorder, he knew the difference between right and wrong.

The Supreme Court invalidated Eddings's death sentence, holding that a capital decision maker could not refuse to consider any relevant mitigating evidence. However, Chief Justice Burger believed that the Oklahoma courts had, in fact, considered and rejected the mitigating evidence. A footnote in his dissent was particularly dismissive of expert testimony related to psychopathy, which "may connote little more than that [Eddings] is egocentric, concerned only with his own desires and unremorseful, has a propensity for criminal conduct, and is unlikely to respond well to conventional psychiatric treatment—hardly significant 'mitigating' factors" (*Eddings v. Oklahoma*, 1982: 126 n.8). According to Burger, the Court's opinion exemplified the untenable legal position that the occurrence of a crime should be met by a judicial search for a theory to explain it.

BAREFOOT V. ESTELLE

Chief Justice Burger's position in *Eddings* has never been embraced by the Supreme Court. To this day, the Eighth Amendment requires that a state allow any relevant mitigating evidence to be presented and that a sentencer consider such evidence, including testimony about mental disorders such as psychopathy. A year after *Eddings* was decided, however, the Justices turned to the other side of the two-edged sword. The Court in *Jurek* had concluded that predictions of future dangerousness could serve as an aggravating factor in capital cases, but the decision's full implications remained unclear. In *Barefoot v. Estelle* (1983), the Justices considered in depth whether the Constitution placed limitations on evidence of future dangerousness, generating the seminal decision for psychopathy in capital sentencing.

Defendant Thomas Barefoot had been convicted of the capital murder of a Texas police officer. During the penalty phase, the prosecution called two psychiatrists, James Grigson and John Holbrook, to testify about future dangerousness. Neither man had examined Barefoot; instead, they were provided an extended hypothetical question, asking them to assume true the events surrounding the killing as well as the defendant's prior convictions, reputation for lawlessness, and previous escape from prison. Dr. Holbrook diagnosed Barefoot as a "criminal sociopath," whose condition was not treatable and would likely get worse in the ensuing years. He then testified that, within reasonable psychiatric certainty, it was probable that the defendant would commit crimes of violence in the future. Dr. Grigson's testimony was even more damning. He diagnosed Barefoot as being in the most severe category of sociopaths, testifying that on a scale of 1 to 10, the defendant was "above ten." According to Grigson, there was a "one hundred percent and absolute" chance that Barefoot would commit future crimes of violence. On cross examination, Holbrook said he was aware of studies showing the unreliability of psychiatric predictions of future dangerousness, but he disagreed with their conclusions. In turn, Grigson said he was not familiar with most of these studies, but whatever conclusion they reached would be accepted by only a small minority of psychiatrists. "It's not the American Psychiatric Association that believes that," Grigson stated (*Barefoot v. Estelle*, 1983: 919–920).

The jury sentenced Barefoot to death, and the reviewing courts affirmed the conviction and punishment. During one court hearing, medical experts testified that it was impossible to give a reliable psychiatric prediction of dangerousness and, in any case, no doctor should give a diagnosis without a full examination of the individual in question. These limits were advocated before the U.S. Supreme Court in an *amicus curiae* brief by the American Psychiatric Association, which argued that predictions of long-term future dangerousness were unreliable and should be excluded from capital sentencing proceedings. Moreover, the Association's principles of medical ethics declare it unethical for a psychiatrist to offer a professional opinion unless he had conducted an examination. In the end, however, these claims were rejected by the Supreme Court.

Writing for the majority in *Barefoot*, Justice Byron White refused to adopt a constitutional rule barring expert testimony on future dangerousness, saying that this is "somewhat like asking us to disinvent the wheel" (*Barefoot v. Estelle*, 1983: 896). *Jurek* had already upheld future dangerousness as an acceptable aggravating circumstance, and lay jurors would have to assess this criterion with or without expert evidence. Not only would it make little sense to prohibit testimony on the issue, White argued, but a contrary position would call into question all other legal contexts in which behavioral predictions were being made. Testimony on future dangerousness was not so unreliable that the adversary system and its professional and lay fact-finders could not be trusted to ferret out implausible claims. As such, the appropriate course was not to prohibit testimony; instead, both sides should be permitted to present their own experts on future dangerousness, allowing the juries to weigh the competing positions.

Likewise, there was no problem with the admission of expert testimony based on hypothetical questions, describing it as a common practice. The fact that neither Grigson nor Holbrook had examined the defendant went to the weight of their testimony, not its admissibility. Nor was it of any constitutional moment that the case involved the ultimate punishment, the *Barefoot* majority argued, refusing to inject a special constitutional limitation into the rules of evidence. In passing, the opinion seemed to suggest that the American Psychiatric Association should respond to its own members rather than asking the judiciary to police the psychiatric profession—which is precisely what the organization would do in the coming years. During the 1980s, the Association twice reprimanded Dr. Grigson for giving unethical and untrustworthy testimony on future dangerousness, and in 1995 it formally expelled Grigson from the organization (Beil, 1995).

In his *Barefoot* dissent, Justice Harry Blackmun found it disconcerting that the Court's death penalty jurisprudence would permit evidence on future dangerousness that was believed to be wrong two-thirds of the time. The possibility that such testimony might mislead the trier of fact was too great, the potential prejudice to the defendant from this evidence was too large, and the consequences of a jury decision on future dangerousness were too grave to meet the constitutional requirements. "In a capital case," Blackmun argued, "the specious testimony of a psychiatrist, colored in the eyes of an impressionable jury by the inevitable untouchability of a medical specialist's words, equates with death itself" (*Barefoot v. Estelle*, 1983: 916).

To some extent, the concerns expressed by the dissent have received greater attention in recent years, as more and more capital defendants challenge the admissibility of predictions on future dangerousness. These arguments draw upon post-*Barefoot* case law that calls for judges to serve as evidentiary gatekeepers, taking into consideration a series of factors to ensure the reliability of expert testimony (*Daubert v. Merrell Dow Pharmaceuticals, Inc.*, 1993; *General Electric Co. v. Joiner*, 1997; *Kumho Tire Co. v. Carmichael*, 1999). Almost all scholars (as well as the occasional jurist) have concluded that predictions of dangerousness fail this standard and should be inadmissible in capital sentencing (Beecher-Monas & Garcia-Rill, 2003; Flores v. Johnson, 2000: 464–466). But like Chief Justice Burger's dissent in *Eddings*, the views expressed by Justice Blackmun in *Barefoot* and echoed in the literature do not reflect the current state of the law. In practice, courts tend to admit expert testimony on future dangerousness, and defendants continue to be sentenced to death as a result of these predictions (*Espada v. State*, 2008; *United States v. Rodriguez*, 2005; Texas Defender Service, 2004).

CAPITAL SENTENCING: RELATED ISSUES

Since *Barefoot*, the Supreme Court has not directly addressed psychopathy in capital sentencing. In 1992, the Court refused to hear a case raising the issue of whether jurors must be provided a special instruction on antisocial personality disorder as a mitigating circumstance (*Demouchette v. Collins*, 1992). Three months later, Justice Clarence Thomas referenced that case as indicative of the baseless mitigation claims that constantly barrage the Court. He found it astonishing that someone would argue the Constitution required a jury instruction about the fact that "the defendant suffers from chronic 'antisocial personality disorder'—that is, that he is a sociopath." To Thomas, "such a business...makes a mockery of the concerns that inspired *Furman*" (*Graham v. Collins*, 1993: 500).

Nonetheless, there have been some developments in the Supreme Court's death penalty jurisprudence that have at least indirect relevance to issues of psychopathy. For instance, the Court has twice referenced antisocial personality disorder in cases limiting the application of capital punishment to juvenile defendants. In the 1988 case, *Thompson v. Oklahoma*, the Court held that the death penalty was unconstitutional for those who were under the age of 16 when they committed the predicate offense. In concluding that juveniles have reduced culpability, the *Thompson* plurality quoted a report suggesting that "homicidal adolescents must cope with brain dysfunction, cognitive limitations, and severe psychopathology" (*Thompson v. Oklahoma*, 1988: 835, 835 n.42). In 2005, the Supreme Court ruled in *Roper v. Simmons* that the death penalty was unconstitutional for all offenders who committed their crimes before the age of 18, concluding that the diminished culpability of juveniles made capital punishment disproportionate for this class of offenders. The case mentioned the American Psychiatric Association's prohibition on diagnosing a juvenile as having antisocial personality disorder, which is "also referred to as psychopathy or sociopathy, and which is characterized by callousness, cynicism, and contempt for the feelings, rights, and suffering of others" (*Roper v. Simmons*, 2005: 573). If psychiatrists may not brand a juvenile with the disorder, the *Roper* majority argued, jurors should not be asked to deliver the "far graver condemnation" of death.[40]

Another significant development on psychopathy and the death penalty has come through a doctrinal backdoor of sorts: the constitutional right to counsel and the concomitant review of defense decision-making in capital cases. As interpreted by the Supreme Court, the Sixth Amendment guarantees not only the presence of counsel but also the effective assistance of that counsel.[41] Some of the more difficult issues in this area involve defense counsel's investigation and exploration of facts relevant to the case, and his choices regarding the presentation of evidence at trial and sentencing. The interaction between the Sixth Amendment and psychopathy first appeared in the 1987 case, *Burger v. Kemp*. Prior to the capital sentencing phase, defense counsel reviewed the pretrial testimony of a mental health expert who had interviewed defendant Burger and had opined that he was a psychopath. If called to the stand, the defendant may demonstrate the related behavior and affect, the psychologist suggested, or he might even brag about the crime. Given this prospect, as well as the potentially unfavorable nature of expert testimony about future dangerousness and the lack of remorse, the Supreme Court concluded that it was reasonable for defense counsel not to call either his client or the psychologist as a witness at sentencing.

Twenty years later, the Supreme Court was presented with a somewhat similar issue. In *Schriro v. Landrigan*, the defendant was a convicted murderer who had escaped from an Oklahoma prison, only to resurface in Phoenix, Arizona, where he strangled and stabbed his sexual partner to death. After being convicted of capital murder, the defendant presented almost no mitigating evidence during the penalty phase and was eventually sentenced to death. Without mentioning the disorder, the trial judge described Landrigan in words that sound of psychopathy: "Mr. Landrigan is a person who has no scruples and no regard for human life and human beings.... [He] appears to be an amoral person" (*Landrigan v. Stewart*, 2001: 1224).

In subsequent proceedings, Landrigan claimed that he had received ineffective assistance of counsel due to his attorney's failure to investigate whether, among other things, he suffered from a mental disease. A psychologist conducted several interviews and multiple clinical tests, diagnosing Landrigan with antisocial personality disorder. According to the psychologist's affidavit, the defendant suffered a life-long diminished capacity that prevented him from making reasoned decisions, "as we have come to define free will," taking into account the impact of his behaviors on other individuals and society (*Schriro v. Landrigan*, 2007: 490–491). Before the Supreme Court, the defense claimed that a competent attorney would have introduced this type of mitigating evidence, which might have made a difference in the ultimate assessment of Landrigan's "moral culpability" and helped to explain his defiant behavior at trial. The government rebuffed these assertions, noting that the psychologist's diagnosis only confirmed character traits that are aggravating in nature and make Landrigan an ideal candidate for capital punishment.

Almost the entire oral argument in the Supreme Court focused on whether the defendant had waived his rights, without once mentioning antisocial personality disorder. The same was true of Justice Thomas's opinion for the Court, which rejected the defendant's Sixth Amendment claim on procedural grounds.[42] Toward the end, however, the opinion appeared to agree with the government that evidence related to Landrigan's psychiatric disorder would have provided poor mitigation, if not providing aggravating evidence. In particular, Thomas quoted from the decisions below and the "chilling" prospect facing the trial court:

> [B]efore he was 30 years of age, Landrigan had murdered one man, repeatedly stabbed another one, escaped from prison, and within two months murdered still another man.... In his comments [to the sentencing judge], defendant not only failed to show remorse or offer mitigating evidence, but he flaunted his menacing behavior. On this record, assuring the court that genetics made him the way he is could not have been very helpful. (*Schriro v. Landrigan*, 2007: 480–481).

In a dissenting opinion delivered by Justice Stevens, four members argued that the Court had trivialized the value of the mitigating evidence and the impact of the attorney's failure to uncover "a serious psychological condition that sheds important light on [Landrigan's] earlier actions" (*Schriro v. Landrigan*, 2007: 482). Testimony regarding the neurological disorder might have affected any assessment of the defendant's blameworthiness and helped explain his in-court behavior, potentially altering the life-or-death judgment of the trial court. "[T]he Court's decision can only be explained by its increasingly familiar effort to guard the floodgates of litigation," Stevens concluded (*Landrigan v. Stewart*, 2001: 499).

Whatever the majority's motivation, it seems plausible that a contrary decision would have given some credence to the theory of psychopathy as mitigation. In all probability, however, criminal justice actors would still view psychopathy as an aggravating rather than mitigating factor in capital trials. As one defense attorney put it, "you might as well throw gasoline on a fire" by presenting "testimony that [the defendant] was a manipulative psychopath" (*Cummings v. Secretary for Dept. of Corrections*, 2010: 1347). If anything, the Supreme Court's decision confirmed that defense attorneys are unlikely to be second-guessed for failing to present evidence of the disorder during the penalty phase. This sentiment received further support in the 2011 case of *Cullen v. Pinholster*, which, like *Landrigan*, focused on the penalty phase of a capital trial involving a brutal murderer with a violent past.

At trial, defense counsel initially sought to prevent the prosecution from presenting aggravating evidence on procedural grounds. When that failed, counsel called defendant Pinholster's mother in an apparent attempt to provoke sympathy for his family. The defense did not present a mental health expert, although it had consulted with a psychiatrist, Dr. John Stalberg, who had reached the following conclusions: "Pinholster did not manifest any significant signs or symptoms of mental disorder or defect other than his antisocial personality disorder"; "he was fully aware of what he was doing at the time of the offenses"; and "it is likely he would be recalcitrant and a security problem while in custody" (*Pinholster v. Ayers*, 2009: 703). A staff psychiatrist at San Quentin State Prison had also diagnosed Pinholster as having antisocial personality disorder, noting that "[t]his man's conduct showed a high degree of cruelty, callousness and viciousness" and "no responsible regard for the reasonable rights of other people" (*Pinholster v. Ayers*, 2009: 719). Ultimately, the court imposed a sentence of death.

In subsequent proceedings, Pinholster claimed ineffective assistance of counsel, accusing his trial attorneys of failing to adequately investigate and present evidence of his psychological abnormalities. To bolster the argument, mental health experts testified that, *inter alia*, Pinholster did not have antisocial personality disorder but instead suffered from an "organic personality syndrome" resulting from a brain injury, which, in turn, explained his aggressive, impulsive, and antisocial behavior. In accepting this argument, a divided federal appellate court concluded that the additional medical evidence would have helped counter the government's case for capital punishment.

> First, evidence that Pinholster's brain damage may have influenced, or even caused, his behavior at the time of the crime may have led jurors to conclude that he was less morally culpable at the time of the offense.... Second, properly presented evidence of Pinholster's brain injury, and its profound effect on his behavior, could have altered the jury's impressions of...his boastful, disrespectful demeanor by indicating an organic basis for his inappropriate expressions and for his tendency to exaggerate his past.... Third, evidence of Pinholster's organic brain injury would have humanized him in the eyes of the jury, even if the jury concluded that his brain injury was not responsible for his actions during his commission of the crime. (*Pinholster v. Ayers*, 2009: 676–677)

In contrast, a pointed dissenting opinion emphasized that trial counsel had hired a competent psychiatric expert, whose diagnosis was simply devastating for the defense.

> Given Dr. Stalberg's expert opinion, trying to develop a psychiatric mitigation case at the time of trial would have been extraordinarily difficult. He told the lawyers that Pinholster was entirely sane and sober on the night of the murder and that, in general, he's a psychopath.... Had Dr. Stalberg testified, Pinholster would be here arguing that his lawyers were incompetent for putting him before the jury. With no realistic possibility of a psychiatric mitigation defense, what could Pinholster's lawyers do?... If [counsel] puts on evidence about his medical or mental problems, this opens the door to evidence that he's a remorseless psychopath. Pinholster's counsel faced a serious risk that a mitigation case could turn out to be aggravating. Unlike habeas counsel, who have years of time, unlimited resources and the power to conjure imaginary mitigation cases with which to mesmerize federal judges, trial counsel are stuck with the hard realities that are the lot of the trial lawyer. (*Pinholster v. Ayers*, 2009: 703, 707)[43]

After describing the defendant's condition in prison—where "[h]e sits in his cell reading Machiavelli, Voltaire 'and all the philosophers,' drawing pictures to sell over the internet," "still stab[bing] people whenever he can, without passion or regret," and "never express[ing] the least remorse for his killings"—the dissent predicted that the Supreme Court would reinstate Pinholster's death sentence (*Pinholster v. Ayers*, 2009: 723–724).

In 2011, this prognostication was proven correct. Writing for the majority, Justice Thomas stressed that trial counsel faced a challenging penalty phase with an unsympathetic client, thereby limiting any mitigation strategy.

> By the end of the guilt phase, the jury had observed Pinholster "glor[y]" in "his criminal disposition" and "hundreds of robberies." During his cross-examination, Pinholster laughed or smirked when he told the jury that his "occupation" was "a crook," when he was asked whether he had threatened a potential witness, and when he described thwarting police efforts to recover a gun he had once used. He bragged about being a "professional robber." (*Cullen v. Pinholster*, 2011: 1405)[44]

Noting Dr. Stalberg's conclusion that the defendant was a legally sane psychopath, the *Pinholster* majority believed it was a reasonable penalty-phase strategy to focus on evoking sympathy for the defendant's family.

In dissent, Justice Sonia Sotomayor argued that "Dr. Stalberg's two-page report, which was based on a very limited record," could not excuse trial counsel's failure to investigate a broader range of potential mitigating circumstances including the possibility that Pinholster's behavior resulted from a brain injury, as well as other factors such as a history of substance abuse and parental neglect (*Cullen v. Pinholster*, 2011: 1429). In the end, however, the majority and dissent may both be right: The defendant might be a psychopath, and his disorder may be related to a brain abnormality, dysfunctional upbringing, and substance abuse. Nonetheless, the Court's holding reaffirmed that counsel will not be deemed incompetent in foregoing evidence of psychopathy during capital sentencing.

SOME REFLECTIONS

This brings the chapter full circle. As a matter of theory and doctrine, pscyhopathy could serve as mitigation at sentencing based on an offender's diminished capacity for rational decision-making. But the disorder might also offer good reason to increase punishment, given the high rate of recidivism among psychopaths and the lack of effective treatment. Psychopathy thus provides a near perfect example of the perpetual conflict between consequentialist and non-consequentialist theories of punishment. The precise attributes that a retributivist might point to as justification for a reduced sentence—the long-term cognitive and affective defects of the psychopath—may offer the utilitarian a good reason for enhanced punishment. The issue epitomizes the more general clash among punishment rationales, the disagreement among scholars, and the theoretical uncertainty in sentencing schemes across the nation.

In practice, however, consequentialism and expressive condemnation of the psychopathic offender have prevailed at sentencing, particular in death penalty proceedings.[45] As one court observed, "a diagnosis as a psychopath is a mental health factor viewed negatively by jurors and is not really considered mitigation" (*Looney v. State*, 2006: 1028–1029).[46] Rather, evidence of psychopathy is typically offered by prosecutors to demonstrate an offender's risk of recidivism and potential dangerousness (DeMatteo & Edens, 2006; Walsh & Walsh, 2006). This evidence can have a substantial impact on the relevant decision makers, as suggested by mock-trial studies on support for capital punishment (Edens, Colwell, Desforges, & Fernandez, 2005; Krauss & Sales, 2001). Conversely, trial judges and juries tend to reject psychopathy evidence as grounds for mitigation, and reviewing courts are unlikely to upset such decisions or second guess defense attorneys who fail to raise the issue at sentencing. One of the few times a defendant successfully employed the PCL-R was to demonstrate that he was *not* a psychopath and therefore was amenable to rehabilitation and unlikely to recidivate (*Muhammad v. State*, 2001).[47]

Obviously, the one-sided nature of psychopathy as aggravation in sentencing does not reflect the theoretical stalemate between retributivism and utilitarianism. Nor is it consistent with the more general opinions of the citizenry, whose views on sentencing seem to comport with notions of desert (Carlsmith, Darley, & Robinson, 2002; Robinson & Kurzban, 2007). But when punishment rationales are diametrically and dramatically opposed, as is the case with the psychopathic offender, decision makers may simply opt for crime prevention over retributive justice. In these circumstances, utilitarian talk about future dangerousness may also provide the appearance of dispassionate analysis, the weighing of social costs and benefits, when in fact the real impetus for heightened punishment can be found in base emotions (e.g., anger, fear, and disgust) and an enduring belief in the existence of human evil (Kahan, 1999).

Needless to say, the behavior and affect of the psychopath can be both terrifying and mind-boggling. In search of explanation for what appears to be coldly calculated criminality, people may prefer to characterize the offender as evil rather than mentally ill. By rejecting psychopathy as an affirmative defense at trial, American criminal law seems to embrace the aptly named strategy of "monster-barring" (Colb, 1999). Along these lines, criminal justice actors may be engaged in a kind of "monster-slaying" by sentencing the psychopathic offender to death or imprisoning him for long periods of time. Social condemnation need not be derived from ascriptions of evil, however. The reactive attitudes of decision makers may turn on the psychopath's appearance of normality, his ostensible rationality, and his comprehensible but reprehensibly selfish acts. He is not a child whose behavior we might easily attribute to immaturity. Nor is he the archetypal raving lunatic of palpable irrationality. Instead, the psychopath *looks* like all other members of society who may be blamed and punished for their crimes.

The most straightforward account of punishment practice is that psychopaths are sentenced based on a commonsensical approach that includes relatively thin conceptions of rationality and volition. Whatever his problems may be, the psychopath grasps reality and has sufficient mental abilities for his actions to be deemed voluntary and deliberate. His self-interested decisions and lack of remorse justify harsh punishment. The works of allied disciplines that could offer insight and add depth to sentencing principles—philosophical tracts on moral agency and reasoning, and scientific studies on the etiology of psychopathy and the associated cognitive and affective dysfunctions—may not be readily accessible or, for that matter, comprehensible to most criminal justice actors.

Then again, these works might be considered irrelevant to sentencing, given that legal principles do not always, or even usually, coincide with those in allied fields. Law may be a parasitic discipline, inevitably drawing upon the humanities and the natural and social sciences (Luna, 2009), but criminal justice is not merely a conduit of philosophical and scientific theory. Models of moral responsibility do not (and, some would argue, should not) define legal responsibility as a matter of doctrine and practice. Most positive law is a political endeavor, of course, and certain polices may be nonstarters in American politics, no matter how well grounded they are in moral theory or scientific study. Moreover, lawmakers can ignore or define things away by fiat, which no decent philosopher or researcher would dare do in his scholarly analysis.

Law and science maintain an especially contentious relationship. As mentioned, criminal law assumes human rationality and volition based on folk psychology,[48] whereas science has historically taken a deterministic worldview of physiological causation. The result is a "cold war between lawyers and psychiatrists," as Karl Menninger (1968) titled it. Mental health experts can diagnose defendants as suffering from a disease, for instance, but judges and juries might reject such conclusions as incorrect or simply beside the point. To be fair to criminal justice actors, a degree of suspicion may be in order for any expert testimony in this area. Such evidence can be cloaked in scientific objectivity, but value judgments (and even raw self-interest) may underlie both the reason to serve as a witness and the shape and content of the resulting testimony (Halleck, 2001; Murrie, Boccaccini, Johnson, & Janke, 2008).

For psychopathy, the problems start with basic taxonomy. Up to this point, the chapter has accepted references to sociopathy and antisocial personality disorder as synonyms for psychopathy. The reason was straightforward: The terms are often used interchangeably in court, and not just by legal professionals (Hare, 2002; Ogloff & Lyon, 1998; Zinger & Forth, 1998). In fact, the bible of mental health assessment, the American Psychiatric Association's *Diagnostic and Statistical Manual*, states that antisocial personality disorder (APD or ASPD) "has also been referred to as psychopathy, sociopathy, or dissocial personality disorder" (American Psychiatric Association, 2000: 702).[49] For the researcher and clinician, the distinction between psychopathy and sociopathy may depend on his views on the determinants of the disorder, whether its origins are properly traced to innate characteristics or socializing institutions. "The same individual therefore could be diagnosed as a sociopath by one expert," Professor Hare notes, "and as a psychopath by another" (Hare, 1993: 24). The choice appears subjective and any alleged difference could be misleading, eliding the complex interrelationship between biological and environmental factors. Interestingly, sociopathy has not existed as a formal psychiatric diagnosis for four decades (Ogloff & Lyon, 1998: 411), yet expert witnesses still refer to defendants as sociopaths and courts continue to believe that "sociopathy is a recognized mental disease" (*U.S. ex rel. Griffin v. Mathy*, 2009: 6 n.9).

Despite considerable overlap between psychopathy and APD, the differences between the disorders can be critical in punishment. Psychopaths tend to meet the criteria for APD, which has a heavy focus on behavior. The reverse is not always true, however, as APD leaves out almost all of the character traits that distinguish the psychopath from nonpsychopathic criminals (Hare, 2002; Kiehl & Hoffman, 2011). It would be a potentially lethal mistake at sentencing, therefore, to associate an APD-diagnosed offender with the most disconcerting propositions about psychopathy (e.g., the lack of effective treatment). But given the interchangeable use of terms by expert witnesses and more than a little confusion within the health profession as a whole, it should not be surprising that legal actors, even Justices of the U.S. Supreme Court, treat psychopathy, sociopathy, and APD as different words for the same disorder. The concern here is not merely the loose language of judicial opinions, but the very real possibility that the terms are at times used to convey an ambiguous lay concept, more or less, the "evil, unredeemable, misanthrope" (Edens, Petrila, & Buffington-Vollum, 2001; Wolfson, 2008).

Whatever term is used—and assuming it is not merely a condemnatory label—a question remains as to the usefulness of psychopathic diagnoses for purposes of sentencing. Predicting future dangerousness is a notoriously difficult endeavor, susceptible to a high number of false positives. At the time the Supreme Court decided *Barefoot*, expert testimony on future dangerousness was expected to be incorrect at least two-thirds of the time (Brief for American Psychiatric Association, 1983). More recent scholarship indicates that risk predictions have improved to better than chance, but they still have a significant error rate (Cunningham, Reidy, & Sorensen, 2008; Edens, Buffington-Vollum, Keilen, Roskamp, & Anthony, 2005; Krauss & Sales, 2001; Mossman, 1994; Otto, 1992; Texas Defender Service, 2004). The sources of faulty forecasting may include inadequate training and experience of examiners, assessments made without sufficient information (e.g., collateral data alone), and the distortive effects on scoring from cultural differences and extended incarceration (Edens & Petrila, 2001; Krauss & Sales, 2001; Nair & Weinstock, 2008). Other problems include the low base rates of the disorder, disagreement over cutoff scores for psychopathy diagnoses, the limited validation of risk assessment instruments in detention facilities, and the differences in predicting violence in the community versus that in jails and prisons (e.g., Cunningham & Reidy, 2002; Dorland & Krauss, 2005; Edens, Buffington-Vollum, Keilen, Roskamp, & Anthony, 2005; Friedman, 2001). Not least of all, provisions on future dangerousness use vague language—for example, "a probability that the defendant would commit criminal acts of violence that would constitute a continuing threat to society"[50]—often without clarification as to how high the probability must be or whether the term *society* refers to those inside or outside of the prison walls (Dorland & Krauss, 2005; Edens, Buffington-Vollum, Keilen, Roskamp, & Anthony, 2005; Schopp, 2006).

For these and other reasons, diagnoses of psychopathy may generate arbitrary outcomes at sentencing akin to lightning strikes. Contemporary scholarship has focused heavily on pscyhopathy as an aggravating factor in death penalty jurisprudence, which is understandable given the role that future dangerousness plays in capital decision-making and the lethal nature of the resulting verdict.[51] But the stakes are large in noncapital sentencing as well. A high PCL-R score could, in the words of some offenders, "flush their lives down the drain" (Hare, 1998, 2002: 203). Recent works (Hare & Neumann, 2010; Skeem & Cooke, 2010) have only heightened the debate and might encourage further challenges to the use of psychopathy diagnoses in court. In light of the potential consequences of misjudgment—a defendant's liberty or even his life lying in the balance—a formidable case can be made against admitting evidence of psychopathy at sentencing.[52]

CONCLUSION

Of course, some of this may change in the coming years. For instance, the American Psychiatric Association recently released a working draft of the fifth edition of the *Diagnostic and Statistical Manual* (DSM-V), with the final version scheduled for publication in May 2013 (DSM-5 Development). The most significant revisions concern Axis II diagnoses and, in particular, personality disorders, which have long been the subject of criticism and proposals for change (Hare, 2006; Krueger, Markon, Patrick, & Iacono, 2005; Westen & Shedler, 2000; Widiger, Simonsen, Krueger, Livesley, & Verheul, 2005). Among other things, the draft offers a new definition of personality disorders, provides a narrative description of each disorder, and creates a multipart, gradated assessment of personality types and traits. Most relevant for present purposes, DSM-V reformulates APD as the "antisocial/psychopathic type" of personality disorder. Unlike its predecessor, the new disorder type incorporates many of the affective and interpersonal components associated with psychopathy, such as callousness and narcissism, drawing heavily upon the criteria specified in the PCL-R. Although the DSM-V approach to antisocial/psychopathic disorder is not identical to that of the PCL-R, the proposed reformulation could help rectify lingering confusion over psychopathy and any resulting misdiagnoses in court.

Moreover, the scientific study of psychopathy is proceeding at a rapid pace, providing a greater understanding of the disorder, its etiology, and its consequences. Basic propositions about the disorder are contestable today—for instance, the claim that a psychopath cannot be treated—and in time, they may turn out to be incorrect.[53] Among other things, the future could bring effective treatment strategies for psychopaths and nonincarcerative means of protecting society from their offending. Compelling evidence may be forthcoming of a causal link between the biological disorder and individual misconduct, adding a cause-and-effect relationship to the powerful work of neuroimaging and genetic testing. This knowledge could provide a viable basis for defense attorneys to seek mitigated sentences for psychopathic offenders.[54]

For better or worse, scientific advancement might also contribute to the double-edged nature of psychopathy in theory and doctrine, and the disorder's damning character in real-world practice. Predictions of future offending may become more accurate as diagnostic tools are honed and more clinicians trained in their appropriate use, thereby buttressing the arguments for increased punishment. It is even conceivable that psychopathy could become an important aggravating factor in noncapital sentencing. The proliferation of civil incarceration schemes for sexual predators could have a spill-over effect, with prosecutors seeking extended incapacitation of psychopathic offenders at the front-end of punishment. Some jurisdictions have already incorporated risk assessment instruments into their sentencing schemes (Ostrom, Kleiman, Cheesman, Hansen, & Kauder, 2002; Demleitner, 2004; Wolff, 2006). Although these instruments focus on criminal records and demographic data, future tools might well incorporate, for instance, the interpersonal and affective deficits associated with psychopathy. For many, the idea of selective incapacitation of genuine psychopaths may be particularly attractive, given the staggering costs of large prison populations and the natural desire of society to reduce the amount of serious and violent crime.

For now, the intellectual debate continues unresolved, stuck in the interstices of science, philosophy, and law. The theories of punishment remain at loggerheads—retributive justice versus utilitarian security, individual liberty clashing with actuarial assessment. Considerations of safety still reign in court, however, where psychopathy tends to engender harsher sentences for the disordered offender. Only time will tell whether this latter practice is unjust, a violation of liberal sentencing principles, or just sound public policy for a society overwhelmed by its own system of crime and punishment.

ACKNOWLEDGMENTS

Many thanks to Deborah Denno, Kent Kiehl, and Walter Sinnott-Armstrong for their thoughtful comments on this chapter.

NOTES

1. See Chapters 15 by Litton and 16 by Pillsbury on responsibility and Chapters 17 by Morse and 18 by Corrado on preventive detention. Another issue that is beyond the scope of this chapter is the relevance of psychopathy for determinations of competency to stand trial. See, for example, *United States v. Gabrion* (2011); *United States v. Mitchell* (2010).
2. But see, for example, *Ward v. State* (2009: 263), questioning psychopathy as mitigation because, inter alia, "[t]he defendant's expert witness acknowledged that 'psychopath' and 'psychopathy' are listed as diagnoses in neither the DSM-IV nor in its subsequent revision." See the section "Some Reflections."
3. For useful introductions and collections on punishment theory, see Ashworth & von Hirsch

(2008); Ezorsky (1972); Honderich (2005); Primoratz (1989).

4. For discussions of different types of retributivism, see Dressler (2009); Robinson (2008).
5. See, for example, Fletcher (2007: 10), noting that the "problem of attributing agency has traditionally been addressed under the label of 'free will'" and describing the history and problems of the concept.
6. See, for example, Morse (2001, 2002, 2007). But see Cal. Penal Code § 207(c), using term "free will" in crime of kidnapping; Colo. Rev. Stat. § 18-3-401, defining consent in terms of "free will" for purposes of sex crimes; Barovick and Seaman (1999), quoting forensic psychologist as insisting that those with antisocial personality disorder "lack free will."
7. See, e.g., Morse (2001, 2002, 2007). See also Dressler (2009), discussing retribution, free will, and rationality. The capacity for rational decision-making is often referred to as "practical reasoning," which itself is a protean concept in moral and legal theory. See, for example, Audi (1982), referring to practical reasoning as "a term of art with little life in ordinary parlance and a multiple personality in philosophical literature."
8. See Chapters 8 by Anderson and Kiehl and 9 by Boccardi.
9. See Chapters 10 by Viding, Fontaine, and Larsson and 11 by Waldman and Rhee.
10. See Chapter 7 by Borg and Sinnott-Armstrong.
11. See Chapter 13 by Rice and Harris. See also Kiehl & Hoffman (2011).
12. But see Chapter 12 by Caldwell in the present volume for promising developments in the treatment of youth with significant psychopathic traits. See also Kiehl & Hoffman (2011).
13. This part and the next offer potential arguments about psychopathy and punishment, rather than my personal opinions on the topic. I continue to reflect upon psychopathy's role (if any) in determinations of criminal liability and sentencing.
14. The term *diminished capacity* actually encompasses two distinct concepts: a "mens rea" variant that is offered to show the defendant lacked the requisite mental state for a given crime, and a "partial responsibility" variant that is presented to reduce the offense level and/or the amount punishment based on a defendant's impaired capacity for rationality. See, for example, Arenella (1977); Dressler (2009); Morse (1984, 2002-b). The term is used here to refer to the diminished capacity variant.
15. In a series of works, Professor Morse has argued in favor of a mitigating excuse of partial responsibility and a new verdict of "guilty but partially responsible," which would call for fixed sentencing reductions that (presumably) could apply to those suffering from psychopathy. See Morse (1999, 2003, 2011). See also von Hirsch and Ashworth (2008); *United States v. Leandre* (1998).
16. See, for example, Morse (2000) (critiquing social deprivation as an excuse); von Hirsch & Ashworth (2005) (same for mitigation); U.S. Sentencing Guidelines Manual §§ 5H1.10, 5H1.12 (rejecting socioeconomic disadvantage and lack of guidance as a youth as bases for reduced punishment). But see, for example, Bazelon (1976) (arguing in favor of social deprivation defense); Delgado (1985) (same); *United States v. Brady* (2005) (holding that childhood abuse could serve as a basis for a sentence reduction).
17. To be clear, Professor Murphy was not advocating this type of legal response to psychopaths. In fact, he concluded with some "very grave objections" to practices that treat psychopaths as less than persons. See pp. 295–298 in Murphy (1972).
18. Moreover, coercion—even a threat of lethal violence—is not always an excusing condition. See, for example, Wash. Rev. Code § 9A.16.060 (duress defense unavailable for homicide or if the defendant placed himself in a situation where he was likely to be subject to coercion).
19. See, for example, Morse and Hoffman, (2007) ("if one believes that the law should not blame and punish those who are not morally responsible, there is a strong case for excusing some psychopaths").
20. For instance, a psychopath who cannot understand the moral wrongfulness of his conduct might conceivably meet the requirements of some insanity laws. See, for example, *People v. Serravo* (1992).
21. 18 U.S.C. § 3553(a)(1).
22. For instance, hybrid (or mixed) sentencing theories adopt a general rationale for punishment constrained on the edges by principles of another justification. See Hart (1968); Morris (1974); Robinson (1987). See also American Law Institute, Model Penal Code: Sentencing,

Tentative Draft No. 1 (April 9, 2007) (espousing Morris's "limiting retributivism" approach).

23. See Utah Sentencing Commission, 2011 Adult Sentencing and Release Guidelines, available at http://www.sentencing.utah.gov/guidelines/Adult/2011%20Adult%20Sentencing%20and%20Release%20Guidelines.pdf; Utah Sentencing Commission, Penalty Distribution for Selected Crimes, available at http://www.sentencing.utah.gov/Penalty%20Distribution/PenaltyDistributionBooklet.pdf.

24. See Cal. Penal Code § 1170.

25. See Cal. Rules of Court, Rules 4.421(b)(1) & 4.423(b)(2). See, for example, *Cunningham v. California* (2007), describing California's sentencing scheme. This chapter does not consider anti-recidivist statutes, epitomized by California's "three strikes and you're out" law, which focus solely on prior offending without any meaningful consideration of mental disorders such as psychopathy.

26. See U.S. Sentencing Guidelines Manual § 4A1.3(a)(1).

27. See U.S. Sentencing Guidelines Manual § 5k2.13. It should be noted, however, that this guideline states that a downward departure is not appropriate when, for instance, the defendant's criminal history indicates a need to incarcerate the defendant to protect the public.

28. See Chapter 14 by Edens, Magyar, and Cox.

29. See also *Wilkerson v. Felker* (2009), upholding parole board denial of defendant's application for parole based on, inter alia, psychological testimony of psychopathy; *Lovelace v. Oregon* (2009), rejecting federal civil rights suit by offender whose parole was revoked based in part on PCL-R testing and a finding of psychopathy.

30. Interestingly, the psychologist in *McNichols* testified that "some people think of [such behaviors] as being psychopath...but the psychology term is Antisocial Personality Disorder" (*McNichols v. State*, 2009: *5)

31. Similarly, an Ohio trial judge said the following when sentencing a violent recidivist: "I had to look up a word, Mr. Coomer, because it came to mind as I also read the numerous letters of support. Some of them indicate you are an honest, hard-working, law-abiding—well, good guy. And the word that came to mind when I read those letters and the disparity between your criminal history and these folks who may not be aware of your criminal history is the word psychopath. A psychopath is defined in the dictionary as a person with a personality disorder, especially one manifested in an aggressively antisocial behavior. That's what I see here, aggressive antisocial behavior" (*State v. Coomer* 2010: *2). The defendant then received a 10-year prison term in order to protect the victims and the general public. On appeal, the reviewing court rejected the claim that the trial judge had lost his ability to remain impartial when he labeled the defendant a psychopath.

32. For discussion about the limitations of a diagnosis for antisocial personality disorder, see Chapter 2 by Forth, Bo, and Kongerslev.

33. "[The defendant's] mental retardation and history of abuse is thus a two-edged sword: it may diminish his blameworthiness for his crime even as it indicates that there is a probability that he will be dangerous in the future" (*Penry v. Lynaugh*, 1989: 324).

34. Specifically, the question was "whether the fact of the location of the body so close to a line that forbids the death penalty"—that is, the jurisdiction of abolitionist Michigan—"allows counsel to try to convince one or more jurors that imposing the death penalty in these circumstances would treat life or death in a random and arbitrary way based on chance" (*United States v. Gabrion*, 2011: 323). Although the jury found that the defendant's significant antisocial personality disorder was a mitigating factor, it ultimately sentenced him to death.

35. See Wyo. Stat. at § 6-2-102(j)(vi), (viii).

36. See Wyo. Stat. at § 6-2-102(h)(xi).

37. See, for example, 21 Okla. Stat. § 701.12(7); Or. Rev. Stat. § 163.150(1)(b)(B); Tex. Code Crim. Proc. art. 37.071, § 2(b)(1); Va. Code § 19.2-264.2(1).

38. More than three decades later, however, Justice Stevens would openly denounce the death penalty. See, for example, Stevens (2010); *Baze v. Rees* (2008: 71–86).

39. *Jurek v. Texas* (1976: 273), discussing *Smith v. State*, 540 S.W.2d 693 (Tex. Crim. App. 1976).

40. Relying upon the reasoning in *Roper*, the Supreme Court subsequently held that the Eighth Amendment prohibits the imposition of a life without parole sentence on a juvenile offender who did not commit homicide. See *Graham v. Florida* (2010).

41. To establish a violation of this right, the defendant must demonstrate that his counsel's performance was constitutionally deficient and that the deficient performance prejudiced his case. See *Strickland v. Washington* (1984). The often complicated doctrine in this area of law is beyond the scope of the present chapter.
42. The precise legal issue in *Landrigan* concerned the discretion of U.S. District Courts to deny evidentiary hearings for petitions seeking a writ of habeas corpus. The intricacies of federal habeas corpus review are extremely complex and beyond the scope of this chapter.
43. See also *Pinholster v. Ayers* (2009: 718–719), describing how the prosecution's cross-examination of Dr. Stalberg might have proceeded.
44. The precise legal issue in *Pinholster* concerned the deference that federal habeas courts must give to state court judgments. Again, habeas corpus review is beyond the scope of this chapter.
45. Although noting that in theory mental abnormality is "a knife that cuts both ways in sentencing," Professor Morse has argued that "the empirical basis for the alternatives of mitigation and aggravation is asymmetrical" in favor of an excuse of partial responsibility (Morse, 2011: 944). However, he acknowledges the unlikelihood that a jurisdiction would preclude aggravating evidence in the form of predictions of future dangerousness. For this reason, Professor Morse proposes a pair of prophylactic measures designed to limit inaccurate assessments: (1) "the state should require use of the most empirically validated prediction methods rather than clinical evaluations or responses to hypothetical questions"; and (2) "the defendant must have access to an independent mental health professional to help him prepare mitigation evidence and to defend against aggravation evidence of future dangerousness" (Morse, 2011: 945).
46. See also *Cummings v. Secretary for Dept. of Corrections* (2010), collecting cases and concluding that APD diagnosis "is not mitigating but damaging").
47. See also *United States v. Stall* (2009), where a defense expert testified that the defendant "did not have a 'criminal mind' or exhibit the qualities of a psychopath"; *United States v. Campbell* (2010), where a forensic psychologist concluded that the defendant was a low risk to recidivate based on, *inter alia*, a PCL-R assessment.
48. "The criminal sanction," Herbert Packer wrote, "does not rest on an assertion that human conduct *is* a matter of free choice; that philosophic controversy is irrelevant. In order to serve purposes far more significant than even the prevention of socially undesirable behavior, the criminal sanction operates *as if* human beings have free choice" (Packer, 1968: 132).
49. But see, for example, *Ward v. State* (2009: 963), questioning psychopathy as mitigation because, *inter alia*, "[t]he defendant's expert witness acknowledged that 'psychopath' and 'psychopathy' are listed as diagnoses in neither the DSM-IV nor in its subsequent revision."
50. See 21 Okla. Stat. Ann. § 701.12(7); Or. Rev. Stat. § 163.150(1)(b)(B); Tex. Crim. Proc. Code. Ann. art. 37.071, § 2(b)(1).
51. Although one might assume that categorical injustice is a relatively infrequent occurrence in this area, the case of Randall Adams offers a reminder of the high stakes involved. He was convicted of a 1976 cop killing and sentenced to death, based in large part on testimony by Dr. Grigson that Adams was a psychopath "at the very extreme, worse or severe end of the scale," who could not be rehabilitated and presented a future danger to society. As it turns out, however, Adams was innocent of the crime and was eventually exonerated after spending a dozen years in prison. See Texas Defender Service (2004: 24–25); Radelet, Bedau, & Putnam (1992). But for Supreme Court intervention on an issue unrelated to Adams's innocence, he might be dead. See *Adams v. Texas* (1980). And but for a filmmaker's interest in his case, Adams might have languished in prison. See *The Thin Blue Line* (Third Floor Productions 1988).
52. In fact, there may be good arguments against allowing parties to refer to a defendant as a "psychopath," just as it would be inappropriate to call him an "animal" that should be "put...down to sleep." See *Wilson v. Sirmons* (2008), distinguishing between these prosecutorial references.
53. See Chapter 12 by Caldwell. See also Kiehl & Hoffman (2011: 357, 391–397).
54. In late 2009, one of the editors of the present volume, Kent Kiehl, testified in the case of Brian Dugan, an admitted murderer facing the death penalty in Illinois. Dugan scored in the 99th percentile on the PCL-R, and functional magnetic

resonance imaging (fMRI) showed abnormalities in his paralimbic system. Professor Kiehl testified regarding much of this information and opined that Dugan suffered from psychopathy when he committed the murder a quarter of a century earlier. Indeed, this may have been the first case in which a psychopathic defendant offered expert testimony on fMRI scans as mitigation evidence. But the trial judge precluded the jury from seeing the actual scans of Dugan's brain, and the jury eventually voted in favor of the death penalty. See Haederle (2010); Rozek, (2009: 24); Gregory & Barnum (2009: 14).

REFERENCES

Adams v. Texas, 448 U.S. 38 (1980).

American Law Institute (April 9, 2007). Model Penal Code: Sentencing, Tentative Draft No. 1.

American Psychiatric Association (2000). *Diagnostic and statistical manual of mental disorders* (4th ed.)–Text Revision. Washington, DC: Author.

Arenella, P. (1977). The diminished capacity and diminished responsibility defenses: Two children of a doomed marriage. *Columbia Law Review*, 77, 827.

Arenella, P. (1992). Convicting the morally blameless: Reassessing the relationship between legal and moral accountability. *UCLA Law Review*, 39, 1511.

Ashworth, A., & von Hirsch, A. (2008). *Principled sentencing: Readings on theory and policy*. Oxford: Hart.

Audi, R. (1982). A theory of practical reasoning. *American Philosophical Quarterly*, 19, 25.

Barefoot v. Estelle, 463 U.S. 880 (1983).

Barovick, H., & Seaman, D. (December 27, 1999). Bad to the bone. *Time*.

Bazelon, D. L. (1976). The morality of the criminal law. *Southern California Law Review*, 49, 385.

Beecher-Monas, E., & Garcia-Rill, E. (2003). Danger at the edge of chaos: Predicting violent behavior in a post-*Daubert* world. *Cardozo Law Review*, 24, 1845.

Beil, L. (July 26, 1995). Groups expel Texas psychiatrist known for murder cases. *Dallas Morning News*, 21A.

Blair, J., Mitchell, D., & Blair, K. (2005). *The psychopath: Emotion and the brain*. Malden, MA: Blackwell.

Blume, J. H., & Garvey, S. P. (2001). Future dangerousness in capital cases: Always "at issue." *Cornell Law Review*, 86, 397.

Boyd v. Brunsman, No. 3:10CV2166, 2012 WL 274745 (N.D. Ohio Jan. 31, 2012).

Brief for the American Psychiatric Association as Amicus Curiae in Support of Petitioner, *Barefoot v. Estelle*, 463 U.S. 880 (1983) (No. 82–6080).

Burger v. Kemp, 483 U.S. 776 (1987).

Carlsmith, K. M., Darley, J. M., & Robinson, P. H. (2002). Why do we punish?: Deterrence and just deserts as motives for punishment. *Journal of Personality & Social Psychology*, 83, 284.

Carter, L. E., Kreitzberg, E. S., & Howe, S. W. (2008). *Understanding capital punishment law* (2d ed.). New York: Lexis-Nexis.

Cleckley, H. (1988). *The mask of sanity: An attempt to clarify some issues about the so-called psychopathic personality* (5th ed.). St. Louis, MO: Mosby.

Colb, S. F. (1999). The character of freedom. *Stanford Law Review*, 52, 235.

Coyne, R., & Entzeroth, L. (2006). *Capital punishment and the judicial process* (3d ed.). Durham, NC: Carolina Academic Press.

Cullen v. Pinholster, 131 S. Ct. 1388 (2011).

Cummings v. Secretary for Dept. of Corrections, 588 F.3d 1331 (11th Cir. 2010).

Cunningham v. California, 549 U.S. 270 (2007).

Cunningham, M. D., & Reidy, T. J. (2002). Violence risk assessment at federal capital sentencing. *Criminal Justice & Behavior*, 29, 512.

Cunningham, M. D., Reidy, T. J., & Sorensen, J. R. (2008). Assertions of "future dangerousness" at federal capital sentencing: Rates and correlates of subsequent prison misconduct and violence. *Law & Human Behavior*, 32, 46.

Dahmer killed by fellow inmate. (November 29, 1994). *Chicago Sun-Times*.

Daubert v. Merrell Dow Pharmaceuticals, Inc., 509 U.S. 579 (1993).

Delgado, R. (1985). "Rotten social background": Should the criminal law recognize a defense of severe environmental deprivation? *Law & Inequality*, 3, 9.

DeMatteo, D., & Edens, J. F. (2006). The role and relevance of the Psychopathy Checklist-Revised in court. *Psychology, Public Policy & Law*, 12, 214.

Demleitner, N. V. (2004). Risk Assessment: Promises and Pitfalls. *Federal Sentencing Reporter*, 16, 161.

Demouchette v. Collins, 972 F.2d 651 (5th Cir.), cert. denied, 505 U.S. 1246 (1992).

Dorland, M., & Krauss, D. (2005). The danger of dangerousness in capital sentencing: Exacerbating the problem of arbitrary and capricious decision-making. *Law & Psychology Review*, 29, 63.

Dressler, J. (2009). *Understanding criminal law* (5th ed.). New York: Lexis-Nexis.

DSM-5 Development. Available at: http://www.dsm5.org.

Duff, R. A. (1977). Psychopathy and moral understanding. *American Philosophical Quarterly*, 14, 189.

Eddings v. Oklahoma, 455 U.S. 104 (1982).

Edens, J. F., & Petrila, J. Legal and ethical issues in the assessment and treatment of psychopathy. In: C. J. Patrick (Ed.), *Handbook of psychopathy*. New York: Guilford.

Edens, J. F., Buffington-Vollum, J. K., Keilen, A., Roskamp, P., & Anthony, C. (2005). Predictions of future dangerousness in capital murder trials: Is it time to "disinvent the wheel"? *Law & Human Behavior*, 29, 55.

Edens J. F., Colwell, L. H., Desforges, D. M., & Fernandez K. (2005). The impact of mental health evidence on support for capital punishment: Are defendants labeled psychopathic considered more deserving of death? *Behavioral Science & Law*, 23, 603.

Edens, J. F., Petrila, J., & Buffington-Vollum, J. K. (2001). Psychopathy and the death penalty: Can the Psychopathy Checklist-Revised identify offenders who represent "a continuing threat to society"? *Journal of Psychiatry & Law*, 29, 433.

Elliott, C. (1992). Diagnosing blame: Responsibility and the psychopath. *Journal of Medicine & Philosophy*, 17, 199.

Espada v. State, No. AP-75219, 2008 WL 4809235 (Tex. Crim. App. Nov. 5, 2008).

Ezorsky, G., Ed. (1972). *Philosophical perspectives on punishment*. Albany, NY: SUNY Press.

Felthous, A., & Saß, H., Eds. (2008). *The international handbook of psychopathic disorder: Law and policies*. Hoboken, NJ: Wiley.

Fine, C., & Kennett, J. (2004). Mental impairment, moral understanding, and criminal responsibility: Psychopathy and the purposes of punishment. *International Journal of Law & Psychiatry*, 27, 425.

Fischer, J. M., & Ravizza, M. (1998). *Responsibility and control: A theory of moral responsibility*. New York: Cambridge University Press.

Fischette, C. (2004). Psychopathy and responsibility. *Virginia Law Review*, 90, 1423.

Fletcher, G. P. (2007). *The grammar of criminal law: American, comparative, and international*. New York: Oxford University Press.

Flores v. Johnson, 210 F.3d 456 (5th Cir. 2000).

Freedman, D. (2001). False prediction of future dangerousness: Error rates and Psychopathy Checklist-Revised. *Journal of the American Academy of Psychiatry & Law*, 29, 89.

Furman v. Georgia, 408 U.S. 238 (1972).

General Electric Co. v. Joiner, 522 U.S. 136 (1997).

Glannon, W. (2008). Moral responsibility and the psychopath. *Neuroethics*, 1, 158.

Graham v. Collins, 506 U.S. 461 (1993).

Graham v. Florida, 130 S. Ct. 2011 (2010).

Gregg v. Georgia, 428 U.S. 153 (1976).

Gregory, T. & Barnum, A. (November 6, 2009). Dugan's brain is focus of testimony; images show killer to be psychopathic, neuroscientist says. *Chicago Tribune*, 14.

Haederle, M. (February 23, 2010). A mind of crime. *Miller-McCune*.

Haji, I. (2003). The emotional depravity of psychopaths and culpability. *Legal Theory*, 9, 63.

Haji, I. (2008). The inauthentic evaluative schemes of psychopaths and culpability. In: J. McMillan & L. Malatesti (Eds.), *Responsibility and psychopathy: Interfacing law, psychiatry, and philosophy* (pp. 261–282). New York: Oxford University Press.

Halleck, S. L. (2001). Psychiatry and the death penalty: A view from the front lines. In: L. E. Frost & R. J. Bonnie (eds.), *The evolution of mental health law* (pp. 181–192). Washington, DC: American Psychological Association.

Hare, R. D. (1993). *Without conscience: The disturbing world of the psychopaths among us*. New York: Guilford.

Hare, R. D. (February 1, 1996). Psychopathy and antisocial personality disorder: A case of diagnostic confusion. *The Psychology Times*.

Hare, R. D. (1998). The Hare PCL-R: Some issues concerning its use and misuse. *Legal & Criminological Psychology*, 3, 99, 114.

Hare, R. D., Clark, D., Grann, M., & Thornton, D. (2000). Psychopathy and the predictive validity of the PCL-R: An international perspective. *Behavioral Science & Law*, 18, 623.

Hare, R. D. (2002). Psychopaths and their nature: Implications for the mental health and criminal justice systems. In: T. Millon, E. Simonsen, R. D. Davis, & M. Birket-Smith (Eds.), *Psychopathy: Antisocial, criminal, and violent behavior* (pp. 188–212). New York; Guilford.

Hare, R. D. (2003). *Manual for the Psychopathy Checklist-Revised* (2d ed.). Toronto, Ontario, Canada: Multi-Health Systems.

Hare, R. D., & Neumann, C. S. (2010). The role of antisociality in the psychopathy construct: Comment on Skeem and Cooke. *Psychological Assessment, 22*, 446.

Hart, H. L. A. (1968). *Punishment and responsibility: Essays in the philosophy of law.* New York: Oxford University Press.

Honderich, T. (2005). *Punishment: The supposed justifications revisited* (rev. ed.). London: Pluto Press.

Jurek v. Texas, 428 U.S. 262 (1976).

Kahan, D. M. (1999). The secret ambition of deterrence. *Harvard Law Review, 113*, 413.

Kaplan, J., Weisberg, R., & Binder, G. (2008). *Criminal law: Cases and materials* (6th ed.). New York: Aspen.

Kiehl, K. A., & Hoffman, M. B. (2011). The criminal psychopath: History, neuroscience, treatment, and economics. *Jurimetrics Journal, 51*, 355.

Krauss, D. A., & Sales, B. D. (2001). The effects of clinical and scientific expert testimony on juror decision-making in capital sentencing. *Psychology, Public Policy, & Law, 7*, 267.

Krueger, R. F., Markon, K. E., Patrick, C. J., & Iacono, W. G. (2005). Externalizing psychopathology in adulthood: A dimensional-spectrum conceptualization and its implications for DSM-V. *Journal of Abnormal Psychology, 114*, 537.

Kumho Tire Co. v. Carmichael, 526 U.S. 137 (1999).

Landrigan v. Stewart, 272 F.3d 1221 (9th Cir. 2001).

Levy, N. (2007). The responsibility of the psychopath revisited. *Philosophy, Psychiatry & Psychology, 14*, 129.

Litton, P. J. (2008). Responsibility status of the psychopath: On moral reasoning and rational self-governance. *Rutgers Law Journal, 39*, 349.

Looney v. State, 941 So.2d 1017 (Fla. 2006).

Lovelace v. Oregon, No. CV-08-3107-PA, 2009 WL 2450298 (D. Or. Aug. 10, 2009).

Luna, E. (2005). Gridland: An allegorical critique of federal sentencing. *Journal of Criminal Law & Criminology, 96*, 25.

Luna, E. (2009). Criminal justice and the public imagination. *Ohio State Journal of Criminal Law, 7*, 71.

Lykken, D. T. (1995). *The antisocial personalities.* New York: Psychology Press.

Maughs, S. (1941). A concept of psychopathy and psychopathic personality: Its evolution and historical development. *Journal of Criminal Psychopathology, 2*, 329.

McNichols v. State, No. 14-08-00125-CR, 2009 WL 196066 (Tex. App. Jan. 29, 2009).

McPhail v. Renico, 412 F. Supp. 2d 647, 652 (E.D. Mich. 2006).

Menninger, K. (1968). *The crime of punishment.* New York: Viking Press.

Millon, T., Simonsen, E., Birket-Smith, M., Davis, R. D. (2003). Historical conceptions of psychopathy in the United States and Europe. In: T. Millon, E. Simonsen, R. D. Davis, & M. Birket-Smith (Eds.), *Psychopathy: Antisocial, criminal, and violent behavior* (pp. 3–31). New York; Guilford.

Morris, N. (1974). *The future of imprisonment.* Chicago: University of Chicago Press.

Morse, S. J. (1984-a). Justice, mercy, and craziness. *Stanford Law Review, 36*, 1485.

Morse, S. J. (1984-b). Undiminished confusion in diminished capacity. *Journal of Criminal Law & Criminology, 75*, 1.

Morse, S. J. (1994). Culpability and control. *University of Pennsylvania Law Review, 142*, 1587.

Morse, S. J. (1999). Excusing and the new excuse defenses: A legal and conceptual review. *Crime & Justice, 23*, 329.

Morse, S. J. (2000-a). Deprivation and desert. In: W. C. Heffernan & J. Kleinig (Eds.), *From social justice to criminal justice* (pp. 114–160). New York: Oxford University Press.

Morse, S. J. (2000-b). Rationality and responsibility. *Southern California Law Review, 74*, 251.

Morse, S. J. (2001). From *Sikora* to *Hendricks*: Mental disorder and criminal responsibility. In: L. E. Frost & R. J. Bonnie (Eds.), *The evolution of mental health law* (pp. 129–166). Washington, DC: American Psychological Association.

Morse, S. J. (2002-a). Psychopathy. In: J. Dressler (Ed.), *Encyclopedia of crime and justice*: Vol. 3 (2nd ed., pp. 1264–1269). New York: Macmillan.

Morse, S. J. (2002-b). Diminished capacity. In *Encyclopedia of crime and justice*: Vol. 2 (2nd ed., pp. 528–533). New York: Macmillan.

Morse, S. J. (2003). Diminished rationality, diminished responsibility. *Ohio State Journal of Criminal Law, 1*, 289.

Morse, S. J. (2004). Reasons, results, and criminal responsibility. *University of Illinois Law Review, 2004*, 363.

Morse, S. J. (2007). The non-problem of free will in forensic psychiatry and psychology. *Behavioral Science & Law, 25*, 203.

Morse, S. J. (2008-a). Psychopathy and criminal responsibility. *Neuroethics, 1*, 205.

Morse, S. J. (2008-b). Thoroughly modern: Sir James Fitzjames Stephen on criminal responsibility. *Ohio State Journal of Criminal Law, 5*, 505.

Morse, S. J. (2011). Mental disorder and criminal law. *Journal of Criminal Law & Criminology, 101*, 885.

Morse, S. J., & Hoffman, M. B. (2007). The uneasy entente between legal insanity and mens rea: Beyond *Clark v. Arizona*. *Journal of Criminal Law & Criminology, 97*, 1071.

Mossman, D. (1994). Assessing predictions of violence: Being accurate about accuracy. *Journal of Consulting & Clinical Psychology, 62*, 783.

Muhammad v. State, 46 S.W.3d 493 (Tex. App. 2001).

Murphy, J. G. (1972). Moral death: A Kantian essay on psychopathy. *Ethics, 82*, 284.

Murrie, D. C., Boccaccini, M. T., Johnson, J. T., & Janke, C. (2008). Does interrater (dis)agreement on Psychopathy Checklist scores in sexually violent predator trials suggest partisan allegiance in forensic evaluations? *Law & Human Behavior, 32*, 352.

Nair, M. S., & Weinstock, R. (2008). Psychopathy, diminished capacity and responsibility. In: Felthous, A., & Saß, H., Eds., *The international handbook of psychopathic disorder: Law and policies* (pp. 291–293). Hoboken, NJ: Wiley.

Newman, J. P., Curtin, J. J., Bertsch, J. D., & Baskin-Sommers, A. R. (2010). Attention moderates the fearlessness of psychopathic offenders. *Biological Psychiatry, 67*, 66.

Nordenfelt, L. (2007). *Rationality and compulsion: Applying action theory to psychiatry*. New York: Oxford University Press.

Ogloff, J. R. P., & Lyon, D. R. (1998). Legal issues associated with the concept of psychopathy. In: D. J. Cooke, Forth, A. E., & Hare, R. D. (Eds.), *Psychopathy: Theory, research, and implications for society* (pp. 401–422). Dordrecht, The Netherlands: Kluwer.

O'Meara, G. (2009). He speaks not, yet he says everything; What of that?: Text, context, and pretext in *State v. Jeffrey Dahmer*. *Denver University Law Review, 87*, 97.

Ostrom, B. J., Kleiman, M., Cheesman, F., Hansen, R. M., & Kauder, N. B. (2002). *Offender risk assessment in Virginia*. Williamsburg, VA: National Center for State Courts.

Otto, R. K. (1992). Prediction of dangerous behavior: A review and analysis of "second-generation" research. *Forensic Reports, 5*, 103.

Packer, H. L. (1968). *The limits of the criminal sanction*. Stanford, CA: Stanford University Press.

Patrick, C. J., Ed. (2006). *Handbook of psychopathy*. New York: Guilford.

Penry v. Lynaugh, 492 U.S. 302 (1989).

People v. Boreham, No. A099043, 2003 WL 21399747 (Cal. Ct. App. June 18, 2003).

People v. Garza, No. B228779, 2012 WL 1115886 (Cal. Ct. App. Apr. 4, 2012).

People v. Nichols, 964 N.E.2d 1190 (Ill. App. Ct. 2012).

People v. Serravo, 823 P.2d 128 (Colo. 1992).

Pillsbury, S. H. (2009). Misunderstanding provocation. *Michigan Journal of Law Reform, 43*, 143.

Pillsbury, S. H. (1992). The meaning of deserved punishment: An essay on choice, character, and responsibility. *Indiana Law Journal, 67*, 719.

Pinholster v. Ayers, 525 F.3d 742 (9th Cir. 2008).

Pinholster v. Ayers, 590 F.3d 651 (9th Cir. 2009).

Primoratz, I. (1989). *Justifying legal punishment* (2d ed.). Amherst, NY: Humanity Books.

Primoratz, I. (1989). Punishment as language. *Philosophy, 64*, 187.

Prichard, J. C. (1835). *A treatise on insanity and other disorders affecting the mind*. New York: Arno Press. Can the man be morally insane in whom is found no insane delusion? (1862). *British Medical Journal, 54*, 38.

Radelet, M. L., Bedau, H. A., & Putnam, C. E. (1992). *In spite of innocence*. Holliston, MA: Northeastern.

Raine, A., & Sanmartin, J., Eds. (2001). *Violence and psychopathy*. New York: Springer.

Reznek, L. (1997). *Evil or ill?: Justifying the insanity defence*. London: Routledge.

Robinson, P. H. (2008). Competing conceptions of modern desert: Vengeful, deontological, and empirical. *Cambridge Law Journal, 67*, 145.

Robinson, P. H. (1994). Harm v. culpability: Which should be the organizing principle of the criminal law? *Journal of Contemporary Legal Issues, 5*, 1.

Robinson, P. H. (1987). Hybrid principles for the distribution of criminal sanctions. *Northwestern University Law Review, 82*, 19.

Robinson, P. H., & Kurzban, R. (2007). Concordance and conflict in intuitions of justice. *Minnesota Law Review, 91*, 1829.

Roper v. Simmons, 543 U.S. 551 (2005).

Rozek, D. (November 6, 2009). Dugan among very worst in psychopathic traits. *Chicago Sun-Times*.

Schopp, R. (2006). Two-edged swords, dangerousness, and expert testimony in capital sentencing. *Law & Psychology Review, 30*, 57.

Schopp, R. T., Scalora, M. J., & Pearce, M. (1999). Expert testimony and professional judgment: Psychological expertise and commitment as a sexual predator after Hendricks. *Psychology, Public Policy & Law, 5*, 120.

Schriro v. Landrigan, 550 U.S. 465 (2007).

Shapiro, M. (2008). An overdose of dangerousness. *American Journal of Criminal Law, 35*, 145.

Skeem, J. L., & Cooke, D. J. (2010). Is criminal behavior a central component of psychopathy?: Conceptual directions for resolving the debate. *Psychological Assessment, 22*, 433.

State v. Bare, No. 00-1497-CR, 2000 WL 1874113 (Wis. Ct. App. Dec. 27, 2000).

State v. Coomer, Nos. CA2009-09-016 & CA2009-09-017, 2010 WL 2891748 (Ohio Ct. App. July 26, 2010).

State v. Evans, 806 P.2d 512 (Mont. 1991).

Strawson, P. (1993). Freedom and resentment. In: J. M. Fischer & M. Ravizza (Eds.), *Perspectives on moral responsibility* (p. 45–66). Ithaca, NY: Cornell University Press.

Strickland v. Washington, 466 U.S. 668 (1984).

Texas Defender Service (2004). *Deadly speculation: Misleading Texas capital juries with false predictions of future dangerousness*. Houston, TX: Author.

Thompson v. Oklahoma, 487 U.S. 815 (1988).

U.S. ex rel. Griffin v. Mathy, No. 98-C-5024, 2009 WL 2252238 (N.D. Ill. July 29, 2009).

U.S. Sentencing Guidelines Manual §§ 5H1.10, 5H1.12.

United States v. Clark, 310 F. App'x 275 (10th Cir. 2009).

United States v. Barnette, 211 F.3d 803 (4th Cir. 2000).

United States v. Brady, 417 F.3d 326 (2d Cir. 2005)

United States v. Campbell, 738 F. Supp. 2d 960 (D. Neb. 2010).

United States v. Chisichlly, 30 F.3d 1144 (9th Cir. 1994).

United States v. Coates, No. 2:09CR00008, 2010 WL 582785 (W.D. Va. Feb. 17, 2010).

United States v. Gabrion, 648 F.3d 307 (6th Cir. 2011).

United States v. Leandre, 132 F.3d 796 (D.C. Cir. 1998).

United States v. McNeil, No. 8:06CR123, 2007 WL 2318548 (D. Neb. Aug. 9, 2007).

United States v. Mitchell, 706 F. Supp. 2d 1148 (D. Utah 2010).

United States v. Motto, 70 F. Supp. 2d 570 (E.D. Pa. 1999).

United States v. Rodriguez, 389 F. Supp. 2d 1135 (D. N.D. 2005).

United States v. Stall, 581 F.3d 276, 279 (6th Cir. 2009).

United States v. Taylor, 320 F. Supp. 2d 790 (N.D. Ind. 2004).

United States v. Torres, 217 F. App'x 540 (7th Cir. 2007).

Vargas, M., & Nichols, S. (2007). Psychopaths and moral knowledge. *Philosophy, Psychiatry & Psychology, 14*, 157.

Vitiello, M. (1993). Reconsidering rehabilitation. *Tulane Law Review, 65*, 1011.

von Hirsch, A., & Ashworth, A. (2005). *Proportionate sentencing: Exploring the principles*. New York; Oxford University Press.

Wallace, R. J. (1994). *Responsibility and the moral sentiments*. Cambridge, MA: Harvard University Press.

Walsh, T., & Walsh, Z. (2006). The evidentiary introduction of Psychopathy Checklist-Revised assessed psychopathy in U.S. courts: Extent and appropriateness. *Law & Human Behavior, 30*, 493.

Ward v. State, 903 N.E.2d 946 (Ind. 2009).

Werlinder, H. (1978). *Psychopathy: A history of the concepts*. Uppsala: Almqvist & Wiksell.

Westen, D., & Shedler, J. (2000). A prototype matching approach to diagnosing personality disorders: Toward DSM-V. *Journal of Personality Disorders, 14*, 109.

Widiger, T. A., Simonsen, E., Krueger, R., Livesley, W. J., & Verheul, R. (2005). Personality disorder research agenda for the DSM-V. *Journal of Personality Disorders, 19*, 315.

Wilkerson v. Felker, No. 2:06-cv-01981-AK, 2009 WL 54901 (E.D. Cal. Jan. 8, 2009).

Wilson v. Sirmons, 536 F.3d 1064 (10th Cir. 2008).

Wolff, M. A. (2006). Missouri's information-based discretionary system. *Ohio State Journal of Criminal Law, 4*, 95.

Wolfson, J. K. (2008). Psychopathy and the death penalty in the United States. In: A. Felthous & H. Saß (Eds.), *The international handbook* of psychopathic disorder: Law and policies (pp. 329–342). Hoboken, NJ: Wiley.

Wootton, B. (1959). *Social science and social pathology*. St. Leonards, NSW, Australia: Allen & Unwin.

Zimring, F. E., & Hawkins, G. (1995). *Incapacitation: Penal confinement and the restraint of crime*. New York: Oxford University Press.

Zimring, F. E., & Hawkins, G. J. (1973). *Deterrence: The legal threat in crime control*. Chicago: University of Chicago Press.

Zinger, I., & Forth, A. E. (1998). Psychopathy and Canadian criminal proceedings: The potential for human rights abuses. *Canadian Journal of Criminology & Criminal Justice, 40*, 237.

INDEX

Note: Page numbers in italics indicate tables or graphs

adolescents
   adolescent psychopathy and the law, 78–89
      court proceedings and psychopathy, 78–79
      new research in adolescent psychopathy, 84–86
      risk assessment, 79–82
      treating psychopathic traits in youthful offenders, 83–84
      using "psychopathy" label, 82–83, 262–263
   community recidivism, 257–258
   genetics and psychopathy
      antisocial behavior and psychopathy in adulthood, 163, 171–172
      environmental influences, 163, 168–169
      and environmental influences, 174–175
      and explaining covariance, 169–170
      legal implications, 175–176
      studies of identical and fraternal twins, 162–163, *164–167*
   life-course and psychopathy, 238–240
   Mendota Juvenile Treatment Center, 205–214
      background of, 205
      conceptual and philosophical basis, 205–209
      treatment outcome studies, 209–214, *212–218*
   psychopathic features
      assessing, 61–63
      developmental origins of affective features, 64–66
      developmental origins of interpersonal features, 66–67
      early interventions, 67–69
      legal and clinical implications, 69–71
      stability and change in, 63–64, 170–171
   psychopathic traits and treatment progress, 219–223, *220–223*
   recidivism following treatment, 214–219, 222–223
   stigma of psychopathy label, 22, *23*, 24, 82–83
   treating those with psychopathic features, 201–228
      study outcomes and implications for clinical practice, 223–225
      study outcomes and implications for policy, 225
      treatment response of adults, 202–204
      treatment studies of adolescents, 204–205
adoption studies, distribution of effect sizes, *182*
affective psychopathic features, developmental origins of, 64–66
aggravation, psychopathy as, 363–365
aggression
   conditional reasoning tests for, 49–50
   examining psychopathy/violence relationship, 259–260
Anderson, Nathaniel E., 131–149
antisocial behavior
   and adoption studies, 181–182
   approaches to researching, 180–198
      assessment of confounding among moderators, 185
      assessment of potential moderators, 182–185
      biometric model-fitting, 181–182
      candidate-genes and endophenotypes, 191–193
      candidate-genes and environment, 188–191
      meta-analyses of genetic studies, 185–186
      results of recent studies, 186–188
   predicting reliably in forensic settings, 257
antisocial personality disorder, 12–13
assessing psychopathy, 5–33
   and comorbidity issues, 153
   contexts for assessing, *15*
   forensic use of psychopathy measures, 250–256

assessing psychopathy (*Cont.*)
   field reliability, 252–255
   reliability in applied settings, 251–252
   Hare Psychopathy Checklist Measures
      and aggression and reoffending, 22
      alternatives to, 34–57
      and antisocial personality disorder, 12–13
      concerns regarding misuse, 14, 16
      conclusions regarding, 24
      cutoff scores, 8
      described, 5–6, 19
      evaluator scores in adversarial contexts, *9*
      factor structure of, 6–8, 20–21
      forensic uses of, 14, 256–262
      generalizability of, 13–14, 21–22
      items and factors in, *7*
      and prevalence of psychopathic traits, 10
      psychometric properties of, 20
      reliability of, 8–10, 17, 236–238
      scores across different population samples, *10*
      scores across settings and sexes, *17*
      screening version, 16–18
      use and administration of, 19–20
      validity of, 10–12
      youth version, 18–22, 24
   implicit measures, 49–50
      and conditional reasoning tests, 49–50
      and projective devices, 49
   observer measures, 47–49
      The B-Scan, 48
      Interpersonal Measure of Psychopathy, 47–48
      Minnesota Temperament Inventory, 48–49
      Psychopathy Q-Sort, 48
      Psychopathy Scan, 48
   psychopathic features in youth, 61–63
   self-report measures, 35–47
      advantages of, 35
      disadvantages of, 35–37
      Levenson Primary and Secondary Psychopathy Scales, 41–42
      longstanding empirical problems, 38–39
      misconceptions and misunderstandings regarding, 37–38
      Psychopathy Personality Inventory, 44–47
      and psychopathy "profiles," 39–40
      psychopathy-specific measures, 40
      Self-Report Psychopathy Scale, 42–43
   See also moral judgments and psychopathy
attention deficits, and psychopathic decision-making, 95–97, 98–99

behavioral science and criminal law, 298–300, vii–ix

Beyond Harm-based morality, and moral judgments, 119–121
biometric model-fitting analyses, 181–182
Bo, Sune, 5–33
Boccardi, Marina, 150–157
brain
   amygdala, 135–137
   basic functional organization of, 131–132
   electrocortical measures of activity, 139–140
   functional connectivity and psychopathy, 138–139
   methods for assessing function, 132–134
   paralimbic and other structures, 137–138
   prefrontal cortex, 134–135
   structural abnormalities and psychopathy, 150–157
      influence of comorbidities, 153
      interpreting imaging findings, 154
B-Scan, The, 48
Byrd, Amy L., 61–77

Caldwell, Michael F., 201–228
capital sentencing, and psychopathy, 369–376
   aggravating and mitigating factors, 370–371
   background of, 369–370
   Barefoot vs. Estelle, 371–372
   recent cases and issues, 372–376
children
   adolescent recidivism following treatment, 214–219, 222–223
   community recidivism in juveniles, 257–258
   genetics and psychopathy
      antisocial behavior and psychopathy in adulthood, 163, 171–172
      environmental influences, 163, 168–169
      and environmental influences, 174–175
      and explaining covariance, 169–170
      legal implications, 175–176
      studies of identical and fraternal twins, 162–163, *164–167*
   life-course and psychopathy, 238–240
   Mendota Juvenile Treatment Center, 205–214
      background of, 205
      conceptual and philosophical basis, 205–209
      treatment outcome studies, 209–214, *212–218*
   psychopathic features
      assessing, 61–63
      developmental origins of affective features, 64–66
      developmental origins of interpersonal features, 66–67
      early interventions, 67–69
      legal and clinical implications, 69–71

INDEX

stability and change in, 63–64, 170–171
psychopathic traits and treatment progress, 219–223, *220–223*
stigma of psychopathy label, 22, *23,* 24, 262–263
treating adolescents with psychopathic features, 201–228
    study outcomes and implications for clinical practice, 223–225
    study outcomes and implications for policy, 225
    treatment response of adults, 202–204
    treatment studies, 204–205
See also adolescents
clinical psychology
    legal and clinical implications of psychopathic features, 69–71
    Mendota Juvenile Treatment Center, 205–214
        background of, 205
        conceptual and philosophical basis, 205–209
        treatment outcome studies, 209–214, *212–218*
    new research in adolescent psychopathy, 84–86
    psychopathic traits and treatment progress, 219–223, *220–223*
    recidivism following treatment, 214–219, 222–223
    treating psychopathic features in adolescents, 201–228
        study outcomes and implications for clinical practice, 223–225
        study outcomes and implications for policy, 225
        treatment response of adults, 202–204
        treatment studies, 204–205
    treating psychopathic traits in youthful offenders, 83–84
See also forensic psychology
cognitive deficits, and psychopathy, 97–99
Columbine High School, 308–310
comorbidity issues, and assessing psychopathy, 153
conduct judgment, and criminal liability, 302–303
Corrado, Michael Louis, 346–357
Cox, Jennifer M., 250–272
criminal responsibility, and psychopathy, 275–296
    behavioral science and criminal law, 298–300
    biological causes and responsibility, 277–278
    case for excusing the psychopath, 278–287
        the law's assumption of moral competence, 286–287
        moral competence and criminal responsibility, 280–286
        psychopathy and moral competence, 278–280
    Eric Harris and Columbine High School, 308–310
    holding psychopaths criminally responsible, 287–294, 297–298
        challenges to implementing an excuse, 290–292
        moral considerations against excusing, 290
        no right to criminal excuse, 288–290
        reasons for holding psychopaths responsible, 292–294
    liability
        and conduct judgment, 302–303
        and the conduct limitation, 306–308
        and the question of humanity, 310–311
        and reason-responsiveness, 301–302
    overview, 275–277
    proof of psychopathy, 303–305
    psychopathy as a test, 311–312
    the rationality of the psychopath, 305–306
See also forensic psychology
See also sentencing, and psychopathy

decision-making and psychopathy
    neurobiological mechanisms for psychopathy, 100–102
    psychological mechanisms, 93–99
        cognitive deficits, 97–99
        psychopathy as a decision-making disorder, 93–97
Defining Issues Test of moral judgment, 111–112
Deserved Punishment Test, of moral judgment, 118–119
detention of psychopaths, 321–345
    classification of offenders, 321–322
    desert-based detention, 325–328
        recidivist offender enhancements, 326–328
        sentencing, 325
    disease-based detention, 329–341
        quasi-criminal commitment, 332–333
        sexual predator and violent offender commitments, 335–341
        sexual predator commitments, 333–335
        and traditional insanity acquittal, 329–332
        traditional involuntary civil commitment, 329
    future science of, 342–343
    justification of, 346–348
    logic of preventive detention, 322–325
    philosophical underpinnings, 354–356
    preventive detention and criminal responsibility, 348–351
    purely preventive detention, 341–342
    rights approach, 351–354
        dignity and preventive detention, 351–352
        undeterrability and preventive detention, 352–353
See also sentencing, and psychopathy

developmental origins
  of affective psychopathic features, 64–66
  of interpersonal psychopathic features, 66–67
Diagnostic and Statistical Manual of Mental Disorders (DSM-IV), 12–13

Edens, John F., 250–272
electrocortical measures, and assessing psychopathy, 139–140
emotional deficits, and psychopathic decision-making, 94–95
empathy, lack of in psychopathy, 108–109
environmental influences, and psychopathy
  adoption studies, 181–182
  approaches to researching, 180–198
    assessment of confounding among moderators, 185
    assessment of potential moderators, 182–185
    candidate-genes and endophenotypes, 191–193
    candidate-genes and environment, 188–191
    meta-analyses of genetic studies, 185–186
    in quantitative genetic studies, 181–182
    results of recent studies, 186–188
  genetic studies informing on, 174–175
  and heritable psychopathic personality traits, 163, 168–169

Fontaine, Natalie M. G., 161–179
forensic psychology
  adolescent psychopathy and the law, 78–89
  behavioral science and criminal law, 298–300
  contexts for assessing psychopathy, *15*
  impact of psychopathic label
    on adults, 263–264
    on juveniles, 262–263
  legal and clinical implications of psychopathic features, 69–71
  recidivism, violent, 231–249
    assessing risk for, 232–236, 241–242
    legal implications of research on, 240–241
    life-course and, 238–240
  recidivism following treatment, 214–219, 222–223
  treatment outcomes for adolescents, *212–218*
  use of Hare Psychopathy Checklist Measures, 14, 236–238
  See also clinical psychology
  See also criminal responsibility, and psychopathy
  See also sentencing, and psychopathy
Forth, Adelle, 5–33
Fowler, Katherine A., 34–57

functional neuroimaging and psychopathy, 131–149
brain
  amygdala, 135–137
  basic functional organization of, 131–132
  electrocortical measures of activity, 139–140
  functional connectivity and psychopathy, 138–139
  methods for assessing function, 132–134
  paralimbic and other structures, 137–138
  prefrontal cortex, 134–135
interpreting structural imaging findings, 154
key considerations and methodological issues, 140–141, 144
structural brain abnormalities, 150–153

genetics and psychopathy
  antisocial behavior and psychopathy in adulthood, 163, 171–172, 182
  approaches to researching, 180–198
    assessment of confounding among moderators, 185
    assessment of potential moderators, 182–185
    candidate-genes and endophenotypes, 191–193
    candidate-genes and environment, 188–191
    meta-analyses of genetic studies, 185–186
    quantitative genetic studies, 181–182
    results of recent studies, 186–188
  and explaining covariance, 169–170
  legal implications for minors, 161–179
  molecular genetic studies, 172–174
  studies of identical and fraternal twins, 162–163, *164–167*, 183
  See also environmental influences, and psychopathy

Hare, Robert D., vii–ix
Hare Psychopathy Checklist Measures, 5–33
  and aggression and reoffending, 22
  alternatives to, 34–57
    future directions, 50–51
    observer measures, 47–49
    self-report measures, 35–47
  and antisocial personality disorder, 12–13
  concerns regarding misuse, 14, 16
  conclusions regarding, 24
  cutoff scores, 8
  described, 5–6, 19
  evaluator scores in adversarial contexts, *9*
  factor structure of, 6–8, 20–21
  forensic uses of, 14
  generalizability of, 13–14, 21–22

items and factors in, 7
and predicting poor outcomes, 256–262
   antisocial conduct, 257
   community recidivism in juveniles, 257–258
   contribution to risk assessment, 260–261
   examining psychopathy/violence relationship, 259–260
   institutional misconduct, 258–259
and prevalence of psychopathic traits, 10
psychometric properties of, 20
reliability of
   in forensic settings, 251–256
   overall reliability, 8–10, 17
   for risk assessment, 236–238
scores across different population samples, 10
scores across settings and sexes, 17
screening version, 16–18
use and administration of, 19–20
validity of, 10–12
youth version, 18–22, 24
Harris, Eric and Columbine High School, 308–310
Harris, Grant T., 231–249

implicit measures, and assessing psychopathy, 49–50
   and conditional reasoning tests, 49–50
   and projective devices, 49
institutional misconduct, 258–259
intentionality, and cognitive deficits in psychopathy, 97–99
Interpersonal Measure of Psychopathy, 47–48
interpersonal psychopathic features, developmental origins of, 66–67
interventions, early, 67–69

Kiehl, Kent A., 1–2, 131–149
Koenigs, Michael, 93–105
Kongerslev, Mickey, 5–33

Larsson, H., 161–179
legal implications
   adolescent psychopathy and the law, 78–89
      new directions in research, 84–86
      treating psychopathic traits in youthful offenders, 83–84
      using "psychopathy" label, 82–83, 262–263
   of genetic research, 175–176
   of psychopathic features, 69–71
   of psychopathic label on adults, 263–264
   of research on violent recidivism, 240–241, 241–242
Levenson Primary and Secondary Psychopathy Scales, 41–42

liability
   and conduct judgment, 302–303
   conduct limitation in, 306–308
   and the question of humanity, 310–311
   and reason-responsiveness, 301–302
life-course and psychopathy, 238–240
Lilienfeld, Scott O., 34–57
Litton, Paul, 275–296, 349–350
Luna, Erik, 358–388

Magyar, Melissa S., 250–272
Mendota Juvenile Treatment Center, 205–214
   background of, 205
   conceptual and philosophical basis, 205–209
      theoretical foundation, 206–207
      treating intractable patients, 206
      treatment components, 207–209
   treatment outcome studies, 209–214
Minnesota Temperament Inventory, 48–49
mitigation, psychopathy as, 361–363
moral judgments and psychopathy, 106–126
   defining moral judgment, 107–109
   defining psychopathy, 106–107
   the law's assumption of moral competence, 286–287
   moral competence and criminal responsibility, 280–286
   moral judgment tests, 109–122
      Beyond Harm-based morality, 119–121
      Defining Issues Test, 111–112
      Moral Judgment Interview, 109, 111
      Moral Judgment Task, 112–113
      moral pictures with brain scans, 121–122
      personal *versus* impersonal dilemmas, 115–118
      philosophical scenarios, 115
      Robinson and Kurzban's Deserved Punishment Test, 118–119
      Turiel's Moral/Conventional Test, 113–115
   published studies testing, *110*
   See also criminal responsibility, and psychopathy
Moral Judgment Task, test of moral judgment, 112–113
moral philosophy and criminal law, 301–302
moral responsiveness, and criminal responsibility, 350–351
Morse, Stephen J., 321–345, 348–349
Murphy, Jeffrie, 351–352

neurobiological mechanisms and psychopathy, 100–102
neuroimaging and psychopathy, 131–149
   brain

neuroimaging and psychopathy (*Cont.*)
    amygdala, 135–137
        basic functional organization of, 131–132
        electrocortical measures of activity, 139–140
        functional connectivity and psychopathy, 138–139
        methods for assessing function, 132–134
        paralimbic and other structures, 137–138
        prefrontal cortex, 134–135
    interpreting structural imaging findings, 154
    key considerations and methodological issues, 140–141, 144
    structural brain abnormalities, 150–153
Newman, Joseph P., 93–105

observer measures, and assessing psychopathy, 47–49
    The B-Scan, 48
    Interpersonal Measure of Psychopathy, 47–48
    Minnesota Temperament Inventory, 48–49
    Psychopathy Q-Sort, 48
    Psychopathy Scan, 48

Pardini, Dustin A., 61–77
PCL-R
    See Hare Psychopathy Checklist Measures
philosophical scenarios, as moral judgment tests, 115
Pillsbury, Samuel H., 297–318
preventive detention, 321–345
    classification of offenders, 321–322
    and criminal responsibility, 348–351
    desert-based detention, 325–328
        recidivist offender enhancements, 326–328
        sentencing, 325
    disease-based detention, 329–341
        quasi-criminal commitment, 332–333
        sexual predator and violent offender commitments, 335–341
        sexual predator commitments, 333–335
        and traditional insanity acquittal, 329–332
        traditional involuntary civil commitment, 329
    and future science, 342–343
    justification of, 346–348
    logic of prevention, 322–325
    philosophical underpinnings, 354–356
    purely preventive detention, 341–342
    rights approach, 351–354
        dignity and preventive detention, 351–352
        undeterrability and preventive detention, 352–353
    See also sentencing, and psychopathy
projective devices, 49

psychopathic features, 61–77
    assessing in youth, 61–63
    developmental origins of affective features, 64–66
    developmental origins of interpersonal features, 66–67
    early interventions, 67–69
    legal and clinical implications, 69–71
    stability and change in, 63–64, 170–171
    treatment of adolescents with, 201–228
        treatment response of adults, 202–204
        treatment studies, 204–205
psychopathic label
    stigma of for adults, 263–264
    stigma of for children and adolescents, 22, *23*, 24, 262–263
psychopathic traits
    inheritability of, 161–179
    prevalence of, 10, 17–18
    and psychopathy "profiles," 39–40
    results of recent genetic studies, 186–188
    stability of, 21, 170–171
    treatment progress, 219–223
Psychopathy Personality Inventory, 44–47
    advantages and disadvantages, 46–47
    construction and format, 44
    psychometric properties, 45–46
Psychopathy Q-Sort, The, 48
Psychopathy Scan, 48
punishment theory, and psychopathy, 360–361

reason-responsiveness, and liability, 301–302
recidivism
    community recidivism, 257–258
    treatment outcomes and, 214–219, 222–223
    using Hare Psychopathy Checklist Measures to assess risk, 236–238
    violent recidivism, 231–249
        assessing risk for, 232–236, 241–242
        legal implications of research, 240–241
        life-course and, 238–240
Rhee, Soo Hyun, 180–198
Rice, Marnie E., 231–249
risk assessment
    and adolescent psychopathy, 79–82
    contribution of Hare Psychopathy Checklist Measures, 260–261
    effectiveness of Hare Psychopathy Checklist Measures, 236–238
    recommendations for adolescents, 81–82
    for violent recidivism, 232–236
Robinson and Kurzban's Deserved Punishment Test, 118–119

INDEX

Salekin, Randall T., 78–89
Schaich Borg, Jana, 106–126
self-promotion, aberrant, 49–50
self-report measures, and assessing psychopathy, 35–47
    advantages of, 35
    disadvantages of, 35–37
    Levenson Primary and Secondary Psychopathy Scales, 41–42
    longstanding empirical problems, 38–39
    misconceptions and misunderstandings regarding, 37–38
    Psychopathy Personality Inventory, 44–47
    and psychopathy "profiles," 39–40
    psychopathy-specific measures, 40
    Self-Report Psychopathy Scale, 42–43
sentencing, and psychopathy, 358–388
    capital sentencing, 369–376
        aggravating and mitigating factors, 370–371
        background of, 369–370
        Barefoot vs. Estelle, 371–372
        recent cases and issues, 372–376
    psychopathy as aggravation, 363–365
    psychopathy as mitigation, 361–363
    punishment theory, 360–361
    sentencing doctrine, 365–367
    sentencing practice, 367–369
Sinnott-Armstrong, Walter P., 1–2, 106–126
Slobogin, Christopher, 352–353

Turiel's Moral/Conventional Test, 113–115

Viding, Essi, 161–179
Vitacco, Michael J., 78–89

Waldman, Irwin D., 180–198